Advanced Oxidation Processes
for Wastewater Treatment

Wastewater Treatment and Research

Series Editor: Maulin P. Shah

Wastewater Treatment: Molecular Tools, Techniques, and Applications
Maulin P. Shah

Advanced Oxidation Processes for Wastewater Treatment: An Innovative Approach
Maulin P. Shah, Sweta Parimita Bera, and Günay Yıldız Töre

Advanced Oxidation Processes for Wastewater Treatment

An Innovative Approach

Edited by
Maulin P. Shah, Sweta Parimita Bera, and Günay Yıldız Töre

CRC Press
Taylor & Francis Group
Boca Raton   London

CRC Press is an imprint of the
Taylor & Francis Group, an **informa** business

First edition published 2022
by CRC Press
6000 Broken Sound Parkway NW, Suite 300, Boca Raton, FL 33487–2742

and by CRC Press
2 Park Square, Milton Park, Abingdon, Oxon, OX14 4RN

CRC Press is an imprint of Taylor & Francis Group, LLC

Library of Congress Cataloging-in-Publication Data
Names: Shah, Maulin P., editor. | Parimita Bera, Sweta, editor. | Yildiz Tore, Günay, editor.
Title: Advanced oxidation processes for wastewater treatment : an innovative approach / edited by Maulin P. Shah, Sweta Parimita Bera, and Günay Yıldız Töre.
Description: First edition. | Boca Raton : CRC Press, 2022. | Series: Wastewater treatment and research | Includes bibliographical references.
Identifiers: LCCN 2021046708 | ISBN 9780367762117 (hardback) | ISBN 9780367762124 (paperback) | ISBN 9781003165958 (ebook)
Subjects: LCSH: Sewage—Purification—Oxidation. | Water—Purification—Oxidation. | Water—Purification—Photocatalysis.
Classification: LCC TD758 .A325 2022 | DDC 628.3—dc23/eng/20211116
LC record available at https://lccn.loc.gov/2021046708

ISBN: 978-0-367-76211-7 (hbk)
ISBN: 978-0-367-76212-4 (pbk)
ISBN: 978-1-003-16595-8 (ebk)

DOI: 10.1201/9781003165958

Contents

Preface

In recent decades, scientific insight into the chemistry of water has increased enormously, leading to the development of advanced wastewater and water purification technologies. However, the quality of freshwater resources has continually deteriorated worldwide, both in industrialized countries and in developing countries. Although traditional wastewater technologies are focused on the removal of suspended solids, nutrients and bacteria, hundreds of organic pollutants occur in wastewater and affected urban surface waters. These new pollutants are synthetic or naturally occurring chemicals that are not often monitored in the environment but have the potential to penetrate the environment and cause known or suspected adverse ecological and/or human health effects. These contaminants are collectively referred to as "emerging contaminants" and are mostly derived from domestic use and occur in trace concentrations ranging from picograms to micrograms per liter. Environmental contaminants are recalcitrant for conventional wastewater treatment processes and most of them remain unaffected, leading to the contamination of receiving water. This scenario leads to the need for an advanced wastewater treatment process that can remove environmental contaminants to safely monitor freshwater sources.

This book explains the technologies of biological and chemical wastewater treatment processes. The biological wastewater treatment processes presented include (1) bioremediation of wastewater that includes aerobic treatment (oxidation ponds, aerating lagoons, aerobic bioreactors, active sludge, percolation or drip filters, biological filters, rotating biological contactors, biological removal of nutrients) and anaerobic treatment (anaerobic bioreactors, anaerobic lagoons); (2) phytoremediation of wastewater consisting of engineered wetlands, rhizofiltration, rhizodegradation, phytodegradation, phytoaccumulation, phytotransformation and hyperaccumulators; and (3) mycoremediation of wastewater. The chemical wastewater treatment processes discussed include chemical precipitation (coagulation, flocculation), ion exchange, neutralization, adsorption and disinfection (chlorination/dechlorination, ozone, UV light). In addition, this volume explains wastewater treatment plants and illustrates them in terms of plant size, plant layout, plant design and installation location.

Editors

Maulin P. Shah
Sweta Parimita Bera
Günay Yıldız Töre

Editor Bio

Maulin P. Shah's major work involves isolation, screening, identification and genetic engineering of high-impact microbes for the degradation of hazardous materials. He has more than 250 research publications in highly reputed national and international journals.

Sweta Parimita Bera is currently working as Assistant Professor at the School of Sciences, P Savani University, Kosamba, India. Her specialization includes bioremediation and waste management. She has worked in textile wastewater treatment during her research studies and has expertise in the field of microbial bioremediation. Her research findings are published in highly reputed international journals.

Günay Yıldız Töre has a PhD in environmental science. Her research interest is related to biological treatment of industrial wastewater, industrial waste management and industrial reclamation and reuse with emerging treatment technologies. Nowadays, she is studying antibiotic removal with emerging treatment technologies from treated urban and industrial wastewater for agricultural reuse.

Contributors

Noshin Afshan
Department of Chemistry
Govt. College Women University
Faisalabad, Pakistan

Feryal Akbal
Environmental Engineering Department, Faculty of
    Engineering
Ondokuz Mayıs University Kurupelit
Samsun, Turkey

Paulina Alulema-Pullupaxi
Escuela de Ciencias Químicas
Pontificia Universidad Católica del Ecuador
Quito, Ecuador

Hithu Anand
Sri Sivasubramanyia Nadar College of Engineering
Kalavakkam, India

Idil Arslan-Alaton
Department of Environmental Engineering, School of
    Civil Engineering
Istanbul Technical University
Istanbul, Turkey

Ambreen Ashar
Department of Chemistry
University of Agriculture
Faisalabad, Pakistan

Tanzila Aslam
Department of Chemistry
University of Agriculture
Faisalabad, Pakistan

Fatos Germirli Babuna
Department of Environmental Engineering, School of
    Civil Engineering
Istanbul Technical University
Istanbul, Turkey

Alina Bari
Anhui Normal University
Wuhu, China

Zeeshan Ahmad Bhutta
Department of Clinical Medicine and Surgery

University of Agriculture
Faisalabad, Pakistan

Anjali Bishnoi
Department of Chemical Engineering
LD College of Engineering
Ahmadabad, Gujarat, India

Hanife Büyükgüngör
Environmental Engineering Department, Faculty of
    Engineering
Ondokuz Mayis University
Samsun, Turkey

Wei Cao
Nano and Molecular Systems Research Unit, Faculty of
    Science
University of Oulu
Finland

Mamta Chahar
Nalanda College of Engineering Chandi
Nalanda, Bihar, India

E Çokgör
Environmental Engineering Department, Faculty of Civil
    Engineering
Istanbul Technical University
Istanbul, Turkey

Praveen Dahiya
Amity Institute of Biotechnology
Amity University Uttar Pradesh
Noida, India

Pranjal P. Das
Department of Chemical Engineering
Indian Institute of Technology Guwahati
Guwahati, Assam, India

Mrunal Deshpande
Department of Electrical and Electronics
    Engineering
Sri Sivasubramanyia Nadar College of Engineering
Kalavakkam, India

Subhasish Dutta
Department of Biotechnology

Haldia Institute of Technology, ICARE Complex,
  Hatiberia
Haldia, West Bengal, India

Guleda Onkal Engin
Civil Engineering Faculty, Environmental Engineering
  Department
Yildiz Technical University
Istanbul, Turkey

Hanife Sari Erkan
Civil Engineering Faculty, Environmental Engineering
  Department
Yildiz Technical University
Istanbul, Turkey

Patricio J. Espinoza-Montero
Escuela de Ciencias Químicas
Pontificia Universidad Católica del Ecuador
Quito, Ecuador

Lenys Fernández
Escuela de Ciencias Químicas
Pontificia Universidad Católica del Ecuador
Quito, Ecuador

Bernardo A. Frontana-Uribea
Centro Conjunto de Investigación en Química Sustentable
  UAEM-UNAM, Carretera Toluca-Ixtlahuaca Km 14.5,
  Toluca, 50200, Estado de México, México; Instituto de
  Química, Universidad Nacional Autónoma de México,
  Circuito Exterior, Ciudad Universitaria, Coyoacán,
  04510 CDMX, México

Tarun Gangar
Department of Biosciences and Bioengineering
Indian Institute of Technology Guwahati
Guwahati, Assam, India

Sougata Ghosh
Department of Microbiology, School of Science
RK University
Rajkot, Gujarat, India

Nomso C. Hintsho-Mbita
Department of Chemistry, Faculty of Science and Agriculture
University of Limpopo
Polokwane, South Africa

Marko Huttula
Environmental and Chemical Engineering Research Unit,
  Faculty of Technology
University of Oulu
Finland

Mika Huuhtanen
Environmental and Chemical Engineering Research Unit,
  Faculty of Technology
University of Oulu
Finland

Gulen Iskender
Department of Environmental Engineering, School of
  Civil Engineering
Istanbul Technical University
Istanbul, Turkey

Ishani Joardar
Department of Biotechnology
Haldia Institute of Technology, ICARE Complex, Hatiberia
Haldia, West Bengal, India

Riitta L. Keiski
Environmental and Chemical Engineering Research Unit,
  Faculty of Technology
University of Oulu
Finland

Sarita Khaturia
Science & Technology
Mody University
Lakshamangarh, Rajasthan, India

Ayşe Kuleyin
Environmental Engineering Department, Faculty of
  Engineering
Ondokuz Mayıs University Kurupelit
Samsun, Turkey

K. Sathish Kumar
Department of Chemical Engineering
Sri Sivasubramanyia Nadar College of Engineering
Kalavakkam, India

Jaya Lakkakula
Amity Institute of Biotechnology
Amity University
Mumbai, Maharashtra, India

CB Mahto
Nalanda College of Engineering Chandi
Nalanda, Bihar, India

Slimane Merouani
Laboratory of Environmental Process Engineering,
  Department of Chemical Engineering, Faculty of
  Process Engineering
University Salah Boubnider Constantine 3
Constantine, Algeria

Samuel S. Mgiba
Department of Chemistry, College of Science,
  Engineering and Technology
University of South Africa
Johannesburg, South Africa
Department of Maths, Science and Technology Education,
  Faculty of Humanities
University of Limpopo
Polokwane, South Africa

Vimbai Mhuka
Department of Chemistry, College of Science,
  Engineering and Technology
University of South Africa
Johannesburg, South Africa

Shilpa Mishra
Department of Civil Engineering
National Institute of Technology A.P.
India

Nomvano Mketo
Department of Chemistry, College of Science,
  Engineering and Technology
University of South Africa
Johannesburg, South Africa

Mohammad Mohsin
Department of Chemistry
University of Agriculture
Faisalabad, Pakistan

Piyal Mondal
Department of Chemical Engineering
Indian Institute of Technology Guwahati
Guwahati, Assam, India

Karla Montenegro-Rosero
Escuela de Ciencias Químicas
Pontificia Universidad Católica del Ecuador
Quito, Ecuador

Nida Naeem
Department of Chemistry
University of Agriculture
Faisalabad, Pakistan

Sonal Nigam
Amity Institute of Microbial Technology
Amity University
Noida, Uttar Pradesh, India

Sadia Noor
Department of Chemistry
Govt. College Women University
Faisalabad, Pakistan

Stalin Andrés Ochoa-Chavez
Centro Conjunto de Investigación en Química Sustentable
  UAEM-UNAM, Carretera Toluca-Ixtlahuaca Km 14.5,
  Toluca
Estado de México, México

Sevde Üstün Odabaşi
Environmental Engineering Department, Faculty of
  Engineering
Ondokuz Mayis University
Samsun, Turkey

Satu Ojala
Environmental and Chemical Engineering Research Unit,
  Faculty of Technology
University of Oulu
Finland

Burcu Özkaraova
Environmental Engineering Department, Faculty of
  Engineering
Ondokuz Mayıs University Kurupelit
Samsun, Turkey

Shweta Patel
Department of Biosciences and Bioengineering
Indian Institute of Technology Guwahati
Guwahati, Assam, India

Amrin Pathan
Department of Microbiology, Parul Institute of Applied
  Sciences
Parul University
Vadodara, Gujarat, India

Sanjukta Patra
Department of Biosciences and Bioengineering
Indian Institute of Technology Guwahati
Guwahati, Assam, India

Mihir K. Purkait
Department of Chemical Engineering
Indian Institute of Technology Guwahati
Guwahati, Assam, India

Jennyffer Martinez Quimbayo
Environmental and Chemical Engineering Research Unit,
  Faculty of Technology
University of Oulu
Finland

R. Rengaraj
Department of Electrical and Electronics Engineering
Sri Sivasubramanyia Nadar College of Engineering
Kalavakkam, India

Muthukumar S
Department of Bioengineering
BIT Mesra
Ranchi, India

Nazia Saleem
Department of Chemistry
Govt. College Women University
Faisalabad, Pakistan

Bishwarup Sarkar
Department of Microbiology and Biotechnology Centre
The Maharaja Sayajirao University of Baroda
Vadodara, Gujarat, India

Reshmi Sasi
School of Biotechnology
National Institute of Technology Calicut
Kozhikode, Kerala, India

Satyam
Department of Biosciences and Bioengineering
Indian Institute of Technology Guwahati
Guwahati, Assam, India

Apoorva Sharma
Amity Institute of Biotechnology
Amity University Uttar Pradesh
Noida, India

Anupama Shrivastav
Department of Microbiology, Parul Institute of Applied Sciences
Parul University
Vadodara, Gujarat, India

Har Lal Singh
Science & Technology
Mody University
Lakshamangarh, Rajasthan, India

Surbhi Sinha
Amity Institute of Biotechnology
Amity University
Noida, Uttar Pradesh, India

T.V. Suchithra
School of Biotechnology
National Institute of Technology Calicut
Kozhikode, Kerala, India

Baranidharan Sundaram
Department of Civil Engineering
National Institute of Technology A.P.
India

Muskan Syed
Department of Biotechnology
Jamia Hamdard
New Delhi, India

Muhammad Babar Taj
Department of Chemistry
Islamia University
Bahawalpur, Pakistan

Günay Yıldız Töre
Environmental Engineering Department of Çorlu Engineering Faculty
Tekirdağ Namık Kemal University
Tekirdag, Turkey

Nouha Bakaraki Turan
Civil Engineering Faculty, Environmental Engineering Department
Yildiz Technical University
Istanbul, Turkey

Samuli Urpelainen
Nano and Molecular Systems Research Unit, Faculty of Science
University of Oulu
Finland

Ronald Vargas
Instituto Tecnológico de Chascomús
Universidad Nacional de San Martín – Consejo Nacional de Investigaciones Científicas y Técnicas
Chascomús, Argentina

G.R. Venkatakrishnan
Department of Electrical and Electronics Engineering
Sri Sivasubramanyia Nadar College of Engineering
Kalavakkam, India

Nilesh S. Wagh
Amity Institute of Biotechnology
Amity University
Mumbai, Maharashtra, India

GE Zengin
Environmental Engineering Department, Faculty of Civil Engineering
Istanbul Technical University
Istanbul, Turkey

1

Treatment of Industrial Wastewater Utilizing Standalone and Integrated Advanced Oxidation Processes

Pranjal P. Das, Piyal Mondal and Mihir K. Purkait

CONTENTS

1.1 Introduction

The most fundamental requirement for life is water, which generally gets employed in various industrial and household activities. However, the rapid urbanization and abrupt expansion of industries, viz. textile, pharmaceuticals, fertilizers, distilleries, tanneries and mining industries, result in the formation of harmful recalcitrant wastewater consisting of various hazardous pollutants and eventually leading to a water crisis. Owing to its low biodegradability index, the recalcitrant content in wastewater becomes impervious to biological process treatment and as such is categorized by high chemical oxygen demand (COD) values. The existence of such nonbiodegradable refractory organic compounds, due to their potential carcinogenic activity in wastewater, may lead to serious health hazards. Such refractory compounds result in an increase in waterborne diseases, thereby calling for highly efficient advanced treatment techniques towards water security [1]. Although several conventional techniques extensively utilized for the elimination of toxic contaminants from wastewater have led to significantly good results viz.

chemical coagulation [2], adsorption [3], biological methods [4] and membrane filtration [5], various limitations linked with each of these methods has become a matter of serious concern. For instance, the treatment time for adsorption process (pH-dependent) is very long. Also, apart from requiring a source of steam or vacuum for adsorbents regeneration, there is a gradual deterioration of the adsorbent potential with respect to an increase in the number of cycles [3]. Likewise, during chemical coagulation, there is a constant requirement for pH adjustment throughout the analysis. The process also requires the inclusion of various chemicals viz. acids and coagulants such as alum, chloride, polymeric or caustic flocculants, lime and ferric sulfate, besides generating a significant quantity of secondary contaminants and a substantial amount of sludge [2]. Moreover, in the membrane separation process, the permeate flux may significantly decline due to the occurrence of fouling on the membrane surface. The separation efficiency of the process may also be hampered due to membranes with wider pore size distribution [5]. In addition, the biological methods are generally utilized for industrial effluent treatment. Also, it requires the mandatory application of pathogenic as well as nonpathogenic microbes during the treatment [4].

DOI: 10.1201/9781003165958-1

However, over the last few decades, the application of various advanced oxidation processes (AOPs), viz. ozonation, Fenton oxidation, electrochemical oxidation, UV irradiation and ultrasonication, have been established as the most significant and effective techniques in removing numerous toxic and harmful pollutants present in wastewater. Glaze et al. (1987) [6] first suggested the conceptualization of AOP through the generation of hydroxyl radicals ($OH^•$) in 1987. The superoxides formed during AOPs promote the *in situ* production of highly reactive oxygen species (ROS) viz. singlet oxygen ($^1O_2$), ozone (O_3), superoxide radicals ($O_2^{•-}$), hydrogen peroxide (H_2O_2), sulfate radical ($SO_4^{•-}$) and hydroxyl radicals ($OH^•$) with subsequent commencement of various oxidative reactions by reactive oxygen species present in water. Apart from the standalone AOPs, hybrid AOPs can be further employed to obtain a higher degree of degeneration for the refractory contaminants, a reduction in treatment time and an intensification of mineralization process. The hybrid AOPs discussed in this chapter may be classified as ozonation-based AOPs, Fenton-based AOPs, UV-based AOPs and ultrasound-based AOPs. However, the mechanism for the oxidation of complexing agents as well as the inactivation of pathogenic microbes by different AOPs and hybrid AOPs still lack a detailed understanding in regard to the elimination of trace organic contaminants (TrOCs), removal of heavy metal complexes, effect of different organic and inorganic compounds present, generation of various by-products, toxicity evaluation as well as consideration of economic aspects during the reduction of various wastewater contaminants.

This chapter outlines the overview of several standalone AOPs published in the literature for the removal of various complexing agents, heavy metal ions, microbes as well as TrOCs present in industrial and synthetic wastewater, with specific attention paid to hybrid AOPs as categorized previously. A brief profile for both the standalone and hybrid AOPs based on a thorough literature survey has been displayed. Moreover, detailed descriptions along with various experimental conditions associated with each of the processes, which eventually influence the performance efficiency of AOPs, are described. In addition, an economic assessment of both the standalone and hybrid AOPs is also demonstrated. This chapter also suggests future guidelines in the field of water chemistry correlation, nature of contaminants, aspects of mineralization, identification of reaction intermediates, development of rate expressions as well as identification of scale-up parameters.

1.2 Advanced Oxidation Processes (AOPs)

Advanced oxidation processes can be considered the most promising and significant techniques for the treatment of synthetic and industrial effluents, owing to the *in situ* generation of different powerful oxidants that helps in the degradation of all types of organic pollutants. Due to the occurrence of several oxidative reactions between the refractory organic compounds and *in situ* generated hydroxyl radicals ($OH^•$, $E^° = 2.8$ V versus NHE), AOPs are able to substantially reduce

TABLE 1.1

Oxidation Potentials of Different Oxidant Species Used in Water Treatment (Modified with permission from Wang et al. (2020) © Elsevier [7].)

Sl No.	Oxidant	Oxidation Potential (V)
1.	Fluorine (F_2)	3.06
2.	Hydroxyl Radical ($OH^•$)	2.80
3.	Persulfate ($S_2O_8^{2-}$)	2.1
4.	Ozone (O_3)	2.07
5.	Peroxymonosulfate (HSO_5^-)	1.80
6.	Hydrogen Peroxide (H_2O_2)	1.77
7.	Hydroperoxyl Radical ($HO_2^•$)	1.70
8.	Permanganate (MnO_4^-)	1.67
9.	Chlorine Dioxide (ClO_2)	1.50
10.	Chlorine (Cl_2)	1.09

all the recalcitrant contaminants present in wastewater and also significantly increase its biodegradability. The oxidation potential of different oxidizers utilized during water treatment is shown in Table 1.1 [7].

Moreover, the principle of AOP is based on the production of several oxidizing radicals, particularly the non-selective hydroxyl radicals ($OH^•$), which are very effective in oxidizing the refractory organic pollutants into their final end products, such as water and carbon dioxide. The generated $OH^•$ radicals strike the refractory organic compounds and as such abstracts a hydrogen atom from the compound, by means of a well-established pathway [1].

$$RH + OH^• \rightarrow H_2O + R^• \tag{1.1}$$

$$R^• + H_2O_2 \rightarrow ROH + OH^• \tag{1.2}$$

$$R^• + O_2 \rightarrow ROO^• \tag{1.3}$$

$$ROO^• + RH \rightarrow ROOH + R^• \tag{1.4}$$

The advanced oxidation processes discussed in this chapter include several standalone and combined AOPs as shown in Figure 1.1.

1.2.1 Ozone Oxidation Process

Ozonation is one of the most promising and effective techniques for water treatment, which produces zero amount of sludge, followed by dissolution of the residual ozone into O_2 and H_2O. Ozone is considered a very strong oxidizing agent, having an oxidation potential of 2.07 V. It comprises one strong double bond and a weak single bond that tends to react with the organic molecules present in water either by a direct ozone (O_3) attack through electrophiles or by an indirectly generated *in situ* hydroxyl radical ($OH^•$) attack. The generation of ozone is established on the principle of the corona discharge method, which discharges very high voltage inside a dried/cooled gaseous phase consisting of oxygen. An enhancement in ozone concentration of the solution increases the rate of oxidation for many organic compounds, whereas it has

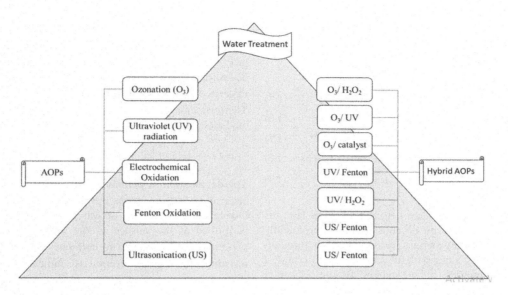

FIGURE 1.1 Overview and classification of AOPs utilized for water treatment.

no significant impact on the deterioration of a few contaminants, depending upon their molecular structure. The reason may be attributed to the interaction of ozone with contaminants as per the following categories: (i) the reaction of ozone with either superoxide or hydroperoxyl radicals (oxidation-reduction reaction) generally proceeds via the phenomenon of electron transfer, (ii) during the ozone reaction with organic compounds, a ozonide structure (five-pattern ring) formation takes place through cycloaddition reaction, (iii) electrophilic attack of ozone on the nucleophilic site and groups of organic and aromatic compounds, as such -Cl, $-OH^-$ and $-NO_2^-$ has a considerable impact on the reactivity of ozone with the aromatic rings, (iv) ozone exhibits nucleophilic characteristics, as a result of which molecules consisting of carbonyl bonds undergoes nucleophilic reaction [8].

Some of the common ozone applications include decolorization, degradation of micropollutants as well as reduction in COD content. The degradation efficiency of the ozonation process depends upon the rate of ozone generation. As the generation rate of ozone in the solution increases, the reduction performance of the process also increases. Chang et al. (2009) [9] reported the effect of ozone generation rates on the process performance during the decolorization of digital textile printing effluent. The study showed that at an ozone generation rate of 17.7 g.s/L, a reduction efficiency of only 12% and 66% for COD and color were achieved respectively, whereas 63% COD and 81% color were removed for an increased ozone generation rate of 255 g.s/L. The reason may be due to the effective and easy cleavage of the chromophore groups present in the dyes at enhanced ozone concentration. Moreover, the size of ozone bubbles plays a significant role in enhancing the performance of the process. Khuntia et al. (2014) [10] have demonstrated the improved oxidative ability of ozone microbubbles due to their ability to effectively generate $OH^{\bullet}$ radicals in bulk during the treatment of wastewaters containing

arsenic. They reported that As(III) content in the effluent can be completely oxidized over a wide pH range by employing a microbubble-aided ozonation process in a pilot plant. In the oxidation of As(III), enhanced $OH^{\bullet}$ radical generation by ozone microbubbles empowered the degradation under acidic pH. However, the degradation of some organic contaminants is relatively low, owing to the nature of selective oxidation by ozone with the contaminants. This leads to the formation of toxic gases as well as intermediate by-products viz. aldehydes and carboxylic acids, which do not interact with ozone, thereby resulting in incomplete mineralization. Such types of complications can be overcome by combining ozonation with other advanced oxidation processes viz. H_2O_2, UV radiation, catalyst and ultrasound to enhance the production of hydroxyl radicals, which eventually leads to an increase in treatment efficiency. The discussion concerning the hybrid ozone-based AOPs are elaborately explained in sections 5.1–5.3 and 5.7.

1.2.2 Fenton Oxidation Process

Among the various AOPs, Fenton oxidation is considered the most prevalent technique, because of its powerful oxidative capacity, economic advantage and significant decontaminant efficiency over a wide variety of wastewater pollutants. It is a complex catalytic system in which $OH^{\bullet}$ radicals are generated due to the interaction of iron (Fe(II) ion) with hydrogen peroxide (H_2O_2) under acidic conditions. Moreover, there are many other processes linked to Fenton oxidation with diverse treatment efficiency and reaction chemistry. For instance, the homogeneous Fenton oxidation process involves the use of soluble Fe(II) as catalyst, and as such requires an acidic pH to sustain the catalyst solubility throughout the reaction for an optimal process performance. On the other hand, heterogeneous Fenton oxidation process depends on the utilization of solid iron catalysts such as iron minerals to intensify the production of $OH^{\bullet}$ radicals under nearly neutral pH conditions. The reaction mechanisms involved during the Fenton process are as follows [11].

$$Fe^{2+} (aq) + H_2O_2 (aq) \rightarrow Fe^{3+} (aq) + OH^{\bullet} (aq) + OH^- \quad (1.5)$$

$$Fe^{3+} (aq) + H_2O_2 (aq) \rightarrow [FeOOH]^{2+} (aq) + H^+ (aq) \quad (1.6)$$

$$[FeOOH]^{2+} (aq) \rightarrow Fe^{2+} (aq) + HO_2^{\bullet} (aq) \quad (1.7)$$

$$Fe^{3+} (aq) + HO_2^{\bullet} (aq) \rightarrow Fe^{2+} (aq) + H^+ (aq) + O_2 (g) \quad (1.8)$$

$$R (aq) + OH^{\bullet} (aq) \rightarrow H_2O (l) + CO_2 (g) \quad (1.9)$$

The electron transfer mechanism between Fe^{2+} and H_2O_2 as depicted by Equation 1.5 can be considered the profoundly accepted reaction, which generates hydroxyl radicals throughout the Fenton process, along with the recovery of iron catalyst between the oxidation states of Fe^{3+} and Fe^{2+}. Besides, Equation (1.9) shows the pollutants degradation mechanism by $OH^{\bullet}$ radicals via catalytic dissolution of H_2O_2. Several parameters that enhance the mineralization efficiency during the Fenton oxidation process include Fe (II)/H_2O_2 molar ratio, pollutant concentration, pH content as well as the proportion of iron salts.

The Fenton process can successfully remediate highly contaminated wastewater from various sources, namely breweries, olive oil mills, sawmills, tanneries and palm oil mill effluent. The solution pH plays a very crucial role in achieving proper oxidation of the organic compounds during the conventional Fenton process. An optimum acidic pH of around 3 is generally reported during Fenton oxidation for the treatment of both synthetic and industrial effluents. During the treatment of benzene dye intermediate effluent, Guo et al. (2018) [12], under an acidic pH of 4.1, obtained a COD removal efficiency of 85% and was able to increase the BOD_5/COD ratio from 0.08 to 0.49 after 1 h of treatment time. Another important parameter that determines the process performance of Fenton oxidation process is the Fe^{2+}:H_2O_2 molar ratio. An optimal range of 1.42 to 20 was considered to be the most effective ratio for [Fe^{2+}:H_2O_2]. Thus, employing a molar ratio [Fe^{2+}:H_2O_2] of 1:2 at pH 3 during the treatment of palm oil refinery effluent, Nithyanandam et al. (2015) [13] reported a COD and BOD removal efficiency of 63.9% and 61.3% respectively, after 30 min of treatment time. The result suggested a considerable mineralization of contaminants present in the effluent, despite the fact that a reasonable quantity of sludge is generated, during the oligomerization of organic compounds. However, there are a few limitations in the Fenton process, namely a rigid pH range during the acidification period, consumption of Fenton reagent during the oxidation period and generation of sludge during the flocculation period. Also, the cost of operation may be significantly increased due to the consecutive supply of acids and Fenton reagents during the treatment of effluents containing high organic content. These type of problems can be overcome by combining the Fenton process with other advanced oxidation processes, such as UV-assisted Fenton and sonication-assisted Fenton to enhance the production of hydroxyl radicals, which eventually leads to an increase in the treatment efficiency. The review of hybrid Fenton-based AOPs are elaborately discussed in sections 5.4 and 5.6.

1.2.3 Electrochemical Oxidation Process

Over the past few decades, electrochemical advanced oxidation processes (EAOPs) have significantly gained much recognition due to their implementation of a clean and efficient approach in generating $OH^{\bullet}$ radicals, without the addition of chemicals. The formation of $OH^{\bullet}$ radicals in EAOPs is based on direct oxidation of water as portrayed by the Equation 1.10 [14]:

$$H_2O \rightarrow OH^{\bullet} + H^+ + e^- \quad (1.10)$$

The degradation of pollutants can be attributed to the direct electron transfer reactions during $OH^{\bullet}$ oxidation in EAOPs. Owing to the oxidation of water, an acidic boundary layer is formed at the surface of the anode. Here, bicarbonate (HCO_3^-) gets protonated to carbonic acid (H_2CO_3), thereby resisting the hindrance from $OH^{\bullet}$ radical scavenging, along with the occurrence of HCO_3^- in the solution. The generation of $OH^{\bullet}$ radical is either due to direct anodic oxidation or by indirect electrochemical production of Fenton's reagent. During the anodic oxidation, $OH^{\bullet}$ radicals are formed heterogeneously, owing to the direct water discharge at the surface of the anode. On the other hand, homogeneous production of $OH^{\bullet}$ radicals occurs, during the electrochemical production of Fenton's reagent [15]. Mineralization of various pharmaceutical compounds, azo dyes, herbicides, pesticides and smaller acids as well as decomposition of contaminants in the form of solid, liquid and gas can be effectively accomplished by EAOPs. Moreover, EAOPs are mainly categorized into (i) electro-Fenton (EF), (ii) photoelectro-Fenton (PEF), (iii) solar photoelectro-Fenton (SPEF), (iv) anodic oxidation (AO) and (v) anodic oxidation with electrogenerated H_2O_2 (AO-H_2O_2). In general, the performance of EAOPs in oxidizing the organic pollutants present in different effluents can be organized as follows: AO ≈ AO-H_2O_2 < EF < PEF-UV(A) < SPEF.

Anodic oxidation (AO) can be described as the process by which the organic pollutants undergo either direct oxidation due to transfer of electrons at the anode surface or through indirect oxidation by $OH^{\bullet}$ radicals physisorbed at the surface of the anode and also via agents at the bulk solution such as $S_2O_8^{2-}$, active chlorine species, H_2O_2 and O_3. The electrogeneration of H_2O_2 at the cathode surface during anodic oxidation process is termed anodic oxidation with electrogenerated H_2O_2 (AO-H_2O_2). Barışçı et al. (2018) [16] studied the anodic oxidation process during the treatment of cytostatic drugs, viz. methotrexate (MTX) and capecitabine (CPC), present in pharmaceutical wastewater and reported a degradation efficiency of 95% and 80% respectively, after 30 min of treatment time, using stainless steel as the cathode and Ti/IrO_2-RuO_2 as anode under a current density of 30 mA/cm². On the other hand, Brillas et al. (1995) [17] carried out experiments utilizing AO-H_2O_2 process for the treatment of 4-chloroaniline and aniline containing synthetic effluent, in which Pb/PbO_2 was used as anode and carbon-polytetrafluoroethylene (PTFE) O_2-fed as cathode. It was observed that at pH 12.7 and a current value of 300 mA, more than 97% of all the generated organic intermediates were mineralized and a complete destruction of both aniline and 4-chloroaniline were achieved.

The incorporation of Fe^{2+} to the bulk solution along with the electrochemical generation of H_2O_2 constitutes the electro-Fenton process, where additional hydroxyl radicals are produced via Fenton's reaction in the bulk. The application of boron-doped diamond (BDD) as anode and carbon felt (CF)/carbon sponge/carbon-PTFE gas (O_2 or air) diffusion electrodes as cathodes can substantially increase the oxidation power of the EF process. As such, Özcan and Özcan (2018) [18] examined the color removal and mineralization efficiency during the treatment of a diazo dye (naphthol blue black) containing effluent in the electro-Fenton system. The study showed that, using BDD as anode and CF as cathode, complete decolorization of naphthol blue black took place after 180 min of reaction time, at a current value of 60 mA. However, the TOC removal efficiency reached 97% at an applied current value of 300 mA, indicating a complete mineralization after 360 min of treatment time. The pollutant decomposition during both PEF and SPEF photo-assisted processes can be accomplished by either photoreduction of Fe(III)-hydroxy complexes with subsequent generation of Fe^{2+} and OH· radicals or through ligand-to-metal charge transfer of complexes produced between the organic pollutants (R-COOH) and Fe^{3+} by direct photolysis, resulting in Fe^{2+} regeneration with simultaneous formation of $O_2^{·-}$ radicals and H_2O_2. Moreira et al. (2015) [19] employed a UV(A) light source in PEF mediated Fe(III)–carboxylate process for the treatment of aqueous solution containing antibiotic such as trimethoprim (TMP). The result obtained using carbon-PTFE air diffusion as cathode and BDD as anode at a current density of 5 mA/cm^2 indicated 100% and 85%–98% removal efficiency of TMP and DOC respectively after 180 min of reaction time. On the other hand, Burgos-Castillo et al. (2018) [20] investigated the SPEF process for the mineralization of bisphenol A solution in 0.050 M Na_2SO_4 electrolyte. The experiments were conducted utilizing carbon-PTFE air diffusion as cathode and BDD as anode material. The study showed a removal efficiency of 100% and 98.2% for bisphenol A and TOC content respectively, at a current density of 100 mA/cm^2. Thus, the feasibility of each EAOP to decompose various kinds of refractory and toxic organic pollutants can be attributed to the accomplishment of high removal rates, low energy consumptions and significant mineralization efficiency.

1.2.4 UV Oxidation Process

The use of ultraviolet radiation (UV) is considered to be the best disinfectant technology with regard to various conventional water treatment methods for the elimination of microbes, pathogens, viruses and other chemical contaminants in drinking water. The wavelength for UV radiation varies between 1 and 380 nm. Thus, it can be categorized into vacuum UV (VUV) (100–200 nm), UV(C) (200–280 nm), UV(B) (280–315 nm) and UV(A) (315–380 nm). Low-pressure mercury lamps (partial pressure = 1 Pa) are the most frequently used UV radiation sources. The emitted wavelength spectrum for low-pressure mercury lamps are focused only at a selected number of distinct and well-defined lines. Thus, the light sources are known as monochromatic. The wavelengths at 184.9 and 253.7 nm are considered to be the most vital resonance lines.

In general, a wavelength of 254 nm is used for disinfection, as it is in close proximity to the wavelength of maximum DNA absorption. Therefore, the major radiation sources used for disinfection process are either low- or medium-pressure mercury lamps. However, several limitations are associated with the UV lamps. Some of the major disadvantages are that they require a substantial amount of energy to function, because of their low wall plug efficiency (15%–35%), they have low resistance to shock characteristics and they have a relatively short lifetime (10,000 h), besides containing toxic mercury, thereby requiring proper disposal. The intensive use of ultraviolet light-emitting diodes (UV-LEDs) over the years, due to their shock-resistant property, ability to provide a wide variety of wavelength, longer lifetime (100,000 h), and lower energy consumption, has significantly reduced the limitations associated with conventional UV. Also, UV-LEDs contain nontoxic substances, namely gallium and aluminum nitride (AlGaN) or aluminum nitride (AlN) [21].

Disinfection technology via UV-LEDs has been applied remarkably well to inactivate various pathogens, both in potable as well as surface water treatment. Hamamoto et al. (2007) [22] applied a 365 nm UV(A)-LED radiation source for the removal of *E. coli* from aqueous solution and reported a UV dose response of 55 and 263 mJ/cm^2 per log inactivation. These responses are found to be significantly higher, compared to the UV dose response obtained (2 mJ/cm^2 per log inactivation) using the conventional 254 nm UV radiation source. Moreover, the use of UV-LEDs employing adjustable pulsed illumination during the disinfection process is one of its distinct characteristics. Wengraitis et al. (2013) [23] employed a pulsed UV(C)-LED radiation source at 272 nm with a frequency of 1 Hz for the inactivation of *E. coli* on agar plates. The study reported a 3.8 times higher log inactivation capacity for the pulsed UV(C)-LED compared to continuous illumination, operated under an identical UV dose, thereby indicating a higher effectiveness of UV radiation process utilizing pulse illumination. In addition to UV-LEDs, various analyses have also been conducted on the utilization of vacuum-UV (VUV) irradiation for the degradation of natural organic matter (NOM) in water. The study conducted by Thomson et al. (2004) [24] using VUV irradiation at 185 nm indicated a significant increase in the case of both mineralization efficiency and biodegradability of NOM, compared to the conventional UV radiation (254 nm). Moreover, the study showed a decrease in DOC concentration by 71% with subsequent increase in the biodegradability efficiency of residual DOC by 68%. The decrease in DOC content can be effectively correlated with biodegradability enhancement and, as such, the efficiency of biologically degradable DOC using VUV irradiation was reported to be much higher than the conventional UV radiation. In addition, the UV oxidation process can also be combined with other AOPs, namely H_2O_2, ozone and Fenton, to further enhance its disinfection performance. The discussion in regard to the integrated UV-based AOPs are elaborately illustrated in sections 5.2, 5.4 and 5.5.

1.2.5 Ultrasound-Based Oxidation Process

The act of sonochemistry involves the implementation of sound energy to aqueous solutions, suspensions as well as solid particles dispersed in liquid for various applications; nanomaterial, biomaterial, biochemistry, pharmaceuticals, organic and environmental chemistry constitute the phenomenon of ultrasonication (US). Operational parameters such as power density, power intensity, frequency, energy dosage and treatment time play a very significant part in governing the ultrasonication process. The formation of cavitation bubbles caused by consecutive compression and rarefaction cycles during the application of ultrasonic radiation eventually leads to the production, growth and implosive bubble collapse in an aqueous solution. The cavitation bubbles get extremely heated during its disintegration period, thereby forming localized hot spots with pressure of around 500 atm, temperature of approximately 5000 °C and a shelf life of a few microseconds. The cavitation phenomenon can be classified as acoustic and hydrodynamic cavitations. Acoustic cavitation (AC) is caused by the use of high-power ultrasonic radiation with wave frequency ranging from tens of kHz to tens of MHz. Meanwhile, hydrodynamic cavitation is caused by a decrease in static pressure due to compression effect during the flow transfer through asymmetrical geometries [25].

The cavitation phenomenon has been extensively utilized in the fields of pharmaceuticals, bacterial inactivation, wastewater treatment and the oil extraction process. A rotational hydrodynamic cavitation reactor was used by Mezule et al. (2009) [26] to generate cavitation bubbles, in which the reproduction potential of *E. coli* based on the influence of cavitation in aqueous solution was studied. The study showed that an energy input and a reaction time of 490 W/L and 3 min respectively could lead to a 75% decrement in the reproduction potential of *E. coli*, thereby confirming the efficiency of the cavitation process on bacterial decomposition in the treated solution. On the other hand, Amirante et al. (2017) [27] analyzed the influence of acoustic cavitation during the extraction of virgin olive oil. The thermal and mechanical characteristics of acoustic cavitation helped in exerting a greater disruption effect on the tissue structures of elaioplasts' membranes, which eventually helped in intensifying the extraction process of the oily phase. They concluded that the use of this novel technique remarkably decreases the treatment time and enhances the overall extraction efficiency of the process. Moreover, the cavitation phenomenon can also be used to enhance the biodegradability of wastewater. The recalcitrant compounds in wastewater undergo complete mineralization due to the non-selective attack by $OH^\bullet$ radicals during cavitation. Most studies report that an increase in the biodegradability of wastewater can be achieved at an optimum inlet pressure of HC reactor between 4 and 10 bars. Padoley et al. (2012) [28], during their treatment of distillery effluent, reported an increase in biodegradability indices from 0.13 to 0.32 with an increase in the inlet pressure from 5 to 13 bars. Thus, the cavitation phenomenon has been widely employed as a clean and efficient technology for the disinfection of water and wastewater as well as for various biological and chemical applications. In addition, besides employing acoustic cavitation and hydrodynamic cavitation reactors, the performance of the sonication process can further be improved by integrating with other AOPs such as sonication-assisted Fenton reagent and sonication-assisted ozonation. Integrated ultrasound-based AOPs are elaborately reviewed in sections 5.6 and 5.7. Furthermore, Table 1.2 [29–38] indicates the application and the various experimental conditions employed during the standalone AOPs for the removal of pollutants from various wastewater sources.

1.3 Sustainability of Advanced Oxidation Process

Among the various contaminant removal processes, AOPS have been extensively studied for obtaining both water security and sustainability. The sustainability of AOPs depends upon various factors such as circular economy, waste minimization, low energy consumption and enhanced catalytic AOP performance. Circular economy can be described as an economic model to keep the resources in a closed loop, through maximum resource consumption and minimum waste generation. It also indicates effective reusability of waste or by-products generated in the process as valuable compounds. The prospects of a circular economy have led to extensive research on the production of hydrogen gas from water along with successful waste management. Solar water oxidation is the procedure of extracting hydrogen from water, thereby resulting in the formation of $OH^\bullet$ radicals. The electrodes utilized during the solar water oxidation process consist of photocatalysts (such as ZnO as visible light catalyst), as they supply an adequate amount of voltage to the PEC cell in order to obtain a desired current density, thereby helping in the acceleration of electrochemical reaction kinetics [39]. Minimization of waste is also a crucial factor that affects the sustainability of different AOPs. As such, the solar photocatalysis process had been thoroughly investigated due to its cost effectiveness and environmental advantages over conventional UV lamps.

Dominguez et al. (2018) [40] carried out a life cycle assessment on various AOPs and reported that a solar photocatalysis process utilizing photovoltaic (PV) cells operated at the minimum waste conditions, that is, atmospheric acidification potential (4.52×10^{-3} kg SO_2), stratospheric ozone depletion (5.36×10^{-10} kg CFC), lowest global warming potential (2.14 kg CO_2) and lowest global warming potential (2.14 kg CO_2). Moreover, the feasibility of AOPs significantly depends upon the consumption of a low amount of energy during its various applications. In order to evaluate and compare different AOPs for the consumption of energy, Miklos et al. (2018) [41] applied the theory of electrical energy per order (E_{EO}) to categorize various AOPs. It was observed that ozone-based AOPs and UV-based AOPs have E_{EO} values lower than 1 kW h/m^3 compared to photo-Fenton-based AOPs (2.6 kW h/m^3), indicating that ozone and UV-based AOPs are more energy efficient for extensive applications. Also, the ability to completely mineralize the refractory organic compounds in wastewater plays a very important role in the sustainability assessment of AOPs. The utilization of enhanced catalytic AOPs has emerged as a promising technology for the degradation of various recalcitrant pollutants. Thus, Wang and Bai (2017) [42] used Fe-based catalysts during the application of heterogeneous catalytic ozonation for the treatment of various emerging contaminants. The study reported that several Fe-based catalysts used in the process (Fe_3O_4-derived, FeO-derived,

TABLE 1.2

Utilization of AOPs for the Degradation of Various Water Contaminants

Sl No.	Process	Wastewater/ Contaminants	Experimental conditions	Removal performance	References
1.	Ozonation	17ß-estradiol (E_2) and bisphenol A (BPA) in aqueous medium	Volume = 250 mL; pH = 6.25 (E_2) and 5.25 (BPA); T = 20 °C; t = 80 min; r_{ox} = 0.016	100% degradation of E_2 and BPA at O_3 dose = 15.78×10^{-3} and 18.67×10^{-3} mmol min^{-1} resp.	[29]
2.	Ozonation	Wastewater containing azo dye reactive brilliant red X-3B	Volume = 1 L; pH = 11; O_3 dose = 34.08 mg/L; T = 20–25 °C; t = 120 min; r_{ox} = 0.085	97% decolorization and 90% COD removal BOD_5 / COD ratio = 0.40	[30]
3.	Fenton (Fe^{2+}/H_2O_2)	Palm oil mill effluent	pH = 3.0–5.0; t = 30 min; H_2O_2 conc. = 3.50 g/L; Fe^{2+} conc. = 1.88 g/L; Agitation rate = 120 rpm	85.1% COD removal and 92.1% TOC removal	[31]
4.	Fenton (Fe^{2+}/H_2O_2)	Winery wastewater	pH = 3.5; t = 30 min; $FeSO_4$ dose = 50–600 mg/L; H_2O_2 dose = 150–1000 mg/L; T = 25 °C	94.4–95.8% COD removal, 96.7–97.7% BOD removal and 91.7% colour removal.	[32]
5.	Electrochemical oxidation	Textile wastewater	pH = 1; t = 180 min Sample volume = 0.12 L; Anode = Si / BDD (15 cm^2); Cathode = Zr; Current density = 8 mA/cm^2;	80% COD removal, 75% DOC removal and 100% colour removal	[33]
6.	Electrochemical oxidation	Pharmaceutical wastewater	pH = 8.5; T = 20 °C; Volume = 0.6 L; Anode: BDD (78 cm^2); Cathode: SS (78 cm^2); Current density = 26 mA cm^{-2}; Flowrate = 0.10–0.56 L min^{-1}	100% degradation of DOC and 60% COD removal	[34]
7.	UV radiation	Water containing bacteriophages Qß, and MS2	Wavelength = 255 nm; UV dose (Qß and MS2) = 30 and 41 mJ/cm^2 resp.; Wavelength = 280 nm; UV dose (Qß and MS2) = 43 and 58 mJ/cm^2 resp.; Log inactivation = 1	UV dose responses at 255 nm = 12.5 (Qß) and 12.8 (MS2) mJ/cm^2 UV dose responses at 280 nm = 28.7 (Qß), and 30.5 (MS2) mJ/cm^2	[35]
8.	UV radiation	Water containing E.coli	Wavelength = 310 nm; UV dose = 56.9 mJ/cm^2; Log inactivation = 0.6	UV dose responses = 94.8 mJ/cm^2	[36]
9.	Ultrasound (US)	Water containing micro-organisms (Microcystis aeruginosa)	Volume (batch and continuous) = 1 L and 3.5 L resp.; t = 10 min and 15 min resp.; Power density = 0.0177 and 0.0256 W/mL; Frequency = 16 and 20 kHz	60% cell reduction (batch mode) and 46% cell reduction (recirculating mode at 1 L/min).	[37]
10.	Ultrasound (US)	Pharmaceutical effluent containing antiepileptic drug (carbamazepine)	Reactor mode = Hydrodynamic and Acoustic reactors; Frequency = 24 kHz; t = 15 min; Power generation = 200 W;	HC reactor = 26.6% removal rate; AC reactor = 32.7% removal rate; Hybrid (HC+AC) reactor = 96% removal rate	[38]

Fe_2O_3-derived and FeOOH-derived catalysts) were capable of significantly decomposing the emerging contaminants such as nitrobenzenes, herbicides, pesticides, phthalic acid, phenols and dyes present in the wastewater.

1.4 Utilization of AOPs in Bioremediation

The integration of the AOP and bioremediation process specifically concentrates on the treatment of industrial effluent, consisting of various emerging bio-resistant pollutants. This combination has been explored because of its economic advantages and the capacity of AOPs to decompose the toxic pollutants chemically into biodegradable compounds. This section describes the classification and streams of wastewaters, along with the influence of AOPs on the combined AOP-biological oxidation process.

The performance of the integrated AOP-biological process significantly depends upon the classification of water composition, consisting of several pollutants of emerging concern. The water classification can be mainly divided into four types such as wastewater containing (i) inhibitory substances, (ii) substantial amount of refractory compounds, (iii) significant biodegradable wastes with low amount of refractory compounds and (iv) intermediate dead-end products. The wastewater containing inhibitory substances requires prior treatment to eliminate the toxic compounds, which would otherwise obstruct the biological treatment process involving the use of fungi and/or bacteria. Similarly, wastewaters having a significant proportion of refractory substances requires pre-treatment by AOPs, while effluents with a low quantity of refractory compounds usually prefers AOPs for post-treatment techniques. Moreover, wastewaters with intermediate dead-end products should circumvent the accumulation of metabolites, which may hinder the growth of microorganisms during the biological treatment step [43]. Also, toxicity analysis is another parameter that can be used to evaluate and optimize the combined AOP-bioremediation technique. The analysis depends upon the required accuracy as well as the environment of the microorganisms (microalgal strains, *P. tricornutum, S. aureus, E. coli* and *V. fischeri*) used. The efficiency of the combined AOP-biological process depends upon the type of AOPs used, as the generation of intermediates varies with different processes. Photo-Fenton and ozonation are the most preferred AOPs, which are often used in combination with the biological process. Zapata et al. (2010) [44], in their pilot-scale study, employed an integrated solar photo-Fenton/biological system for the treatment of pesticide-contaminated wastewater and reported a significant increase in the biodegradability from 50% to 95% and a reduction in toxicity index from 96% to 50%. Also, the performance of the combined system in terms of mineralization was found to be 94%, of which 35.5% indicates the efficiency of solar photo-Fenton process and 58.5% corresponds to the aerobic biological process. The industrial-scale application of photo-Fenton/biological process resulted in 84% mineralization, of which 35% corresponded to the photo-Fenton process and 49% indicated the efficiency of biological treatment. Moreover, Thanekar et al. (2019) [45] applied hydrodynamic cavitation in combination with ozone and H_2O_2 as a pre-treatment process for biological oxidation. The study focused on the removal of naproxen (NAP) and COD content from aqueous solution. The maximum degradation of NAP achieved using a combination of HC+H_2O_2 and HC+O_3 were reported as 80% after 120 min and 100% after 40 min of reaction time respectively. The optimal process (HC+O_3) obtained from the pre-treatment approach was further treated by biological oxidation process. It was reported that the combined process (HC+O_3/aerobic oxidation) resulted in a COD removal of ~89.5%. Thus, the study showed significant benefits concerning the utilization of ozone and hydrodynamic cavitation in combination with aerobic oxidation process. Also, time taken by the hybrid process was much less compared to the conventional process, thereby lowering the cost of operation for the entire system.

1.5 Integrated Advanced Oxidation Processes

Although the decomposition and decolorization of effluents containing various contaminants and dyes can be successfully achieved by the application of AOPs, several drawbacks are associated with each of the AOPs. For instance, the Fenton oxidation process may be evaluated as economically unviable, due to the high quantity of H_2O_2 and catalyst used. The operating pH range for the process is <3.5, especially during the application of homogeneous catalyst. Also, small-size flocs produced during the process eventually results in a huge amount of sludge generation, thereby requiring an additional treatment step. On the other hand, the ozonation process may not result in complete mineralization of wastewaters, leading to the generation of toxic by-products in the solution. The decolorization efficiency during ozonation primarily depends upon the initial concentration of dye. The higher the dye concentration, the higher is the amount of ozone required for decolorization, thereby leading to an increase in the operational cost. Also, the ultrasonication process consists of a high energy requirement as well as a low decomposition efficiency when applied as a standalone process. The radicals generated during the process may interact with one another to form H_2O_2. A loss of oxidation efficiency may result due to the low oxidation potential of H_2O_2 compared to the $OH^·$ radicals. Moreover, ultraviolet radiation is only effective for microorganisms, which means that if the effluent contains minerals such as magnesium and calcium, the rays will not be able to eliminate them. Also, the photochemical damage caused by UV may be repaired, depending upon the types of organisms present, and the effluent must be clear enough so that UV radiation can pass through and destroy the pathogens [46]. Thus, it can be inferred that each process functioning individually may not always be beneficial for the elimination of complex pollutants containing effluents. Therefore, the combination of such techniques can significantly subjugate the limitations of the individual processes. Following are the various integrated advanced oxidation processes discussed in this chapter.

1.5.1 Ozone/Hydrogen Peroxide (O_3/H_2O_2) Process

The combination of O_3 and H_2O_2 is termed the peroxone process. It is one of the most widely used hybrid AOPs and generates non-selective $OH^·$ radicals for pollutant degradation at a comparatively low cost. During the peroxone process, the interaction of ozone with peroxide anion (HO_2^-) results in the formation of $OH^·$ precursors, which eventually react with $OH^·$. However, the residual H_2O_2 needs to be eliminated from the treated water before being released into the environment. The ideal molar ratio and typical ozone dose range for the combined process are 0.5 mol/mol and 1–20 mg respectively. The reaction of ozone with the water matrix can also result in the generation of a small amount of peroxide. The peroxone process yields better degradation of organic pollutants, due to the capacity of H_2O_2 to significantly accelerate the ozone decomposition, thereby leading to the formation of $OH^·$ radicals via the mechanism of electron transfer. Though the rate of reaction and the efficiency of peroxone process are remarkably higher compared to the standalone ozonation process, the pH and H_2O_2/O_3 ratio of the combined process should be controlled correctly, as a surplus of H_2O_2 in the solution can scavenge the $OH^·$ radicals, resulting in low process performance [1].

Jiao et al. (2016) [47] applied the peroxone process for the treatment of nitrobenzene (NB) contaminated wastewater and observed that the NB reduction efficiency initially increases with an increase in H_2O_2 concentration and then starts decreasing. The reduction in NB concentration during the peroxone process was solely attributed to the oxidative degradation of organic compounds by the highly reactive $OH^{\cdot}$ radicals. Moreover, Chueca et al. (2015) [48] observed the decomposition of *E. coli* in a peroxone system and found that the inactivation results from an indirect pathway, given that the synergy of O_3 and H_2O_2 amplifies the decomposition of O_3 by H_2O_2, thereby favoring the generation of non-selective $OH^{\cdot}$ radicals. The presence of H_2O_2 significantly promotes the decomposition of ozone to produce $OH^{\cdot}$ radicals, thereby enhancing the biodegradability and mineralization efficiency of effluents containing refractory compounds.

1.5.2 Ozone/UV Radiation (O_3/UV) Process

The hybrid AOP consisting of O_3 and UV radiation is initiated by ozone photolysis. The generation of H_2O_2 and $OH^{\cdot}$ radicals results from the photodecomposition of ozone (at $\lambda < 310$ nm). Thus, UV radiation leads to the splitting of dissolved ozone, with subsequent reaction between the atomic oxygen and water to produce a thermally excited H_2O_2. The generated H_2O_2 then breaks down into two $OH^{\cdot}$ radicals. At $\lambda = 254$ nm, the molar extinction coefficient (ε) of ozone is 3300 M^{-1} cm^{-1}, which is considerably higher than H_2O_2 at similar wavelength. Only a slight amount of H_2O_2 breaks down into $OH^{\cdot}$ radicals due to cage recombination, thereby leading to a quantum yield of just 0.1 [1]. However, direct oxidation by the hybrid O_3/UV process can cover a wide variety of trace organic chemicals reactivity, which contributes to the main benefit of this combined process.

Shang et al. (2007) [49] applied the hybrid O_3/UV process for the treatment of methyl methacrylate (MMA) from semiconductor wastewater. The study reported a reduction efficiency of 96% and 22% after a treatment time of 30 min, for MMA and COD respectively. The COD degradation efficiency is further enhanced to 51% after a reaction time of 120 min. The reason may be attributed to the decomposition of dissolved ozone in the solution along with the simultaneous formation of $OH^{\cdot}$ radicals due to the presence of UV radiation. As such, the mineralization efficiency significantly increases due to the greater reactivity of $OH^{\cdot}$ radicals towards the organic compounds. In addition, Pillai et al. (2009) [50] treated terephthalic acid contaminated wastewater and observed a slight decrease in the solution pH from 8.5 to 7 during the application of O_3/UV process, compared to the standalone ozonation system, which results in a high acidic pH of the treated solution (around pH 4). However, the treatment of citrus wastewater by Guzmán et al. (2016) [51] reported a decolorization efficiency of 68.4% at pH 7, after 60 min of reaction time. Nevertheless, in both cases, the pH of the treated solution during the hybrid O_3/UV process was well within the favorable pH scale of 6–9, for adequate contribution in radical generation path.

1.5.3 Ozone/Catalysts Process

The combination of catalyst and ozonation is termed the catalytic ozonation process. It can be broadly categorized as a homogeneous and heterogeneous catalytic ozonation process, in which solid catalysts and transition metal ions are extensively used to accelerate the ozone decomposition as well as $OH^{\cdot}$ radical production. Homogeneous catalytic ozonation process can be classified into a two-phase catalytic cycle, namely (i) ozone decomposition by metal ions, resulting in the production of $OH^{\cdot}$ radicals and (ii) generation of complexes between the catalyst and organic compounds, followed by oxidation of the complex. Some of the most commonly employed metal ions as a potent catalyst during the homogeneous catalytic ozonation process are Cu(II), Mn(II), Cr(III), Co(II), Fe(II), Fe(III) and Zn(II). The formation of $OH^{\cdot}$ radicals significantly increases due to the presence of metal ions as an effective catalyst, thereby leading to an enhanced efficiency of the ozonation process. Moreover, a metal ion such as Fe(II), due to its low cost, results in a cost-effective treatment, and as such it can be regarded as the most favorable homogeneous metal catalyst [1]. During the treatment of terephthalic acid contaminated wastewater, Pillai et al. (2009) [50] employed Fe^{2+} doses of 0.25, 0.50 and 0.75 mM to accelerate the ozonation process, thereby achieving a COD removal efficiency of 73%, 78% and 81% respectively, after a reaction time of 240 min. Similarly, Sameena et al. (2019) [52] applied catalytic ozonation process for the treatment of pharmaceutical wastewater. The study reported that the biodegradability index (BI) of the wastewater was increased by 3.5 times the initial value, due to the addition of Fe^{2+} to the ozonation process. However, a high number of Fe^{2+} doses may lead to the consumption of a substantial amount of $OH^{\cdot}$ radicals, eventually resulting in a poor process performance. Although the metal ions used during the ozonation process show high catalytic activity, their removal from the treated solution becomes essential. Also, the catalytic activity of metal ions in the solution may be restricted due to their addition in limited concentration. Such limitations persuaded the researchers to conduct their research in the field of heterogeneous catalytic ozonation process.

The mechanism involved in heterogeneous catalytic ozonation process is mediated by the application of metal oxides (MnO_2, TiO_2, CeO_2, Al_2O_3 and FeOOH) on supports (zeolites modified with metals, activated carbon, SiO_2 and CeO_2). These are the most extensively used catalysts, leading to more complex reaction paths and establishing the mechanism for the multiple-phase transport phenomenon. The effectiveness of the process depends upon operational parameters such as the solution pH and catalyst along with its surface properties. Moreover, it can be classified into several factors, namely activated carbon, photocatalyzed ozonation, metal oxides and metals on supports. Bai et al. (2017) [53] treated carbamazepine contaminated municipal secondary effluent by utilizing Fe_3O_4-CeO_2/MWCNTs as a potent catalyst and observed a degradation efficiency of 50.3% and 64.1% in standalone and heterogeneous catalytic ozonation process respectively, after a reaction time of 4 min. The result indicates that the incorporation of catalyst during the process instigated the generated $OH^{\cdot}$ radicals to significantly enhance the removal rate. Thus, the quantity of catalyst added plays a major role in enhancing the OH˙ radical production during the heterogeneous catalytic ozonation process. Moreover, during the application of the catalytic ozonation process for the treatment of tannery wastewater, Huang et al. (2016) [54] observed that an increase in Mn-Cu/Al$_2$O$_3$ catalyst loading considerably increases the COD removal efficiency. The reason may be attributed to the overall increase in active surface area with respect to an increase in the catalyst quantity. Thus, the presence of supplementary active sites on the surface of catalysts leads to an increase in the generation rate of OH˙ radicals/unit of ozone conversion. The study further concluded that an increase in the amount of catalyst significantly helps in the degradation of the organic contaminants. As such, deep insight into the application of various catalysts during ozonation is crucial to fabricate an ideal catalyst for proper mineralization of the organic contents.

1.5.4 UV Radiation/Fenton Reagent Process

The combination of UV radiation and the Fenton process is commonly known as photo-Fenton oxidation. The efficiency of the hybrid process is increased owing to the presence of UV irradiation, which in turn enhances the overall oxidation rate of the process. Also, due to UV radiation, Fe^{2+} is oxidized to Fe^{3+} while simultaneously producing OH˙ radicals, followed by the formation of Fe^{3+}-hydroxy complexes in the solution. These complexes are further broken down with subsequent regeneration of Fe^{2+} ions, caused by the additional OH˙ radical formation. Moreover, in the hybrid process H$_2$O$_2$ is broken down into two hydroxyl ions, along with the formation of Fe^{2+} ions and OH$^-$ molecules due to the decomposition of Fe(OH)$^{2+}$. The photo-Fenton process can generate a substantial amount of OH˙ radicals via photoreduction of Fe^{3+} to Fe^{2+} ions and also through direct UV/H$_2$O$_2$ reaction. Also, subsequent reaction of Fe^{2+} ions with H$_2$O$_2$ further generates OH˙ radicals in the system. In addition, the photo-Fenton process can be categorized as a homogeneous or heterogeneous process, based on the types of catalyst utilized in the system [46].

Pérez-Estrada et al. (2005) [55] examined the decomposition of diclofenac in aqueous solution by the application of the photo-Fenton process. The result indicated a degradation efficiency of 50% and 100% by Fenton and photo-Fenton process, respectively. Moreover, treatment of 2-chlorophenol by Fenton and photo-Fenton process led to 60% and 100% removal after 30 min of reaction time respectively. The study also reported an increase in mineralization efficiency during the Fenton process from 10% to 95% with the application of ultraviolet radiation in the system for both diclofenac and 2-chlorophenol. Thus, the collaborative effect of combining Fenton with ultraviolet radiation enhanced the overall degradation efficiency of the process. Nevertheless, the homogeneous photo-Fenton process consist of a very firm acidic pH range, apart from generating secondary contaminants like Fe^{2+} ions, and as such requires an extra removal system. Consequently, several heterogeneous catalysts (activated carbons, clays, zeolites and aluminas) have been extensively developed for both Fenton and photo-Fenton processes with properties such as cost effectiveness, long-term

stability and photocatalytic activity. In heterogeneous photo-Fenton oxidation, the treatment can be carried out at neutral pH and room temperature and does not require neutralization of effluents after the operation. Palas et al. (2017) [56] applied both ultraviolet and visible light radiation to analyze its effect on the catalytic activity of LaCuO$_3$ during the photo-Fenton treatment of tartrazine (food dye) in aqueous solution. The influence of catalyst loading, solution pH, H$_2$O$_2$ concentration and irradiation source on the degradation of tartrazine and decolorization efficiency was investigated. The study reported a decolorization and tartrazine degradation efficiency of 83.9% and 46.6% respectively for visible light irradiation, whereas the degradation efficiency significantly increases to 90.2% for color and 64.4% for tartrazine during the utilization of UV irradiation. Thus, it was concluded that the prepared perovskite catalysts showed satisfactory results under both UV and visible light irradiation.

1.5.5 UV Radiation/Hydrogen Peroxide (UV/H$_2$O$_2$) Process

The hybrid UV/H$_2$O$_2$ process is the phenomenon of adding H$_2$O$_2$ to the system, thereby leading to the formation of two OH˙ radicals. The formed OH˙ radicals eventually unite to further produce H$_2$O$_2$ or interact with other chemical species. As such, the chance for the availability of free OH˙ radicals is approximately 50% in the system. These radicals are primarily generated by the utilization of UV(C) light irradiation. UV irradiation of $\lambda > 254$ nm is not sufficient for the breakdown of H$_2$O$_2$. Moreover, at $\lambda = 254$ nm, H$_2$O$_2$ has a relatively low molar absorption coefficient (ε) of 18.6 M^{-1}cm^{-1}, thereby leading to a H$_2$O$_2$ turnover of <10%. As such, in order to generate an adequate amount of OH˙ radicals, a high concentration of H$_2$O$_2$ (5–20 mg/L) is needed during the application of low-pressure UV lamps, which results in the addition of a subsequent step for the removal of residual H$_2$O$_2$ from the system. Therefore, H$_2$O$_2$ photolysis requires the use of UV(C) light irradiation. Other parameters such as the optical characteristics of the UV reactor and the reaction transparency also influence the UV photolysis of H$_2$O$_2$ [41]. In addition, the photodegradation of H$_2$O$_2$ generates a substantial amount of OH˙ radicals, which can effectively degrade a wide variety of pollutants.

Perisic et al. (2016) [57] compared the efficiency of stand-alone UV photolysis with the hybrid UV/H$_2$O$_2$ process for the degradation of diclofenac in aqueous solution. The result indicated a degradation efficiency of 90% and 100% for UV photolysis and UV/H$_2$O$_2$ process respectively. Also, the mineralization efficiency obtained for UV photolysis (2%) and hybrid UV/H$_2$O$_2$ process (30%) varied significantly. They inferred that the overall reaction kinetics was significantly enhanced by the addition of H$_2$O$_2$ in the UV photolysis system, due to the production of a substantial amount of OH˙ radicals by both direct and indirect photolysis. Similarly, during the application of the UV/H$_2$O$_2$ process, Alharbi et al. (2017) [58] observed that the degradation efficiency of four TrOCs – diclofenac, sulfamethoxazole, trimethoprim and carbamazepine – significantly increases with a higher H$_2$O$_2$ concentration in the system. Both diclofenac and sulfamethoxazole were completely degraded after a reaction time of 8 min, whereas the degradation efficiency of trimethoprim and carbamazepine were reported as 71% and 25% respectively. Even though the reaction time was increased to 60 min for the treatment of carbamazepine, it displayed minimal reactivity with regard to direct standalone UV irradiation. However, it was concluded that the additional H_2O_2 in the UV irradiation system completely degrades both carbamazepine and trimethoprim after a reaction time of 30 min.

1.5.6 Sonolysis/Fenton Process

The combination of ultrasound treatment and Fenton oxidation is commonly referred to as the sono-Fenton oxidation process. The hybrid process utilizes the advantages of both individual techniques to generate a substantial amount of $OH^{\cdot}$ radicals and as such intensifies the degradation efficiency of organic pollutants. The cavitation phenomenon in terms of pyrolysis plays a very important role in eliminating the chemicals that are resistant to $OH^{\cdot}$ radicals. Moreover, the turbulent condition created during cavitation helps in preventing external mass transfer resistance. As such, the mechanisms of microstreaming and microturbulence in ultrasonication (US) process are responsible for rigorous micromixing in the reaction mixture, besides disrupting the sludge particles effectively. Consequently, the hybrid sono-Fenton oxidation process results in the production of a substantial amount of $OH^{\cdot}$ radicals, provides efficient mixing followed by better interaction of $OH^{\cdot}$ radicals with the contaminants as well as increases the regeneration of Fe^{2+} ions, thereby leading to a more effective and faster mineralization of the contaminants. Thus, the combination of Fenton and US treatment process has emerged as a promising technology for the degradation of refractory organic compounds in wastewater [46].

Cetinkaya et al. (2018) [59] compared the classic Fenton and ultrasound-assisted Fenton oxidation process during the treatment of textile wastewater. The decolorization and COD removal efficiency were reported as 95% and 70% after a reaction time of 90 min for the classic Fenton process (Fe^{2+} = 0.10 g/L and H_2O_2 = 2.2 g/L), whereas in the ultrasound-assisted Fenton oxidation system (Fe^{2+} = 0.05 g/L and H_2O_2 = 1.65 g/L), complete decolorization and 80% COD removal was achieved after 60 min of reaction time. Thus, they concluded that the Fenton process when combined with ultrasonication treatment results in better decolorization efficiency while requiring a lesser quantity of chemicals and shorter reaction time. Moreover, Gholami et al. (2020) [60] reported that the application of the sono-Fenton oxidation process demonstrated a very positive synergistic effect, in terms of total nitrogen and TOC degradation efficiencies. The study reported very high TOC and nitrogen removal efficiencies of 91.6% and 87.4% respectively. They inferred that the performance of the sono-Fenton oxidation process remarkably increases with an increase in ultrasonic power as well as H_2O_2 and Fe^{2+} concentrations during the treatment.

1.5.7 Sonolysis/Ozonation Process

The combination of ultrasound treatment and ozonation is widely termed the sonolytic ozonation process. The addition of the sonolytic treatment with ozonation considerably enhances the removal rate of refractory pollutants in wastewater compared to the standalone ozonation process. The reason can be attributed to the improvement in decomposition efficiency of ozone due to the collapsing of bubbles in the system, thereby leading to a significant production of $OH^{\cdot}$ radicals. Also, the decomposition of ozone in the solution is enhanced, owing to the ultrasonic dispersion generated by the sonication wave, which eventually leads to a synergistic effect in terms of removal efficiency. Application of the ultrasonication process with frequency ranging from 20 to 100 kHz is utilized for the production of high-intensity bubbles, which are promoted by the phenomenon of compression (high pressure) and rarefaction (low pressure) in alternative cycles. In addition, high shearing effects are formed in the liquid phase, while simultaneous generation of $H^{\cdot}$ and $OH^{\cdot}$ radicals results in an effective oxidation of the organic compounds present in the effluent [61–64]. Also, the arrangement of the ultrasonic transducer (either ultrasonic bath or ultrasonic horn) significantly influences the configuration of the batch sonolytic ozonation setup.

He et al. (2007) [65] coupled the ultrasonic treatment with ozonation process for the mineralization of p-aminophenol (PAP) in aqueous solution. The study indicated an increase in the mineralization efficiency of PAP from 88% to 99% with an increase in treatment time from 10 min to 30 min respectively. Moreover, Yavaş and Ince (2018) [66] investigated the influence of sonication on the catalytic ozonation treatment of paracetamol. The catalyst utilized in the system was Pt-supported nanocomposites of Al_2O_3. They observed that the mineralization efficiency of the process was considerably enhanced by the addition of ultrasonic treatment. The transport of reaction site from the bulk solution to the solid surface, formation of gaseous cavity bubbles through increased hydrophobicity, enhanced ozone decomposition in collapsing bubbles and high ozone mass transfer rate are the main reasons behind an increased mineralization efficiency of the hybrid process. Also, integrating the sonication process with ozonation helps in preventing the poisonous intermediate by-products formed during the ozonation process.

1.6 Economical Aspects of Individual and Integrated AOPs

The application of AOPs involves several expenses, such as that of adding H_2O_2, the electricity required for anodic oxidation and UV lamp operation as well as the energy needed for ozone generation, activation of catalyst and high-intensity sonication. Bolton et al. (2001) [67] introduced the electrical energy per order factor to investigate the consumption of electrical energy for various AOPs. E_{EO} is specified as the amount of energy needed in degrading the effluent pollutants by a single order magnitude in a unit volume of water or air. E_{EO} can be evaluated as follows:

$$E_{EO} = \frac{P \times t \times 1000}{V \times 60 \times \log \frac{C_i}{C_f}} \quad (11)$$

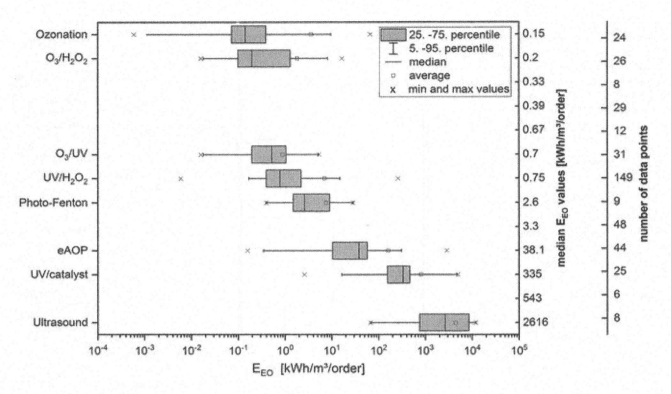

FIGURE 1.2 Overview of published E_{EO} values for AOPs sorted according to median values. (Modified with permission from Miklos et al. (2018) © Elsevier [41].)

Here, C_i and C_f = initial and final concentrations of the pollutant (mg/L); V = volume of the solution (liters); P = power generated (kW) and t = treatment time (min). An outline for the energy assessment, based on E_{EO} values of several AOPs, is shown in Figure 1.2.

During the treatment of surface water by ozonation process, Gerrity et al. (2012) [68] achieved an E_{EO} value of <0.005 and 0.004 kWh/m³ for the degradation of trimethoprim and carbamazepine respectively. Also, the removal of diclofenac from membrane reactor filtrate required a low E_{EO} value of 0.022 kWh/m³, as reported by Pisarenko et al. (2012) [69]. However, during the application of ultrasonication treatment, Güyer and Ince (2011) [70] reported that the degradation of diclofenac from aqueous solution required a much higher E_{EO} value of 60 kWh/m³ and as such may not be considered an energy efficient process.

The energy demand for various contaminant removal is significantly reduced during the application of hybrid AOPs. The mean value of E_{EO} for the combined UV/chemical-based processes is approximately 10 kWh, indicating that the photochemical AOPs are more energy efficient in terms of pollutant degradation compared to photoelectrochemical AOPs [71]. During the application of the $UV/S_2O_8^{2-}$ and UV/H_2O_2 processes, Xiao et al. (2016) [68] and Sichel et al. (2011) [72] evaluated the influence of various operational parameters on the energy consumption of both processes. The results indicated an E_{EO} value of 0.39 and 0.62 kWh/m³ for $UV/S_2O_8^{2-}$ and UV/H_2O_2 process respectively. They concluded that for low pollutant concentrations in the solution, the value of E_{EO}

was also found to be low. However, the $S_2O_8^{2-}$ concentration needs to be increased as the pollutant concentration in the solution increases, thereby resulting in an enhanced E_{EO} value for $UV/S_2O_8^{2-}$ compared to UV/H_2O_2 process. Pisarenko et al. (2012) [69] reported an E_{EO} value < 1 kWh for the ozone-based AOPs. They observed a low E_{EO} value of 0.26 kWh/m³ during the removal of ibuprofen from secondary effluents using the O_3/H_2O_2 process. Also, during the degradation of atrazine from surface water, Katsoyiannis et al. (2011) [73] carried out a comparison analysis study, based on the energy requirement by both UV/H_2O_2 and O_3/H_2O_2 processes. They observed an E_{EO} value of 0.25 kWh/m³ for the O_3/H_2O_2 process, compared to a much higher E_{EO} value of 0.82 kWh/m³ for the UV/H_2O_2 process. Hence, ozone-based AOPs are much more energy efficient for the degradation of pollutants than are the UV-based AOPs. In addition, effective degradation capacity and less energy requirement makes the hybrid AOPs a promising option for water treatment.

1.7 Conclusion and Future Aspects

This chapter provides an elaborative review of the principles, benefits, limitations, performances and successful applications of different AOPs and hybrid AOPs for decolorization, degradation of organic and inorganic contaminants as well as disinfection of water and wastewater. Effluents containing dyes, pharmaceutical drugs, TrOCs, heavy metals and various other industrial pollutants is a very serious problem, which requires distinctive and effective treatments before being discharged into the environment. Several advanced oxidation processes, namely ozonation, Fenton oxidation, UV radiation, sonication and electrochemical oxidation, have emerged as the most attractive options to decompose and mineralize the organic and inorganic contaminants as well as to decolorize the effluents. However, standalone AOPS may result in the generation of toxic by-products, incomplete mineralization, a high energy requirement and a rigid pH condition as well as operational ineffectiveness, owing to the existence of possible radical scavengers in the solution. Thus, an integrated approach is adopted in order to overcome such limitations. The effect of combined processes results in an enhanced formation of OH˙ radicals and various reactive species in the system, thereby significantly improving the degradation efficiency of pollutants over a wide range of reaction conditions. The application of hybrid processes also provides high mineralization efficiency as a result of low effluent toxicity. Moreover, utilization of the combined processes reduces the cost of operation, intensifies the OH˙ radical production and as such enhances the overall treatment efficacy of the process.

Future studies in regards to both standalone and hybrid AOPs must concentrate on the recognition of reaction intermediates and scale-up variables, apart from developing various rate expressions. With the utilization of chromatographic techniques, detail analysis of the reactions occurring during the oxidation processes should be investigated. At times, the intermediate products formed during the treatment exhibit more toxic properties than do the parent compounds. Moreover, insufficient mineralization results in the generation of toxic by-products in the solution. Therefore, mineralization and effluent toxicity need to be examined closely at every treatment step for a wide range of pollutants. Also, it is very important to conduct frequent pilot-plant scale experiments employing real industrial effluents, in order to identify the parameters required for the scale-up process. In addition, the values for electrical energy per order must be calculated and analyzed in detail to compare the energy requirement of each AOP. Very few works have been reported on the operational cost estimation of standalone and hybrid AOPs. As such, a more elaborate study on cost estimation for different AOPs must be conducted to obtain deep insight regarding the interconnection of water chemistry, the nature of the pollutants as well as the process variables with associated cost.

REFERENCES

1. Malik, S. N., Ghosh, P. C., Vaidya, A. N., & Mudliar, S. N. (2020). Hybrid ozonation process for industrial wastewater treatment: Principles and applications: A review. *Journal of Water Process Engineering*, 35, 101193.
2. Padmaja, K., Cherukuri, J., & Reddy, M. A. (2020). A comparative study of the efficiency of chemical coagulation and electrocoagulation methods in the treatment of pharmaceutical effluent. *Journal of Water Process Engineering*, 34, 101153.
3. Purkait, M. K., Mondal, P., & Chang, C. T. (2019). *Treatment of Industrial Effluents: Case Studies*. CRC Press.
4. Oller, I., Malato, S., & Sánchez-Pérez, J. (2011). Combination of advanced oxidation processes and biological treatments for wastewater decontamination – a review. *Science of the Total Environment*, 409(20), 4141–4166.
5. Deepti, Sinha, A., Biswas, P., Sarkar, S., Bora, U., & Purkait, M. K. (2020). Utilization of LD slag from steel industry for the preparation of MF membrane. *Journal of Environmental Management*, 259, 110060.
6. Glaze, W. H., Kang, J. W., & Chapin, D. H. (1987). The chemistry of water treatment processes involving ozone, hydrogen peroxide and ultraviolet radiation. *Ozone: Science and Engineering*, 9(4).
7. Wang, J., & Chen, H. (2020). Catalytic ozonation for water and wastewater treatment: Recent advances and perspective. *Science of the Total Environment*, 704, 135249.
8. Hoigné, J., & Bader, H. (1983). Rate constants of reactions of ozone with organic and inorganic compounds in water – II: Dissociating organic compounds. *Water Research*, 17(2), 185–194.
9. Chang, I. S., Lee, S. S., & Choe, E. K. (2009). Digital textile printing (DTP) wastewater treatment using ozone and membrane filtration. *Desalination*, 235(1–3), 110–121.
10. Khuntia, S., Majumder, S. K., & Ghosh, P. (2014). Oxidation of As (III) to As (V) using ozone microbubbles. *Chemosphere*, 97, 120–124.
11. Sani, S., Dashti, A. F., & Adnan, R. (2020). Applications of Fenton oxidation processes for decontamination of palm oil mill effluent: A review. *Arabian Journal of Chemistry*, 13, 7302–7323.
12. Guo, Y., Xue, Q., Zhang, H., Wang, N., Chang, S., Wang, H., Pang, H., & Chen, H. (2018). Treatment of real benzene dye intermediates wastewater by the Fenton method: Characteristics and multi-response optimization. *RSC Advances*, 8(1), 80–90.
13. Rajesh, N., Lai, C. S., Veena, D., & Nguyen, H. T. T. (2015). Treatment of palm oil refinery effluent using advanced oxidation process. *Journal of Engineering Science and Technology*, 10, 26–34.
14. Moreira, F. C., Boaventura, R. A., Brillas, E., & Vilar, V. J. (2017). Electrochemical advanced oxidation processes: A review on their application to synthetic and real wastewaters. *Applied Catalysis B: Environmental*, 202, 217–261.
15. Feng, L., Van Hullebusch, E. D., Rodrigo, M. A., Esposito, G., & Oturan, M. A. (2013). Removal of residual anti-inflammatory and analgesic pharmaceuticals from aqueous systems by electrochemical advanced oxidation processes: A review. *Chemical Engineering Journal*, 228, 944–964.
16. Barışçı, S., Turkay, O., Ulusoy, E., Şeker, M. G., Yüksel, E., & Dimoglo, A. (2018). Electro-oxidation of cytostatic drugs: Experimental and theoretical identification of by-products and evaluation of ecotoxicological effects. *Chemical Engineering Journal*, 334, 1820–1827.
17. Brillas, E., Bastida, R. M., Llosa, E., & Casado, J. (1995). Electrochemical destruction of aniline and 4-chloroaniline for wastewater treatment using a carbon-PTFE O_2-fed cathode. *Journal of the Electrochemical Society*, 142(6), 1733.
18. Özcan, A. A., & Özcan, A. (2018). Investigation of applicability of electro-Fenton method for the mineralization of naphthol blue black in water. *Chemosphere*, 202, 618–625.

19. Moreira, F. C., Boaventura, R. A., Brillas, E., & Vilar, V. J. (2015). Degradation of trimethoprim antibiotic by UVA photoelectro-Fenton process mediated by Fe (III) – carboxylate complexes. *Applied Catalysis B: Environmental*, 162, 34–44.

20. Burgos-Castillo, R. C., Sirés, I., Sillanpää, M., & Brillas, E. (2018). Application of electrochemical advanced oxidation to bisphenol A degradation in water: Effect of sulfate and chloride ions. *Chemosphere*, 194, 812–820.

21. Chatterley, C., & Linden, K. (2010). Demonstration and evaluation of germicidal UV-LEDs for point-of-use water disinfection. *Journal of Water and Health*, 8(3), 479–486.

22. Hamamoto, A., Mori, M., Takahashi, A., Nakano, M., Wakikawa, N., Akutagawa, M., Ikehara, T., Nakaya, Y., & Kinouchi, Y. (2007). New water disinfection system using UVA light-emitting diodes. *Journal of Applied Microbiology*, 103(6), 2291–2298.

23. Wengraitis, S., McCubbin, P., Wade, M. M., Biggs, T. D., Hall, S., Williams, L. I., & Zulich, A. W. (2013). Pulsed UV-C disinfection of *Escherichia coli* with light-emitting diodes, emitted at various repetition rates and duty cycles. *Photochemistry and Photobiology*, 89(1), 127–131.

24. Thomson, J., Roddick, F. A., & Drikas, M. (2004). Vacuum ultraviolet irradiation for natural organic matter removal. *Journal of Water Supply: Research and Technology – AQUA*, 53(4), 193–206.

25. Sun, X., Liu, J., Ji, L., Wang, G., Zhao, S., Yoon, J. Y., & Chen, S. (2020). A review on hydrodynamic cavitation disinfection: The current state of knowledge. *Science of the Total Environment*, 139606.

26. Mezule, L., Tsyfansky, S., Yakushevich, V., & Juhna, T. (2009). A simple technique for water disinfection with hydrodynamic cavitation: effect on survival of *Escherichia coli*. *Desalination*, 248(1–3), 152–159.

27. Amirante, R., Distaso, E., Tamburrano, P., Paduano, A., Pettinicchio, D., & Clodoveo, M. L. (2017). Acoustic cavitation by means ultrasounds in the extra virgin olive oil extraction process. *Energy Procedia*, 126, 82–90.

28. Padoley, K. V., Saharan, V. K., Mudliar, S. N., Pandey, R. A., & Pandit, A. B. (2012). Cavitationally induced biodegradability enhancement of a distillery wastewater. *Journal of Hazardous Materials*, 219, 69–74.

29. Irmak, S., Erbatur, O., & Akgerman, A. (2005). Degradation of 17β-estradiol and bisphenol A in aqueous medium by using ozone and ozone/UV techniques. *Journal of Hazardous Materials*, 126, 54–62.

30. Lu, X., Yang, B., Chen, J., & Sun, R. (2009). Treatment of wastewater containing azo dye reactive brilliant red X-3B using sequential ozonation and up flow biological aerated filter process. *Journal of Hazardous Materials*, 161, 241–245.

31. Saeed, M. O., Azizli, K., Isa, M. H., & Bashir, M. J. K. (2015). Application of CCD in RSM to obtain optimize treatment of POME using Fenton oxidation process. *Journal of Water Process Engineering*, 8, e7–e16.

32. Yang, J., Xing, M.Y., & Zhou, X.B. (2008). Degradation of recalcitrant organics from winery wastewater by Fenton's reaction. *Environmental Engineering Science*, 25, 1229–1234.

33. Tsantaki, E., Velegraki, T., Katsaounis, A., & Mantzavinos, D. (2012). Anodic oxidation of textile dyehouse effluents on boron-doped diamond electrode. *Journal of Hazardous Materials*, 207–208, 91–96.

34. Domínguez, J. R., González, T., Palo, P., Sánchez-Martín, J., Rodrigo, M. A., & Sáez, C. (2012). Electrochemical degradation of a real pharmaceutical effluent. *Water, Air, & Soil Pollution*, 223(5), 2685–2694.

35. Aoyagi, Y., Takeuchi, M., Yoshida, K., Kurouchi, M., Yasui, N., Kamiko, N., Araki, T., & Nanishi, Y. (2011). Inactivation of bacterial viruses in water using deep ultraviolet semiconductor light-emitting diode. *Journal of Environmental Engineering*, 137(12), 1215–1218.

36. Oguma, K., Kita, R., Sakai, H., Murakami, M., & Takizawa, S. (2013). Application of UV light emitting diodes to batch and flow-through water disinfection systems. *Desalination*, 328, 24–30.

37. Wu, X., & Mason, T. J. (2017). Evaluation of power ultrasonic effects on algae cells at a small pilot scale. *Water*, 9(7), 470.

38. Braeutigam, P., Franke, M., Schneider, R. J., Lehmann, A., Stolle, A., & Ondruschka, B. (2012). Degradation of carbamazepine in environmentally relevant concentrations in water by hydrodynamic-acoustic-cavitation (HAC). *Water Research*, 46(7), 2469–2477.

39. Amat, A. M., Arques, A., Lopez, F., & Miranda, M. A. (2005). Solar photo-catalysis to remove paper mill wastewater pollutants. *Solar Energy*, 79(4), 393–401.

40. Dominguez, S., Laso, J., Margallo, M., Aldaco, R., Rivero, M. J., Irabien, Á., & Ortiz, I. (2018). LCA of greywater management within a water circular economy restorative thinking framework. *Science of the Total Environment*, 621, 1047–1056.

41. Miklos, D. B., Remy, C., Jekel, M., Linden, K. G., Drewes, J. E., & Hübner, U. (2018). Evaluation of advanced oxidation processes for water and wastewater treatment – a critical review. *Water Research*, 139, 118–131.

42. Wang, J., & Bai, Z. (2017). Fe-based catalysts for heterogeneous catalytic ozonation of emerging contaminants in water and wastewater. *Chemical Engineering Journal*, 312, 79–98.

43. Chatzitakis, A., Berberidou, C., Paspaltsis, I., Kyriakou, G., Sklaviadis, T., & Poulios, I. (2008). Photocatalytic degradation and drug activity reduction of chloramphenicol. *Water Research*, 42(1–2), 386–394.

44. Zapata, A., Malato, S., Sánchez-Pérez, J. A., Oller, I., & Maldonado, M. I. (2010). Scale-up strategy for a combined solar photo-Fenton/biological system for remediation of pesticide-contaminated water. *Catalysis Today*, 151(1–2), 100–106.

45. Thanekar, P., Garg, S., & Gogate, P. R. (2019). Hybrid treatment strategies based on hydrodynamic cavitation, advanced oxidation processes, and aerobic oxidation for efficient removal of naproxen. *Industrial & Engineering Chemistry Research*, 59(9), 4058–4070.

46. Atalay, S., & Ersöz, G. (2020). Hybrid application of advanced oxidation processes to dyes' removal. In *Green Chemistry and Water Remediation: Research and Applications* (pp. 209–238). Elsevier.

47. Jiao, W., Yu, L., Feng, Z., Guo, L., Wang, Y., & Liu, Y. (2016). Optimization of nitrobenzene wastewater treatment with O_3/H_2O_2 in a rotating packed bed using response surface methodology. *Desalination and Water Treatment*, 57(42), 19996–20004.

48. Rodríguez-Chueca, J., Melero, M. P. O., Abad, R. M., Finol, J. E., & Narvión, J. L. O. (2015). Inactivation of *Escherichia*

coli in fresh water with advanced oxidation processes based on the combination of O_3, H_2O_2, and TiO_2. Kinetic modelling. *Environmental Science and Pollution Research*, 22(13), 10280–10290.

49. Shang, N. C., Chen, Y. H., Ma, H. W., Lee, C. W., Chang, C. H., Yu, Y. H., & Lee, C. H. (2007). Oxidation of methyl methacrylate from semiconductor wastewater by O3 and O3/UV processes. *Journal of Hazardous Materials*, 147(1–2), 307–312.

50. Pillai, K. C., Kwon, T. O., & Moon, I. S. (2009). Degradation of wastewater from terephthalic acid manufacturing process by ozonation catalyzed with Fe^{2+}, H_2O_2 and UV light: Direct versus indirect ozonation reactions. *Applied Catalysis B: Environmental*, 91(1–2), 319–328.

51. Guzmán, J., Mosteo, R., Sarasa, J., Alba, J. A., & Ovelleiro, J. L. (2016). Evaluation of solar photo-Fenton and ozone based processes as citrus wastewater pre-treatments. *Separation and Purification Technology*, 164, 155–162.

52. Malik, S. N., Khan, S. M., Ghosh, P. C., Vaidya, A. N., Kanade, G., & Mudliar, S. N. (2019). Treatment of pharmaceutical industrial wastewater by nano-catalyzed ozonation in a semi-batch reactor for improved biodegradability. *Science of the Total Environment*, 678, 114–122.

53. Bai, Z., Wang, J., & Yang, Q. (2017). Advanced treatment of municipal secondary effluent by catalytic ozonation using Fe 3 O 4-CeO 2/MWCNTs as efficient catalyst. *Environmental Science and Pollution Research*, 24(10), 9337–9349.

54. Huang, G., Pan, F., Fan, G., & Liu, G. (2016). Application of heterogeneous catalytic ozonation as a tertiary treatment of effluent of biologically treated tannery wastewater. *Journal of Environmental Science and Health, Part A*, 51(8), 626–633.

55. Pérez-Estrada, L. A., Malato, S., Gernjak, W., Agüera, A., Thurman, E. M., Ferrer, I., & Fernández-Alba, A. R. (2005). Photo-Fenton degradation of diclofenac: Identification of main intermediates and degradation pathway. *Environmental Science & Technology*, 39(21), 8300–8306.

56. Palas, B., Ersöz, G., & Atalay, S. (2017). Photo Fenton-like oxidation of Tartrazine under visible and UV light irradiation in the presence of LaCuO3 perovskite catalyst. *Process Safety and Environmental Protection*, 111, 270–282.

57. Perisic, D. J., Kovacic, M., Kusic, H., Stangar, U. L., Marin, V., & Bozic, A. L. (2016). Comparative analysis of UV-C/ H_2O_2 and UV-A/TiO_2 processes for the degradation of diclofenac in water. *Reaction Kinetics, Mechanisms and Catalysis*, 118(2), 451–462.

58. Alharbi, S. K., Kang, J., Nghiem, L. D., Van De Merwe, J. P., Leusch, F. D., & Price, W. E. (2017). Photolysis and UV/H_2O_2 of diclofenac, sulfamethoxazole, carbamazepine, and trimethoprim: Identification of their major degradation products by ESI – LC – MS and assessment of the toxicity of reaction mixtures. *Process Safety and Environmental Protection*, 112, 222–234.

59. Cetinkaya, S. G., Morcali, M. H., Akarsu, S., Ziba, C. A., & Dolaz, M. (2018). Comparison of classic Fenton with ultrasound Fenton processes on industrial textile wastewater. *Sustainable Environment Research*, 28(4), 165–170.

60. Gholami, P., Khataee, A., & Bhatnagar, A. (2020). Environmentally superior cleaning of diatom frustules using sono-Fenton process: Facile fabrication of nanoporous silica with homogeneous morphology and controlled size. *Ultrasonics Sonochemistry*, 64, 105044.

61. Anandan, S., Ponnusamy, V. K., & Ashokkumar, M. (2020). A review on hybrid techniques for the degradation of organic pollutants in aqueous environment. *Ultrasonics Sonochemistry*, 105130.

62. Shah, M. P. (2021). *Removal of Emerging Contaminants Through Microbial Processes*. Springer.

63. Shah, M. P. (2020). *Advanced Oxidation Processes for Effluent Treatment Plants*. Elsevier.

64. Shah, M. P. (2020). *Microbial Bioremediation and Biodegradation*. Springer.

65. He, Z., Song, S., Ying, H., Xu, L., & Chen, J. (2007). p-Aminophenol degradation by ozonation combined with sonolysis: Operating conditions influence and mechanism. *Ultrasonics Sonochemistry*, 14(5), 568–574.

66. Ziylan-Yavaş, A., & Ince, N. H. (2018). Catalytic ozonation of paracetamol using commercial and Pt-supported nanocomposites of Al_2O_3: The impact of ultrasound. *Ultrasonics Sonochemistry*, 40, 175–182.

67. Bolton, J. R., Bircher, K. G., Tumas, W., & Tolman, C. A. (2001). Figures-of-merit for the technical development and application of advanced oxidation technologies for both electric-and solar-driven systems (IUPAC Technical Report). *Pure and Applied Chemistry*, 73(4), 627–637.

68. Gerrity, D., Gamage, S., Jones, D., Korshin, G. V., Lee, Y., Pisarenko, A., . . . Snyder, S. A. (2012). Development of surrogate correlation models to predict trace organic contaminant oxidation and microbial inactivation during ozonation. *Water Research*, 46(19), 6257–6272.

69. Pisarenko, A. N., Stanford, B. D., Yan, D., Gerrity, D., & Snyder, S. A. (2012). Effects of ozone and ozone/peroxide on trace organic contaminants and NDMA in drinking water and water reuse applications. *Water Research*, 46(2), 316–326.

70. Güyer, G. T., & Ince, N. H. (2011). Degradation of diclofenac in water by homogeneous and heterogeneous sonolysis. *Ultrasonics Sonochemistry*, 18(1), 114–119.

71. Xiao, Y., Zhang, L., Zhang, W., Lim, K. Y., Webster, R. D., & Lim, T. T. (2016). Comparative evaluation of iodoacids removal by UV/persulfate and UV/H_2O_2 processes. *Water Research*, 102, 629–639.

72. Sichel, C., Garcia, C., & Andre, K. (2011). Feasibility studies: UV/chlorine advanced oxidation treatment for the removal of emerging contaminants. *Water Research*, 45(19), 6371–6380.

73. Katsoyiannis, I. A., Canonica, S., & von Gunten, U. (2011). Efficiency and energy requirements for the transformation of organic micropollutants by ozone, O_3/H_2O_2 and UV/ H_2O_2. *Water Research*, 45(13), 3811–3822.

2

Electro- and Photo-Fenton-Based Techniques in Wastewater Treatment for Advanced Oxidation of Recalcitrant Pollutants

Satyam, Tarun Gangar, Shweta Patel and Sanjukta Patra

CONTENTS

2.1 Introduction

Industrial, municipal and agricultural waste are the prime sources of water pollution across the globe. Pollutants from industrial processes impart a huge organic load to natural water bodies. Industrial wastewater contains pollutants that are rigid and do not degrade readily by conventional wastewater remediation plants. A high load of recalcitrant pollutants renders water unfit for agricultural, livestock and domestic use. Municipal and industrial wastewater is generally treated by physical, chemical, biological or a combination of these processes. The primary goal of wastewater treatment is to remove organic compounds, solids and nutrients from effluents. Current wastewater treatment technologies are diverse and can effectively target one pollutant at a time. Conventional remediation methods of wastewater from municipal and industrial sources include primary and secondary treatment. Primary wastewater treatment entails settling down solid waste in a sedimentation tank. This step is done just after filtering out large insoluble pollutants. Primary treatment water

is routed through a series of tanks and filters, which separate water from solid contaminants. The subsequent "sludge" is then transferred into a digester, where it is further processed. The first batch of sludge comprises approximately half of the suspended solids found in wastewater. Secondary wastewater treatment includes biofiltration, aeration and oxidation ponds where microbes act on the pollutants and reduce the organic load. Phosphates and nitrates left in wastewater are treated in the last step. All the steps from primary to tertiary treatment of wastewater require plenty of time and effort. Advanced oxidation process (AOP)-based treatment technologies, on the other hand, use highly reactive hydroxyl radicals to achieve complete mineralization of organic compounds into water and carbon dioxide. Fenton process is one of the AOP-based wastewater remediation techniques that can rapidly act upon all organic compounds. Electro- and photo-Fenton processes are the most advanced form of Fenton techniques (Bokare & Choi, 2014; Sharma et al., 2018). This chapter delineates electro- and photo-Fenton-based techniques for wastewater remediation.

Photo-Fenton processes use ultraviolet (UV) or visible light in conjunction with the standard Fenton reaction, which boosts catalytic efficiency, increases biodegradation rate and limits iron sludge production. The photo-Fenton processes utilize light energy to enhance the reduction of Fe^{3+} to Fe^{2+}. At lower pH (2.8–3.5), Fe^{2+} quickly reacts with H_2O_2 to produce Fe^{3+}, primarily existing as $[Fe(OH)]^{2+}$. $[Fe(OH)]^{2+}$ initiates metal charge transfer excitation when exposed to light, which regenerates Fe^{2+} that catalyzes the breakdown of H_2O_2 and releases additional •OH radicals represented by Equation 2.1. These radicals degrade organic contaminants. Furthermore, active photolysis of H_2O_2 generates •OH radicals, which can be used to degrade the organic load in wastewater (Equation 2.2). Complementary catalytic impact of Fe^{2+} and light creates more •OH radicals, which increases the oxidation efficiency in photo-Fenton processes. Sunlight or UV light is used as the primary light source in photo-Fenton processes (M. hui Zhang et al., 2019).

$$[Fe(OH)]^{2+} + h\upsilon \rightarrow \bullet OH \qquad (2.1)$$

$$H_2O_2 + h\upsilon \rightarrow 2 \bullet OH \qquad (2.2)$$

The electro-Fenton process has been designed to address the limitation of the conventional Fenton processes. Conventional Fenton processes have many drawbacks like deposition of iron sludge, regular transport, storage of reagents, high operating cost and risk of handling hazardous chemicals. Incorporating electrochemistry in conventional Fenton can significantly enhance mobility, efficiency, convenience and compatibility. Thus, a fusion of the traditional Fenton process and electrochemistry is known as the electro-Fenton process. In this method, hydrogen peroxide is generally generated *in situ* by electrochemical reduction of O_2 on the cathode. This method reduces the hazard associated with handling H_2O_2. Fe^{3+} produced by the Fenton reaction is reduced to Fe^{2+} on the cathode. This process leads to the production of •OH radicals reducing iron sludge production (Trellu et al., 2018; Zhou et al., 2012). Based on the addition or formation of Fenton reagents, electro-Fenton processes are divided into four groups:

- Cathode electro-Fenton process
- Cathode and Fe^{2+} cycling electro-Fenton process
- Galvanic anode electro-Fenton process
- Fe^{2+} cycling electro-Fenton process

In the cathode electro-Fenton process, Fe^{2+} is added to the system, and production of H_2O_2 is achieved by electrochemical reduction of O_2 on the cathode and is shown in Equation 2.3 (Oturan et al., 2010). In the galvanic anode, electro-Fenton process hydrogen peroxide is added to the system, and Fe^{2+} is electrochemically generated at the galvanic anode. This process is depicted by Equation 2.4 (Kurt et al., 2007). In the Fe^{2+} cycling electro-Fenton process, hydrogen peroxide and Fe^{2+} are added to the system, and Fe^{3+} produced by Fenton reaction is reduced to Fe^{2+} at the cathode that reduces the production of iron sludge and commencing Fe^{2+} input concentration represented by Equation 2.5 (Isarain-Chávez et al., 2011; Rahim Pouran et al., 2015). The cathode and Fe^{2+} cycling electro-Fenton process is a combination of the cathode electro-Fenton process and the Fe^{2+} cycling electro-Fenton process. The

four types of electro-Fenton reaction pathways are shown in Figure 2.1.

$$O_2 + 2H^+ + 2\bar{e} \rightarrow H_2O_2 \qquad (2.3)$$

$$Fe \rightarrow Fe^{2+} + 2\bar{e} \qquad (2.4)$$

$$Fe^{3+} + \bar{e} \rightarrow Fe^{2+} \qquad (2.5)$$

2.2 Type of Fenton Reactors

There are several types of electro/photo-Fenton reactors. The design of these reactors depends on the nature of wastewater to be remediated. Apart from the specific nature of producing hydroxyl radicals by UV lamp or electrodes in the photo- and electro-Fenton processes, all other probes and mechanical parts in both reactors remain the same. In the next section, we will discuss the basic design of electro- and photo-Fenton reactors.

2.2.1 For Electro-Fenton Reaction

Electro-Fenton reactors have a metallic (generally stainless steel) body that acts as a cathode. The centrally placed anode (IrO_2/Ta_2O_5) electrode and steel body is connected to a DC power supply. However, based on the type of wastewater used and significant constituents, the electrode used in electro-Fenton reactors varies. Real-time dissolved oxygen (DO), pH and oxidation reduction potentiometer (ORP) sensors attached to the reactor continuously monitor pH, DO, and oxidation reduction values. A sparger and stirrer continuously supply air and mix wastewater thoroughly. Automatic dosing pumps dispense Fe^{2+}/H_2O_2 as per the requirement. All dosing pumps and probes are connected and controlled by a local computer. A quintessential electro-Fenton reactor has shown promising results by degrading 80% of the initial 100 mg/L of dye-contaminated wastewater in 20 minutes (Bañuelos et al., 2014). A schematic diagram of the electro-Fenton reactor is shown in Figure 2.2.

2.2.2 For Photo-Fenton Reaction

Photo-Fenton reactors have a medium-pressure Hg lamp at the center of the reactor. This lamp is housed in quartz glass for proper light distribution and reduces the heat generated by the Hg lamp. These reactors support both batch and continuous (less efficient) modes of water remediation. There is a dedicated sparger connected to an air pump for constant aeration of the reactor chamber. A temperature control unit attached to the reactor's body provides an appropriate continuous temperature for the Fenton reaction. Photo-Fenton reactors are also equipped with a tunable stirrer to mix wastewater with Fenton reagents. Advance photo-Fenton reactors have many sensors such as the dissolved oxygen sensor, pH sensor, temperature sensor, automatic reagent dispenser and water-pumping unit attached to the main reactor. Photo-Fenton reactors with the aforementioned equipment have been successfully tested to remove 83.2% of COD from landfill leachate from municipal solid waste under optimal conditions (Bañuelos et al., 2014).

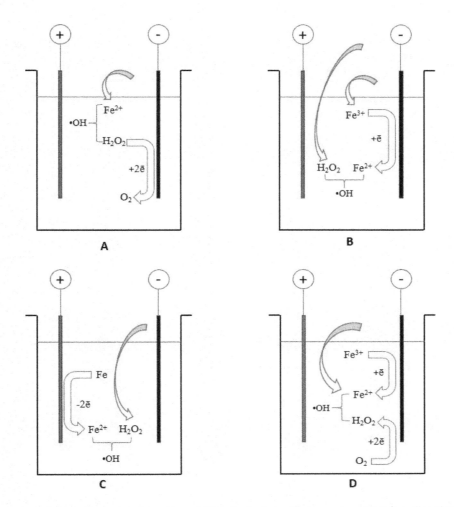

FIGURE 2.1 The four types of electro-Fenton reaction pathways: (A) cathode electro-Fenton process, (B) Fe^{2+} cycling electro-Fenton process, (C) galvanic anode electro-Fenton process and (D).Cathode and Fe^{2+} cycling electro-Fenton process.

A schematic diagram of a photo-Fenton reactor is shown in Figure 2.3.

2.3 Kinetics of Fenton Reaction

Reduction of organic pollutants can be described by the Fenton process, which can be represented by Equation 2.6:

$$RH + \bullet OH \rightarrow \text{Oxidation products} \tag{2.6}$$

Here, •OH is highly reactive and does not pile up in solution. It maintains an almost constant concentration during wastewater treatment. The decay rate of organic pollutant in water can be expressed as:

$$\frac{d[RH]}{dt} = K_{abs}[RH][\bullet OH] \tag{2.7}$$

As [•OH] is static in a steady state, K_{abs} [•OH] is equal to K_{app}. Where K_{abs} is the absolute rate constant, and K_{app} is the apparent rate constant. Now, the kinetic competition method can be used to calculate the value of K_{abs} using a standard. Generally,

benzoic acid is used as a standard, with a K_{abs} value of 4.3×10^9 L mol^{-1} s^{-1}. Now, Equation 2.7 can be represented as:

$$d\frac{[RH]}{dt} = K_{app}[RH] \tag{2.8}$$

Integration of Equation 2.8 yields a first-order kinetic equation of Fenton process, which can be represented as:

$$\frac{\ln[RH]_0}{[RH]_t} = K_{app}t \tag{2.9}$$

Here, $[RH]_0$ is the initial concentration of organic pollutant, and $[RH]_t$ is the concentration of an organic pollutant at a time "*t*". The slope of concentration vs. time plot in compliance with the Equation 2.9 can be used to calculate K_{app} analytically. The equation for the second-order kinetic equation of Fenton reaction can be represented as:

$$\frac{[RH]_t}{[RH]_0} = \frac{1}{1 + Kt[RH]_0} \tag{2.10}$$

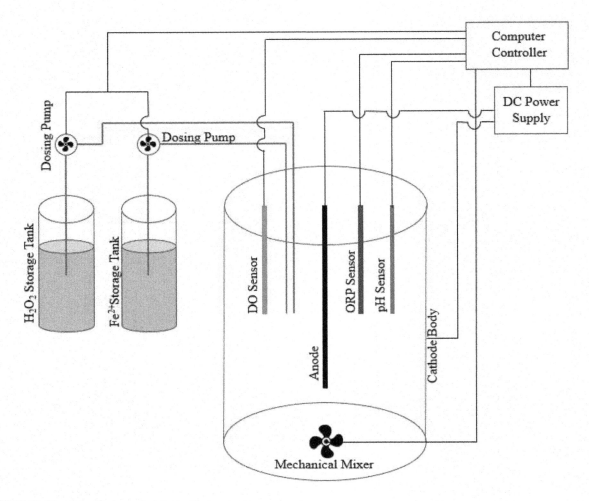

FIGURE 2.2 Illustrative diagram of an electro-Fenton reactor.

2.4 Application of Fenton Processes in Different Industrial Sectors

A lot of emphasis has been placed on the application of efficient Fenton reaction-based techniques for the remediation of toxic and recalcitrant compounds from wastewater. With low energy and capital requirements, the Fenton process can degrade a wide range of refractory organic compounds. This ability makes Fenton-based processes suitable for the remediation of wastewater from various industries. Some significant industries where Fenton-based remediation techniques were successfully tested for wastewater remediation are discussed next (Liu et al., 2012).

2.4.1 Remediation of Pollutants from Textile Industries Using Electro/ Photo-Fenton Processes

The textile industry is a huge global market that has an impact on every country. Most of the 7×10^5 tons of commercial dyes produced globally make their way to textile industries. Wastewater discharges from textile industries impart

enormous damage to water bodies. They block sunlight and cause irreversible damage to aquatic life-forms (Crini, 2006; S. Wang, 2008). Some of the dyes used in textile industries, such as CI Disperse Blue 373, CI Disperse Orange and nitroaminoazobenzene, are mutagenic and carcinogenic. Even less than one part per million of some dyes are visible with the naked eye and are undesirable in natural water bodies as they can severely affect aquatic ecosystems (Alves de Lima et al., 2007; Eren & Acar, 2006).

Fenton process-based remediation methods are a powerful tool for remediation of dye-contaminated water. Dyes like p-methyl red, azobenzene and methyl orange were successfully treated using Fenton reaction with 80% chemical oxygen demand (COD) removal efficiency. The primary reaction mechanism involved was azo bond cleavage followed by aromatic ring hydroxylation (Guivarch et al., 2003). A synthetic dye mixture with yellow drimaren, methylene blue and Congo red was also remediated by Lahkimi et al. with a COD removal efficiency of 89% (Lahkimi et al., 2007). Similarly, efficacious removal of dyes like yellow 36 (Ruiz et al., 2011), indigo carmine (Flox et al., 2006), rhodamine B (Ai et al., 2008) and levafix blue (C. T. Wang et al., 2008) using Fenton reaction is well researched, showing the

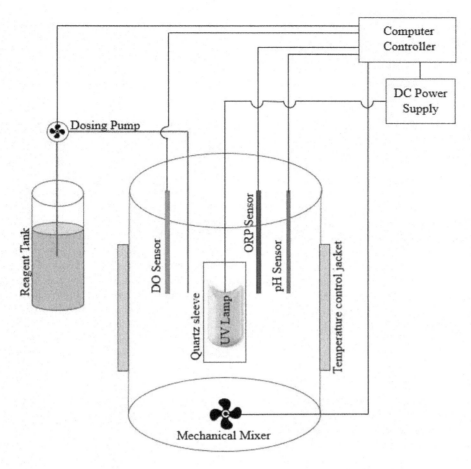

FIGURE 2.3 Illustrative diagram of a photo-Fenton reactor.

potential application of Fenton processes in textile industry wastewater remediation.

2.4.2 Remediation of Wastewater from Agricultural Runoff Using Electro/Photo-Fenton Processes

Pesticides and herbicides are the two commonly detected contaminants in most natural water bodies because of their widespread use in agricultural operations (Kowalkowski et al., 2007). Most of these pesticides are used in quantities greater than 50,000 kg per year. Chlorinated pesticides like HCH (hexachlorocyclohexane) isomers, DDT (dichlorodiphenyltrichloroethane) and their derivatives are predominantly present in groundwater near farmlands. This is regarded as a widespread issue since these compounds typically have direct adverse effects on living organisms. Fenton process-based remediation techniques are effective in degrading toxic herbicides and pesticides. Significant removal of monocrotophos was observed in acidic pH by using electro-Fenton reaction supplemented with H_2O_2. An herbicide 4-chlorophenoxyacetic acid was reduced 80% by using the Fenton process. Similarly, propham (Özcan et al., 2009), diuron (Edelahi et al., 2003), imazapyr (Kaichouh et al., 2004), chlortoluron (Abdessalem

et al., 2008), picloram (Özcan, Şahin, Koparal et al., 2008), imazaquin (Kaichouh et al., 2008), amitrole (Da Pozzo et al., 2005), carbofuran (Abdessalem et al., 2010) and bentazon (Abdessalem et al., 2010) were successfully degraded with high total organic carbon (TOC) and COD removal efficiency. The reduction time taken by electro/photo-Fenton reaction in the cases described earlier was much less (10 minutes to 8 hours depending upon the nature of the pollutant) compared to the conventional method of organic load reduction using microbes.

2.4.3 Remediation of Pollutants from the Pharmaceutical Industry Using Electro/Photo-Fenton Processes

In recent years, many substantial pharmaceutical drugs have been reported in natural water bodies across the globe. Chlorophene, a widely used strong antimicrobial drug, has been recently found in many water bodies in a range of 10–50 mg/L. Direct municipal disposal and wastewater discharge from pharmaceutical industries are the primary sources of chemical drugs in natural water bodies. Some of the drugs identified in water bodies are carcinogenic and mutagenic. Such a compound in wastewater can seriously impact the whole aquatic food chain. Oxidation methods like Fenton process-based water remediation technique can effectively combat a wide range of pharma waste. Photo- and electro-Fenton processes can effectively degrade pharmaceutical waste with aromatic compounds, chlorophene and β blocker atenolol (Isarain-Chávez et al., 2010; Sirés et al., 2007).

2.5 Factors Affecting Fenton-Based Remediation

In advanced oxidation processes, the homogeneous Fenton reaction is a major event for the production of hydroxyl radicals. Fenton's oxidation efficiency is induced by hydroxyl radicals produced during the iron-catalyzed breakdown of H_2O_2 in the acidic medium. The efficiency and rate of degradation of various compounds are dependent on multiple factors like temperature, oxygen sparging rate, pH, iron concentration, applied current density (in electro-Fenton), the distance between the electrodes (in electro-Fenton), light intensity, and reach (in photo-Fenton) and nature of supporting electrolyte/chemicals used. Some of these crucial factors are discussed in the following subsections.

2.5.1 Role of pH in Fenton Processes

One of the most critical factors influencing the overall efficiency of the Fenton process is the pH value. The pH has a powerful impact on free radical formation. The Fenton process is typically performed in an acidic condition because it promotes the production of H_2O_2. According to Ghoneim et al., the overall efficiency of the Fenton process rises at pH 3 (Ghoneim et al., 2011). The iron species tend to precipitate at higher pH values as ferric hydroxides in a typical Fenton process. Conversely, low pH values lead to the deactivation of the catalysts due to the production of stable complexes of iron species with H_2O_2, resulting in a lower oxidation efficiency. Low pH usually lowers the number of active sites for the production of hydrogen peroxide, resulting in the evolution of hydrogen. At pH above 5, the effectiveness of the Fenton reaction decreases significantly because of the unstable nature of H_2O_2 in a basic solution (C. T. Wang et al., 2010; Zhou et al., 2007). The effective pH of Fenton processes slightly varies with the nature of pollutant and type of Fenton reaction. As per the comparative analysis done by Nidheesh and Gandhimathi, the optimum level of pH for the Fenton reaction was estimated to be around 2 to 4.

2.5.2 Role of Temperature in Fenton Processes

In comparison with the other variables, the contribution of temperature in improving the efficiency of a Fenton reaction is relatively lower. An extremely high or low temperature, however, has a negative impact on the Fenton reaction. An increase in temperature in some studies has shown enhanced COD removal efficiency. Elevating temperature above 30 °C decreases the oxygen solubility, which negatively impacts the Fenton reaction as it reduces the concentration of hydrogen peroxide. Since H_2O_2 is less stable at high temperatures, the degradation efficiency of organic pollutants decreases drastically at higher temperatures due to the self-decomposition of H_2O_2. A comparative analysis of various independent studies suggests that the optimal range of temperature for the Fenton process is around 20 to 30 °C (Özcan, Şahin, Savaş Koparal et al., 2008; Umar et al., 2010; C. T. Wang et al., 2010).

2.5.3 Role of Ferrous ion Concentration in Fenton Processes

The Fenton process is highly dependent on optimal ferrous ion concentration. The level of hydroxyl radical increases with an increase in Fe^{2+} concentration; consequently, the efficiency of the Fenton process increases. Hydrogen peroxide's oxidizing potential is insufficient to degrade recalcitrant pollutants when there are no ferrous ions in the Fenton process. Pollution degradation and COD removal efficiency increase with the addition of ferrous ions. TOC removal rate also increases with an increase in ferrous ions. However, an excessive rise in ferrous ions can initiate a competitive reaction between hydroxyl radicals and ferrous ions, which interfere with pollution degradation processes. Therefore, an optimal level of ferrous ion concentration is essentially required for an efficient Fenton process (Sankara Narayanan et al., 2003; C. T. Wang et al., 2010; Zhou et al., 2007).

2.5.4 Peroxide Concentration and Feedback Control in Fenton Processes

The concentration of H_2O_2 plays a crucial role in the Fenton process. Generally, the efficiency of pollution degradation increases with an increase in hydroxyl radical concentration. The concentration of hydroxyl radicals is directly proportional to hydrogen peroxide concentration. Due to the radical-scavenging effect and recombination of hydroxyl radicals of H_2O_2, a very high concentration of peroxide can decrease the efficiency of the Fenton process. A gradual but limiting supply of peroxide is necessary for the effective breakdown of pollutants in a Fenton process (Ting et al., 2009; H. Zhang et al., 2006). Any residual hydroxyl radicals can critically affect aquatic life-forms if left untreated. To ensure the limiting amount of hydrogen peroxide discharge in the Fenton process, many reactors have a feedback control that limits the dosage pump from dispensing excess hydrogen peroxide in the reactor.

2.5.5 Role of Electrolyte, Electrode Distance and Light Source in Fenton Processes

The electrolyte is essential in the E-Fenton process because it boosts solution conductivity by facilitating electron transfer. For the solution with insufficient conductivity, the optimal supporting electrolyte concentration is required. In the E-Fenton process, sodium sulfate (Na_2SO_4) is commonly used as a supporting electrolyte. A high concentration of Na_2SO_4 leads to faster and increased hydrogen peroxide production, which increases the efficiency of electro-Fenton processes. However, highly conducting electrolytes cause heating, which reduces dissolved oxygen levels in the reaction (Jiang & Zhang, 2007).

Another critical factor influencing pollutant removal in the E-Fenton process is the distance between electrodes. When the distance between the electrodes is reduced, the ohmic drop through the electrolyte decreases, resulting in reduced cell voltage and energy consumption. Greater distance between electrodes results in the restricted mass transfer of ferric ion to the cathode surface, regulating ferrous ion regeneration. Thus, optimal distance should be maintained between the electrode for efficient Fenton reaction (Fockedey & Van Lierde, 2002; Qiang et al., 2003; H. Zhang et al., 2006).

In photo-Fenton processes, hydroxyl radicals are generated using light. Thus, the intensity and light source significantly control the photo-Fenton reaction. Generally, UV lamps are used in photo-Fenton reactors. Continuous mixing of reactants and wastewater is therefore required for light energy to act and produce hydroxyl radicals. As excessive heating can negatively impact the Fenton reaction, a temperature control unit is supplemented with a photo-Fenton reactor to control heat generated by the lamp (O'Dowd & Pillai, 2020).

2.6 Iron Recovery after Fenton Reaction

One of the critical disadvantages of the Fenton process is the use of iron as a homogeneous catalyst. As a result, there is a large amount of iron left in treated water at the end of oxidation. High metal concentrations in water bodies can severely impact aquatic life-forms (Martins et al., 2017). Before effluent disposal, Fenton-based remediation techniques must be accompanied by an iron separation technology. The most popular iron recovery method after the Fenton processes is magnetic separation, alkalinization precipitation and ion exchange. Magnetic separation is carried out using a strong magnetic field to separate iron traces after the Fenton reaction. However, this type of separation is only feasible when bioremediation is carried out in a batch reactor, as continuously moving water can still carry iron traces. Alkalinization precipitation is carried out to precipitate out residual metal during the Fenton reaction. This method uses additional chemicals to raise the pH, which later causes reduced Fenton efficiency. The chemicals used in this method later require further management and are harmful to the aquatic environment if left untreated (Ostroski et al., 2009). Ion exchange is one of the most efficient techniques for the removal/recovery of metal particles (Dąbrowski et al., 2004). This method can easily be applied and is distinguished by relatively inexpensive cost and low reactant requirement (Abdel-Ghani et al., 2007). Recent research has shown strong acid cation coupled with weak base anion resin can efficiently reduce iron concentration from 5 mg/L to 200 µg/L (Víctor-Ortega et al., 2014, 2016).

2.7 Advantages and Challenges of the Fenton Process

One of the main advantages of using Fenton reaction-based remediation techniques is the short reaction time. Advanced oxidation processes readily perform the reduction of organic compounds in wastewater, which saves time and effort. In Fenton-based processes, there is no energy requirement as a catalyst or for activation of hydrogen peroxide. The homogeneous catalytic nature of Fenton processes provides limitless mass transfer during remediation steps. Easy operation and maintenance and cost-effective reagents are the major advantages of using Fenton reaction-based water remediation techniques (Nidheesh & Gandhimathi, 2012).

Despite the advantages listed previously, there is some limitation associated with Fenton reaction-based remediation of wastewater. One major disadvantage of Fenton processes is the high consumption rate of ferrous ion than the regeneration rate. The Fenton reaction requires a narrow pH range to act and degrade organic compounds. The presence of complex compounds such as phosphate anions and intermediate oxidation products in wastewater can also lead to iron ions' deactivation. One of the most challenging tasks in the Fenton processes is removing residual Fe ions after the reaction, which requires capital, chemical and human resources (Nidheesh & Gandhimathi, 2012).

2.8 Conclusion

There has been promising development in wastewater treatment technologies in the past decade. The advanced oxidation method of wastewater treatment is one such technology that has a powerful potential to attack almost all organic compounds present in wastewater. So far, ozonation and electro- and photo-Fenton-based processes are widely researched and used to remediate organic pollutants, which significantly reduces TOC, COD, dyes and phenolic compounds typically found in municipal and industrial wastewater. The initial concentration of pollutants, quantity of catalysts and oxidizing agents, illumination strength, irradiation period and wastewater composition (presence of ions and solid compounds, pH) are significant factors controlling Fenton processes. Most of the experiments based on AOP are critically examined to understand unknown chemical reactions so that process optimization can be done accordingly. Process optimization enhances efficacy by limiting residual reagents in the effluents and ensures minimal production of toxic by-products. Fenton processes can be the first choice for remediation of point-source wastewater from industries due to advantages like short reaction time, no energy requirement for catalyst activation, easy operation, low energy and low capital requirement. However, progressive research and development are required to address the iron recovery issue after the remediation process. Also, it has been proven that undertaking pilot plant studies is essential for estimating capital expenditure, production expenses and maintenance cost. These pilot plant tests are effectively capable of providing a close environment to evaluate realistic estimates.

2.9 Acknowledgment

Satyam, Tarun Gangar, Shweta Patel and Sanjukta Patra acknowledge the Department of Biosciences and Bioengineering, Indian Institute of Technology, Guwahati, for providing the infrastructure to carry out research work.

Satyam, Tarun Gangar and Shweta Patel acknowledge the Indian Institute of Technology, Guwahati, for providing research work fellowship. The authors also acknowledge the Indo-EU Horizon 2020 project (BT/IN/EU-WR/60/SP/2018) funded by the Department of Biotechnology, New Delhi.

2.10 Conflict of Interest

The authors declare no conflict of interest.

REFERENCES

Abdel-Ghani, N. T., Hefny, M., & El-Chaghaby, G. A. F. (2007). Removal of lead from aqueous solution using low cost abundantly available adsorbents. *International Journal of Environmental Science & Technology*, 4(1), 67–73.

Abdessalem, A. K., Bellakhal, N., Oturan, N., Dachraoui, M., & Oturan, M. A. (2010). Treatment of a mixture of three pesticides by photo- and electro-Fenton processes. *Desalination*, 250(1), 450–455. https://doi.org/10.1016/j.desal.2009.09.072

Abdessalem, A. K., Oturan, N., Bellakhal, N., Dachraoui, M., & Oturan, M. A. (2008). Experimental design methodology applied to electro-Fenton treatment for degradation of herbicide chlortoluron. *Applied Catalysis B: Environmental*, 78(3–4), 334–341. https://doi.org/10.1016/j.apcatb.2007.09.032

Ai, Z., Xiao, H., Mei, T., Liu, J., Zhang, L., Deng, K., & Qiu, J. (2008). Electro-Fenton degradation of rhodamine B based on a composite cathode of Cu2O nanocubes and carbon nanotubes. *Journal of Physical Chemistry C*, 112(31), 11929–11935. https://doi.org/10.1021/jp803243t

Alves de Lima, R. O., Bazo, A. P., Salvadori, D. M. F., Rech, C. M., de Palma Oliveira, D., & de Aragão Umbuzeiro, G. (2007). Mutagenic and carcinogenic potential of a textile azo dye processing plant effluent that impacts a drinking water source. *Mutation Research – Genetic Toxicology and Environmental Mutagenesis*, 626(1–2), 53–60. https://doi.org/10.1016/j.mrgentox.2006.08.002

Bañuelos, J. A., Rodríguez, F. J., Manríquez, J., Bustos, E., Rodríguez, A., & Godínez, L. A. (2014). A review on arrangement and reactors for Fenton-based water treatment processes. *Evaluation of Electrochemical Reactors as a New Way to Environmental Protection*, 1, 95–135.

Bokare, A. D., & Choi, W. (2014). Review of iron-free Fenton-like systems for activating H2O2 in advanced oxidation processes. In *Journal of Hazardous Materials* (Vol. 275, pp. 121–135). Elsevier. https://doi.org/10.1016/j.jhazmat.2014.04.054

Crini, G. (2006). Non-conventional low-cost adsorbents for dye removal: A review. In *Bioresource Technology* (Vol. 97, Issue 9, pp. 1061–1085). Elsevier. https://doi.org/10.1016/j.biortech.2005.05.001

Dąbrowski, A., Hubicki, Z., Podkościelny, P., & Robens, E. (2004). Selective removal of the heavy metal ions from waters and industrial wastewaters by ion-exchange method. In *Chemosphere* (Vol. 56, Issue 2, pp. 91–106). Elsevier. https://doi.org/10.1016/j.chemosphere.2004.03.006

Da Pozzo, A., Merli, C., Sirés, I., Garrido, J. A., Rodríguez, R. M., & Brillas, E. (2005). Removal of the herbicide amitrole from water by anodic oxidation and electro-Fenton.

Environmental Chemistry Letters, 3(1), 7–11. https://doi.org/10.1007/s10311-005-0104-0

Edelahi, M. C., Oturan, N., Oturan, M. A., Padellec, Y., Bermond, A., & El Kacemi, K. (2003). Degradation of diuron by the electro-Fenton process. *Environmental Chemistry Letters*, 1(4), 233–236. https://doi.org/10.1007/s10311-003-0052-5

Eren, Z., & Acar, F. N. (2006). Adsorption of reactive black 5 from an aqueous solution: equilibrium and kinetic studies. *Desalination*, 194(1–3), 1–10. https://doi.org/10.1016/j.desal.2005.10.022

Flox, C., Ammar, S., Arias, C., Brillas, E., Vargas-Zavala, A. V., & Abdelhedi, R. (2006). Electro-Fenton and photoelectro-Fenton degradation of indigo carmine in acidic aqueous medium. *Applied Catalysis B: Environmental*, 67(1–2), 93–104. https://doi.org/10.1016/j.apcatb.2006.04.020

Fockedey, E., & Van Lierde, A. (2002). Coupling of anodic and cathodic reactions for phenol electro-oxidation using three-dimensional electrodes. *Water Research*, 36(16), 4169–4175. https://doi.org/10.1016/S0043-1354(02)00103-3

Ghoneim, M. M., El-Desoky, H. S., & Zidan, N. M. (2011). Electro-Fenton oxidation of sunset yellow FCF azo-dye in aqueous solutions. *Desalination*, 274(1–3), 22–30. https://doi.org/10.1016/j.desal.2011.01.062

Guivarch, E., Trevin, S., Lahitte, C., & Oturan, M. A. (2003). Degradation of azo dyes in water by electro-Fenton process. *Environmental Chemistry Letters*, 1(1), 38–44.

Isarain-Chávez, E., Arias, C., Cabot, P. L., Centellas, F., Rodríguez, R. M., Garrido, J. A., & Brillas, E. (2010). Mineralization of the drug β-blocker atenolol by electro-Fenton and photoelectro-Fenton using an air-diffusion cathode for H2O2 electro-generation combined with a carbon-felt cathode for Fe2+ regeneration. *Applied Catalysis B: Environmental*, 96(3–4), 361–369. https://doi.org/10.1016/j.apcatb.2010.02.033

Isarain-Chávez, E., Garrido, J. A., Rodríguez, R. M., Centellas, F., Arias, C., Cabot, P. L., & Brillas, E. (2011). Mineralization of metoprolol by electro-Fenton and photoelectro-Fenton processes. *Journal of Physical Chemistry A*, 115(7), 1234–1242. https://doi.org/10.1021/jp110753r

Jiang, C. C., & Zhang, J. F. (2007). Progress and prospect in electro-Fenton process for wastewater treatment. *Journal of Zhejiang University: Science A*, 8(7), 1118–1125. https://doi.org/10.1631/jzus.2007.A1118

Kaichouh, G., Oturan, N., Oturan, M. A., El Kacemi, K., & El Hourch, A. (2004). Degradation of the herbicide imazapyr by Fenton reactions. *Environmental Chemistry Letters*, 2(1), 31–33. https://doi.org/10.1007/s10311-004-0060-0

Kaichouh, G., Oturan, N., Oturan, M. A., El Hourch, A., & El Kacemi, K. (2008). Mineralization of herbicides imazapyr and imazaquin in aqueous medium by, Fenton, photo-Fenton and electro-Fenton processes. *Environmental Technology*, 29(5), 489–496. https://doi.org/10.1080/09593330801983516

Kowalkowski, T., Gadzała-Kopciuch, M., Kosobucki, P., Krupczyńska, K., Ligor, T., & Buszewski, B. (2007). Organic and inorganic pollution of the Vistula River basin. *Journal of Environmental Science and Health, Part A*, 42(4), 421–426. https://doi.org/10.1080/10934520601187336

Kurt, U., Apaydin, O., & Gonullu, M. T. (2007). Reduction of COD in wastewater from an organized tannery industrial region by electro-Fenton process. *Journal of Hazardous Materials*, 143(1–2), 33–40. https://doi.org/10.1016/j.jhazmat.2006.08.065

Lahkimi, A., Oturan, M. A., Oturan, N., & Chaouch, M. (2007). Removal of textile dyes from water by the electro-Fenton process. *Environmental Chemistry Letters*, 5(1), 35–39. https://doi.org/10.1007/s10311-006-0058-x

Liu, Y., Li, J., Zhou, B., Lv, S., Li, X., Chen, H., Chen, Q., & Cai, W. (2012). Photoelectrocatalytic degradation of refractory organic compounds enhanced by a photocatalytic fuel cell. *Applied Catalysis B: Environmental*, 111–112, 485–491. https://doi.org/10.1016/j.apcatb.2011.10.038

Martins, P. J. M., Reis, P. M., Martins, R. C., Gando-Ferreira, L. M., & Quinta-Ferreira, R. M. (2017). Iron recovery from the Fenton's treatment of winery effluent using an ion-exchange resin. *Journal of Molecular Liquids*, 242, 505–511. https://doi.org/10.1016/j.molliq.2017.07.041

Nidheesh, P. V., & Gandhimathi, R. (2012). Trends in electro-Fenton process for water and wastewater treatment: An overview. In *Desalination* (Vol. 299, pp. 1–15). Elsevier. https://doi.org/10.1016/j.desal.2012.05.011

O'Dowd, K., & Pillai, S. C. (2020). Photo-Fenton disinfection at near neutral pH: Process, parameter optimization and recent advances. In *Journal of Environmental Chemical Engineering* (Vol. 8, Issue 5, p. 104063). Elsevier. https://doi.org/10.1016/j.jece.2020.104063

Ostroski, I. C., Barros, M. A. S. D., Silva, E. A., Dantas, J. H., Arroyo, P. A., & Lima, O. C. M. (2009). A comparative study for the ion exchange of Fe(III) and Zn(II) on zeolite NaY. *Journal of Hazardous Materials*, 161(2–3), 1404–1412. https://doi.org/10.1016/j.jhazmat.2008.04.111

Oturan, N., Zhou, M., & Oturan, M. A. (2010). Metomyl degradation by electro-Fenton and electro-Fenton-like processes: A kinetics study of the effect of the nature and concentration of some transition metal ions as catalyst. *Journal of Physical Chemistry A*, 114(39), 10605–10611. https://doi.org/10.1021/jp1062836

Özcan, A., Şahin, Y., Koparal, A. S., & Oturan, M. A. (2008a). Degradation of picloram by the electro-Fenton process. *Journal of Hazardous Materials*, 153(1–2), 718–727. https://doi.org/10.1016/j.jhazmat.2007.09.015

Özcan, A., Şahin, Y., Koparal, A. S., & Oturan, M. A. (2008b). Carbon sponge as a new cathode material for the electro-Fenton process: Comparison with carbon felt cathode and application to degradation of synthetic dye basic blue 3 in aqueous medium. *Journal of Electroanalytical Chemistry*, 616(1–2), 71–78. https://doi.org/10.1016/j.jelechem.2008.01.002

Özcan, A., Şahin, Y., Koparal, A. S., & Oturan, M. A. (2009). A comparative study on the efficiency of electro-Fenton process in the removal of propham from water. *Applied Catalysis B: Environmental*, 89(3–4), 620–626. https://doi.org/10.1016/j.apcatb.2009.01.022

Qiang, Z., Chang, J. H., & Huang, C. P. (2003). Electrochemical regeneration of Fe2+ in Fenton oxidation processes. *Water Research*, 37(6), 1308–1319. https://doi.org/10.1016/S0043-1354(02)00461-X

Rahim Pouran, S., Abdul Aziz, A. R., & Wan Daud, W. M. A. (2015). Review on the main advances in photo-Fenton oxidation system for recalcitrant wastewaters. In *Journal of Industrial and Engineering Chemistry* (Vol. 21, pp. 53–69). Korean Society of Industrial Engineering Chemistry. https://doi.org/10.1016/j.jiec.2014.05.005

Ruiz, E. J., Arias, C., Brillas, E., Hernández-Ramírez, A., & Peralta-Hernández, J. M. (2011). Mineralization of Acid Yellow 36azo dye by electro-Fenton and solar photoelectro-Fenton processes with a boron-doped diamond anode. *Chemosphere*, 82(4), 495–501. https://doi.org/10.1016/j.chemosphere.2010.11.013

Sankara Narayanan, T. S. N., Magesh, G., & Rajendran, N. (2003). Degradation of O-chlorophenol from aqueous solution by electro-Fenton process. *Fresenius Environmental Bulletin*, 12(7), 776–780.

Sharma, A., Ahmad, J., & Flora, S. J. S. (2018). Application of advanced oxidation processes and toxicity assessment of transformation products. In *Environmental Research* (Vol. 167, pp. 223–233). Academic Press Inc. https://doi.org/10.1016/j.envres.2018.07.010

Sirés, I., Garrido, J. A., Rodríguez, R. M., Brillas, E., Oturan, N., & Oturan, M. A. (2007). Catalytic behavior of the Fe3+/Fe2+ system in the electro-Fenton degradation of the antimicrobial chlorophene. *Applied Catalysis B: Environmental*, 72(3–4), 382–394. https://doi.org/10.1016/j.apcatb.2006.11.016

Ting, W. P., Lu, M. C., & Huang, Y. H. (2009). Kinetics of 2,6-dimethylaniline degradation by electro-Fenton process. *Journal of Hazardous Materials*, 161(2–3), 1484–1490. https://doi.org/10.1016/j.jhazmat.2008.04.119

Trellu, C., Oturan, N., Keita, F. K., Fourdrin, C., Pechaud, Y., & Oturan, M. A. (2018). Regeneration of activated carbon fiber by the electro-Fenton process. *Environmental Science and Technology*, 52(13), 7450–7457. https://doi.org/10.1021/acs.est.8b01554

Umar, M., Aziz, H. A., & Yusoff, M. S. (2010). Trends in the use of Fenton, electro-Fenton and photo-Fenton for the treatment of landfill leachate. In *Waste Management* (Vol. 30, Issue 11, pp. 2113–2121). Pergamon. https://doi.org/10.1016/j.wasman.2010.07.003

Víctor-Ortega, M. D., Ochando-Pulido, J. M., Hodaifa, G., & Martinez-Ferez, A. (2014). Final purification of synthetic olive oil mill wastewater treated by chemical oxidation using ion exchange: Study of operating parameters. *Chemical Engineering and Processing: Process Intensification*, 85, 241–247. https://doi.org/10.1016/j.cep.2014.10.002

Víctor-Ortega, M. D., Ochando-Pulido, J. M., & Martínez-Ferez, A. (2016). Thermodynamic and kinetic studies on iron removal by means of a novel strong-acid cation exchange resin for olive mill effluent reclamation. *Ecological Engineering*, 86, 53–59. https://doi.org/10.1016/j.ecoleng.2015.10.027

Wang, C. T., Chou, W. L., Chung, M. H., & Kuo, Y. M. (2010). COD removal from real dyeing wastewater by electro-Fenton technology using an activated carbon fiber cathode. *Desalination*, 253(1–3), 129–134. https://doi.org/10.1016/j.desal.2009.11.020

Wang, C. T., Hu, J. L., Chou, W. L., & Kuo, Y. M. (2008). Removal of color from real dyeing wastewater by electro-Fenton technology using a three-dimensional graphite cathode. *Journal of Hazardous Materials*, 152(2), 601–606. https://doi.org/10.1016/j.jhazmat.2007.07.023

Wang, S. (2008). A Comparative study of Fenton and Fenton-like reaction kinetics in decolourisation of wastewater. *Dyes and Pigments*, 76(3), 714–720. https://doi.org/10.1016/j.dyepig.2007.01.012

Zhang, H., Zhang, D., & Zhou, J. (2006). Removal of COD from landfill leachate by electro-Fenton method. *Journal of Hazardous Materials*, 135(1–3), 106–111. https://doi.org/10.1016/j.jhazmat.2005.11.025

Zhang, M. hui, Dong, H., Zhao, L., Wang, D. xi, & Meng, D. (2019). A review on Fenton process for organic wastewater treatment based on optimization perspective. In *Science of the Total Environment* (Vol. 670, pp. 110–121). Elsevier B.V. https://doi.org/10.1016/j.scitotenv.2019.03.180

Zhou, M., Tan, Q., Wang, Q., Jiao, Y., Oturan, N., & Oturan, M. A. (2012). Degradation of organics in reverse osmosis concentrate by electro-Fenton process. *Journal of Hazardous Materials, 215–216*, 287–293. https://doi.org/10.1016/j.jhazmat.2012.02.070

Zhou, M., Yu, Q., Lei, L., & Barton, G. (2007). Electro-Fenton method for the removal of methyl red in an efficient electrochemical system. *Separation and Purification Technology, 57*(2), 380–387. https://doi.org/10.1016/j.seppur.2007.04.021

3

Emerging Contaminants

Nouha Bakaraki Turan, Hanife Sari Erkan and Guleda Onkal Engin

CONTENTS

3.1 Introduction

The increased consumption of chemicals all around the world has resulted in chemical pollution of the aquatic environment. The chemical compounds available at trace levels of $\mu g\ L^{-1}$ to $ng\ L^{-1}$ are called micropollutants (MPs). These emerging pollutants pose a public concern due to unknown side effects to aquatic and human life [1–2]. Micropollutants are released from different industrial activities, runoff from different livestock and agricultural activities and wastewater effluents. Currently available regulations around the world do not implement strict regulations for the management of industrial effluents. Besides, conventional wastewater treatment plants are not suitable for the complete removal of these trace organic compounds [3]. However, based on the pollutant properties, removal is carried out extensively through biological treatment processes, mainly the conventional activated sludge process, oxidation ditch, anaerobic/anoxic/oxic process, membrane bioreactor or advanced treatment process such as ozonation, UV photolysis and chlorination [4]. Membrane bioreactors and the conventional activated sludge process are reported as the most employed processes [3]. Advanced oxidation processes (AOPs) garnered great interest for their possible application in micropollutant removal. AOPs are mostly applied as a post-treatment process after the biological process in both water and wastewater treatment plants. The application of AOPs, as a post-treatment process, aims to reduce the toxicity of micropollutants by transforming them into less harmful compounds. However, there is a possibility to apply AOPs as a pre-treatment step especially when the effluent is highly toxic, as they increase effluent biodegradability and reduce its toxicity [5]. AOPs utilize highly reactive radicals, mainly hydroxyl radicals and sulfate radicals, for the degradation of a wide range of carbon-based pollutants. Fenton, UV, ozone, electrochemical, sulfate and nanomaterials based AOPs are of interest for micropollutant removal [6]. The generation of toxic by-products and high operating cost and energy requirement are the main disadvantages of AOPs [7].

This chapter aims to define micropollutants based on their different categories, production sources and adverse effects. Various analytical techniques used for the detection of micropollutants will be discussed as well. Furthermore, this chapter will also present different AOPs used in the removal of micropollutants, providing the reader with different example studies from the literature. Finally, a brief overview of the cost operation of these AOPs will be presented, supporting the literature data with a comprehensive chapter related to micropollutant detection by AOPs.

DOI: 10.1201/9781003165958-3

3.2 Classification of MPs

MPs are classified based on their properties and their application areas such as pharmaceuticals (PhACs), personal care products (PCPs), industrial products and pesticides. Further classification can be performed as seen in Figure 3.1. For instance, fragrances, antimicrobial agents, detergents, sunscreens and insect repellants are grouped under the class named "personal care products." Whereas hormones, ß-blockers, analgesics, antiepileptic drugs and anti-inflammatory drugs are examples from pharmaceuticals [8].

3.2.1 Pharmaceuticals

Pharmaceutical compounds are largely produced worldwide for the treatment of many disorders and diseases. The number of commercially produced pharmaceuticals in Europe is estimated to be around 3000. The consumption of these medicaments and their entrance into the human body will lead to their metabolization and their further release into urine and feces. Thus, parent pharmaceutical compounds and their metabolites are known to be available in wastewaters. Lipid regulators, β-blockers, analgesics, anti-inflammatory drugs, psychiatric drugs and antibiotics, antihistamines, iodinated contrast media and antidiabetics are some examples of pharmaceutical drugs [1]. The concentration of pharmaceuticals in raw wastewater varies according to the country and the consumption habits, from less than 1 ng L^{-1} to more than 100 µg L^{-1} [9]. The removal efficiency of pharmaceuticals in wastewater treatment plants (WWTPs) differ based on their characteristics, with an average efficiency of less than 50%. Removal of pharmaceuticals by sorption is only efficient for hydrophobic and positively charged compounds such as ciprofloxacin, ofloxacin,

fenofibrate, norfloxacin and mefenamic acid, with a removal efficiency of between 10% and 80%. However, hydrophilic and negatively charged compounds can be removed by biotransformation and biodegradation processes [1]. For instance, several drugs were reported to be better removed in WWTPs having a sludge retention time higher than 10 days, increasing in turn the possible degradation of these compounds by bacterial enzymatic and metabolic activity [9]. Human drug metabolites are estimated to be a serious issue in WWTPs due to their high polarity and hydrophilic characteristics caused by their transformation in the kidney and liver [1].

3.2.2 Personal Care Products

Personal care products consist of fragrances, insect repellents, sunscreens, shampoos, cosmetics and preservative-like products. Paraben is one of the antimicrobial agents used in deodorants and shampoos found at varying concentrations between 2 ng L^{-1} and 2500 ng L^{-1} in raw wastewaters of Spain. Paraben was reported to be successfully removed in WWTPs up to more than 95%. On the other hand, *N,N*-diethyl-*m*-toluamide (DEET) compound used as an insect repellent showed a removal efficiency between 10% and 99% based on the WWTP type, noting that this compound has low volatility with low sorption characteristics and that biodegradation is the only effective removal process [1]. These products enter the human body by their direct application to the human body and not by ingestion, as pharmaceutical compounds do. Disinfectants such as triclosan and chloroprene are categorized under personal care products. They are released into the aquatic environments through showering, bathing and other recreational activities. A large number of these products have been detected in wastewater treatment plants. Some of them are harmless and degrade into carbon dioxide and water,

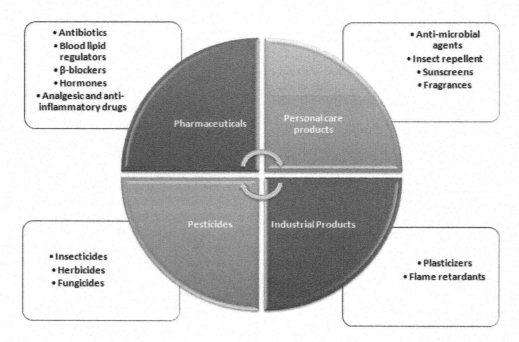

FIGURE 3.1 Classification of the main categories of micropollutants.

while others known as endocrine-disrupting compounds persist in the environment, posing a serious risk to human health [10]. Papageorgiou et al. (2016) reported low and high removal efficiencies for different pharmaceuticals and personal care products in WWTPs, reflecting the fact that these products are not efficiently removed and may persist untreated in the effluent with a concentration varying between 18 ng L^{-1} and 9965 ng L^{-1}, according to their study [11].

3.2.3 Industrial Products

The industrial compounds are mainly divided into food and beverage additives, plasticizers and plastic additives, anticorrosives, chelating agents and retardants. A plasticizer is commonly added to plastic material to increase its flexibility and softness. About 85% of the added plasticizers are phthalates known as phthalic acid esters (PAE). This last was detected in different environmental compartments such as air, seawater, sediments, freshwater and soil samples. The extensive usage of PAEs is associated with various negative impacts such as metabolic disorders, tumor formation, fluctuations in hormone levels and birth defects [12]. Flame retardants are mainly grouped into brominated, chlorinated paraffin and organophosphorus compounds that are widely included in different household equipment to inhibit fire occurrence [1]. Kim et al. (2017) investigated the occurrence of 14 organophosphate flame retardants (OPFRs), 18 plasticizers and their metabolites in a wastewater treatment plant in the New York State. The studied OPFR compounds were detected in wastewater with an average concentration that fluctuated between 20.1 ng L^{-1} and 30,100 ng L^{-1}, for tris(methylphenyl) phosphate (TMPP) and tris(2-butoxyethyl)phosphate (TBOEP), respectively. However, their concentrations reduced in effluents registering a concentration of 7.68 ng L^{-1} for TMPP and 12,600 ng L^{-1} for TBOEP. The fate of these compounds was investigated in sludge as well declaring their presence in the WWTP sludge. Thus, the removal of OPFRs in WWTPs is suggested to be incomplete, as the removal efficiencies of these compounds were only around 60% for TMPP, TBOEP and tris(2–35 ethylhexyl)phosphate (TEHP) and lower than 40% for the other OPFRs [13].

3.2.4 Pesticides

Pesticides mainly grouped into herbicides, fungicides and insecticides are used in agriculture to prevent, mitigate and destroy the growth of unwanted microorganisms, plants, insects and animals. Furthermore, pesticides are applied for the regulation of the plant growth, nitrogen production, the drying of plant tissue and their defoliance [14]. Persistent organic pollutants (POPs) are a group of pollutants characterized by their long-term persistence, bioaccumulation, toxicity and high potential to be transported into different environmental compartments. Twelve of the "old" generation group of pesticides were classified within the group of POPs being thus restricted from usage [15]. The European Commission listed 1331 types of pesticides where 667 are not approved and 48 of them are still on review. The extensive application of these pesticides leads to the contamination of soil, rivers and different environmental compartments in distal regions [16]. The concentration of pesticides in WWTPs is highly affected by precipitation and seasonal changes, reaching up to several µg L^{-1} during high rain loads. However, the average concentration of pesticides is usually lower than 1 µg L^{-1}. The removal of pesticides in WWTPs do not exceed 50% of removal efficiency, leading thus to a high concentration of pesticides in the effluents [1].

3.3 Sources of Micropollutants

MPs are ubiquitous in the water environment, having various origins mainly from domestic and hospital households, industrial activities, animal farming, agricultural activities, aquaculture and untreated sewage [7]. These origins can be classified into point sources such as wastewater treatment plant effluents and non-point or diffuse sources such as the runoff from roads and fields. For instance, pesticides, listed as one of the main classes of MPs, are widely applied to farmland and livestock-rearing areas in order to control the appearance of insects, weeds, pests and various disease-carrying living organisms and to increase crop productivity. The intensive use of these pesticides results in further runoff and release of these chemicals to the aquatic environment. The same is true for the antibiotics and hormonal steroids used in the maintenance of livestock. Furthermore, the consumption of different pharmaceuticals by humans and animals leads to their complete metabolization into residual pharmaceuticals and metabolites that will be excreted in urine and feces, finally reaching wastewater and the aquatic environment. Contaminated surface water infiltration and hyporheic interaction leads to the entrance of MPs into the aquifer layers. Artificial recharge of wastewater, leaky pipes or sewer usage and accidents such as spillages and leakages may result as well in the contamination of groundwater that later reaches the drinking water system [17].

3.4 Behavior and Effects of MPs in the Environment

Five main factors can affect the presence of MPs in natural waters: (1) physicochemical properties, (2) environmental conditions, (3) transformation, (4) transport and retention and (5) accumulation. Other factors such as the water solubility, volatility and stability of the chemical affect the behavior of MPs in water. The movement of MPs between environmental compartments is related to their chemical properties such as the octanol-water partitioning coefficient (K_{ow}) and the logarithmic acid dissociation constant (pKa). MPs can be either retained in water by precipitation, sorption and complexation-like mechanisms or transported by dispersion, diffusion and advection processes [18]. Thus, the physicochemical properties of MPs and their metabolites affect their interaction with the biotic and abiotic environment. However, this interaction can help in getting information about the flow ways, coexisting redox reactions, field temperature and gradients. The use of micropollutants may vary depending on time and place, which complicates making meaningful interpretations on their occurrence [17]. Unfortunately, there is still a lack in implementing a routine monitoring process for the MPs in the water environment. Released MPs may belong to the group of persistent mobile organic compounds (PMOCs) known to last longer in the environment, posing a serious risk for drinking water. Other MPs can consistently be released to the environment with a high possibility to track them all over the world [17].

It is important to know that the lifetime of a product in the environment is related to its chemical and biological characteristics. The attenuation of MPs in the environment can be caused through different biodegradation and nonbiological degradation processes [19]. For instance, sotalol, diclofenac and sulfamethoxazole-like micropollutants submit an abiotic degradation, mainly photolysis, in the water. An example on the biotic attenuation of MPs can be noticed in the WWTPs caused by microbial activity. However, the attenuation and degradation processes can change over time due to ecosystem characteristics [17].

MPs were reported to have adverse effects on human health and ecosystems [8]. The exposure to MPs can be associated with different short-term and long-term toxic effects as well as endocrine-disrupting effects resulting in the occurrence of cancerous tumor and newborn deficits. Thyroid and testicular cancers in males, breast cancer and reproductive failure were reported in connection to possible adverse effects of MPs. A study performed on 12 obese and 12 underweight persons revealed the accumulation of n-propyl-paraben, benzophenone-3, bisphenol A, triclocarban, and methyl paraben in the brain tissue, while methyl-paraben, bisphenol A, ethyl-paraben, triclosan, triclocarban, benzyl-paraben and n-propyl-paraben accumulated in the hypothalamus [20]. It was reported that endocrine disruptors are associated with various metabolic disorders such as diabetes, cardiovascular diseases, insulin resistance and hepatic injury [21]. Furthermore, acute and chronic exposure to endocrine disruptors was reported to have an effect on the reproductive system and histopathological changes in mammals, fishes, birds and snails [8].

3.5 Analytical Methods for the Detection of Micropollutants

Micropollutants are present at concentrations that vary between ng L^{-1} and mg L^{-1} in different environmental compartments such as groundwater, seawater, river, ponds and surface water [22]. The detection of these micropollutants in different matrices in their complex form is a serious issue that requires selective, sensitive and specific instruments [23]. However, a sample preparation step is required prior to measurement in the aim to remove any possible interferences from the matrix. This step consists of increasing the concentration of the target analytes followed by the cleaning up of the samples for further analysis. Different extraction and microextraction methods are used nowadays, such as liquid-liquid extraction, solid-phase extraction, Soxhlet extraction, single drop microextraction, dispersive liquid-liquid microextraction and continuous flow microextraction. Microextraction methods overcome the limitations of the conventional extraction methods that usually demand high organic solvents volume and extended extraction time. Different analytical methods are listed in the literature for the qualitative/quantitative analysis of micropollutants such as Fourier transform infrared spectroscopy (FTIR), UV-Vis spectroscopy, NMR spectroscopy, Raman spectroscopy, HPLC/UV/fluorescence, gas chromatography (GC) and liquid chromatography (LC). Among the listed techniques, column-based chromatography is preferred due to its high selectivity and identification potential of analytes from a complex mixture. Based on the partition of the compounds, GC is used for the thermostable volatile and semi-volatile compounds, while LC is not very effective for the volatile compounds [24]. Liquid chromatography–mass spectrometry (LC-MS) is effective in the measurement of most of the micropollutants, such as bisphenol A, 2-naphthol, ethinyloestradiol and propranolol hydrochloride [25]. Barbosa et al. (2016) analyzed 12 pharmaceuticals in drinking water using a solid-phase extraction procedure followed by ultra-high-performance liquid chromatography coupled to tandem mass spectrometry (offline SPE – UHPLC – MS/MS). A detection limit of 0.20 ng L^{-1} was reached using the previously mentioned method. The highest frequency was obtained for diclofenac while the lowest was for citalopram [26]. GC-MS is another analytical method applied for pharmaceuticals detection. However, in most cases it may need a derivatization step that inhibits any analyte loss that affects the measurement process. Thus, the use of LC-MS may be preferred over GC-MS because it does not require the derivatization step and ensures a rapid analysis for a wide range of micropollutants, reducing in turn the analysis timeframe [27].

3.6 Removal of MPs Using Advanced Oxidation Processes

Different treatment technologies were reported for the removal of MPs from water and wastewater. Both chemical and biological treatment methods were used in the treatment of water and wastewater containing MPs. Adsorption, coagulation and sedimentation and ozonation are some of the applied chemical methods. Reverse osmosis, microfiltration and nanofiltration were some of the membrane technologies used for MP removal [8]. The activated sludge process, biotrickling filter, biofilm reactors and nitrification/denitrification processes are examples of conventional biological treatment methods. However, the application of these conventional methods was found to be limited for MP removal, and advanced biological treatment such as two-phase partitioning bioreactor, membrane-based bioreactor, immobilized cell bioreactor and moving-bed biofilm reactor were reported to have better removal efficiencies [7]. In this chapter, the advanced oxidation processes will be listed for their efficiencies on MP removal.

3.6.1 Hydroxyl Radical–Based AOPs

3.6.1.1 Ozone-Based AOPs

Ozonation is an environmentally friendly process having an oxidation potential of 2.07 V used widely in the treatment of wastewater. Ozonation can act directly or indirectly for the degradation of organic pollutants. During indirect oxidation, ozone enters into reaction with water molecules leading to the formation of hydroxyl radicals known to have a stronger oxidation capacity for the degradation of pollutants [28]. This reaction can be enhanced with the help of homogeneous and heterogeneous catalyzers. Mn^{2+}, Zn^{2+}, Ni^{2+} and Ag^+ are some of the liquid catalysts used during homogeneous catalytic ozonation. On the other hand, activated carbon and metal oxides are examples of the solid catalysts used during heterogeneous catalytic ozonation [29]. However, during the direct ozone-based oxidation processes, ozone is used as a powerful oxidant that attacks amines, aromatic rings and alkenes. Thus, ozonation is applied for the removal of organic micropollutants from water and wastewater effluents and is able to reach a removal percentage higher than 80%. Ozonation does not lead to the mineralization of organics; however, it transforms organics to ozonation transformation products that are less toxic and have a higher resistance to biodegradation. For this reason, activated carbon filters or sand filtration is preferred as a post-ozonation treatment for the reduction of these by-products. The formation of oxidation by-products is mainly related to the dose of ozone applied [30]. Besides, it was proved that the decomposition of antibiotics increases with an increase in the applied dose [31]. For instance, erythromycin, ofloxacin, tetracycline, azithromycin, ampicillin and trimethorprim antibiotics removal increased as the dosage of ozone increased. Besides this, an increase in the initial concentration of ozone was proved to affect the removal of levofloxacin, ciprofloxacin, clarithromycin and nalidixic. Another factor that affects the degradation of organics is pH. For instance, it was reported that antibiotics degradation performance is affected by the pH of the solution. In acidic conditions, pollutants are degraded by direct oxidation of ozone, while, at basic conditions, hydroxyl radicals and ozone molecules intervene in the pollutants' degradation [31]. Supportive results were obtained by Feng et al. (2016a), who reported an increase in the reaction rate of the degradation of flumequine when the pH values increased from 3.0 to 11.0. An increase in pH leads to the transformation of ozone to OH·, which triggers the indirect oxidation of flumequine [32]. Taliawy et al. (2017) tested the removal of different micropollutants using the ozonation process. The removal efficiency of compounds such as diclofenac, carbamazepine, azithromycin, sulfamethoxazole and propanolol was higher than 70%. It was also proved that the removal efficiency may vary depending on the characterization of wastewater tested, especially the pH and the nitrite concentration [33]. Furthermore, Lu et al. (2019) proved the removal of trimethoprim by 70% after the application of ozone at a dose of 2 mg L^{-1} [34].

The main restrictions of the ozonation process can be listed as limited mineralization of dissolved organic compounds (DOCs), incomplete oxidation of the compounds and formation of unknown by-products such as toxic nitrosamines, bromate and formaldehyde. The application of a post-biological filtration is usually proposed for the elimination of the biodegradable by-products, whereas the addition of catalysts or H_2O_2 can be used for further mineralization of DOCs [35].

Hence, hydroxyl radical generation yield can be improved by the presence of other oxidants or irradiation. For this reason, the ozone process can be combined with H_2O_2, which is called peroxone (O_3/H_2O_2), or with ultraviolet irradiation (O_3/UV irradiation) [36]. For the peroxone process, generation of hydroxyl radicals from ozone is accelerated by reducing the time required for the degradation of MPs. The applied ozone concentration during the peroxone process is between 1 and 20 mg L^{-1}, noting that the H_2O_2/O_3 ratio should also be adjusted to around 0.5. A higher H_2O_2 dose is not recommended because it leads to the formation of weak hydroperoxyl radicals (HO_2) [35]. It is important to note that residual H_2O_2 needs to be removed prior to discharge of treated water to the environment. However, this method shows satisfactory reduction in the formation of bromate [37]. The same increase in hydroxyl radical formation can be ensured with the use of ultraviolet light lower than 300 nm. The irradiation disintegrates the O_3 molecule leading to the generation of a thermally excited H_2O_2, which will be later decomposed into two hydroxyl radicals. The limitation of this method is related to the large amount of electricity needed for the UV lamps and the ozone generator [37]. The application of O_3/H_2O_2 and O_3/H_2O_2/UV was tested for the removal of MPs such as atrazine and isoproturon spiked to a pretreated surface water. A removal percentage of 90% for isoproturon was ensured after the application of O_3/H_2O_2, while about 70% of atrazine was removed through the O_3/H_2O_2/UV process [38].

3.6.1.2 UV-Based AOPs

UV-based AOPs consist of the use of UV light to initiate hydroxyl radical generation in the presence of different radical promoters such as catalysts or oxidants. The UV dose applied during AOP treatment is higher than 200 mJ cm^{-2}, exceeding the dose needed for the 4-log inactivation of pathogens [39]. UV-based AOPs can be used in combination with H_2O_2 or with persulfate and chlorine for the generation of radicals, such as sulfate and hydroxyl and chlorine radicals, respectively. The photocatalytic reaction takes place under light irradiation among a photocatalyst, its surface substrate such as H_2O_2 and O_2 and target pollutants [40]. During the photocatalytic reaction, the photocatalyst is excited by the effect of energy leading to the production of oxidative positive holes and reductive negative electrons. The organic pollutant is then decomposed by superoxide radical O_2 and hydroxide radicals HO generated during the reduction of O_2 and the oxidation of H_2O by the effect of reduced electrons and oxidative holes, respectively. Metal oxides (ZnO, TiO_2), metal semiconductors ($BiVO_4$, $GdVO_4$, $SmVO_4$, Ag_3O_4, BiOBr, BiOCl), nonmetallic semiconductors and metal sulfides are different kinds of catalysts applied during photochemical oxidation reactions [40]. The photocatalyst reaction may show performance variations depending on the pH of the solution. For instance, UV-based AOPs using TiO_2 were applied for amoxicillin degradation. This last removal efficiency increased from 75% to 95% when pH was decreased from 7.5 to 5 [41]. In another study, cefotaxime removal was performed by the effect of sunlight in association with TiO_2 and ZnO in solution. The removal efficiency of cefotaxime increased with an increase of pH from 4 to 6.2. However, for pH higher than 6.2, a decrease in the degradation rate of cefotaxime was observed [42]. Besides, photocatalyst dosage applied is a crucial factor in the performance of the reaction. For instance, an applied dosage between 0 to 1 g L^{-1} was sufficient for the degradation of chloramphenicol (CAP). However, higher TiO$_2$ dosage results in a decrease in the light efficiency [43]. The photocatalytic process was tested for the removal of five different pesticides, chlorotoluron, diuron, fluometuron, isoprotouron and linuron, in the presence of ZnS, TiO$_2$, WO$_3$, ZnO and SnO$_2$ catalysts at a fixed dosage of 200 mg L^{-1}. The highest removal efficiency of 99% was obtained in the presence of ZnO at a pH of 8.2 during a reaction time of 240 min [44]. Similar satisfactory results were obtained for the removal of antibiotics using the photocatalytic process. Sarkhosh et al. (2019) applied the UV/ZnO process for the removal of ciproflaxacin. A removal efficiency of 100% was registered for the applied method [45]. Furthermore, MPs such as atenolol, caffeine, bisphenol-A, carbamazepine, diclofenac, ibuprofen and sulfamethoxazole removal efficiencies were investigated in the presence of UV/peroxymonosulfate (UV/PMS), UV/persulfate (UV/PS) and UV/H$_2$O$_2$. Atenolol and caffeine were easily removed at even low UV contact time of 9 seconds and low oxidant dosage of 0.5 mM. Obtained removal efficiency varied between 84% and 100%. However, the rest of the MPs showed variable removal efficiencies based on the dosage, contact time and process applied. UV/PMS shows higher removal efficiencies, between 63% and 83%, than do UV/PS and UV/H$_2$O$_2$ processes [46].

3.6.1.3 Fenton-Based AOPs

Fenton oxidation processes are widely used in wastewater treatment for the degradation of a variety of pharmaceuticals based on the reaction between iron salts (Fe^{2+}) and H$_2$O$_2$. The oxidation of these will lead to the release of (Fe^{3+}) and the generation of hydroxyl radicals (HO$^{\bullet}$) with a high oxidant capacity [47]. The performance of the oxidation process is usually affected by several operational parameters such as pH, H$_2$O$_2$ and Fe^{2+} concentration. High degradation efficiency and the simplicity of the processes are the main advantages of the Fenton processes. However, the effectivity is limited to the acidic conditions of the medium. Besides, the disposal of the treatment sludge from Fenton processes is a serious concern due to the large quantity of produced sludge containing Fe^{2+}. Fenton-like processes are used to overcome these disadvantages using catalysts other than Fe^{2+}. Iron minerals, zerovalent iron, single metal, metallic oxide and metal-organic frameworks are some examples of the catalysts used in heterogeneous Fenton catalysts [31]. The application of Fenton and Fenton-like processes for the removal of pharmaceuticals has been tested in different studies. For instance, the removal of sulfamethoxazole at an initial concentration of 158 µM was studied in the presence of 100 µM of H$_2$O$_2$ and 179 µM of Fe^{2+}. About 99.7% of sulfamethoxazole was removed at a pH of 3.0, reaction time of 50 min and temperature of 25 °C [48]. Furthermore, Gupta et al. (2018) studied the removal of 100 mg L^{-1} of ciprofloxacin at an acidic pH of 3 and a reaction time of 60 min and in the presence of [H$_2$O$_2$]/[Fe^{2+}] with a ratio of 10. Obtained removal efficiency was found to be around 70% [49]. Catalyst dosage is crucial in Fenton and Fenton-like processes affecting the degradation performance

of organic pollutants. The effect of catalyst dosage was tested by Qi et al. (2019), who analyzed the effect of CuCo$_2$O$_4$ nanocatalyst dosage on metacycline removal. The catalyst dosage was increased from 5.0 to 12.0 mg, leading to an increase in the removal efficiency from 38.4% to 89.1%. However, the removal efficiency slightly increased to 92.5% when the catalyst dosage increased to 15.0 mg [50]. On the other hand, the concentration of H$_2$O$_2$ is also important in the effectiveness of Fenton oxidation, as H$_2$O$_2$ is the dominant source of hydroxyl radicals. Insufficient or excessive amount of hydroxyl radicals can negatively affect pollutant degradation [51]. Nasseh et al. (2019) analyzed the effect of H$_2$O$_2$ dosage on the removal of metronidazole in the presence of FeNi$_3$/SiO$_2$. It was found that the removal efficiency increased with H$_2$O$_2$ dosage when increased from 50 mg L^{-1} to 150 mg L^{-1}. However, an excessive increase in H$_2$O$_2$ dosage to 200 mg L^{-1} led to a decrease in the pollutants' degradation [52].

3.6.2 Electrochemical AOPs

Electrochemical oxidation processes (EAOPs) are another class of advanced oxidation processes that have raised interest over the last two decades [2]. Anodic, electro-Fenton, photoelectro-Fenton and solar photoelectro-Fenton are the main types of the EAOPs. The anodic oxidation known as the former and simplest kind of EAOPs consist of the direct oxidation of organics on the anode surface or the indirect oxidation by O$_3$, H$_2$O$_2$ and persulfate-like agents in the solution. However, the addition of Fe^{2+} accompanied with the generation of the H$_2$O$_2$ leads to an additional production of hydroxyl radicals in the solution. This kind of oxidation is called the electro-Fenton process. The combination of artificial or sunlight with the electro-Fenton processes was defined by the Brillas group as the photoelectro-Fenton (PEF) and solar photoelectro-Fenton (SPEF) processes [53]. Initial concentration of organics, nature and quantity of the electrolyte used, applied current density, mixing rate, temperature, pH and initial iron concentration are the main parameters that affect the performance of EAOPs. The role of EAOPs is prominent for the removal of dyes, pharmaceuticals and pesticides from wastewater [2]. Table 3.1 summarizes some of the studies performed on micropollutant removal in the last five years.

3.6.3 Sulfate Radical–Based AOPs

The efficiency of hydroxyl radicals is limited to a narrow pH range between 2 and 4. It is affected by the instability of H$_2$O$_2$, in addition to the limited effectiveness to complex organics such as alkanes. Sulfate-based oxidation using peroxymonosulfate (PMS) and peroxydisulfate (PDS) came to overcome the limitation of hydroxyl radical–based oxidation. SO$_4^-$-based oxidation acts on a wide range of pH with a higher oxidation potential and half-life span [58]. A higher half-life span is usually important in the interaction and mass transfer between sulfate radicals and pollutants [59]. Sulfate-based radicals were evaluated as inert and non-pollutant according to the United States Environmental Protection Agency (USEPA). Besides, acceptable taste and odor values have been listed, allowing for the use of sulfate radicals in wastewater treatment

TABLE 3.1

Electrochemical Studies Conducted on the Removal of Various Micropollutants (Coverage: 2018–2021)

Process	Micropollutant	Operating Conditions	Removal Efficiency (%)	Reference
Electro-Fenton	Metformin (10 mg L^{-1})	Reaction time: 10 min pH: 3 Current density: 6 mA cm^{-2} H_2O_2 dosage: 250 μL L^{-1}	98.57%	[54]
Heterogeneous electro-Fenton using activated carbon fibers (ACFs) supported ferric citrate (Cit-Fe/ACFs) as the cathode, with RuO_2/Ti was used as the anode	Ibuprofen	pH: circumneutral Reaction time: 120 min Current density: 7 mA cm^{-2}	97%	[55]
Photoelectro-Fenton	Ampicillin, sulfamethazine, tetracycline, diclofenac and salicylic acid	Fe^{2+}: 1 mmol L^{-1} Na_2SO_4: 0.05 mol L^{-1} pH: 3 Temperature: 35 °C Reaction time: 100 min Current densities: 10, 25 and 50 mA cm^{-2}	99.2% of TOC	[56]
Anodic oxidation using boron-doped diamond (BDD) as anode and stainless-steel electrode as cathode	30 different pharmaceuticals	pH: natural Current densities: 6, 20 and 40 mA cm^{-2}	High than 85%	[57]

[60]. Ushani et al. (2020) discussed the details of the effect of pH, temperature and transition metals in the effectiveness of sulfate-based radicals in advanced oxidation processes. Briefly, it was reported that PDS decomposition rate to SO_4^- increases in strongly acidic and alkaline solution (3 < pH < 13). Besides, high temperature degrees are more favorable for the formation of reactive SO_4^- from persulfate. Finally, the transition metals such as Mn^{2+}, Co^{2+}, Mn^{2+} and Cu^{2+} was proved to activate the oxidation process by acting as electron donors, which will be used for the formation of sulfate radicals [58]. Wang et al. (2020) investigated the effectiveness of a PMS-amended iron coagulation process for the removal of carbamazepine. The process applied showed improved carbamazepine removal rate from 50% to 80% in comparison to the conventional iron coagulation method [61].

3.6.4 Nanomaterials-Based AOPs

Nanoparticles (NPs) are defined as particles with a size between 1 and 100 nm. NPs are optically active, mechanically strong and chemically reactive particles with a large surface area that gives them unique properties. Based on these characteristics, NPs are applied widely in different fields of science, mainly in drugs and medications, manufacturing materials, electronics and environmental remediation processes [62]. The combination of AOPs with other treatment technologies shows higher hydroxyl radical generation and high energy efficiency in comparison to individual AOP techniques. Nano-based AOPs such as photo-Fenton with nanophotocatalysts, UV/O_3 with nanophotocatalysts, and sono-Fenton catalytic processes involving nanoparticles were applied for the removal of different dyes, pharmaceuticals and pesticides and other toxic pollutants from wastewater [63].

In a study performed on the removal of 4-chlorophenol, the photocatalysis treatment was combined with three different kinds of NPs: the commercial TiO_2, Zr^{4+} doped TiO_2 and nano TiO_2. After the optimization of the parameters, Zr^{4+} doped TiO_2 was found to give higher degradation efficiency for 4-chlorophenol removal in comparison with the other two nanoparticles [64].

The photocatalytic treatment combination with Pt modified nano-sized tungsten trioxides (Pt/WO_3) showed an excellent photocatalytic activity for the removal of ciproflaxacin antibiotic [65]. Furthermore, the hybrid sono-Fenton process combined with Fe_3O_4 magnetic nanoparticles (MNPs) was applied for the degradation of bisphenol-A. About 95% of bisphenol-A was degraded and half of the total organic carbon (TOC) was reduced. Besides, a good stability of the applied Fe_3O_4 MNPs for a pH range changing between 3 and 9 was noticed even after five recycles [66]. Similar hybrid AOPs combined with nanomaterials were applied for the removal of the pharmaceutical drug diclofenac sodium. The hydrodynamic cavitation in conjunction with $UV/TiO_2/H_2O_2$ registered a removal of 95% of diclofenac and a reduction of 76% of the TOC [67]. Mohseni et al. (2021) reported a study on the removal of the pharmaceutical drug carbamazepine using carbon microtubes (CMTs) with entrapped Fe_3O_4 combined with the heterogeneous electro-Fenton process. This technology shows a superiority over other treatment methods that require acidic conditions, as it works under neutral pH values. Besides, the mineralization of carbamazepine is related to unselective radicals that take place once the H_2O_2 reacted with the entrapped Fe_3O_4 nanomaterials [68]. Another study performed by Wei et al. (2020) implemented plasmonic nanoparticles (Au NRs) with silica that work under a visible-wavelength solar degradation. This treatment system drives a persulfate-based advanced oxidation for the removal of the model micropollutant known as phenol. This technology seems to be a promising system for water treatment especially in the places where solar energy is abundant [69]. Mohsin and Mohammed (2021) investigated the efficiency of ozonation in the presence of zinc oxide nanoparticles as catalytic agents for the removal of oxytetracycline antibiotics. The results showed high effectiveness of the catalytic ozonation, as 94% of the oxytetracycline was removed after 35 minutes of the reaction process at a pH of 7, ozone dosage rate of 1.38 mg s^{-1} and ZnO dosage of 100 mg L^{-1} [70].

3.7 Conclusion

This chapter presents emerging contaminants classes, sources and effects on human health and the environment. Pharmaceuticals, hormones, personal care products, industrial products and pesticides are the main sources of micropollutants, and although relatively small amounts have been found in water bodies, the threat caused by them is quite serious. The introductory part is followed by the analytical methods used for detection and advanced oxidation methods used for treatment. The advantages and disadvantages of the treatment methods were also emphasized. It is known that conventional methods have limitations for the removal of micropollutants and, therefore, there is a tendency to use advanced oxidation methods. It is evident that all these advanced oxidation processes are quite effective in removing the micropollutants, with removal efficiencies of higher than 85%. Among other AOPs, nanomaterial-based AOPs stand out due to their high potential for applicability for water and wastewater treatment, high surface area and relatively low production costs.

REFERENCES

1. Margot, J., Rossi, L., Barry, D.A., Holliger, C.: A review of the fate of micropollutants in wastewater treatment plants. *Wiley Interdisciplinary Reviews: Water* **2**(5), 457–487 (2015).
2. Moreira, F.C., Boaventura, R.A., Brillas, E., Vilar, V.J.: Electrochemical advanced oxidation processes: A review on their application to synthetic and real wastewaters. *Applied Catalysis B: Environmental* **202**, 217–261 (2017).
3. Barbosa, M.O., Moreira, N.F., Ribeiro, A.R., Pereira, M.F., Silva, A.M.: Occurrence and removal of organic micropollutants: An overview of the watch list of EU decision 2015/495. *Water Research* **94**, 257–279 (2016).
4. Ben, W., Zhu, B., Yuan, X., Zhang, Y., Yang, M., Qiang, Z.: Occurrence, removal and risk of organic micropollutants in wastewater treatment plants across China: Comparison of wastewater treatment processes. *Water Research* **130**, 38–46 (2018).
5. Ribeiro, A.R., Nunes, O.C., Pereira, M.F., Silva, A.M.: An overview on the advanced oxidation processes applied for the treatment of water pollutants defined in the recently launched directive 2013/39/EU. *Environment International* **75**, 33–51 (2015).
6. Vagi, M.C., Petsas, A.S.: Recent advances on the removal of priority organochlorine and organophosphorus biorecalcitrant pesticides defined by directive 2013/39/EU from environmental matrices by using advanced oxidation processes: An overview (2007–2018). *Journal of Environmental Chemical Engineering* **8**(1), 102940 (2020). https://doi.org/10.1016/j.jece.2019.102940

7. Kanaujiya, D.K., Paul, T., Sinharoy, A., Pakshirajan, K.J.C.P.R.: Biological treatment processes for the removal of organic micropollutants from wastewater: A review. *Current Pollution Reports* **5**(3), 112–128 (2019).
8. Khanzada, N.K., Farid, M.U., Kharraz, J.A., Choi, J., Tang, C.Y., Nghiem, L.D., Jang, A., An, A.K.: Removal of organic micropollutants using advanced membrane-based water and wastewater treatment: A review. *Journal of Membrane Science* **598**, 117672 (2020). doi:https://doi.org/10.1016/j.memsci.2019.117672
9. Margot, J., Kienle, C., Magnet, A., Weil, M., Rossi, L., De Alencastro, L.F., Abegglen, C., Thonney, D., Chèvre, N., Schärer, M.: Treatment of micropollutants in municipal wastewater: Ozone or powdered activated carbon? *Science of the Total Environment* **461**, 480–498 (2013).
10. Tijani, J.O., Fatoba, O.O., Babajide, O.O., Petrik, L.F.: Pharmaceuticals, endocrine disruptors, personal care products, nanomaterials and perfluorinated pollutants: A review. *Environmental Chemistry Letters* **14**(1), 27–49 (2016).
11. Papageorgiou, M., Kosma, C., Lambropoulou, D.: Seasonal occurrence, removal, mass loading and environmental risk assessment of 55 pharmaceuticals and personal care products in a municipal wastewater treatment plant in Central Greece. *Science of the Total Environment* **543**, 547–569 (2016).
12. Armstrong, D.L., Rice, C.P., Ramirez, M., Torrents, A.: Fate of four phthalate plasticizers under various wastewater treatment processes. *Journal of Environmental Science and Health, Part A* **53**(12), 1075–1082 (2018).
13. Kim, U.-J., Oh, J.K., Kannan, K.: Occurrence, removal, and environmental emission of organophosphate flame retardants/plasticizers in a wastewater treatment plant in New York State. *Environmental Science & Technology* **51**(14), 7872–7880 (2017).
14. Sabarwal, A., Kumar, K., Singh, R.P.: Hazardous effects of chemical pesticides on human health – cancer and other associated disorders. *Environmental Toxicology and Pharmacology* **63**, 103–114 (2018).
15. Sander, L.C., Schantz, M.M., Wise, S.A.: Environmental analysis: Persistent organic pollutants. *Liquid Chromatography*, 401–449 (2017).
16. de Souza, R.M., Seibert, D., Quesada, H.B., de Jesus Bassetti, F., Fagundes-Klen, M.R., Bergamasco, R.: Occurrence, impacts and general aspects of pesticides in surface water: A review. *Process Safety and Environmental Protection* **135**, 22–37 (2020).
17. Warner, W., Licha, T., Nödler, K.: Qualitative and quantitative use of micropollutants as source and process indicators: A review. *Science of the Total Environment* **686**, 75–89 (2019).
18. Kim, M.-K., Zoh, K.-D.: Occurrence and removals of micropollutants in water environment. *Environmental Engineering Research* **21**(4), 319–332 (2016).
19. Jekel, M., Dott, W., Bergmann, A., Dünnbier, U., Gnirß, R., Haist-Gulde, B., Hamscher, G., Letzel, M., Licha, T., Lyko, S., Miehe, U., Sacher, F., Scheurer, M., Schmidt, C.K., Reemtsma, T., Ruhl, A.S.: Selection of organic process and source indicator substances for the anthropogenically influenced water cycle. *Chemosphere* **125**, 155–167 (2015). doi:https://doi.org/10.1016/j.chemosphere.2014.12.025

20. Van der Meer, T.P., Artacho-Cordón, F., Swaab, D.F., Struik, D., Makris, K.C., Wolffenbuttel, B.H.R., Frederiksen, H., Van Vliet-Ostaptchouk, J.V.: Distribution of non-persistent endocrine disruptors in two different regions of the human brain. *International Journal of Environmental Research and Public Health* **14**(9), 1059 (2017).

21. Desai, M., Jellyman, J., Ross, M.J.: Epigenomics, gestational programming and risk of metabolic syndrome. *International Journal of Obesity* **39**(4), 633–641 (2015).

22. Ahmed, M.B., Zhou, J.L., Ngo, H.H., Guo, W., Thomaidis, N.S., Xu, J.: Progress in the biological and chemical treatment technologies for emerging contaminant removal from wastewater: A critical review. *Journal of Hazardous Materials* **323**, 274–298 (2017). doi:https://doi.org/10.1016/j.jhazmat.2016.04.045

23. Núñez, M., Borrull, F., Pocurull, E., Fontanals, N.: Sample treatment for the determination of emerging organic contaminants in aquatic organisms. *TrAC Trends in Analytical Chemistry* **97**, 136–145 (2017). doi:https://doi.org/10.1016/j.trac.2017.09.007

24. Ahmed, M.B., Johir, M.A.H., Ngo, H.H., Guo, W., Zhou, J.L., Belhaj, D., Moni, M.A.: Chapter 4 – Methods for the analysis of micro-pollutants. In: Varjani, S., Pandey, A., Tyagi, R.D., Ngo, H.H., Larroche, C. (eds.) *Current Developments in Biotechnology and Bioengineering.* pp. 63–86. Elsevier (2020).

25. Lucci, P., Saurina, J., Núñez, O.: Trends in LC-MS and LC-HRMS analysis and characterization of polyphenols in food. *TrAC Trends in Analytical Chemistry* **88**, 1–24 (2017). doi:https://doi.org/10.1016/j.trac.2016.12.006

26. Barbosa, M.O., Ribeiro, A.R., Pereira, M.F., Silva, A.M.J.A.: Eco-friendly LC – MS/MS method for analysis of multi-class micropollutants in tap, fountain, and well water from northern Portugal. *Analytical and Bioanalytical Chemistry* **408**(29), 8355–8367 (2016).

27. Pérez-Fernández, V., Mainero Rocca, L., Tomai, P., Fanali, S., Gentili, A.: Recent advancements and future trends in environmental analysis: Sample preparation, liquid chromatography and mass spectrometry. *Analytica Chimica Acta* **983**, 9–41 (2017). doi:https://doi.org/10.1016/j.aca.2017.06.029

28. Wang, J., Bai, Z.: Fe-based catalysts for heterogeneous catalytic ozonation of emerging contaminants in water and wastewater. *Chemical Engineering Journal* **312**, 79–98 (2017). doi:https://doi.org/10.1016/j.cej.2016.11.118

29. Kasprzyk-Hordern, B., Ziółek, M., Nawrocki, J.: Catalytic ozonation and methods of enhancing molecular ozone reactions in water treatment. *Applied Catalysis B: Environmental* **46**(4), 639–669 (2003). doi:https://doi.org/10.1016/S0926-3373(03)00326-6

30. Guillossou, R., Le Roux, J., Brosillon, S., Mailler, R., Vulliet, E., Morlay, C., Nauleau, F., Rocher, V., Gaspéri, J.: Benefits of ozonation before activated carbon adsorption for the removal of organic micropollutants from wastewater effluents. *Chemosphere* **245**, 125530 (2020). doi:https://doi.org/10.1016/j.chemosphere.2019.125530

31. Wang, J., Zhuan, R.: Degradation of antibiotics by advanced oxidation processes: An overview. *Science of The Total Environment* **701**, 135023 (2020). doi:https://doi.org/10.1016/j.scitotenv.2019.135023

32. Feng, M., Yan, L., Zhang, X., Sun, P., Yang, S., Wang, L., Wang, Z.: Fast removal of the antibiotic flumequine from aqueous solution by ozonation: Influencing factors, reaction pathways, and toxicity evaluation. *Science of the Total Environment* **541**, 167–175 (2016). doi:https://doi.org/10.1016/j.scitotenv.2015.09.048

33. El-taliawy, H., Ekblad, M., Nilsson, F., Hagman, M., Paxeus, N., Jönsson, K., Cimbritz, M., la Cour Jansen, J., Bester, K.: Ozonation efficiency in removing organic micro pollutants from wastewater with respect to hydraulic loading rates and different wastewaters. *Chemical Engineering Journal* **325**, 310–321 (2017). doi:https://doi.org/10.1016/j.cej.2017.05.019

34. Lu, J., Sun, J., Chen, X., Tian, S., Chen, D., He, C., Xiong, Y.: Efficient mineralization of aqueous antibiotics by simultaneous catalytic ozonation and photocatalysis using MgMnO3 as a bifunctional catalyst. *Chemical Engineering Journal* **358**, 48–57 (2019). doi:https://doi.org/10.1016/j.cej.2018.08.198

35. Lado Ribeiro, A.R., Moreira, N.F.F., Li Puma, G., Silva, A.M.T.: Impact of water matrix on the removal of micropollutants by advanced oxidation technologies. *Chemical Engineering Journal* **363**, 155–173 (2019). doi:https://doi.org/10.1016/j.cej.2019.01.080

36. Deng, Y., Zhao, R.: Advanced oxidation processes (AOPs) in wastewater treatment. *Current Pollution Reports* **1**(3), 167–176 (2015).

37. Miklos, D.B., Remy, C., Jekel, M., Linden, K.G., Drewes, J.E., Hübner, U.: Evaluation of advanced oxidation processes for water and wastewater treatment – a critical review. *Water Research* **139**, 118–131 (2018). doi:https://doi.org/10.1016/j.watres.2018.03.042

38. Lekkerkerker-Teunissen, K., Knol, A.H., van Altena, L.P., Houtman, C.J., Verberk, J.Q.J.C., van Dijk, J.C.: Serial ozone/peroxide/low pressure UV treatment for synergistic and effective organic micropollutant conversion. *Separation and Purification Technology* **100**, 22–29 (2012). doi:https://doi.org/10.1016/j.seppur.2012.08.030

39. EPA, E.P.A.J.: National primary drinking water regulations: Long term 1 enhanced surface water treatment rule. *Final Rule* **67**(9), 1811–1844 (2002).

40. Wang, J., Zhuan, R.: Degradation of antibiotics by advanced oxidation processes: An overview. *Science of the Total Environment* **701**, 135023 (2020). doi:10.1016/j.scitotenv.2019.135023

41. Dimitrakopoulou, D., Rethemiotaki, I., Frontistis, Z., Xekoukoulotakis, N.P., Venieri, D., Mantzavinos, D.: Degradation, mineralization and antibiotic inactivation of amoxicillin by UV-A/TiO2 photocatalysis. *Journal of Environmental Management* **98**, 168–174 (2012). doi:https://doi.org/10.1016/j.jenvman.2012.01.010

42. León, D.E., Zúñiga-Benítez, H., Peñuela, G.A., Mansilla, H.D.: Photocatalytic removal of the antibiotic cefotaxime on TiO 2 and ZnO suspensions under simulated sunlight radiation. *Water, Air, Soil Pollution* **228**(9), 1–12 (2017).

43. Zhang, Y., Shao, Y., Gao, N., Gao, Y., Chu, W., Li, S., Wang, Y., Xu, S.: Kinetics and by-products formation of chloramphenicol (CAP) using chlorination and photocatalytic oxidation. *Chemical Engineering Journal* **333**, 85–91 (2018). doi:https://doi.org/10.1016/j.cej.2017.09.094

44. Fenoll, J., Martínez-Menchón, M., Navarro, G., Vela, N., Navarro, S.: Photocatalytic degradation of substituted phenylurea herbicides in aqueous semiconductor suspensions exposed to solar energy. *Chemosphere* **91**(5), 571–578 (2013). doi:https://doi.org/10.1016/j.chemosphere.2012.11.067

45. Sarkhosh, M., Sadani, M., Abtahi, M., Mohseni, S.M., Sheikhmohammadi, A., Azarpira, H., Najafpoor, A.A., Atafar, Z., Rezaei, S., Alli, R., Bay, A.: Enhancing photodegradation of ciprofloxacin using simultaneous usage of eaq– and OH over UV/ZnO/I- process: Efficiency, kinetics, pathways, and mechanisms. *Journal of Hazardous Materials* **377**, 418–426 (2019). doi:https://doi.org/10.1016/j.jhazmat.2019.05.090

46. Rodríguez-Chueca, J., Garcia-Cañibano, C., Sarro, M., Encinas, Á., Medana, C., Fabbri, D., Calza, P., Marugán, J.: Evaluation of transformation products from chemical oxidation of micropollutants in wastewater by photoassisted generation of sulfate radicals. *Chemosphere* **226**, 509–519 (2019). doi:https://doi.org/10.1016/j.chemosphere.2019.03.152

47. Taoufik, N., Boumya, W., Achak, M., Sillanpää, M., Barka, N.: Comparative overview of advanced oxidation processes and biological approaches for the removal pharmaceuticals. *Journal of Environmental Management* **288**, 112404 (2021). doi:https://doi.org/10.1016/j.jenvman.2021.112404

48. Martínez-Costa, J.I., Rivera-Utrilla, J., Leyva-Ramos, R., Sánchez-Polo, M., Velo-Gala, I., Mota, A.J.: Individual and simultaneous degradation of the antibiotics sulfamethoxazole and trimethoprim in aqueous solutions by Fenton, Fenton-like and photo-Fenton processes using solar and UV radiations. *Journal of Photochemistry and Photobiology A: Chemistry* **360**, 95–108 (2018). doi:https://doi.org/10.1016/j.jphotochem.2018.04.014

49. Gupta, A., Garg, A.: Degradation of ciprofloxacin using Fenton's oxidation: Effect of operating parameters, identification of oxidized by-products and toxicity assessment. *Chemosphere* **193**, 1181–1188 (2018). doi:https://doi.org/10.1016/j.chemosphere.2017.11.046

50. Qi, Y., Mei, Y., Li, J., Yao, T., Yang, Y., Jia, W., Tong, X., Wu, J., Xin, B.: Highly efficient microwave-assisted Fenton degradation of metacycline using pine-needle-like CuCo2O4 nanocatalyst. *Chemical Engineering Journal* **373**, 1158–1167 (2019). doi:https://doi.org/10.1016/j.cej.2019.05.097

51. Wang, N., Zheng, T., Zhang, G., Wang, P.: A review on Fenton-like processes for organic wastewater treatment. *Journal of Environmental Chemical Engineering* **4**(1), 762–787 (2016). doi:https://doi.org/10.1016/j.jece.2015.12.016

52. Nasseh, N., Taghavi, L., Barikbin, B., Nasseri, M.A., Allahresani, A.: FeNi3/SiO2 magnetic nanocomposite as an efficient and recyclable heterogeneous Fenton-like catalyst for the oxidation of metronidazole in neutral environments: Adsorption and degradation studies. *Composites Part B: Engineering* **166**, 328–340 (2019). doi:https://doi.org/10.1016/j.compositesb.2018.11.112

53. Brillas, E., Sirés, I., Oturan, M.A.: Electro-Fenton process and related electrochemical technologies based on Fenton's reaction chemistry. *Chemical Reviews* **109**(12), 6570–6631 (2009).

54. Dolatabadi, M., Ahmadzadeh, S.: A rapid and efficient removal approach for degradation of metformin in pharmaceutical wastewater using electro-Fenton process; optimization by response surface methodology. *Water Science Technology* **80**(4), 685–694 (2019).

55. Liu, D., Zhang, H., Wei, Y., Liu, B., Lin, Y., Li, G., Zhang, F.: Enhanced degradation of ibuprofen by heterogeneous electro-Fenton at circumneutral pH. *Chemosphere* **209**, 998–1006 (2018). doi:https://doi.org/10.1016/j.chemosphere.2018.06.164

56. Bugueño-Carrasco, S., Monteil, H., Toledo-Neira, C., Sandoval, M.Á., Thiam, A., Salazar, R.: Elimination of pharmaceutical pollutants by solar photoelectro-Fenton process in a pilot plant. *Environmental Science and Pollution Research* **28**(19), 23753–23766 (2021). doi:10.1007/s11356-020-11223-y

57. Calzadilla, W., Espinoza, L.C., Diaz-Cruz, M.S., Sunyer, A., Aranda, M., Peña-Farfal, C., Salazar, R.: Simultaneous degradation of 30 pharmaceuticals by anodic oxidation: Main intermediaries and by-products. *Chemosphere* **269**, 128753 (2021). doi:https://doi.org/10.1016/j.chemosphere.2020.128753

58. Ushani, U., Lu, X., Wang, J., Zhang, Z., Dai, J., Tan, Y., Wang, S., Li, W., Niu, C., Cai, T., Wang, N., Zhen, G.: Sulfate radicals-based advanced oxidation technology in various environmental remediation: A state-of-the – art review. *Chemical Engineering Journal* **402**, 126232 (2020). doi:https://doi.org/10.1016/j.cej.2020.126232

59. Guo, W., Su, S., Yi, C., Ma, Z.: Degradation of antibiotics amoxicillin by Co3O4-catalyzed peroxymonosulfate system. *Environmental Progress Sustainable Energy* **32**(2), 193–197 (2013).

60. Venkatesan, A.K., Done, H.Y., Halden, R.U.: United States national sewage sludge repository at Arizona State University – a new resource and research tool for environmental scientists, engineers, and epidemiologists. *Environmental Science and Pollution Research* **22**(3), 1577–1586 (2015).

61. Wang, Y., Pan, T., Yu, Y., Wu, Y., Pan, Y., Yang, X.: A novel peroxymonosulfate (PMS)-enhanced iron coagulation process for simultaneous removal of trace organic pollutants in water. *Water Research* **185**, 116136 (2020). doi:https://doi.org/10.1016/j.watres.2020.116136

62. Khan, I., Saeed, K., Khan, I.: Nanoparticles: Properties, applications and toxicities. *Arabian Journal of Chemistry* **12**(7), 908–931 (2019).

63. Bethi, B., Sonawane, S.H., Bhanvase, B.A., Gumfekar, S.P.: Nanomaterials-based advanced oxidation processes for wastewater treatment: A review. *Chemical Engineering Processing-Process Intensification* **109**, 178–189 (2016).

64. Min, S., Wang, F., Han, Y.: An investigation on synthesis and photocatalytic activity of polyaniline sensitized nanocrystalline TiO 2 composites. *Journal of Materials Science* **42**(24), 9966–9972 (2007).

65. Shi, W., Yan, Y., Yan, X.: Microwave-assisted synthesis of nano-scale BiVO4 photocatalysts and their excellent visible-light-driven photocatalytic activity for the degradation of ciprofloxacin. *Chemical Engineering Journal* **215–216**, 740–746 (2013). doi:https://doi.org/10.1016/j.cej.2012.10.071

66. Huang, R., Fang, Z., Yan, X., Cheng, W.: Heterogeneous sono-Fenton catalytic degradation of bisphenol A by Fe3O4 magnetic nanoparticles under neutral condition. *Chemical*

Engineering Journal **197**, 242–249 (2012). doi:https://doi.org/10.1016/j.cej.2012.05.035

67. Bagal, M.V., Gogate, P.R.: Degradation of diclofenac sodium using combined processes based on hydrodynamic cavitation and heterogeneous photocatalysis. *Ultrasonics Sonochemistry* **21**(3), 1035–1043 (2014). doi:https://doi.org/10.1016/j.ultsonch.2013.10.020

68. Mohseni, M., Demeestere, K., Du Laing, G., Yüce, S., Keller, R.G., Wessling, M.: CNT microtubes with entrapped Fe3O4 nanoparticles remove micropollutants through a heterogeneous electro-Fenton process at neutral pH. *Advanced Sustainable Systems* **5**(4), 2100001 (2021).

69. Wei, H., Loeb, S.K., Halas, N.J., Kim, J.-H.: Plasmon-enabled degradation of organic micropollutants in water by visible-light illumination of Janus gold nanorods. *Proceedings of the National Academy of Sciences* **117**(27), 15473–15481 (2020).

70. Mohsin, M.K., Mohammed, A.A.: Catalytic ozonation for removal of antibiotic oxy-tetracycline using zinc oxide nanoparticles. *Applied Water Science* **11**(1), 1–9 (2021).

4

Application of Advanced Oxidation Processes to Treat Industrial Wastewaters: Sustainability and Other Recent Challenges

Idil Arslan-Alaton, Fatos Germirli Babuna and Gulen Iskender

CONTENTS

4.1 Introduction

Industrial wastewater treatment has always been a challenging task because unlike domestic or urban wastewater (sewage), industrial effluent varies considerably in flow rate and chemical composition (Imai et al., 1995; Tchobanoglous and Burton, 2003). These two factors have triggered the necessity of a sustainable industrial wastewater management strategy featuring in-depth knowledge and evaluation of the industry/industrial sector, the effluent arising from different wet processes of the industry, a detailed effluent characterization in terms of collective and specific environmental quality parameters, waste minimization, recycling and recovery strategies as well as advanced treatment processes for the efficient removal of difficult (problematic) parameters (Aziz et al., 2011). In particular, some industry-specific parameters such as color, toxicity, heavy metals and thermal pollution have to be considered and managed carefully (Shah, 2020a).

Upon brief comparison of domestic and industrial wastewater characteristics and flow rates, one can come to the following general conclusions:

I. Industrial wastewater is extremely variable in flow rates; some industrial effluent treatment plants do not operate regularly (7 days/week, 24 hours/day), which renders the sole biological treatment option unfeasible (Shah, 2020b).

II. Variations even within a day are not unusual for industrial effluents; constant T, pH and other physicochemical characteristics are not typical in the textile industry, for example.

III. Besides, characteristics can be unpredictable and difficult to anticipate and control. At sewage treatment works, however, capacity and loading are typically constant and predictable.

DOI: 10.1201/9781003165958-4

IV. Industrial wastewater contains not only more complex but also a higher number of chemicals, which renders their treatment difficult. Usually, the well-known and -established conventional sewage treatment processes do not work properly here and require at least some modification and integration with advanced, alternative treatment options – if these exist.

From these conclusions it is evident that industrial wastewater is by far more difficult to treat than domestic/urban wastewater due to its inherent complexity and irregular flow rate (Shah, 2020a). Hence, case-specific and special treatment processes have to be developed, applied and verified for efficient industrial wastewater management (Hocaoğlu and Orhon, 2013; Keerthi et al., 2019).

4.2 Principles of Advanced Industrial Wastewater Treatment

For advanced industrial wastewater treatment, biological processes (aerobic, anaerobic and fungal treatments), physicochemical methods (flocculation, coagulation, precipitation), advanced oxidation processes (AOPs, such as photocatalytic, photochemical, Fenton and photo-Fenton processes), electrochemical treatment processes (involving electrocoagulation) and membrane technologies are well-known treatment options (Matilainen et al., 2010). However, some drawbacks resulting in poor treatment performance and high associated costs have been reported (Ngujen et al., 2011). In this section, the most important and efficient chemical treatment processes, namely ozonation and electrocoagulation, have been reported on and detailed as follows.

4.2.1 Ozonation of Industrial Wastewater

A long time after ozone was effectively used for water disinfection and disinfection by-product (trihalomethanes (THM), haloacetic acids (HAA)) control, ozone was applied for the treatment to treat industrial pollutants, including (i) industrial dyes, (ii) pesticides, (iii) surfactants, (iv) industrial solvents (volatile organic compounds (VOCs)) and (v) pharmaceuticals (particularly antibiotics from the antibiotic formulation industry). Many other industrial sectors have incorporated ozonation facilities as a treatment stage prior to or following main biological (activated sludge) treatment (CCOT, 1996). Ozone production is highly energy intensive, and ozonation facilities have capital costs starting from one million USD. Ozone is produced by passing oxygen gas through an electric field/corona discharge, and for the production of 1 kg ozone, approximately 10–15 kWh of electrical energy is required (CCOT, 1996). Ozone is produced from dry oxygen or air; however, its production from pure dry oxygen is two to three times more efficient than production from dry air.

It has been established that the specific ozone dose required for organic carbon removal from industrial wastewater is around ten times higher than that required for organic matter removal from raw (fresh)water; around 2 g ozone is required for the removal of 1 g of organic carbon (total organic carbon [TOC], dissolved organic carbon [DOC]) in industrial wastewater. It should be noted that the ozone demand is very case specific and depends on the type and nature of the industrial wastewater. Generally speaking, a longer treatment time – a higher specific ozone dose – is required for post-treatment (polishing) than for pre-treatment of industrial wastewater. The most frequent candidates for ozonation have been textile, tannery, and pulp and paper mill effluents. On the other hand, the selectivity of ozone (1.8–2.0 eV vs. standard electrode hydrogen, or SHE) has limited its application for more inert (difficult-to-oxidize) polyaromatic, halogenated organic industrial pollutants (Fang et al., 2016). Hence, alternative, advanced treatment systems and their integration with conventional industrial wastewater treatment systems have been proposed in the recent past (Ollis, 2001; Tang et al., 2010).

4.2.2 Electrocoagulation of Industrial Wastewater

Electrocoagulation appears to be an ideal treatment process for industrial wastewater and other complex effluents with irregular chemical composition and flow rates. It is a (physico)chemical treatment method where a complex series of removal mechanisms such as coagulation, flotation, precipitation and redox reactions occur that are initiated by using "sacrificial" anode and cathode materials (for instance, Al, cast Fe, stainless steel [SS]). Electrocoagulation is an alternative and effective chemical treatment option particularly for industrial wastewater types that have a high chemical oxygen demand (COD), TOC and electrical conductivity (Gursoy-Haksevenler et al., 2014; Wang et al., 2017; Wang et al., 2018, Shah, 2021).

Stages of the electrocoagulation process are listed as follows:

I. Dissolution (oxidation) of the electrode material at the anode providing the coagulant

II. Formation of a metal hydroxo complex in the reaction bulk

III. Destabilization of pollutants

IV. Flocculation of pollutants

V. Precipitation and flotation occurring simultaneously

VI. Other pollutant removal mechanisms (adsorption, particle entrapment, oxidation, reduction), which may also become important depending upon the reaction, are pH, electrolyte concentration, oxygen concentration, and type and nature of wastewater

Advantages of electrocoagulation processes are simple equipment and operating conditions; compact reactor units; effective treatment for different industrial effluent types and organic loads, including oil and grease removal; formation of a relatively dense, easily dewaterable, low-in-volume chemical sludge at the top and bottom of the reaction unit; reduction in TDS concentration due to efficient coprecipitation-adsorption-absorption-particle entrapment mechanisms; no additional chemicals requirement (Fe, Al); and simultaneous reactions occurring at the same time in one reactor (Hanafi et al., 2009; Shah, 2021).

Electricity can be generated by integrating solar collectors or hydrogen fuel cells to the electrocoagulation unit (ideal application in countries with appropriate climate conditions and high electrical energy costs).

Disadvantages of electrocoagulation are that the electrode materials have to be periodically replaced (the sacrificial Al, Fe, SS electrodes); electrical energy requirements and hence operating costs can be high (sometimes > 5 kWh/m^3); costs associated with chemical sludge handling; only partial treatment can be achieved (pre-treatment); in some cases an electrolyte (e.g., NaCl) has to be added increasing the salinity; and treatment costs and the possibility to form organochlorines (AOX, RCl) in the effluent during electrochemical treatment, and as shown in Equation 4.1:

$$R + Cl^- \rightarrow (R^+ + Cl^{\bullet-}) \rightarrow RCl \qquad (4.1)$$

Further, upon extension of treatment time and/or applied current density, high sludge volumes are produced especially when Al electrodes are used. The most common and cheapest electrode materials are cast Fe, 304 and 316 grade stainless steel and Al. Fe and SS are preferred electrode materials mainly due to sludge production at relatively lower quantities and relatively low toxicity of released metal ions as compared with Al electrodes (toxic Al release).

4.2.3 UV-C Photolysis of Industrial Pollutants

Under specific conditions and depending on the structural properties of the target pollutant, it is possible to remove industrial pollutants by UV photolysis, which is also called UV fragmentation. UV photolysis can results in homolytic or heterolytic fission of the pollutant causing its fragmentations into low molecular weight moieties. The dissociation of a molecule into fragments requires a photon energy that is higher than the bond energy of the molecules. Breaking a bond usually requires an energy corresponding to the ultraviolet (UV) wavelength region. Hence, UV light in the range of 180–400 nm is necessary to break pollutant bonds by electromagnetic radiation. For example, comparing the bond energy of -C=C- with -C–C- bonds indicates that the dissociation of double bonds requires less photonic energy than those of saturated (single) bonds. Saturated bonds are always more difficult (require photon energy) than double or triple bonds. The wavelength of light in the UV spectral region, which is practically used in photochemical treatment applications, is 200–300 nm. Near-UV (long UV, UV-A) with a spectrum in the 300–390 nm range is seldom used without an activator (catalyst). When UV-A is to be used due to geographical, safety or economic reasons, a catalyst (homogenous or heterogeneous) has to be added into the reaction solution to activate the catalyst with near-UV light. Short-UV on the other hand is 180–250 nm (also called UV-C spectral region) and is mainly used for disinfection (sterilization) of pathogens in treated water/wastewater. The visible light spectrum, that is, in the 400–700 nm wavelength range, can only activate colored (visible light-absorbing) chemicals such as dyes.

The following pollutants are strong UV absorbers and therefore good candidates for direct UV photolysis, treatment with UV light. These are:

- NDMA (nitrosodimethylamine)
- Some chlorinated alkenes like DCE, TCE (dichloro-, trichloroethylene)
- Some nitroaromatics
- Some phenolic contaminants (mono- and polyphenols) being used as solvents, insulators and pesticides

Some major applications of direct UV photolysis are as follows:

- Treatment of groundwater contaminated with industrial wastewater such as NDMA, MTBE
- Photolysis of some haloaliphatic solvents (CCl$_4$, DCE, TCE, chlorinated alkanes and alkenes, chloroform)

It should be pointed out here that no additional oxidants or (photo)catalysts are needed for effective NDMA removal. Hence, NDMA can be treated directly with UV-C radiation and the specific electrical energy consumption for direct UV-C photolysis (photodegradation) is 2.5 kWh/m^3 for 90% removal efficiency (i.e., one (1) log removal).

4.3 Advanced Oxidation Processes (AOPs) for Industrial Wastewater

4.3.1 Basic Principles and Modes of Application

Decades ago, chemical or photochemical oxidation of industrial wastewater was considered to be impractical because most oxidants used in the (photo)chemical oxidation are quite expensive and difficult to handle. Ozone, chlorine dioxide and hydrogen peroxide are some examples for common oxidants used for disinfection and in chemical treatment applications. Moreover, in chemical oxidation applications, complete oxidation of biologically resistant and toxic/hazardous pollutants/micropollutants found in industrial wastewaters is targeted, rendering chemical oxidation very energy intensive and hence too expensive for full-scale applications (Ameta and Ameta, 2018). More recently, however, it has been postulated that full chemical treatment is not always required and partial oxidation of industrial effluent prior to conventional biological treatment or, alternatively, final treatment for polishing purposes is economically and technically more feasible and hence more realistic (Fatta-Kassinos et al., 2015). Other factors that have triggered chemical and photochemical oxidation of industrial wastewaters are (i) discharge requirements and standards becoming stricter every day; (ii) the opportunity to follow more environmental pollution parameters down to ppm–ppb concentration levels in water, thanks to recent advances in analytical instrumentation techniques allowing greater precision in contamination measurements; (iii) the increasing water stress and water scarcity problems that have prompted the need to recycle and reuse wastewater after extensive treatment (Fatta-Kassinos et al., 2015).

Considering all the aforementioned factors, it has become important to treat industrial wastewater thoroughly by employing "destructive" (not phase-transfer-based) advanced chemical/photochemical oxidation processes (Bolton, 2001). As mentioned, chemical oxidation offers several technical and economic advantages over biological treatment processes. These are listed as follows:

- Chemical oxidation is appreciably faster than biological oxidation, allowing the use of more compact treatment facilities (reduced holding times).
- Toxic and/or refractory chemicals found in a variety of industrial wastewaters or segregated industrial effluent streams cannot be treated biologically.
- Recent advances in industrial wastewater management practices have increased the applicability and necessity of using chemical/photochemical treatment processes.

Since treatment costs nowadays have become secondary after protecting human health, water quality and the environment, is it not surprising that more and more chemical oxidation processes have been developed as environmentally friendly, sustainable and ecotoxicological safe treatment alternatives.

As mentioned earlier, applying a variety of chemical oxidation methods yields favorable outcomes by lessening toxic and/or inert characteristics, removing color of these effluents in rather smaller treatment units. The possible positions of AOPs in industrial wastewater treatment are illustrated in Figure 4.1.

The main focus is not letting the advanced oxidation processes compete with biological treatment in removing the biodegradable organics. A solid example of this issue can be found in literature, as the application of ozone prior to biological treatment for a tannery wastewater is stated to yield a similar unwanted outcome (Dogruel et al., 2006). It is well known that lower costs are involved in eliminating biodegradable fractions of wastewaters by biological treatment. In this respect, the characteristics of the industrial wastewater dictate the location of advanced oxidation processes in the whole treatment train. When the raw industrial wastewater has a high inert organic content and/or an excessive toxic nature, a partial pre-treatment with advanced oxidation processes is required prior to a biological treatment units, as given in Figure 4.1a. Application of advanced oxidation previous to biological treatment for the agrochemical industry (Pariente et al., 2013), the pharmaceutical sector (Changotra et al., 2019, 2020) and textile mills (Sathya et al., 2019) are some examples of this practice. Various other studies can be found in the literature, such as on pulp and paper effluents where ozone, ozone plus UV and ozone plus UV plus H_2O_2 all followed by activated sludge treatment trains are investigated (Ledakowicz et al., 2006). Treatment of ozonated and perozonated penicillin formulation effluents by activated sludge (Arslan-Alaton et al., 2004a) and the solar-assisted photo-Fenton process as a preliminary treatment of activated sludge for winery wastewaters (Mosteo et al., 2007) can be considered as cases dealing with the pre-location of advanced oxidation processes.

On the other hand, in some cases advanced oxidation methods are coupled with biological units for polishing the already biologically treated industrial effluents. This alternative location is called post-treatment (Figure 4.1b). The purpose of this application is to remove the residual organics to meet stringent discharge standards. As indicated earlier, the characteristics of the effluent are of importance in deciding the location of advanced oxidation within the whole treatment train. For an industry where elemental chlorine-free bleaching of wood pulp is conducted, a treatment train composed of UV radiation after the activated sludge process is stated to yield substantial reductions in recalcitrant compounds and color (Silva et al., 2007).

Even in some industrial sectors, different locations for advanced oxidation appear to be the most appropriate ones, such as in the case of two different pulp and paper mills: pre-treatment (Gopalakrishnan et al., 2019) and post-treatment (Merayo et al., 2013). In a study on tannery wastewaters, ozonation at a stage in biological treatment where readily biodegradable organics are eliminated by biological means is addressed as the most favorable outcome (Dogruel et al., 2004). Another study on tannery effluents indicates post-treatment with ozone application is suitable to achieve the required effluent discharge standards (Dogruel et al., 2006).

Certain industrial sectors are known to generate segregated wastewater streams with different characteristics. The textile industry can be referred to as a good example for such sectors. Due to the complex nature of textile dye houses where processes, equipment and auxiliary inputs widely differ, segregated effluent streams with various toxicity levels, biodegradability degrees, colors and so on are created. A sound wastewater treatment strategy for such industrial sectors can only be developed by considering the aforementioned variations in segregated streams. If a problematic segregated effluent stream (in terms of biodegradability, toxicity, color, etc.) exists in a facility, the best approach is to handle this stream separately. Otherwise, mixing all the effluents might result in enlarging the problem to be solved for these installations. A schematic illustration of this approach is given in Figure 4.1c. Instead of adding a specific segregated industrial wastewater that might cause difficulties in combined treatment, advanced oxidation processes can be applied to this very stream for easing the problem. AOPs are widely used to segregate textile effluents for this purpose (Paździor et al., 2019). Example cases include the decolorization of a textile dye through Fenton followed by aerobic biological treatment (Lucas et al., 2007; Lodha and Chaudhari, 2007), UV/H_2O_2 and subsequent aerobic biodegradation for another textile dye (Sudarjanto et al., 2006) and the application of immobilized biomass after solar photo-Fenton for a biorecalcitrant precursor used in the synthesis of pharmaceuticals (Malato et al., 2007).

General principles adopted to choose the proper location of AOPs within biological treatment are outlined in the flowchart given in Figure 4.2.

As can be easily seen from the first loop of Figure 4.2, the industrial wastewaters having either a high fraction of refractory (nonbiodegradable) organics must be partially treated with advanced oxidation processes before passing through

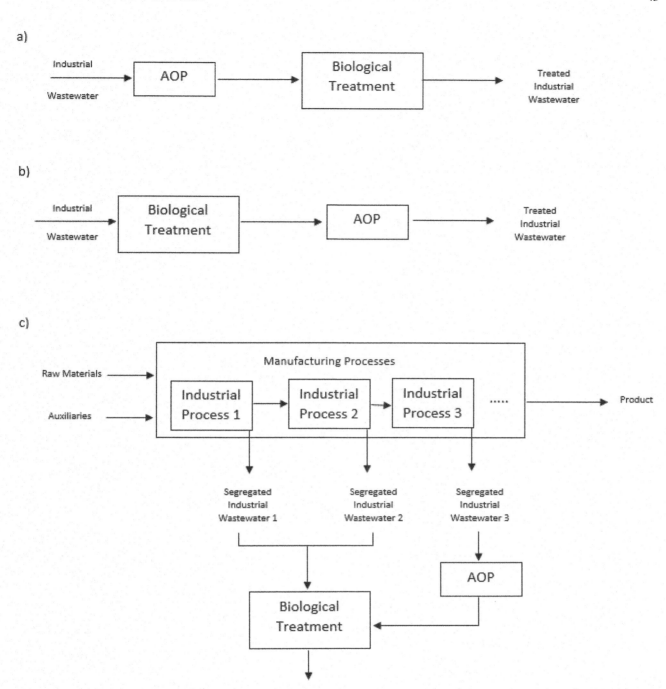

FIGURE 4.1 Application of AOPs in the treatment of industrial wastewaters (a) as pre-treatment, (b) as post-treatment, (c) for segregated effluent streams.

biological treatment units. Similarly, industrial effluents that have high levels of inhibition on biological treatment should be directed towards partial advanced oxidation processes previous to biological treatment. The nature of industrial wastewaters varies substantially. Different wastewater characteristics are noted even for the same sectors or subsectors. Therefore, it is recommended to run a detailed wastewater characterization and treatability studies before deciding on the treatment units and their sequence. Such an effort should include parameters indicating the biodegradability level of the industrial effluent. Refractory fractions dictating also the biodegradability should be defined depending on the type of biotreatment. In other words, constituents of an industrial wastewater that can be perceived as refractory by the microbial consortia of suspended

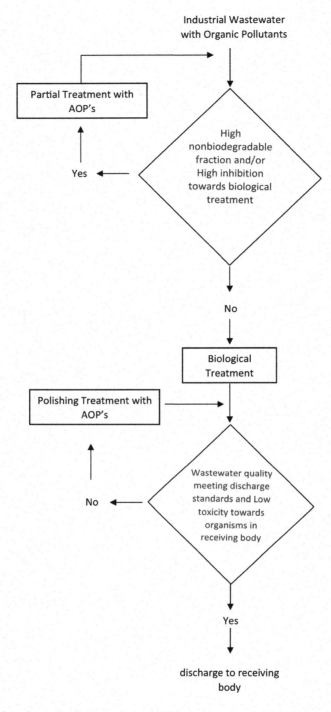

FIGURE 4.2 Flowchart of integrating advanced oxidation processes in biological treatment for industrial wastewaters.

growth aerobic systems can be degraded well by the micro-bial community of an attached growth anaerobic system. Therefore, the concept of biodegradability actually depends on the microbial consortia. Besides, industrial effluents might also cause inhibition on biological activity. The preliminary studies on wastewater characterization and treatability must also cover such inhibitory effects on the microbial community

of biotreatment systems. An activated sludge inhibition test can be given as an example used for this purpose (ISO, 1986).

Once the hurdle related to the first loop is passed and bio-logical treatment is applied, another check is required that tar-gets meeting the discharge standards. Further color removal or organic reduction can be achieved via advanced oxida-tion methods at this second loop. In cases where discharge standards are not complied with in the previous biological treatment, a polishing with advanced oxidation processes is needed. Moreover, it is recommended to run an investigation on the toxicity of treated wastewaters towards the organisms that can be found in the receiving water bodies. For this purpose, the nature of the discharge media gains importance. When the treated wastewaters are discharged into salty waters, toxicity on marine organisms such as *Phaeodactylum tricornutum* and *Vibrio fischeri* are used, whereas toxicity on freshwater bodies can be measured with freshwater organisms, for instance *Daphnia magna*. Table 4.1 presents examples of evaluation bases for advanced oxidation processes applied to various wastewaters generated by industries. Together with industrial

wastewaters arising from the whole production facility, segregated effluents that carry certain auxiliary chemicals added during a specific part of the manufacturing are tabulated. Only after appraising the dataset on biodegradability, inhibition and/or toxicities towards various organisms representing various receiving media can a proper treatment train be proposed.

The role of toxicity must be emphasized in appraisals as in some cases application of advanced oxidation processes result in elevated toxicity levels. Such increase in toxicity can be explained by the generation of low molecular weight fractions of the parental compounds.

As mentioned earlier, some pollutants in industrial effluent cannot be treated effectively by ozone/ozonation process.

TABLE 4.1

Evaluation basis for advanced oxidation processes applied to effluents originating from selected industries.

Industrial Sector	Source	Type of AOP(s) Applied	Evaluations Based on	References
Textile	Segregated dyebath discharge containing Tannin 1	Ozonation	* Biodegradability * Activated sludge inhibition * towards *Phaeodactylum tricornutum*	Iskender et al., 2016; Koyunluoglu et al., 2006; Germirli Babuna et al., 2007
Textile	Segregated dyebath discharge containing Tannin 2	Ozonation	* Biodegradability * Activated sludge inhibition * Toxicity towards *Phaeodactylum tricornutum*	Iskender et al., 2016; Koyunluoglu et al., 2006; Germirli Babuna et al., 2007
Textile	Segregated dyebath discharge containing Biocide 1	Ozonation	* Biodegradability	Iskender et al., 2016
Textile	Segregated dyebath discharge containing Biocide 2	Ozonation	* Biodegradability	Iskender et al., 2016
Textile	Segregated dyebath discharge containing Biocide 1	O_3/OH^-, TiO_2/UV-A Fe_2^+/H_2O_2	* Toxicity towards *Daphnia magna* * Activated sludge inhibition	Arslan-Alaton et al., 2005
Textile	Segregated dyebath discharge containing Biocide 2	O_3/OH^-, TiO_2/UV-A Fe_2^+/H_2O_2	* Toxicity towards *Daphnia magna* * Activated sludge inhibition	Arslan-Alaton et al., 2005
Textile	Segregated dyebath discharge containing Carrier 1	Ozonation	* Biodegradability * Toxicity towards *Daphnia magna* * Toxicity towards *Phaedactylum tricornutum* * Activated sludge inhibition	Arslan-Alaton et al., 2004b
Textile	Segregated dyebath discharge containing Carrier 2	Ozonation	* Biodegradability * Toxicity towards *Daphnia magna* * Toxicity towards *Phaedactylum tricornutum* * Activated sludge inhibition	Arslan-Alaton et al., 2004b
Textile	Segregated dyebath discharge containing naphthalene sulfonic acid	Ozonation	* Biodegradability * Toxicity towards *Phaedactylum tricornutum*	Germirli Babuna et al., 2009
Textile	Segregated dyebath discharge containing lignosulfonate	Ozonation	* Biodegradability * Toxicity towards *Phaedactylum tricornutum*	Germirli Babuna et al., 2011
Textile	Mill effluent	Ozonation	* Biodegradability * Toxicity towards *Phaedactylum tricornutum*	Selcuk et al., 2006
Pharmaceutical	Cefazolin sodium formulation effluent	Ozonation and H_2O_2/O_3 (perozonation)	* Biodegradability * Toxicity towards *Phaedactylum tricornutum* * Activated sludge inhibition	Iskender et al., 2007
Pharmaceutical	Ceftriaxone sodium antibiotic formulation effluent	Ozonation and H_2O_2/O_3 (perozonation)	* Biodegradability * Toxicity towards *Phaedactylum tricornutum* * Activated sludge inhibition	Tezgel et al., 2011

In other words, in some cases even the oxidation potential of ozone is not high enough to completely oxidize industrial pollutants found in water or wastewater such as metal complex dyes, nitrobenzenes, chlorinated phenols, pesticides, surfactants and halogenated biocides (Parsons, 2004). Under these circumstances, chemical oxidation processes have to be improved to increase the oxidation rates and efficiencies. Fortunately, common oxidants such as ozone and peroxides (hydrogen peroxide, persulfate, peroxy monosulfate, peracetic acid, etc.) can be chemically, thermally or photochemically activated to form free radicals. These free radicals (HO$^\bullet$, SO$_4^{\bullet-}$, O$_2^{\bullet-}$, etc.) are very strong, non-selective oxidants and can initiate radical chain reactions that enhance oxidation rates appreciably. For the treatment of refractory and/or toxic, mainly organic pollutants, so-called advanced oxidation processes – AOPs – have been developed and refer to the enhanced catalytic, photocatalytic, chemical or photochemical oxidation of otherwise difficult-to-oxidize pollutants. AOPs use catalysts (Fe, Co, Cu, Ni) and/or ultraviolet light radiation (UV-A or -C or even solar light) in combination with common oxidants (persulfate – S$_2$O$_8^{2-}$, hydrogen peroxide – H$_2$O$_2$, ozone – O$_3$) to produce free radicals. AOPs are based on the "*in situ*" formation of "active oxidizing species" (Peyton, 1993). In this way, pollutant removal rates and efficiencies can be enhanced even at ambient temperatures. In the last two to three decades, AOPs have been developed as effective and economically affordable treatment technologies for full-scale application to industrial wastewaters (Parsons, 2004).

4.3.2 Classification and Applicability of AOPs

AOPs are basically classified into photochemical (involving photocatalysts, oxidants and UV light at short wavelengths) and non-photochemical (involving catalysts, oxidants and power ultrasound) treatments. Most well-known photochemical/photocatalytic AOPs are O$_3$/UV-C, H$_2$O$_2$/UV-C and Fe^{2+}/Fe^{3+}/H$_2$O$_2$/UV-C (photo-Fenton and photo-Fenton-like oxidation), TiO$_2$/UV-A (titanium dioxide–mediated heterogeneous photocatalysts and other semiconductors/chalcogenides) and ultrasound-activated oxidation processes involving UV light radiation, whereas the most frequently reported chemical/catalytic AOPs are Fenton, Fenton-like, O$_3$/H$_2$O$_2$ (called perozone or peroxone) and ultrasound-activated oxidation processes involving peroxides and metal catalysts (Pignatello et al., 2006). The development of new (photo)catalysts, catalyst activators and chemical/photochemical treatment combinations is continuing to be a challenging, hot topic.

Because AOPs are rather energy-intensive, difficult-to-control treatment process with high operating costs, it is highly recommended to prepare a checklist before deciding whether to use/implement AOPs for water/wastewater treatment. The following checklist featuring five critical questions is proposed by experts in the field of AOPs (CCOT, 1996):

1. Is a destruction technology preferred to phase-transfer treatment technology?
2. Are there restrictions for air/soil discharge?
3. Does the target industrial contaminant air strip poorly (for example, phenol)?

4. Does the target industrial contaminant load/sorb poorly on activated carbon (AC; for example, industrial dyes)?
5. Are there disposal concerns associated with loaded AC?

If the answer is "Yes" to three or more of the questions, AOPs should definitely be considered an alternative treatment application.

4.3.3 System Sizing and Cost Calculations for AOPs

In the past, it has been elucidated that treatment costs of AOP are appreciably higher than those of other advanced treatment processes mainly because photochemically driven AOPs are energy intensive and mostly identified with high electric energy expenses. Electric energy costs are associated with the use of UV lamps and lamp replacement costs that account for approximately 30% of the total electricity costs. In fact, two major process design variables are critical in deciding if oxidant/UV treatment applications (photochemical AOPs) are economically/technically feasible. The following have to be carefully optimized before real-scale treatment applications involving AOPs are realized:

1. The UV power (or UV dose) emitted per unit volume of treated wastewater (in kw/m^3)
2. The oxidant (mostly peroxide) consumption (in kg/m^3)
3. For treatment combinations involving ozone, the electric energy required to produce 1 kg of ozone (around 10–15 kWh/kg ozone)

The applied UV dose is an empirical parameter that combines wastewater flow rate, residence (treatment or holding) time and UV light intensity into one single expression for (a) batch and (b) flow-through treatment systems.

a. Batch systems: For batch mode treatment systems, the UV dose is simply the lamp power times treatment time per volume of wastewater to be treated.
b. Flow-through systems: In continuous-mode arrangements, the UV dose is the lamp power per wastewater flow rate. Hence, the unit of UV dose is identical for both treatment modes.

It is important to note that preliminary design and optimization tests should be performed to determine the UV dose required to achieve the desired (targeted) effluent (final) concentration and/or removal efficiency. The target can be the discharge limit value of a critical pollutant parameter (adsorbable organically bound halogens [AOX], COD, color, cyanide, etc.) or a prescribed removal performance (1-log, 2-log, etc.) to achieve this target value. Simple first-order kinetics is applied to support energy calculations. In other words, UV dose is determined by combining operating expenses with kinetic expressions based on the targeted pollutant parameter.

The EE/O (the electric energy required per order and volume of a specific or collective pollutant parameter, in kWh/m^3/order) concept enables the comparison of cost and performance of a variety of photochemical AOPs using a single and simple equation. For instance, by measuring changes in atrazine or 1,4-dioxane in an industrial effluent as a function of photochemical treatment (peroxide/UV-C) time, EE/O values can be established and evaluated to decide for the proposed systems' feasibility check; for atrazine, this universal value has been reported in the range of 2–6 kWh/m^3/order, whereas for 1,4-dioxane it was calculated in the range of 10–30 kWh/m^3/order. The explanation for these values is that the pollutant atrazine (a herbicide) requires approximately five times less electrical energy than 1,4-dioxane (a stabilizer) does for the same removal efficiency/level of destruction.

Although UV-C disinfection (sterilization) is considered to be safer than chemical use (chlorination, ozonation, etc.), all conventional UV lamps typically contain 20–200 milligrams of mercury and are susceptible to breakage during transportation, handling and operation. Conventional UV lamps hold their mercury either in a liquid form (more common in medium-pressure lamps) or in an amalgam (more common in low-pressure, high-output arc lamps). Accidents and improper procedures increase the risk of exposure. The United Nations Environmental Programme (UNEP) initiated the "Minamata Convention on Mercury" to protect human health and the environment from anthropogenic emissions and releases of mercury. The UNEP has set a goal for mercury to be phased out of production by the year 2020. Years ago, organizations and governments were asked to discourage the use of mercury starting immediately, and 127 countries signed to ratify the removal of all mercury by the year 2020 (Pikuda et al., 2020). While it is unclear whether UV lamps for water disinfection will receive an exemption from this regulation, it is clear that alternatives such as light emitting diode (LED) technology should be actively considered and developed in the nearest future. In other words, the conventional UV-based oxidation processes typically use low-, medium- or high-pressure mercury arc lamps for the treatment of water and wastewater. The high power requirements, along with poor service life and low photonic efficiency, render the use of these conventional lamps a very energy-intensive process. Besides, mercury vapor lamps also impose a serious hazardous waste concern and risk to the environment. Hence, the production and use of this kind of light source have been reduced recently. On the other hand, recent advances in LED technology have made it possible to use LEDs at low wavelengths (UV range; λ = 240–260 nm). LEDs have found their applications in the degradation of a variety of industrial pollutants such as cyanotoxins, phenols and industrial dyes. LED UV lamps are currently used either alone or in combination with photocatalysts and oxidants.

LED technology provides several advantages over mercury vapor lamps (World Water, 2019):

- Short/low warm-up time
- No problem of mercury disposal
- Reduced replacement frequency
- Mechanical stability
- Easy handling due to compact design of LEDs
- Low operation voltage resulting in less power requirement
- Monochromatic emission at a particular wavelength

Although the use of LED technology in applications for visible light radiation has become routine, its use for treatment and hence at short wavelengths is still limited due to several economic and technical reasons. More recently and in lab-scale applications, LED technology has become an effective, alternative treatment solution and is still at its development stage.

4.3.4 Preliminary Design Tests for AOPs

Design tests have to be carried out with representative wastewater samples to confirm preliminary capital and operating cost estimations. Preliminary tests also enable the optimization other AOPs (e.g., ozonation, electrochemical oxidation, sonolysis, electrocoagulation, etc.) that do not involve UV light radiation. Preliminary design tests are case specific for a particular application to treat industrial wastewater.

The following issues need to be considered before planning/conducting preliminary tests:

- Adequate volume/representative sample
- Discharge criteria and treatment objectives
- Variations in flow rate/effluent characteristics
- Geographical location of treatment plant
- Local electric power costs
- Operating costs estimation (EE/O values)
- Pre-treatment requirements (filtration, coagulation, pH adjustment, etc.)

4.3.4.1 Organic Carbon

For industrial wastewater treatment with AOPs, a COD < 1000 mg/L is recommended and the optimum (ideal) treatment range is 100–500 mg/L COD (Pignatello et al., 2006). For high COD samples (500–1000 mg/L), the Fenton and photo-Fenton oxidation processes are highly recommended. For CODs > 1000 mg/L, photochemical and photocatalytic treatment is not recommended or a pre-treatment step is required before the application of photochemical/photocatalytic treatment.

4.3.4.2 UV Absorbance

An absorbance of <0.2 cm^{-1} is desired; but <0.5 cm^{-1} can be tolerated. High UV absorbance values result in competition for UV light radiation with non-target species slowing down the photochemical reaction (removal of the target pollutant). If the UV absorbance is >10 cm^{-1}, all light is absorbed within the first mm or less from the quartz sleeve that surrounds the lamp and the UV light energy cannot reach the oxidant. Mixing and good, direct contact is recommended if the UV absorbance in the sample is high (Bolton, 2001).

4.3.4.3 Suspended Solids (SS)

Suspended solids (SS) have two effects on UV/oxidation systems, namely (i) they reflect, scatter and absorb UV light (less photochemical energy reaches the contaminant and oxidant) and (ii) SS can also adsorb organics (Bolton, 2001). Only dissolved organics can effectively react with HO$^\bullet$ in the solution bulk, such that the treatment efficiency slows down for solids and colloidal matter due to mass transfer limitations. Hence, the rate limiting step is the desorption of adsorbed pollutants. This effect is most severe for hydrophobic compounds (PCBs, PAHs, etc.). Generally speaking, filtration prior to the application of AOPs is always strongly suggested, in particular prior to application of photochemical AOPs. For SS > 50 mg/L, filtration with a 10–30 micron cutoff range is highly recommended. By the removal of colloids and oil droplets, partial removal of the organic load (COD) is also achieved and the UV absorbance (competing with the oxidant for UV light) is thereby also reduced.

4.3.4.4 Oil and Grease

Oil and grease slow down the treatment process by competing with the contaminant for $HO^\bullet$ and with the oxidant for UV light, thereby increasing the UV absorbance of the wastewater.

4.3.4.5 Chloride, Nitrate, Sulfate, Phosphate

These inorganic anions compete with contaminants for $HO^\bullet$ acting as $HO^\bullet$ scavengers at acidic pHs (2–3) (Riberia et al., 2019). Bicarbonate and carbonate ions that become important after pH > 8 and >10, respectively, are effective $HO^\bullet$ scavengers at alkaline pHs and an inhibit oxidation reactions appreciably (Petrie et al., 2015; Rastogi et al., 2009).

4.3.5 Treatment Examples and Case Studies for AOPs

4.3.5.1 MTBE (Methyl Tert-Butyl Ether)

MTBE is a gasoline additive, solvent and common groundwater contaminant. Due to its relatively high solubility in water, concentrations of >5 mg/L are not uncommon. At these concentrations, advanced treatment technologies including AC adsorption become very expensive. For MTBE treatment, H_2O_2/UV-C oxidation is recommended (CCOT, 1996). EE/O for H_2O_2/UV-C treatment of 10 mg/L MTBE is in the range of 10 kWh/m^3/order (order = 90% or 1-log removal of target contaminant of group parameter).

4.3.5.2 Volatile Organic Compounds

VOCs are industrial solvents used for cleaning, extraction and purification purposes. VOCs, including DCE, TCE and vinyl chloride, are major groundwater pollutants and treatment options for VOCs, with concentrations above 10 mg/L, are UV/O_3, UV/H_2O_2, UV/Fenton (photo-Fenton) and O_3/H_2O_2 (Pignatello et al., 2006).

4.3.5.3 Cyanides

Cyanide (CN^-) originates from mining activities and electroplating processes. Free cyanide (KCN, NaCN) is easily oxidizable, but complex cyanides (FeCN) cannot be oxidized by conventional oxidants (H_2O_2, HOCl, ClO_2, O_3). Recommended AOPs for cyanide treatment are H_2O_2/UV (40–80 kWh/m^3/

order for full H_2O_2/UV oxidation), photo-Fenton treatment (10–20 kWh/m^3/order; but requires acidification to pH = 2–5). Free CN^- is further oxidized to cyanate (CNO^-) and ultimately to CO_2, NO_2, or N_2 (g) by H_2O_2 or HOCl.

4.3.5.4 Phenols

Many industrial pollutants contain phenolic groups in their structure. Phenol and phenol derivatives are used as solvents and wood preservatives with concentration levels up to 100–800 mg/L being present in effluent discharge. At high pollutant concentrations, photo-Fenton or non-photochemical AOPs are good/better options than photochemical AOPs. Operating costs are usually high for H_2O_2/UV treatment (> 20–30 kWh/m^3/order). When total phenolics are >100–200 mg/L and the wastewater's COD is also high (> 1000 mg O_2/L), operating costs will rise to >50 kWh/m^3/order, rendering AOPs not feasible for treatment of phenol-containing wastewater (example: olive oil mill effluent).

4.3.5.5 Adsorbable Organically Bound Halogens

AOX can be found in pesticide production, pulp and paper, textile and tannery wastewater. As with the previous parameters, AOX is also a discharge parameter for specific industries. It is a relatively difficult-to-treat pollutant parameter due to its halogen content. Operating costs are only feasible for O_3/UV and iron-based, non-photochemical AOPs.

4.4 Conclusions

Industrial wastewaters are known for their variable difficult-to-predict characteristics and complexity that render their effective and at the same time economically/technically feasible treatment sometimes a rather difficult task. Integration of advanced chemical and photochemical oxidation processes with biological treatment processes has often been practiced for the efficient, full treatment of industrial wastewater; however, sometimes with limited success. In this chapter, the advantages, implications and different modes of chemical/photochemical treatment applications to enhance/improve conventional biological treatment processes have been summarized. Apparently, the industrial sector and the industry's process and pollution profiles have to be known very well and studied in detail to decide where and how to apply chemical or photochemical advanced oxidation processes. Ultimately, process economics, technical limitations, and ecotoxicological and safety issues determine whether a treatment process integration is sustainable.

REFERENCES

Ameta, S. C. & Ameta, R. (2018). *Advanced Oxidation Processes for Wastewater Treatment: Emerging Green Chemical Technology*. Academic Press, Oxford, 348 p.

Arslan-Alaton, I., Dogruel, S., Baykal, E. & Gerone, G. (2004a). Combined chemical and biological oxidation of penicillin formulation effluent. *Journal of Environmental*

Management, 73(2), 155–163. https://doi.org/10.1016/j. jenvman.2004.06.007.

Arslan-Alaton, I., Eremektar, G., Germirli Babuna, F., Selçuk, H. & Orhon, D. (2004b). Chemical pre-treatment of textile dye carriers with ozone: Effects on acute toxicity and activated sludge inhibition. *Fresenius Environmental Bulletin*, 13(10), 1040–1044.

Arslan-Alaton, I., Eremektar, G., Germirli-Babuna, F., Insel, G., Selcuk, H., Ozerkan, B. & Teksoy, S. (2005). Advanced oxidation of commercial textile biocides in aqueous solution: Effects on acute toxicity and biomass inhibition. *Water Science & Technology*, 52(10–11), 309–316. https://doi. org/10.2166/wst.2005.0707.

Aziz, S. Q., Aziz, H. A., Yusoff, M. S. & Bashir, M. J. (2011). Landfill leachate treatment using powdered activated carbon augmented sequencing batch reactor (SBR) process: Optimization by response surface methodology. *Journal of Hazardous Materials*, 189(1–2), 404–413. https://doi. org/10.1016/j.jhazmat.2011.02.052.

Bolton, J. R. (2001). *Ultraviolet Applications Handbook*. Bolton Photosciences Inc., Edmonton.

Calgon Carbon Oxidation Technologies (1996). *The AOP Handbook*. CCOT, Ontario.

Changotra, R., Rajput, H. & Dhir, A. (2019). Treatment of real pharmaceutical wastewater using combined approach of Fenton applications and aerobic biological treatment. *Journal of Photochemistry and Photobiology A: Chemistry*, 376, 175–184. https://doi.org/10.1016/j.jphotochem.2019.02.029.

Changotra, R., Rajput, H., Guin, J. P., Khader, S. A. & Dhir, A. (2020). Techno-economical evaluation of coupling ionizing radiation and biological treatment process for the remediation of real pharmaceutical wastewater. *Journal of Cleaner Production*, 242, 118544. https://doi.org/10.1016/j. jclepro.2019.118544.

Dogruel, S., Ateş Genceli, E., Germirli Babuna, F. & Orhon, D. (2004). Ozonation of nonbiodegradable organics in tannery wastewater. *Journal of Environmental Science and Health Part A*, 39(7), 1705–1715. https://doi.org/10.1081/ ESE-120037871.

Dogruel, S., Ateş Genceli, E., Germirli Babuna, F. & Orhon, D. (2006). An investigation on the optimal location of ozonation within biological treatment for a tannery wastewater. *Journal of Chemical Technology and Biotechnology*, 81, 1877–1885. https://doi.org/10.1002/jctb.1620

Fang, S., Wang, C. & Chao, B. (2016). Operating conditions on the optimization and water quality analysis on the advanced treatment of papermaking wastewater by coagulation/ Fenton process. *Desalination and Water Treatment*, 57(27), 12755–12762. https://doi.org/10.1080/19443994.2015.1052 568.

Fatta-Kassinos, D., Manaia, C., Berendonk, T. U., Cytryn, E., Bayona, J., Chefetz, B., . . . Lundy, L. (2015). Cost action ES1403: New and emerging challenges and opportunities in wastewater reuse (NEREUS). *Environmental Science and Pollution Research*, 22(9), 7183–7186. doi: 10.1007/ s11356-015-4278-0.

Germirli Babuna, F., Camur, S., Arslan-Alaton, I., Okay, O. & Iskender, G. (2009). The application of ozonation for the detoxification and biodegradability improvement of a textile auxiliary: Naphtalene sulphonic acid. *Desalination*, 249, 682–686. https://doi.org/10.1016/j.desal.2008.10.031.

Germirli Babuna, F., Oructut, N., Arslan-Alaton, I., Iskender, G. & Okay, O. (2011). Reducing the toxicity and recalcitrance of a textile xenobiotic through ozonation. In: *Survival and Sustainability*. Gokcekus, H., Turker, U. & LaMoreaux, J. W. (Eds.). Environmental Earth Sciences Publisher, Springer, Heidelberg, 955–965.

Germirli Babuna, F., Yilmaz, Z., Okay, O., Arslan-Alaton, I. & Iskender, G. (2007). Ozonation of synthetic versus natural textile tannins: Recalcitrance and toxicity towards Phaeodactylum tricornutum. *Water Science & Technology*, 55(10), 45–52. https://doi.org/10.2166/wst.2007.305.

Gopalakrishnan, G., Somanathan, A. & Jeyakumar, R. B. (2019). Combination of solar advanced oxidation processes and biological treatment strategy for the decolourization and degradation of pulp and paper mill wastewater. *Desalination and Water Treatment*, 158, 87–96. doi: 10.5004/dwt.2019.24180.

Gursoy-Haksevenler, B. H., Dogruel, S. & Arslan-Alaton, I. (2014). Effect of ferric chloride coagulation, lime precipitation, electrocoagulation and the Fenton's reagent on the particle size distribution of olive mill wastewater. *International Journal of Global Warming*, 6(2–3), 194–211. https://doi. org/10.1504/IJGW.2014.061010.

Hanafi, F., Assobhei, O. & Mountadar, M. (2009). Detoxification and discoloration of Moroccan OMWW by electrocoagulation. *Journal of Hazardous Materials*, 174(1–3), 807–812. https://doi.org/10.1016/j.jhazmat.2009.09.124.

Hocaoglu, S. M. & Orhon, D. (2013). Particle size distribution analysis of chemical oxygen demand fractions with different biodegradation characteristics in black water and gray water. *Clean-Soil, Air & Water*, 41(11), 1044–1051. https:// doi.org/10.1002/clen.201100467.

Imai, A., Onuma, K., Inamori, Y. & Sudo, R. (1995). Biodegradation and adsorption in refractory leachate treatment by the biological activated carbon fluidized bed process. *Water Research*, 29(2), 687–694. https://doi. org/10.1016/0043-1354(94)00147-Y.

International Organization for Standardization (ISO) 8192 (1986). *Water Quality-Oxygen Demand Inhibition Assay in Activated Sludge*. ISO, Geneva.

Iskender, G., Arslan-Alaton I., Koyunluoğlu, S., Yilmaz, Z. & Germirli Babuna, F. (2016). Ozonation of common textile auxiliaries. *World Multidisciplinary Earth Sciences Symposium (WMESS'2016). IOP Conf. Series: Earth and Environmental Science*, 44, 052037. doi:10.1088/1755-1315/44/5/052037.

Iskender, G., Sezer, A., Arslan-Alaton, I., Germirli Babuna, F. & Okay, O. (2007). Treatability of cefazolin antibiotic formulation effluent with O_3 & O_3/H_2O_2 processes. *Water Science & Technology*, 55(10), 217–225. https://doi.org/10.2166/ wst.2007.325.

Keerthi, U. S., Nithya, M. & Balasubramanian, N. (2019). Evaluation of advanced oxidation processes (AOPs) integrated membrane bioreactor (MBR) for the real textile wastewater treatment. *Journal of Environmental Management*, 246, 768–775. https://doi.org/10.1016/j.jenvman.2019.06.039.

Koyunluoglu, S., Arslan-Alaton, I., Eremektar, G. & Germirli Babuna, F. (2006). Pre-ozonation of commercial textile tannins effect on biodegradability and toxicity. *Journal of Environment Science and Health Part A*, A41(9), 1873–1886. https://doi.org/10.1080/10934520600779083.

Ledakowicz, S., Michniewicz, M., Jagiella, A., Stufka-Olczyk, J. & Martynelis, M. (2006). Elimination of resin acids by advanced oxidation processes and their impact on subsequent biodegradation. *Water Research*, 40(18), 3439–3446. https://doi.org/10.1016/j.watres.2006.06.038.

Lodha, B. & Chaudhari, S. (2007). Optimization of Fenton-biological treatment scheme for the treatment of aqueous dye solutions. *Journal of Hazardous Materials*, 148(1–2), 459–466. https://doi.org/10.1016/j.jhazmat.2007.02.061.

Lucas, M. S., Dias, A. A., Sampaio, A., Amaral, C. & Peres, J. (2007). Degradation of a textile reactive azo dye by a combined chemical-biological process: Fenton's reagent-yeast. *Water Research*, 4(5), 1103–1109. https://doi.org/10.1016/j.watres.2006.12.013.

Malato, S., Blanco, J., Maldonado, M. I., Oller, I., Gernjak, W. & Pérez-Estrada, L. (2007). Coupling solar photo-Fenton and biotreatment at industrial scale: Main results of a demonstration plant. *Journal of Hazardous Materials*, 146(3), 440–446. https://doi.org/10.1016/j.jhazmat.2007.04.084.

Matilainen, A., Vepsäläinen, M. & Sillanpää, M. (2010). Natural organic matter removal by coagulation during drinking water treatment: A review. *Advances in Colloid and Interface Science*, 159(2), 189–197. https://doi.org/10.1016/j.cis.2010.06.007.

Merayo, N., Hermosilla, D., Blanco, L., Cortijo, L. & Blanco. A. (2013). Assessing the application of advanced oxidation processes, and their combination with biological treatment, to effluents from pulp and paper industry. *Journal of Hazardous Materials*, 262, 420–427. https://doi.org/10.1016/j.jhazmat.2013.09.005.

Mosteo, R., Ormad, M. P. & Ovelleiro, J. L. (2007). Photo-Fenton processes assisted by solar light used as preliminary step to biological treatment applied to winery wastewaters. *Water Science & Technology*, 56(2), 89–94. https://doi.org/10.2166/wst.2007.476.

Nguyen, L. N., Hai, F. I., Kang, J., Price, W. E. & Nghiem, L. D. (2011). Removal of trace organic contaminants by a membrane bioreactor – granular activated carbon (MBR – GAC) system. *Bioresource Technology*, 113, 169–173. https://doi.org/10.1016/j.biortech.2011.10.051.

Ollis, D. F. (2001). On the need for engineering models of integrated chemical and biological oxidation of wastewaters. *Water Science and Technology*, 44(5), 117–123. https://doi.org/10. 2166/wst.2001.0265.

Pariente, M. I., Siles, J.A., Molina, R., Botas, J.A., Melero, J.A. & Martinez, F. (2013). Treatment of an agrochemical wastewater by integration of heterogeneous catalytic wet hydrogen peroxide oxidation and rotating biological contactors. *Chemical Engineering Journal*, 226, 409–415. https://doi.org/10.1016/j.cej.2013.04.081.

Parsons, S. (Ed.) (2004). *Advanced Oxidation Processes for Water and Wastewater Treatment*. IWA Publishing, London.

Paździor, K., Bilińska, L. J. & Ledakowicz, S. (2019). A review of the existing and emerging technologies in the combination of AOPs and biological processes in industrial textile wastewater treatment. *Chemical Engineering Journal*, 376, 120597. https://doi.org/10.1016/j.cej.2018.12.057.

Petrie, B., Barden, R. & Kasprzyk-Hordern, B. (2015). A review on emerging contaminants in wastewaters and the environment: Current knowledge, understudied areas and recommendations for future monitoring. *Water Research*, 72(4), 3–27. doi: 10.1016/j.watres.2014.08.053.

Peyton, G. R. (1993). The free-radical chemistry of persulfate-based total organic carbon analyzers. *Marine Chemistry*, 41(1–3), 91–103.

Pignatello, J. J., Oliveros, E. & MacKay, A. (2006). Advanced oxidation processes for organic contaminant destruction based on the Fenton reaction and related chemistry. *Critical Reviews in Environmental Science and Technology*, 36(1), 1–84. doi: 10.1080/10643380500326564.

Pikuda, O., De Luca, G., Di Salvo, J. L. & Chakraborty, S. (2020). A review of emerging trends in membrane science and technology for sustainable water treatment. *Journal of Cleaner Production*, 266, 121867. https://doi.org/10.1016/j.jclepro.2020.121867.

Rastogi, A., Al-Abed, S. R. & Dionysiou, D. D. (2009). Effect of inorganic, synthetic and naturally occurring chelating agents on Fe (II) mediated advanced oxidation of chlorophenols. *Water Research*, 43(3), 684–694. doi: 10.1016/j.watres.2008.10.045

Ribeiro, A. R. L., Moreira, N. F., Puma, G. L. & Silva, A. M. (2019). Impact of water matrix on the removal of micropollutants by advanced oxidation technologies. *Chemical Engineering Journal*, 363(5), 155–173. https://doi.org/10.1016/j.cej.2019.01.080.

Sathya, U., Keerthi, A., Nithya, M. & Balasubramanian, N. (2019). Evaluation of advanced oxidation processes (AOPs) integrated membrane bioreactor (MBR) for the real textile wastewater treatment. *Journal of Environmental Management*, 246, 768–775. https://doi: 10.1016/j.jenvman.2019.06.039. Epub 2019 Jun 19.

Selcuk, H., Eremektar, G. & Meric, S. (2006). The effect of pre-ozone oxidation on acute toxicity and inert soluble cod fractions of a textile finishing industry wastewater. *Journal of Hazardous Materials*, B137(1), 254–260. https://doi.org/10.1016/j.jhazmat.2006.01.055.

Shah, M. P. (2020a). *Advanced Oxidation Processes for Effluent Treatment Plants*. Elsevier, Amsterdam.

Shah, M. P. (2020b). *Microbial Bioremediation and Biodegradation*. Springer, Singapore.

Shah, M. P. (2021). *Removal of Emerging Contaminants through Microbial Processes*. Springer, Singapore.

Silva, F. T., Mattos, L. R. & Paiva, T. C. B. (2007). Treatment of an ECF effluent by combined use ofactivated sludge and advanced oxidation process. *Water Science & Technology*, 55(6), 151–156. https://doi.org/10.2166/wst.2007.223.

Sudarjanto, G., Keller-Lehmann, B. & Keller, J. (2006). Optimization of integrated chemical-biological degradation of a reactive azo dye using response surface methodology. *Journal of Hazardous Materials*, 138(1), 160–168. https://doi.org/10.1016/j.jhazmat.2006.05.054.

Tang, S., Wang, Z., Wu, Z. & Zhou, Q. (2010). Role of dissolved organic matters (DOM) in membrane fouling of membrane bioreactors for municipal wastewater treatment. *Journal of Hazardous Materials*, 178(1–3), 377–384. https://doi.org/10.1016/j.jhazmat.2010.01.090.

Tchobanoglous, G. & Burton, F. (2003). *Wastewater Engineering: Treatment, Disposal, and Reuse*. 4th ed. McGraw-Hill, Singapore, 1846–1850.

Tezgel, T., Germirli Babuna, F., Arslan-Alaton, I., Iskender, G. & Okay, O. (2011). Pretreatment of ceftriaxone formulation effluents: Drawbacks and benefits. In: *Survival and Sustainability*. Gokcekus, H., Turker, U. & LaMoreaux, J. W. (Eds.). Environmental Earth Sciences Publisher, Springer, Heidelberg, 943–954.

Wang, M., Meng, Y., Ma, D., Wang, Y., Li, F., Xu, X., Xia, C. & Gao, B. (2017). Integration of coagulation and adsorption for removal of N-nitrosodimethylamine (NDMA) precursors from biologically treated municipal wastewater. *Environmental Science and Pollution Research*, 24(13), 12426–12436. doi: 10.1007/s11356-017-8854-3.

Wang, Y., Shia, S., Wang, C. & Fanga, S. (2018). Degradation of pulp mill wastewater by a heterogeneous Fenton-like catalyst Fe/Mn supported on zeolite. *Environment Protection Engineering*, 44(2), 131–145. doi:10.5277/epe180209.

Water World (2019). www.waterworld.com/drinking-water/treatment/article/14072360/.

5

Photoelectrocatalysis: Principles and Applications

Patricio J. Espinoza-Montero, Ronald Vargas, Paulina Alulema-Pullupaxi and Lenys Fernández

CONTENTS

5.1 Introduction

Recalcitrant compounds are a serious environmental and health problem mainly due to their toxicity and potential hazardous effects on living organisms, including human beings. Conventional wastewater treatments have not been able to remove pollutants from water efficiently; however, advanced oxidation processes (AOPs) are able to solve this environmental concern. One of the most recent AOP technology is photoelectrocatalysis (PEC). This consists of applying an external potential bias to photocatalyst (PC) supported on a conductive substrate to avoid the recombination of photogenerated electron/hole (e$^-$/h$^+$) pairs and increase the lifetime of photogenerated holes to promote an efficient degradation and mineralization of organic pollutants in aqueous medium. This chapter is dedicated to reinforcing knowledge in the field of semiconductor electrodes used in PEC systems; the kinetic aspects of PEC mechanisms; the fundamentals of PEC applied to remove pollutants from water, including an analysis of photocatalysts and modified materials to synthesize photoelectrodes; the PEC systems and the influence of operating parameters; and finally the application of PEC to remove contaminants of emerging concern and microorganisms from wastewater.

5.2 Photoelectrochemistry Using Semiconductor Electrodes: A First Glance

The first thing to discuss about light-induced redox reactions on the surface of a semiconductor electrode is the electron flux and charge balance. When illuminating an n-type semiconductor, the electrons (e$^-$) – majority carriers – can be extracted from the conduction band (CB) as photocurrent (i_{ph}) by an external circuit, in which case, the deficiency of electrons or "holes" (h^+) – as minority carriers – in the valence band (VB) are transported to the interface to oxidize a donor species present in the electrolyte. The charge transfer is completed when the reduction of some electron acceptor species occurs in the secondary electrode. In equilibrium the potential of the bands are flat (E_{FB}), while, far from equilibrium, the band bending is sustained by unbalanced charge carriers [1]. The intensity of the illumination (I) and the electric potential (E) are perturbations that promote band bending. Then, the charge carriers can be involved in the redox processes and the effective chemical potential of the h^+ and e^- changes, yielding to the so-called quasi-Fermi levels ($E_{F,e-}$ and E_{F,h^+}). Precisely in Figure 5.1, these magnitudes are presented on the experimental response, specifically Figure 5.1a shows an illustration of the photoelectrochemical processes that take place on this type of interface (n-type semiconductor | electrolyte), and Figure 5.1b illustrates the photocurrent vs. potential typical response. Now, it should be noted that the majority carriers in a p-type semiconductor are the holes, therefore the band bending is outlined in the opposite direction to the n-type semiconductor. In addition, molecule reduction reactions occur in the semiconductor electrode surface. In this case, the charge balance is completed with oxidation processes in an appropriate anode [2]. Figure 5.1c illustrates the respective electron flux (p-type semiconductor | electrolyte) and Figure 5.1d shows the photocurrent *vs.* potential response.

The most recognized limitation of the processes described is the recombination of the charge carriers (k_R) [1]. If a higher gradient of electrochemical potential is applied under lighting condition, the higher efficiency is defined for the charge carrier's separation and extraction. The optoelectrical and surface properties of the electrode material are decisive to drive appropriated photo-redox reactions [3, 4]. In any case, a delicate balance between electron transfer and recombination fluxes in the semiconductor interface | electrolyte dominates the experimental response [2].

5.3 Kinetic Aspects of Photoelectrocatalytic Systems

5.3.1 Charge Transfer and Recombination: Phenomenology

At first approximation, the phenomenological response of the photocurrent depending on the potential can be understood according to the considerations made by Gartner in 1959 [5]. The theory of Gartner involves fundamental concepts, which link the flux of minority carriers with the energy conditions

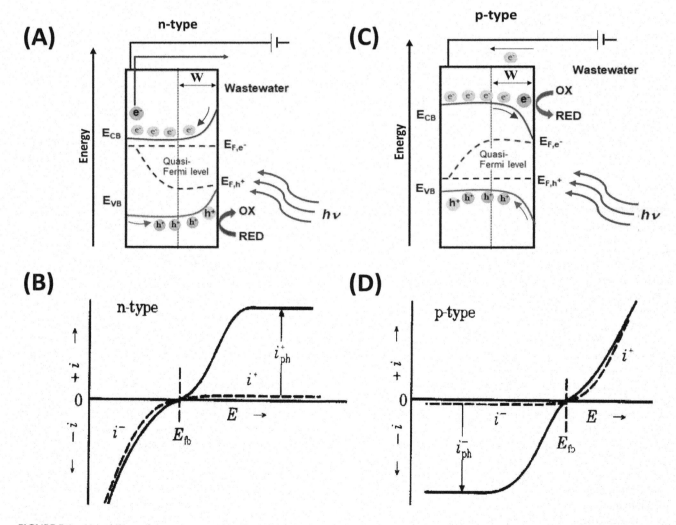

FIGURE 5.1 (a) and (b) are for an n-type semiconductor photoanode, and (c) and (d) correspond to a p-type semiconductor photocathode. (a) and (c) Photoelectrochemical processes that take place on the interface. Reproduced with modifications from [22], Open Access. (b) and (d) photocurrent vs. potential response. Reproduced from [12] with permission from Elsevier.

defined by the light intensity (I_0) and potential (E). In general, the radiation defines an extinction profile in the electrode: $G(x)$ and the experimental response will depend on the characteristic properties of the semiconductor, such as carrier density (N), radiation absorption coefficient (α) and minimum carrier transport length (L_{min}) within the space charge zone (W) [5].

The fundamental postulates of this ideal model are:

i. The rate of generation of carriers depends on the photon flux and absorption coefficient of the semiconductor.

ii. All carriers generated within the W reach the surface when traveling a characteristic length.

iii. At the interface, the photocurrent is supported by a redox process with a very fast charge transfer rate.

iv. There are no recombination losses.

Now, the maximum possible theoretical value can be defined for the photoelectrochemical response in a semiconductor predicted by the Gartner current density (j_G) (Equation 5.1), where I_0 represents the light intensity on the surface.

$$j_G = I_0 \left[1 - \frac{e^{-(\alpha W)}}{1 + \alpha L_{min}} \right] \qquad (5.1)$$

where W is defined by the potential according to Mott and Schottky's theory [1]:

$$W = \left[\frac{2\varepsilon\varepsilon_0 \left(E - E_{FB} \right)}{e_0 N} \right]^{\frac{1}{2}} \qquad (5.2)$$

where: ε and ε_0 are the permittivity of the semiconductor and vacuum, respectively, and e_0 is the electron charge. It is important to note that minority carriers will accumulate in the W in cases of slow electron transfer, changing the bending of the bands and promoting recombination [2]. Then, not all the applied potential will contribute to improving charge separation [2, 6]. This is the case for slow reactions with several steps, such as those involved in the water oxidation reaction (WOR) on photoanodes [2, 6]. Therefore, the general kinetic situation has been treated in a simplified way by considering two apparent kinetic parameters: (i) the direct charge transfer rate constant (k_{CT}) and (ii) the recombination rate constant (k_R) [6]. The practical photocurrent density (j_{ph}) is corrected according to Equation 5.3), where k_{CT} and k_R are potential dependent and are usually determined experimentally by transient photocurrent and/or intensity modulated photocurrent spectroscopy [2].

$$j_{ph} = j_G \left[\frac{k_{CT}}{k_{CT} + k_R} \right] \qquad (5.3)$$

In order to apply the phenomenological Gartner approach, it is fundamental that the data obtained be at potentials much greater than the equilibrium value, where the previously discussed postulates are satisfied [4, 6]. In addition, the situation is analogous in the case of a p-type semiconductor, highlighting

that the minority carriers are the electrons and the current in the interface is supported by a reduction process.

As already mentioned, the equilibrium condition is equivalent to the flat band potential, so determining its value is advantageous in defining criteria for proper evaluation of new semiconducting electrode materials. Its value is usually determined using differential capacitance (C_S) measurements in the W and correlating this data with Mott and Schottky Equation 5.4):

$$\frac{1}{C_S^2} = \left(\frac{2}{\varepsilon\varepsilon_0 A^2 e_0 N_D} \right) E + \left(E_{FB} - \frac{k_B T}{e_0} \right) \qquad (5.4)$$

where A is the area, k_B the Boltzmann's constant and T the temperature. Thus, a plot of C_S^{-2} vs. E will allow one to identify the linear region and then perform a fitting to obtain N_D and E_{FB} values [1]. The experimental determination implicitly considers the structural effects of the electrode. For an n-type semiconductor, the slope of C_S^{-2} vs. E plot is positive and $E_{FB} = E_{CB}$, while for a p-type semiconductor a negative slope is defined and $E_{FB} = E_{VB}$. Knowing that the energetic difference between the valence and conduction bands is the band gap energy ($E_{bg} = E_{VB} - E_{CB}$), then the use of the E_{bg} value experimentally determined by spectroscopic methods in combination with the band potential estimated by Mott-Schottky plot (E_{CB} for n-type semiconductor and E_{VB} for p-type semiconductor), allows one to calculate the position of each energy band of the semiconductor electrode [7].

Now, a first approximation to assess the consequences of recombination processes has to do with elucidating the effects of modifying some key factor involved in the electron flux and charge balance (systematic modification of electrode material from the synthesis process, promotion of adsorbed states, doping, presence or absence of chemical species in the electrolyte, pretreated semiconductor | electrolyte interface, etc.), resulting in changes in the forces that drive photo-redox processes [8]. In any case, the open circuit potential (E_{OC}) measurement indicates the steady-state concentration of photogenerated charge carriers [9]. The lifetime (τ) of photogenerated charge carriers can be estimated by photovoltage measurements as a function of time (t) at open circuit condition with a light flux interruption. The lifetime of charge carriers with respect to E_{OC} can then be evaluated using Equation 5.5):

$$\tau = -\frac{k_b T}{e_0} \left(\frac{dE_{OC}}{dt} \right)^{-1} \qquad (5.5)$$

The decay of photovoltage depends upon the recombination rate of electron-hole pairs, which ultimately defines the lifetime of charge carriers. Due to this, analyzing lifetime trends based on a systematic change in electrode synthesis and doping, among others, would allow elucidation if there are effects related to carrier deactivation processes [3, 4].

5.3.2 Deviations to Ideal and Need for Dynamic Evaluation

In semiconductors with low or moderate doping, the differential capacitance is controlled by the charge carriers accumulated in the W [10]. Reichman incorporated the effects of carrier recombination into the depletion layer in the analysis of the appropriate modifications by establishing the differential balance of minority carriers, as well as the boundary conditions of the differential equation for transport phenomena [11].

The condition where the Fermi level (E_F) is fixed is also possible, especially for surface state densities greater than 1% relative to the density of the doping atom on the surface. In these conditions, changes in applied potential can then develop in the Helmholtz layer [8, 12]. Thus, the total capacitance (C_T) shall be defined by the contribution of the two consecutive capacitances: the space charge zone (C_W) and the Helmholtz layer (C_H): $C_T^{-1} = C_W^{-1} + C_H^{-1}$ [3].

In general, the doping, different crystalline phases, adsorbed states, and material defects introduce modifications to the density of electronic states (DOS), which are manifested between the VB and CB in nanostructures. These defects – known as surface states – affect the dynamics of the interface [3, 8]. In these cases, C_S relates to the intrinsic capacitance of the semiconductor photoactive film or the chemical capacitance ($C\mu$), which is usually much less than C_H and dominates the final electrochemical behavior. The chemical capacitance depends on the geometric characteristics of the nanostructured film, as well as on the variation of the electronic population in the DOS due to a change in Fermi level (E_F). In general, contributions come from the dynamics of CB states, surface states and deep trap states [3]. In any case, it must be emphasized that a way to modify the DOS of a semiconductor has to do with doping, the formation of new crystalline phases and their synthesis to form a nanostructured material.

Regarding the nanostructure effect, Peter's group postulates an interesting analysis in which (invoking only geometric considerations) it is possible to pursue the improvement of quantum efficiencies for the transformation of solar energy by using nanostructured electrodes [10]. It must be recognized that nanostructured electrodes are more complex and involve changes in both the surface state population and recombination kinetics [3, 8]. In addition, the transport phenomena of majority carriers also require attention. For example, it has been demonstrated that kinetics of trapping and detrapping of both superficial and deep trap states in nanostructured electrodes affects the transient response of the photocurrent [13].

The study of these phenomena can be carried out using electrochemical impedance spectroscopy (EIS), intensity modulated photocurrent spectroscopy (IMPS), time-resolved optical spectroscopy measurements and electrochemical methods [2, 8]. In recent literature, a compendium of the electrochemical response of nanostructured semiconductor oxides and compact layers is available, where not only the fundamentals are discussed, but also details of electronic structure of nanoporous electrodes as shallow and deep trap states are provided, emphasizing the nature of the different capacitive phenomena that take place in semiconductor electrodes [3].

5.3.3 Photocatalytic Cell and Electrochemically Assisted Photocatalysis

A photocatalytic process can be understood as the light-induced combustion reaction of organic matter; therefore, if carried out in a separated compartment, it allows the oxidation of organic compounds yielding the corresponding electron flux as an added value. This is analogous to a fuel cell and is called a photo fuel cell [14]. In fact, the oxygen reduction reaction (ORR) must be carried out in a cell using an appropriate electrocatalytic material, and determining factors are the control of electric losses during operation of the cell as well as the selection of appropriate electrodes.

The so-described process is spontaneous, and its kinetics is defined at the experimental condition that flux-matched the oxidation reaction with the reduction reaction. First, the light arrives to the photocatalyst surface (S) and promotes an electron from VB to CB:

$$h\nu \overset{S}{\to} h^+_{VB} + e^-_{CB} \tag{5.6}$$

Then, the water is oxidized by a hole in the VB yielding a hydroxyl radical ($^•OH$) as a first step ($H_2O + h^+_{VB} \to {}^•OH + H^+$) in the pathway to form O_2:

$$2H_2O + 4h^+_{VB} \to O_2 + 4H^+ \tag{5.7}$$

The $^•OH$ attacks the adjacent organic matter (R) to yield the product (P):

$${}^•OH + R \to P \tag{5.8}$$

The charge balance is reached at the cathode by the ORR:

$$O_2 + 4e^-_{CB} + 4H^+ \to 2H_2O \tag{5.9}$$

Now, to interpret the flux-matching condition that results from coupling these redox processes (reactions 5.7 and 5.8 for oxidation and reaction 5.9 for reduction), it will resort to analyzing the photocurrent vs. potential curves. Figure 5.2a illustrates the current density vs. potential plot for a pair of semi-reactions that are coupled with positive $\Delta E = E_{r,c} - E_{r,a}$, where $E_{r,c}$ and $E_{r,a}$ are the reversible potentials of cathodic and anodic processes, respectively. The current response of the photoanode (curve j_a) and the cathode (curve j_c) is shown. Redox flux matching is represented as the point of intersection between j_a and $|j_c|$, this is the operating point: I_{OP} (0) [15]. Indeed, this described condition is in which the photocatalytic process occurs spontaneously, being, for example, the energetic condition that results when photocatalysts are illuminated with sunlight and used to degrade organic compounds with the concomitant electricity production.

On the other hand, a very convenient possibility is to electrochemically assist photocatalysis. ORR usually limits processes kinetically due to its slow kinetics, so attaching a second convenient reduction reaction is interesting. For example, if one wants to oxidize water containing organic matter to achieve effluent purification, then a reduction process such as metal ions (M^{z+}) reduction (reaction 5.10) or protons reduction (reaction 5.11) can be coupled. In all cases, a double benefit will be obtained: (*i*) the organic matter will be removed by oxidation (j_a), and (*ii*) the metal ions will be removed from the

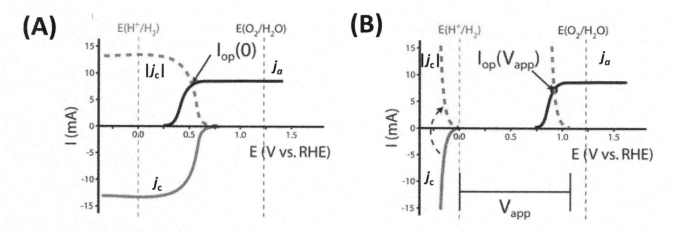

FIGURE 5.2 Current vs. potential plot and operation point (I_{OP}) in photoelectrochemistry. (a) spontaneous photoelectrochemical cell ($E_{r,c} - E_{r,a} > 0$), e.g., photo fuel cell and (b) electrochemical assist photoelectrochemical cell ($E_{r,c} - E_{r,a} < 0$ and $E_{r,c} - E_{r,a} + V_{app} > 0$). (Reproduced from ref. [15] with permission of Royal Society of Chemistry.)

solution by reduction or H_2 will be formed as an added value product (j_c).

$$M^{z+} + ze^-_{CB} \rightarrow M^0 \qquad (5.10)$$

$$2H^+ + 2e^-_{CB} \rightarrow H_2 \qquad (5.11)$$

In this case, reactions 5.7 and 5.8 for oxidation reactions are coupled to reactions 5.10 or 5.11 as reduction reactions, and the photoelectrochemical cell implies $E_{r,c} - E_{r,a} < 0$. So for redox processes to occur, a specific amount of energy must be supplied by electrochemical assistance. This is exemplified in Figure 5.2b, which indicates that the energy supply (V_{app}) must comply with the following thermodynamic constraint: $E_{r,c} - E_{r,a} + V_{app} > 0$. This can be interpreted as a displacement of the curve $|j_c|$ in order to intercept with the curve j_a, defining the point of operation as the one that maximizes the power density of the electrochemical cell, that is: $I_{OP}(V_{app})$ [15].

It should be noted that for practical applications, it is assumed that when in the presence of organic compounds (R), the measured photocurrent is greater than the one obtained in the support electrolyte, that is: j_{ph} (R + H_2O) > j_{ph} (H_2O). Therefore, the appropriate condition for a coupled photoredox process is to operate at the maximum power density [16]. In any case, all the aforementioned physicochemical arguments are valid, but to make adequate quantitative predictions it must be ensured that the losses due to mass transfer and ohmic overpotentials are minimal; otherwise, such losses should be considered [17]. The change in the concentration of the reacting species should be followed by analytical methods such as total organic carbon (TOC), chromatography and/or spectrophotometry. Furthermore, this information can be used to perform kinetic studies using classical chemical kinetic models, calculate the efficiency of the removal process and define the energy consumption [18, 19].

5.4 Photoelectrocatalysis (PEC) Applied to Remove Pollutants from Water

5.4.1 Summarizing the Concepts to Implement Photoelectrocatalysis in Effluent Treatment

The applicability and efficiency of PEC for organic pollutant degradation is directly associated with the nature of photoelectrodes (photoanode) and their intrinsic photocatalytic properties as previously addressed [20]. The application of an adequate potential disturbance during irradiation to the photoanode will cause an efficient separation of charges and will avoid their recombination, which will generate a high concentration of h_{BV}^+ on the surface of the catalyst. The generated h_{BV}^+ have a strong oxidation ability and can react with water to promote the formation of reactive oxygen species such as ·OH and others, which due to their high oxidizing power can then react with organic pollutants up to mineralization [20].

Besides that, the photogenerated electrons e_{CB}^- driven towards the cathode by the external electrical circuit could help to induce some reduction reactions at the cathode, for example, (i) react with dissolved oxygen to form H_2O_2 or react with adsorbed water to produce H_2; (ii) reduce the nitrogen that may be present in water into ammonia or hydrazine; (iii) participate in the reduction of heavy metals, which is advantageous to sufficient usage of electricity for the recovery of heavy metals; and (iv) reduce the CO_2 into CO, CH_4 or longer chain carbon species/another interesting reaction, which is also shown in Figure 5.3 and could occur in a PEC system [21]. Furthermore, as discussed in the previous section, a band bending takes place in the W when the semiconductor is in contact with the electrolyte, which is a really important phenomenon to avoid recombination because it favors the separation of (e^-/h^+) pairs and lets electrons circulate throughout the system and be transferred to the cathode. In other words, the

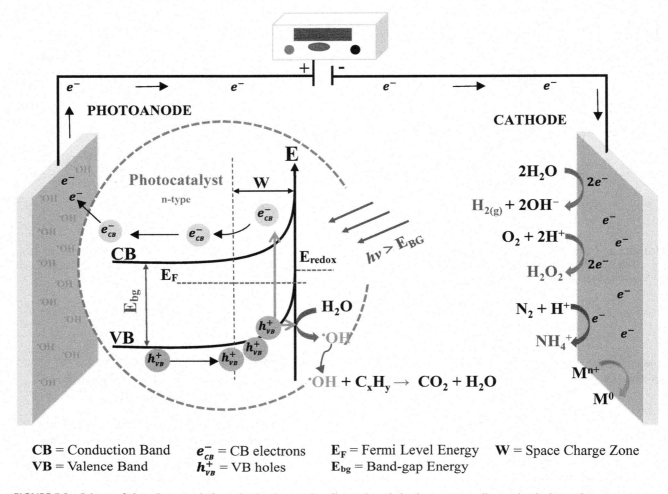

CB = Conduction Band e_{CB}^- = CB electrons E_F = Fermi Level Energy W = Space Charge Zone
VB = Valence Band h_{VB}^+ = VB holes E_{bg} = Band-gap Energy

FIGURE 5.3 Scheme of photoelectrocatalytic mechanism in organic pollutant degradation in aqueous medium and cathode reactions.

larger the band bending, the better charge separation. A complete scheme that sums up the PEC applied to remove organic pollutants is presented in Figure 5.3.

5.4.2 Semiconductor-Based Photocatalysts and Photoelectrodes

Generally, the semiconductor materials used in the PEC process are photoanodes (n-type semiconductors), when oxidation reactions occur in the interface, and photocathodes (p-type semiconductors) when the reduction takes place as the principal process [22]. These electrodes can be coupled to work together (photoanode with photocathode) or connected to a suitable electrocatalyst in order to drive the complementary reaction: an n-type semiconductor with a metal cathode or a p-type semiconductor with a metal anode. For instance, in the case of widely investigated PEC water splitting, the h+ oxidizes water to form oxygen at an n-type semiconductor photoanode, while the electrons reduce protons at a p-type semiconductor or a metallic photocathode to form hydrogen. A similar process occurs with organic pollutants in aqueous medium [23].

Important efforts in the design of efficient photoelectrocatalytic systems have been reported in the literature. The synergy

of the photocatalyst/electrode system must be correctly coupled to the applied external electric polarization. For that, three important processes should be noticed in PEC systems for wastewater treatment [22].

i. Light absorption by semiconductor to generate e-/h+ pairs
ii. Charge separation and migration in the surface semiconductor
iii. Interfacial oxidation and reduction reaction of pollutants

The different methodologies for improving PEC systems are based on the three processes mentioned earlier, principally centered on:

i. The development of semiconductor materials of narrow band gap, which can absorb visible light
ii. Synthesis of materials in which crystalline structure, particle size and so on should be oriented to obtain an efficient charge separation and migration with highest redox reaction rates of pollutants

iii. Good electric conductivity and sufficient stability of semiconductor material for being applied in water treatment process in practical applications to big scale

Visible light is an inexhaustible source of clean energy when compared with the more energetic and harmful UV light. Visible light can be obtained by harvesting solar light that constitutes a safe and unlimited source. However, the photoexcitation of photoelectrodes by low-energy photons requires the design of special semiconductors (coupled or doped semiconductors) that are able to absorb solar light to degrade organic pollutants from wastewater. Even if the photogenerated pairs (e^-/h^+) have thermodynamically sufficient potentials for the degradation of pollutants, the recombination could not be avoided, especially if there are no suitable and available active reaction sites on the surface of semiconductor [22].

This means that the semiconductor must have an adequate position of both the valence and conduction bands in relation to the oxidation and reduction potentials of the molecules of pollutants. In this context, semiconductors modified with suitable catalysts could accelerate the photocatalytic processes by means of having more available reaction sites that are able to promote charge separation and transport across the interface junctions formed between the catalyst and the semiconductor. Furthermore, a high degradation efficiency and a low cost must characterize the semiconductors used as electrodes. Due to this, developing materials with characteristics such as long-term stability, high mechanical resistance, high corrosion resistance, high catalytic activity, suitable band energies and sustainability and low cost are the main challenges within this field of research [22].

5.4.2.1 Photoelectrodes

As mentioned previously, a photoelectrode is prepared through immobilizing a semiconductor material (photocatalyst) on a conductive substrate (electrode). Though this setup contributes to avoiding the fast recombination of the e^-/h^+ pairs, the position of CB and VB of semiconductors in photocatalytic systems must be taken into account, as relatively wide-band gap-semiconductors require the application of UV radiation – which is undesirable – in contrast to visible light – or sunlight – which is with more accessible and less harmful. Additionally, a control of the redox reactions that occur on photoelectrodes by means of the anodic potential (E_a) result in

an improvement of the reaction's selectivity and further control of the Fermi level in a semiconductor, which consequently improve the charge separation [22]. A clear example of this case is the water decomposition by a PEC system, where the position of the CB and VB have to be respectively more negative and positive than the potentials at which H_2 and O_2 are produced [24].

According to the previously reported literature about photoelectrodes, the more commonly used photocatalysts in PEC systems are TiO_2, WO_3, ZnO, CdS, Fe_2O_3 and SnO_2. Table 5.1 summarizes the main characteristics of these semiconductors, where the TiO_2 stands out as the most used photocatalyst due to its well-known photocatalytic properties, which were ascertained in 1972 [20].

According to Table 5.1, ZnO is a semiconductor with similar photocatalytic properties to TiO_2. Nevertheless, the chemical instability of the first in both acid and basic media has limited its application for the degradation of pollutants from water by PEC [24]. Similarly, Fe_2O_3 has good properties such as chemical stability, abundancy in nature, low cost, negligible toxicity, optic penetration ($\delta_p = 118$ a 550 nm) and band gap with low energy [25]. However, its low conductivity is inconvenient for its application in PEC because its real efficiency is hampered by the low mobility of load carriers in its structure, which leads to the accumulation of h^+ on the surface of the semiconductor and a high rate of surface charge recombination. Hopefully, such low conductivity has been solved by doping this oxide with more conductive metals [17]. Likewise, $BiVO_4$ possess a relatively narrow band gap so it can be excited under solar light, its optical penetration is $\delta_p = 100$ a 550 nm, and it is also nontoxic and thermodynamically favorable for water oxidation. However, it requires extremely high potentials for redox reactions on its surface, it is stable only at neutral pH, and it is characterized by low photon efficiency and susceptibility to photocorrosion, which is why this oxide is often coupled with other semiconductors [22]. Similarly, CdS possess an optical penetration of $\delta p = 62$ a 550 nm, which theoretically is active under visible light. Nevertheless, it leads to the accumulation of h^+ at the surface, has low oxidation kinetics and is susceptible to photocorrosion. For that reason, it has been used coupled to sacrificial anodes eliminating h^+, which yields a faster oxidation rate involved in the reaction of interest [26].

According to the aforementioned, special attention must be paid to the development of highly stable semiconductor materials that can be activated by solar irradiation, leading to the required reduction of the costs for the effective application of

TABLE 5.1

Semiconductor Photocatalyst Most Used in PEC Applied in Photoanodes [20, 22, 24]

Photocatalyst	Crystalline Phase	Band Gap (eV)	Adsorption Band (nm)
TiO_2	Rutile	3.05	405
	Anatase	3.25	385
WO_3	Tungstite	2.5 – 2.7	500
ZnO	Wurtzite	3.2 – 3.4	~400
CdS	---------	2.4	~345
α-Fe_2O_3	Hematite	2.2	Visible light
$BiVO_4$	-	2.4–2.5 eV	Visible light

this technology. Additionally, the semiconductor band gap and the efficiency of the surface reactions are parameters related to the electronic structure of the first and to the photon absorption efficiency; therefore, achieving modified materials with a low energy band gap has been the main focus in order to reach better and efficient PEC systems [27].

Additionally, the conductive substrate used for the preparation of photoanodes plays an important role in the electron transfer through the semiconductor film and must comply with several criteria, such as strong adherence, chemical and electrochemical stability, relatively high conductivity, high specific surface area and strong adsorption affinity towards pollutants. Among the most common substrates that have been applied in PEC are fluorine-doped thin oxide glass (FTO), indium thin oxide glass (ITO), titanium (Ti), stainless steel (SS), graphite (G) and carbon fibers; recently, boron-doped diamond (BDD) has also been studied [22, 28].

Both physical (sputtering and evaporation) and chemical methods (sol-gel process combined with dip coating or spin coating, electrophoretic deposition, and anodization processes, among others) have been studied to fabricate photoanodes with high stability and excellent properties [21, 28]. While mechanical methods often require a vacuum environment – which implies especial equipment and high cost – chemical methods could be more convenient and cost effective; but until now, both physical and mechanical methods have been widely used in laboratory-scale thin-film electrode preparation [21–22]. Thus, the preparation of photoanode materials for applications as photoelectrodes still needs improvement to achieve photoelectrocatalytic treatment of wastewater at an industrial scale. Other important aspects of photoelectrodes are the film thickness and stability. A thick film could be beneficial because it has a higher photocatalytic capacity; however, it becomes less stable and would inhibit the action of the support material, whereas a very thin film could have a low photocatalytic capacity [28, 29]. Therefore, a proper control of the film thickness and roughness is mandatory to achieve the highest possible efficiency in a water treatment process. For this reason, the synthesis method and/or the construction of the photoelectrode is of utter importance, since it will have a direct impact on the kinetic aspects of the process, as discussed previously in section 5.3.

5.4.2.2 Modified Semiconductors

A large part of the research about pollutant degradation in wastewater by PEC assisted by sunlight (visible light) is focused on optimizing the efficiency of n-type semiconductor photoanodes and thus improving the performance of oxidation reactions on their surface due to the high overpotentials required in photocatalytic reactions on conventional non-illuminated anodes. The photocatalytic activity at UV or visible light wavelengths has an inverse relation with the width of the band gap; this means that a relatively wide band gap is not desirable for adsorption of solar light in a PEC system [30]. Therefore, some modification methodologies to improve the performance of photocatalyst materials have been explored. To improve the photoelectrocatalytic efficiency on semiconductors, approaches such as the use of nano-sized semiconductors,

the combination of two or more semiconductor materials, the doping of the semiconductors or the creation of thin-layer heterojunctions by deposition of semiconductors have been employed. The interested reader can refer to the review published by Kusmierek in 2020 for more information [22, 31].

The performance of semiconductors in PEC systems is closely related to their crystal size because the band gap is inversely related to the size of the semiconductor crystal. Therefore, some nonviable superficial reactions are allowed to occur as the size is reduced, since the percentage of atoms or ions exposed on the photocatalyst's surface increases enormously as the surface to volume ratio increases, leading to a greater number of surface-arranged available active sites for catalytic reactions [26]. For instance, a 100% degradation of organic pollutants can be achieved when using nanometric TiO_2, which is a substantial increase when compared with the 70% degradation obtained with TiO_2 supported on Ti electrodes [32]. Thus, the surface effects related to the reduction of the size of a semiconductor are greatly important in PEC systems.

A smart approach for obtaining appreciably efficient photoanodes with excellent light-harvesting property can be done by coupling two or more semiconductors appropriately selected for their band gap, constituting a key factor for using the solar light in the photoexcitation of photoelectrodes and consequently avoiding the recombination of photogenerated e^-/h^+ pairs [27]. In this context, a combined material can be obtained by putting together a semiconductor that presents a good degradation efficiency due to its wide band gap but requires photoexcitation with UV radiation, and another with a narrower band gap than the first one. According to Peleyeju et al., combining appropriate semiconductors is thought to create heterojunctions that generate sufficient potential that increase the separation of the e^-/h^+ pairs. Such heterojunctions – formed by semiconductors (p-type and n-type) – are classified according to the positions of the band energies in p-n, n-n, and p-p types. As an example, several heterojunctions with TiO_2 (3.2 eV) coupled to WO_3 (2.6 eV), Cu_2O (2.0 – 2.2 eV), CdS (2.4 eV) or other combined materials as g-C_3N_4:Ag:AgCl:BiVO and Ag:AgCl:BiVO$_4$ have been reported in the literature [22, 27]. These combinations have resulted in a visible-light-promoted more efficient electron transport, a lower recombination rate of charge carriers and the formation of high-energy reactive species.

Semiconductor doping is also related to the semiconductor band gap engineering, and it has been proven to enhance its photocatalytic activity, facilitate photoexcitation by visible light and improve the electrical conductivity [27]. The doping elements used are alkali metals, metalloids, transition metals, nonmetals (C, N, F, S, B, and I) and lanthanides. Peleyeju et al. (2018) presented a compendium of doped semiconductor materials reported in recent decades for PEC systems [27]; Ag nanoclusters, N-TiO_2/Ti, N-TNTs/Ti, Au/TiO_2NTs/Ti, N-S-TiO_2NCs/TNTAs, La-N-TiO_2/Nio, B-TiO_2NTs/Ti, and Cu-WO_3/FTO are the most relevant.

The deposition of metals on a semiconductor surface is another way to improve the photocatalytic activity of a semiconductor material in PEC systems [22]. In this case, the photoelectrode is designed by electrodeposition of a metal on a semiconductor electrode or by dip coating, followed by photo-assisted reduction of the metal ions in aqueous solution under UV irradiation. This surface modification on semiconductors is also called "decoration." For example, the ZnO semiconductor decorated with Ag nanoparticles is a clear example of this kind of modified photocatalyst, where an improvement of photocatalytic activity has been confirmed [31].

In summary, a considerably high pollutant degradation efficiency has been achieved when modification methodologies are applied to improve the photocatalyst performance in the design of photoelectrodes to PEC systems.

5.5 Photoelectrocatalytic Systems

The design of photoelectrochemical systems is an important subject to develop and optimize this technology for its use in water treatment. In this sense, the reactor configuration, operation mode and some operational requirements such as electrode materials, light collection/absorption, potential distribution inside the reactor, proper modeling (physical, diffusion and mass transfer) and chemically resistant materials have been evaluated in order to scale up PEC systems [33]. In this context, several photoelectrocatalytic reactor configurations oriented for wastewater treatment applications have been studied. Among the most widely used reactor configurations are sequential and hybrid reactors, divided and undivided cells, flow cells and stirred tank reactor. Additionally, the configuration of reactor can be determined by the position of light source (UV, visible or solar light) [33].

The sequential reactors (with or without recirculation of the effluent) consist of two linked reactors where the water effluent of the first becomes the substrate of the second reactor. In this way, a photochemical process can be positioned either as a primary treatment followed by an electrochemical process or in the reverse order (Figure 5.4a). On the contrary, hybrid photoelectrochemical reactors engage both processes in only one reactor, offering the possibility of synergy between photochemical and electrochemical reactions (Figure 5.4b) [33]. This last configuration is the option of choice for a high number of reports [20]. However, because of the complexity of aqueous matrixes, for example real wastewater containing a large variety of both organic and inorganic pollutants, some coupled systems or sequential reactors and hybrid systems have been designed. Furthermore, PEC can be coupled with other advanced oxidation processes or even conventional treatment methods (Figure 5.4c) such as flocculation, membrane filtration, microbial fuel cells, electroenzymatic catalysis, photoelectro-Fenton (PEF) or solar photoelectro-Fenton (SPEF) [34], and ozonation or sonolysis [35]. Additionally, hybrid systems could use ozone, hydrogen peroxide and Fe^{+3} species to improve the production of •OH (Figure 5.4c) [34]. Therefore, these kinds of coupled and hybrid systems could be successfully applied for the treatment of a variety of pollutants because of the additional generation of •OH radicals from other reactions as Fenton's reaction [34].

There are two main possibilities of a reactor's operation, one compartment or undivided cell (batch mode) and two compartments or divided cell (Figures 5.4d and 5.4e).

A single-compartment reactor is an electrolytic tank with parallel electrodes in the same chamber [33]. Meanwhile, in a two-compartment reactor, the anode and the cathode are separated inside individual chambers by an ion-exchange membrane to avoid decomposition of oxidants at the anode or cathode [36]. Although the two-compartment reactors can also be attractive for fuel generation (hydrogen production), the high cost of ion-exchange membranes makes its scalability difficult. In addition, the rate of organic matter degradation by PEC is very sensitive to pH changes during electrolysis. For a batch reactor without pH control, it has been shown that a faster degradation can occur when compared with a divided cell reactor. This happens when the cathode is suitable to promote oxidizing oxygenated species such as H_2O_2 or others. However, when the pH of the media is controlled with a buffer, there is no appreciable difference between the achieved degradation in batch reactors and two-compartment reactors [36]. For that reason, undivided cells have been the most studied among PEC reactors, especially for their easier implementation and scale-up at industrial scale [33].

On the other hand, based on its operation mode, there are stirred tank and flow reactors (Figures 5.4e and 5.4f). In the first ones, the flux is forced to go continuously between parallel electrodes. This design could be a plausible design for scaling up applications since the electrodes can be arranged in series or in parallel, in order to increase the treatment capacity and the removal efficiency. On the contrary, in the stirred tank reactor, the electrodes are placed in parallel with mechanically stirred. A more detailed description of the characteristics of flow cells versus stirred tank can be found in the work by Mousset and Dionysiou (2020) [33].

All of the aforementioned designs can incorporate the light source outside or inside the reactor, in such a position that allows minimizing the distance between the light and the photoelectrode [33]. In this sense, the cylindrical PEC reactor is a promising design as it has the potential to be scaled up because of its uniform geometry; also, it can achieve a high photon flux illumination associated with a proper light source positioning (annular reactor – inside) that provides irradiation from every direction [36]. Avoiding the use of artificial electrical lighting systems in favor of using sunlight is another studied improvement [27], which requires modifications at the anode level and the reactor architecture. Within the framework of the reactor structure, some authors have studied designs that include solar collectors to improve the efficiency of light collection by taking advantage of sunlight, successfully reducing the use of artificial light sources, which is translated in a significant reduction of the operating cost, providing great promise for large-scale operation [27].

5.5.1 Operating Parameters in the Performance of PEC Reactors

In addition to the PEC reactor design, some important operating variables such as the external potential (E_{anod}) or the current density (j) applied to the photoanode, the light source and its intensity, the pH of the solution, and the kind of cathode and supporting electrolyte, the O_2 concentration, the stirring velocity and the temperature of the solution, are involved in the performance

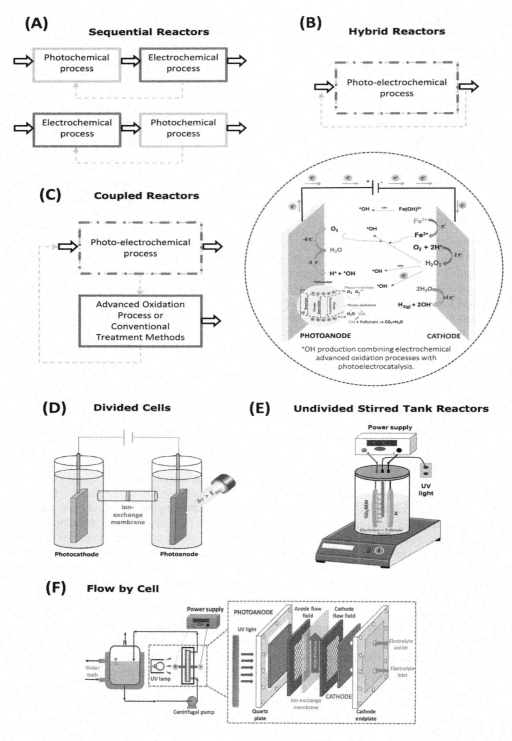

FIGURE 5.4 Sequential (a) versus hybrid (b) reactors for photochemical and electrochemical processes combination; (c) coupled process and mechanism of coupled reactors; (d) divided and (e) undivided cells mixed with stirred tank reactors; and (f) continuous flow cell. (Illustrations based on [20, 33].)

of PEC systems, influencing their scalability as a water treatment technology [37]. For that reason, these parameters were evaluated and optimized in batch systems in most of the studies, like evaluating a reactor prototype design for the kinetic analysis before the modeling and scaling up of the system [20].

In a general way, the pollutant degradation rate increases by increasing j, E_{anod} or E_{cell} because more oxidizing species are formed at a given time. This makes them important variables to take into account as they affect the energy consumption and treatment costs in PEC. Nevertheless, there is an optimal value of j and of E_{anod} to achieve the maximum generation of $^{\cdot}OH$ and the highest degradation rate; above these values there is no additional improvement of the process. Additionally, applying the optimum potential to a semiconductor electrode is of utter importance to avoid its damage – which is the main drawback of using PEC – and to avoid undesired reactions occurring at specified potentials and consuming the electrical charge – for example, the oxygen evolution reaction at anode surface, which competes with the generation of $^{\cdot}OH$ radicals [22, 37].

In most works, UV or visible light has been used to promote PEC reactions, and the criterion used for the selection of the light source are the band gap and the adsorption wavelength of each photocatalyst [21] (see Table 5.1). In this context, artificial sources such as UV light lamps have mostly been used to irradiate the photocatalyst to achieve the removal of pollutants: Only few research works have been reported using visible or solar light because of its inefficiency with the majority of the photocatalysts. For that reason, modified photocatalysts are being studied in order to reduce their band gap energy and to obtain photocatalytic activity using visible light as the excitation source (e.g., semiconductors such as $BiVO_4$, BiOX and Fe_2O_3) [22, 27].

In addition, the fast recombination of electron-hole pairs is avoided by selecting an appropriate material of cathode that works as an electron acceptor. The photoelectrocatalytic configuration (Figure 5.1) opens the possibility of coupling a convenient reaction that completes the charge balance; the extracted electrons from the photoanode to cathode can participate in the reduction reaction, for example water molecules would be reduced into H_2, O_2 into H_2O_2, and CO_2 into CO, CH_4 or longer chain carbon species; reduction of nitrates, reduction of metallic ions (decrease toxicity $Cr^{6+} \rightarrow Cr^{3+} + 3e^-$; $Fe^{3+} \rightarrow Fe^{2+} + 1e^-$) in PEC-photoelectro-Fenton (PEF) hybrid systems [34].

Finally, other parameters can also influence the efficiency of PEC systems. For example, the effect of the amount of dissolved O_2 can help to prevent the recombination of the photogenerated electron-hole pairs, because it acts as an electron acceptor to form reactive species such as superoxides [37]. In addition, a change in temperature has a slight influence on the electrochemical processes under the action of $^{\cdot}OH$. Other important factors to be consider for scaling up PEC as a treatment process are the stirring rate, the supporting electrolyte and the pH of the solution, the concentration of pollutants and the ratio between the real active area of the electrode and the volume of the treated sample, which affects the rate of degradation of the organic pollutants in aqueous medium [37].

5.6 PEC Applied to Remove Pollutants from Water

In recent years, diverse contaminants of emerging concern (CECs) have been detected in wastewater treatment effluents, surface and ground waters, rainwater and even drinking waters, posing potentially new and serious challenges and growing environmental health concerns worldwide [39]. CECs include pesticides, industrial chemicals, pharmaceuticals, personal care products, hormones, synthetic dyes and other categories of contaminants, which are mainly of industrial, municipal or agricultural origin and may cause disorders of human nervous, hormonal and reproductive systems [20, 39]. Because of the harmful effects of these substances on human health and ecosystems worldwide, advanced treatment (biological, chemical and physical, etc.) processes have been investigated, especially due to inefficient degradation of CECs by conventional processes. Although the applicability of PEC to remove contaminants from water has only just begun to be studied, it has shown to be an efficient process for decontaminating water [20, 21, 27]. The following sections provide a summary of the most recent applications of PEC; among these are mineralizing CECs, removing inorganic pollutants as metals, and inactivating or killing some microorganisms such as bacteria, fungi and viruses.

5.6.1 Contaminants of Emerging Concern, CECs

Among CECs that have been studied by PEC degradation are pharmaceutical compounds, pesticides and industrial chemicals (phenolic compounds, synthetic dyes). In this sense, Huda et al. (2019) reported a 95.0% remotion of the synthetic dye azo Acid Yellow using $Sn_3O_4/TiO_2/Ti$ as the photoanode within 1 hour of treatment under UV light irradiation [40]. Similarly, Li et al. (2016) and Ojani et al. (2012) used Rhodamine B to explore the effectiveness of nano-G/TiO_2/Ti-mesh and TiO_2/C (graphite) as photoanodes reaching a removal of 99.99% and 60%, respectively [41, 42].

Similarly, an innovative photoanode activated by visible light (α-Fe_2O_3/ZnO/FTO) has been applied by Cheng et al. (2017) to reach 84.6% of diclofenac degradation after 10 hours by PEC treatment [43]. Similarly, Orimolade et al. (2019) have reported the PEC behavior of FTO/$BiVO_4$/BiOI photoanode to remove acetaminophen (APAP) and ciprofloxacin (CIP) from water matrix using UV light (100 W, V = 50 mL, E_{anod} = 1.5V) [44], reaching 68.0% (APAP) and 62.0% (CIP) after 120 min. Likewise, the pharmaceutical compound mitoxantrone (MTX) – which is hazardous to the aquatic environment – was degraded at 75.0% using CuO/Si as the photoanode under UV-A light irradiation using a Xe lamp (UV-Vis), working with a volume of 50 mL, a time of 120 min., 0.1 mol L^{-1} Na_2SO_4 and E_{anod} = 0.8 V) [45]. In the same context, Xu et al. (2020) studied PEC degradation of the organometallic pollutant triphenyltin chloride (TPTCl), with a transparent photoanode of FTO-TiO_2. The PEC efficiency was enhanced by coupling with a photoelectro-Fenton process (PEC-PEF), which resulted in a complete degradation of TPTCl in 15 minutes and a total mineralization in 2 hours [46]. Similarly, Cheng et al. (2019) used $BiVO_4$/FTO as a photoanode to remove 60.6% of tetracycline (TC) using UV light within 120 min [47]. Also, and using the same treatment time, Cheng et al. (2019) investigated the combination of PEC with electro-enzymatic catalysis (EEC), reaching a 93.6% of removal efficiency. In the same context, other authors have developed combined processes that work using sunlight as an energy source (SPEC) joined with other approaches such as photoelectro-Fenton under solar light. The coupled SPEC/SPEF treatment showed a larger ability than SPEC to degrade pharmaceutical solutions, reaching 100% of drug removal at a short treatment time [20, 34].

The degradation of some persistent and toxic pesticides considered CECs has also been investigated using nanostructured WO_3 photoanodes due to its very good photocatalytic properties and its ability to absorb a part of the visible spectrum and UV light. Roselló-Marquéz et al. (2019) reported a process to obtain optimized WO_3 nanostructures [48], which act as photoanodes, and its effectiveness in PEC degradation of pesticides fenamiphos, chlorfenvinphos and bromacil in aqueous solution under irradiation of simulated solar light. The fenamiphos degradation was achieved at 2 h of experiment and the formation of intermediate species was noticed at 1 h of PEC experiment. The authors obtained a mineralization near 50% TOC of chlorfenvinphos and bromacil solution – as well as their mixture solution – within 360 min. These results demonstrate the good properties of WO_3 nanostructures as visible-light photoelectrocatalysts. Finally, Sánchez-Montes et al. (2020) and Alulema-Pullupaxi et al. (2021) carried out the degradation of the herbicide glyphosate (GLP) using BDD anode and photoanode TiO_2/BDD under UV-C irradiation in the presence of NaCl and Na_2SO_4 as supporting electrolytes [49], achieving a complete GLP mineralization in 1 h and 3 h with forming stoichiometric quantities of NO_3^-, PO_4^{3-} and NH_4^+ ions as final degradation products when using a using a BDD electrode and TiO_2/BDD photoanode, respectively.

5.6.2 Inorganic Pollutants

Generally, wastewater contains recalcitrant organic pollutants and toxic inorganic pollutants. Among inorganic pollutants are heavy metals, which are nonbiodegradable and tend to accumulate in living organisms, making them difficult to remove from sewage. In PEC systems, the oxidation of organic pollutants is carried out on the photoanode while at the cathode the photogenerated electrons are driven by the bias potential to reduce heavy metals. In recent years, the simultaneous removal of organic pollutants and heavy metals by PEC has received more attention. The studied carried out by Ye et al. (2020) summarized the research advances in simultaneous removal of organic-heavy metal mixed pollutants by photoelectrocatalytic treatment under UV light, visible light and sunlight [21]. For instance, Chen et al. (2017) studied the removal efficiency for a mixed solution of 0.05 mM Cu-EDTA in a photoelectroreactor with a rotating cathode under 9W-UVC light. The PEC treatment yielded 74.18% Cu-complex degradation and 75.54% Cu(II) recovery, demonstrating that it is much more efficient for photocatalysis (PC) and electrochemical oxidation (EC) [50]. In the same context, some authors have considered the limited opportunities for industrialization of PEC using UV light sources, so more researchers recently turned their attention to visible light, simulated sunlight and even real sunlight. This is the case for Wang et al. (2019), who reported that by using visible-light-driven PEC, the simultaneous removal of phenol and Cr(VI) has been achieved by using Bi/BiOI-Bi$_2$O$_3$ [51] as photoanodes; high removal percentages of Cr(VI) and phenol degradation in the phenol/Cr(VI) system was achieved as well. The authors also concluded that the removal of Cr(VI) is directly related to phenol concentration from 5 to 20 mg L^{-1} because more phenol molecules can act as hole-trapping agents, enhancing the separation of electrons and holes.

Even though sunlight is a clean energy and inexhaustible source that can be widely utilized in water treatment, there are few reports about PEC under sunlight. Some recent studies that reported the removal of organic or organic-heavy metal mixed pollutants by PEC still used simulated sunlight in the laboratory. For example, a PEC system with high efficiency was constructed to simultaneously remove 2,4-dichlorophenoxyacetic acid (2,4-D) and Cr(VI) under simulated sunlight (light intensity 100 mW cm^{-2}, 280 < λ < 2000 nm) [52]. As a result, Cr(VI) and 2,4-D could be removed faster under simulated sunlight at lower E_{cell} compared with a conventional PEC process. Moreover, the electric energy consumption decreased by nearly 50% in such a proposed PEC system.

5.6.3 PEC Disinfection

There are few reports about photoelectrocatalytic disinfection to inactivate or kill microorganisms, such as some bacteria, fungi and virus species. The complete inactivation of bacteria such as *Escherichia coli*, *Enterococcus faecalis*, *Mycobacterium*, *Pseudomonas aeruginosa* and cyanobacteria (*Microcystin aeruginosa*) have been reported using thin-film TiO_2 electrodes, in dark and UV light conditions, prepared by methods such as electrochemical anodization, sol-gel method with TiO_2 (Degussa P-25) and sol-gel suspensions with Ti(IV) isopropoxide as precursor [20]. Pirsaheb et al. (2021) studied the application of a photoanode composed of TiO_2 nanotubes decorated with Ni nanoparticles (Ni-TiO$_2$-NTs), reaching a high disinfection rate [53]. Similarly, Rather and Lo (2020) studied the application of a composed photoanode (g-C$_3$N$_4$/Ag/AgCl/BiVO$_4$) for the simultaneous degradation of emerging pollutants and hydrogen production and the disinfection of *E. coli*, reaching an effective disinfection in sewage with a final discharge of ≤ 1000 CFU/mL, which was within the permissible discharge limits (≤ 1500 CFU/mL) [54].

On the other hand, Montenegro and collaborators (2020) studied the implementation of photoelectrocatalytic technologies in portable devices [55]. In their research, they presented the electrophotocatalytic disinfection reactor in a kup (e-DRINK) that is based on TiO_2 photoelectrocatalytic disinfection principles for pathogen inactivation. This device was designed to treat 350 mL of water, with a titanium (Ti) reservoir cup as cathode, a Ti meshed inner cup with TiO_2 nanotubes as photoanode and a submergible two-sided UV-LED lamp (365 nm). The e-DRINK lid contains an embedded small rechargeable battery that feeds the UV lamp and provides a small current for photoelectrocatalytic operation. It has been shown also that photoelectrocatalytic treatment provides a 5-log removal of *E. coli* in 10 s, 2.6-log removal of *Legionella* in 60 s and 1-log removal of *E. coli* in 30 s in natural waters due to matrix effects [55].

In addition, the photoelectrocatalytic disinfection of water contaminated with fungi species, unicelular algae (Chlorella), and viruses (Leviviridae Fago MS2) have been studied [20]. However, still required is deeper research into the PEC efficiency on the destruction of viruses, for example poliovirus 1, hepatitis B virus, rotavirus, severe acute respiratory syndrome coronavirus (SARS-CoV), astrovirus, bacteriophages and toxins that present a risk to human health.

5.6.4 Real Wastewater Samples and Scaling Up Designs

There are only a few studies about PEC systems with real wastewater (municipal wastewater, and industrial water as textile and pharmaceutical effluents). Additionally, some pre-pilot systems have been designed for scaling up a photoelectrocatalytic treatment [20]. Studies carried out by Selcuk et al. (2004) [56] and Paschoal et al. (2009) [57] have reported the application of TiO$_2$/Ti as photoanode. In this context, Selcuk et al. (2004) reported the application of a photoelectrocatalytic system of a two-compartment reactor isolated by a Nafion 117 membrane to treat river water with low inorganic ion and high humic matter content, obtaining a 90% removal of color and 58%–80% of TOC removal [56]. In addition, Paschoal et al. (2009) studied a prospective treatment system for textile industry wastewater using a PEC reactor equipped with water refrigeration using a UV lamp as the light source and an applied potential of +1.0 V [57], leading to a reduction from 52.6% to 69.0% of COD, 80% to 89% of discoloration and 47% to 50% removal of TOC. However, all of the aforementioned results correspond only to small volumes of wastewater (50–350 mL). Thus, for industrialization of this very promising technique, the performance of PEC when using higher volumes of wastewater samples are still being studied.

There are only few reports about pre-pilot and pilot scale of PEC systems using larger working volumes between 350 mL and 12 L. Before modeling and scaling up a PEC system, some parameters must be evaluated and optimized in batch systems, like a reactor prototype design for the kinetic analysis [20]. In this context, cylindrical and tubular reactors working in continuous flow are the typical configurations that have been studied in most scaling-up research [33]. The study of Xu et al. (2008) developed a flowing aqueous film PEC reactor (circulating flux = 7.7 L h^{-1}) to enhance PEC degradation of a textile effluent using a TiO$_2$/Ti electrode prepared by sol-gel method with tetrabutyl titanate as the precursor [58]. After 120 min, the color and TOC removal efficiency reached 75% and 50%, respectively. Similarly, Zhao et al. (2010) studied the treatment of a leachate from a municipal landfill site (COD = 560 mg L^{-1}, TOC = 190 mg L^{-1}) in a pilot-scale flow reactor (6.5 L) [59], using Ti/TiORu$_2$ anode and UV light irradiation. At a current density of 67.1 mA cm^{-2} and 2.5 h reaction time, the removal rates achieved were 74.1% for COD, 41.6% for TOC and 94.5% for ammonium. In the same context, Cardoso et al. (2016) reported the development of a versatile system aimed at the treatment of real textile wastewater (COD = 153 ± 1.8 mg L^{-1}) [60]. In their approach, a bubbling annular reactor (8.5 L) of 145 cm of length and 10.2 cm of diameter with a UV-B lamp inserted in the center of the reactor inside a glass tube was used.

Furthermore, in the recent review about applications of PEC, a summary of photoelectrocatalytic disinfection in real wastewater was addressed [20].

5.7 Conclusions and Perspectives

The study of PEC as a water treatment technology has allowed increasing the mineralization efficiency and reducing the water treatment time. This chapter has summarized the principal kinetic aspects of the PEC mechanism, the fundamentals of PEC applied to remove pollutants from water including an analysis of photocatalysts and modified materials to synthesize photoelectrodes, and the influence of operating parameters for the scale-up of PEC to degrade contaminants of emerging concern and microorganisms from wastewater. When revising the available literature, a common conclusion is that PEC is a promising technology and the continuous evaluation of its operating parameters allows scaling up and developing improved designs for its ultimate application on real, industrial-scale wastewater treatment. Nevertheless, the factors that influence the degradation of pollutants by PEC processes must be examined for each particular case. Among the few studies that have been carried out on a real wastewater system, most were oriented to the treatment of textile effluents and disinfection. Moreover, most of the studies were focused on the development of innovative photoelectrodes and scaling-up reactor designs; however, it is necessary to put more effort into the study of hybrid alternatives that could take advantage of renewable energy, in order to avoid using an external power supply and artificial light sources. This could result in some cost-effective approaches by reducing energy consumption, achieving an economic and operational viability for their implementation on a larger scale in real wastewater treatment. Therefore, more studies on the applicability and scale-up of PEC to remove different recalcitrant organic pollutants of greater environmental concern are essential. In this way, it will be possible to develop an applicable, efficient and affordable technology for the treatment of industrial, and even municipal and domestic, wastewater.

REFERENCES

[1] Memming, R. *Semiconductor Electrochemistry*. Weinheim: Wiley, VCH Verlag GmbH & Co. KGaA, 2015. doi: 10.1002/9783527688685.

[2] Peter, L. "Kinetics and mechanisms of light-driven reactions at semiconductor electrodes: Principles and techniques," in *Photoelectrochemical Water Splitting: Materials, Processes and Architectures*, vol. 9, H.-J. Lewerenz and L. Peter, Eds. Royal Society of Chemistry, 2013, pp. 19–51. doi: 10.1039/9781849737739-00019.

[3] Monllor-Satoca, D., M. I. Díez-García, T. Lana-Villarreal and R. Gómez. "Photoelectrocatalytic production of solar fuels with semiconductor oxides: Materials, activity and modeling," *Chemical Communications*, vol. 56, no. 82, pp. 12272–12289, 2020. doi: 10.1039/d0cc04387g.

[4] Hojamberdiev, M., R. Vargas, V. S. Bhati, D. Torres, Z. C. Kadirova and M. Kumar. "Unraveling the photoelectrochemical behavior of Ni-modified ZnO and TiO2 thin films fabricated by RF magnetron sputtering," *Journal of Electroanalytical Chemistry*, vol. 882, Feb. 2021. doi: 10.1016/j.jelechem.2021.115009.

[5] Gärtner, W. W. "Depletion-layer photoeffects in semiconductors," *Physical Review*, vol. 116, no. 1, pp. 84–87, Oct. 1959. doi: 10.1103/PhysRev.116.84.

[6] He, Y., R. Chen, W. Fa, B. Zhang and D. Wang. "Surface chemistry and photoelectrochemistry – case study on tantalum nitride," *Journal of Chemical Physics*, vol. 151,

no. 13. American Institute of Physics Inc., Oct. 7, 2019. doi: 10.1063/1.5122996.

[7] Rajeshwar, K. "Fundamentals of semiconductor electrochemistry and photoelectrochemistry," in *Encyclopedia of Electrochemistry*, vol. 6, A. Bard and et al., Eds. Wiley, 2007. doi: 10.1002/9783527610426.BARD060001.

[8] Bisquert, J. *Nanostructured Energy Devices : Foundations of Carrier Transport*, 1st ed. Boca Raton: CRC Press, Taylor & Francis Group, 2017. doi: 10.1201/9781315117805.

[9] Bisquert, J., A. Zaban, M. Greenshtein and I. Mora-Seró. "Determination of rate constants for charge transfer and the distribution of semiconductor and electrolyte electronic energy levels in dye-sensitized solar cells by open-circuit photovoltage decay method," *Journal of the American Chemical Society*, vol. 126, no. 41, pp. 13550–13559, Oct. 2004. doi: 10.1021/JA047311K.

[10] Peter, L., Gurudayal, L. H. Wong and F. Abdi. "Understanding the role of nanostructuring in photoelectrode performance for light-driven water splitting," *Journal of Electroanalytical Chemistry*, vol. 819, pp. 447–458, June 2018. doi: 10.1016/J.JELECHEM.2017.12.031.

[11] Reichman, J. "The current-voltage characteristics of semiconductor-electrolyte junction photovoltaic cells," *Applied Physics Letters*, vol. 36, no. 7, pp. 574–577, Apr. 1980. doi: 10.1063/1.91551.

[12] Sato, N. *Electrochemistry at Metal and Semiconductor Electrodes*. Elsevier, 1998. doi: 10.1016/b978-0-444-82806-4.x5000-4.

[13] Zhang, Q. et al. "Density of deep trap states in oriented TiO2 nanotube arrays," *Journal of Physical Chemistry C*, vol. 118, no. 31, pp. 18207–18213, Aug. 2014. doi: 10.1021/jp505091t.

[14] Lianos, P. "Review of recent trends in photoelectrocatalytic conversion of solar energy to electricity and hydrogen," *Applied Catalysis B: Environmental*, vol. 210, pp. 235–254, 2017. doi: 10.1016/j.apcatb.2017.03.067.

[15] Coridan, R. H. et al. "Methods for comparing the performance of energy-conversion systems for use in solar fuels and solar electricity generation," *Energy and Environmental Science*, vol. 8, no. 10, pp. 2886–2901, Oct. 2015. doi: 10.1039/c5ee00777a.

[16] Madriz, L. et al. "Photocatalysis and photoelectrochemical glucose oxidation on Bi2WO6: Conditions for the concomitant H2 production," *Renewable Energy*, vol. 152, pp. 974–983, 2020. doi: 10.1016/j.renene.2020.01.071.

[17] Bedoya-Lora, F. E., A. Hankin and G. H. Kelsall. "En route to a unified model for photo-electrochemical reactor optimisation. I – Photocurrent and H$_2$ yield predictions," *Journal of Materials Chemistry A*, vol. 5, no. 43, pp. 22683–22696, Nov. 2017. doi: 10.1039/C7TA05125E.

[18] Leon, D. et al. "Unraveling kinetic effects during photoelectrochemical mineralization of phenols. Rutile: Anatase TiO2 nanotube photoanodes under thin-layer conditions," *Journal of Physical Chemistry C*, vol. 125, no. 1, pp. 610–617, 2021, doi: 10.1021/acs.jpcc.0c09890.

[19] Vargas, R., D. Carvajal, L. Madriz and B. R. Scharifker. "Chemical kinetics in solar to chemical energy conversion: The photoelectrochemical oxygen transfer reaction," *Energy Reports*, vol. 6, pp. 2–12, 2020. doi: 10.1016/j.egyr.2019.10.004.

[20] Alulema-Pullupaxi, P. et al. "Fundamentals and applications of photoelectrocatalysis as an efficient process to remove pollutants from water: A review," *Chemosphere*, vol. 281, Oct. 2021. doi: 10.1016/j.chemosphere.2021.130821.

[21] Ye, S., Y. Chen, X. Yao and J. Zhang. "Simultaneous removal of organic pollutants and heavy metals in wastewater by photoelectrocatalysis: A review," *Chemosphere*, no. xxxx, p. 128503, 2020. doi: 10.1016/j.chemosphere.2020.128503.

[22] Kusmierek, E. "Semiconductor electrode materials applied in photoelectrocatalytic wastewater treatment-an overview," *Catalysts*, vol. 10, no. 439, pp. 1–49, 2020.

[23] Gouda, A., C. Santato and P. Montréal. "Best practices in photoelectrochemistry towards explaining de-wetting phenomena of eumelanin films at surfaces view project is melanin an amorphous semiconductor? View project," *Article in Journal of Power Sources*, 2020, doi: 10.1016/j.jpowsour.2020.228958.

[24] Zhang, H., G. Chen and D. W. Bahnemann. "Environmental photo(electro)catalysis: Fundamental principles and applied catalysts," in *Electrochemistry for the Environment*, New York: Springer, 2010, pp. 371–442. doi: 10.1007/978-0-387-68318-8_16.

[25] Young, K. M. H., B. M. Klahr, O. Zandi and T. W. Hamann. "Photocatalytic water oxidation with hematite electrodes," *Catalysis Science & Technology*, vol. 3, no. 7, pp. 1660–1671, June 2013. doi: 10.1039/C3CY00310H.

[26] Prabakar, K., S. Minkyu, S. Inyoung and K. Heeje. "CdSe quantum dots co-sensitized TiO2 photoelectrodes: Particle size dependent properties," *Journal of Physics D: Applied Physics*, vol. 43, no. 1, p. 012002, Dec. 2009. doi: 10.1088/0022-3727/43/1/012002.

[27] Peleyeju, M. G. and O. A. Arotiba. "Recent trend in visible-light photoelectrocatalytic systems for degradation of organic contaminants in water/wastewater," *Environmental Science: Water Research & Technology*, vol. 4, no. 10, pp. 1389–1411, 2018. doi: 10.1039/C8EW00276B.

[28] Alulema-Pullupaxi, P. et al. "Photoelectrocatalytic degradation of glyphosate on titanium dioxide synthesized by sol-gel/spin-coating on boron doped diamond (TiO2/BDD) as a photoanode," *Chemosphere*, vol. 278, Sep. 2021. doi: 10.1016/j.chemosphere.2021.130488.

[29] Espinola, F., R. Navarro, S. Gutiérrez, U. Morales, E. Brillas and J. M. Peralta. "A simple process for the deposition of TiO2onto BDD by electrophoresis and its application to the photoelectrocatalysis of Acid Blue 80 dye," *Journal of Electroanalytical Chemistry*, vol. 802, no. July, pp. 57–63, 2017. doi: 10.1016/j.jelechem.2017.08.041.

[30] Egerton, T. A. and P. A. Christensen. "Photoelectrocatalysis processes," in *Advanced Oxidation Processes for Water and Wastewater Treatment*, Simons Parsons, Ed. Tundbridge Wells. London: IWA Publishing, 2002, pp. 272–275. doi: 10.1016/s0026-0576(97)86125-9.

[31] González-Rodríguez, J. et al. "Enhanced photocatalytic activity of semiconductor nanocomposites doped with ag nanoclusters under UV and visible light," *Catalysts*, vol. 10, no. 1, p. 31, Dec. 2019, doi: 10.3390/CATAL10010031.

[32] fan Su, Y., G. B. Wang, D. T. F. Kuo, M. ling Chang and Y. hsin Shih. "Photoelectrocatalytic degradation of the antibiotic sulfamethoxazole using TiO2/Ti photoanode," *Applied Catalysis B: Environmental*, vol. 186, pp. 184–192, June 2016. doi: 10.1016/J.APCATB.2016.01.003.

[33] Mousset, E. and D. D. Dionysiou. "Photoelectrochemical reactors for treatment of water and wastewater: A review,"

Environmental Chemistry Letters, vol. 18, no. 4, pp. 1301–1318, July 2020. doi: 10.1007/s10311-020-01014-9.

[34] Brillas, E. "A review on the photoelectro-Fenton process as efficient electrochemical advanced oxidation for wastewater remediation: Treatment with UV light, sunlight, and coupling with conventional and other photo-assisted advanced technologies," *Chemosphere*, vol. 250, p. 126198, 2020. doi: 10.1016/j.chemosphere.2020.126198.

[35] Kim, J. Y. U., G. G. Bessegato, B. C. de Souza, J. J. da Silva and M. V. B. Zanoni. "Efficient treatment of swimming pool water by photoelectrocatalytic ozonation: Inactivation of Candida parapsilosis and mineralization of Benzophenone-3 and urea," *Chemical Engineering Journal*, vol. 378, p. 122094, 2019. doi: 10.1016/j.cej.2019.122094.

[36] Meng, X., Z. Zhang and X. Li. "Synergetic photoelectrocatalytic reactors for environmental remediation: A review," *Journal of Photochemistry and Photobiology C: Photochemistry Reviews*, vol. 24, pp. 83–101, Sept. 2015. doi: 10.1016/J.JPHOTOCHEMREV.2015.07.003.

[37] Zarei, E. and R. Ojani. "Fundamentals and some applications of photoelectrocatalysis and effective factors on its efficiency: A review," *Journal of Solid State Electrochemistry*, vol. 21, no. 2, pp. 305–336, Feb. 2017. New York: Springer. doi: 10.1007/s10008-016-3385-2.

[38] Alulema-Pullupaxi, P. et al. "Photoelectrocatalytic degradation of glyphosate on titanium dioxide synthesized by sol-gel/spin-coating on boron doped diamond (TiO$_2$/BDD) as a photoanode," *Chemosphere*, p. 130488, Apr. 2021. doi: 10.1016/j.chemosphere.2021.130488.

[39] Costa, E. P., M. C. V. M. Starling and C. C. Amorim. "Simultaneous removal of emerging contaminants and disinfection for municipal wastewater treatment plant effluent quality improvement: A systemic analysis of the literature," *Environmental Science and Pollution Research*, 2021, doi: 10.1007/s11356-021-12363-5.

[40] Huda, A. et al. "Visible light-driven photoelectrocatalytic degradation of acid yellow 17 using Sn$_3$O$_4$ flower-like thin films supported on Ti substrate (Sn$_3$O$_4$/TiO$_2$/Ti)," *Journal of Photochemistry and Photobiology A: Chemistry*, vol. 376, pp. 196–205, 2019. doi: 10.1016/j.jphotochem.2019.01.039.

[41] Li, D., J. Jia, Y. Zhang, N. Wang, X. Guo and X. Yu. "Preparation and characterization of Nano-graphite/TiO2 composite photoelectrode for photoelectrocatalytic degradation of hazardous pollutant," *Journal of Hazardous Materials*, vol. 315, pp. 1–10, Sept. 2016. doi: 10.1016/j.jhazmat.2016.04.053.

[42] Ojani, R., J. B. Raoof and E. Zarei. "Electrochemical monitoring of photoelectrocatalytic degradation of rhodamine B using TiO$_2$ thin film modified graphite electrode," *Journal of Solid State Electrochemistry*, vol. 16, no. 6, pp. 2143–2149, 2012. doi: 10.1007/s10008-011-1634-y.

[43] Cheng, L., L. Liu, R. Li and J. Zhang. "Liquid phase deposition of α-Fe$_2$O$_3$/ZnO heterojunction film with enhanced visible-light photoelectrocatalytic activity for pollutant removal," *Journal of The Electrochemical Society*, vol. 164, no. 12, pp. H726–H733, 2017. doi: 10.1149/2.0241712jes.

[44] Orimolade, B. O., B. A. Koiki, G. M. Peleyeju and O. A. Arotiba. "Visible light driven photoelectrocatalysis on a FTO/BiVO$_4$/BiOI anode for water treatment involving emerging pharmaceutical pollutants," *Electrochimica Acta*, vol. 307, pp. 285–292, 2019. doi: 10.1016/j.electacta.2019.03.217.

[45] da Rosa, A. P. P. et al. "H$_2$O$_2$-assisted photoelectrocatalytic degradation of mitoxantrone using CuO nanostructured films: Identification of by-products and toxicity," *Science of the Total Environment*, vol. 651, pp. 2845–2856, 2019. doi: 10.1016/j.scitotenv.2018.10.173.

[46] Xu, J., H. Olvera-Vargas, B. J. H. Loh and O. Lefebvre. "FTO-TiO2 photoelectrocatalytic degradation of triphenyltin chloride coupled to photoelectro-Fenton: A mechanistic study," *Applied Catalysis B: Environmental*, vol. 271, no. Dec. 2019, 2020. doi: 10.1016/j.apcatb.2020.118923.

[47] Cheng, L., T. Jiang, K. Yan, J. Gong and J. Zhang. "A dual-cathode photoelectrocatalysis-electroenzymatic catalysis system by coupling BiVO$_4$ photoanode with hemin/Cu and carbon cloth cathodes for degradation of tetracycline," *Electrochimica Acta*, vol. 298, pp. 561–569, 2019. doi: 10.1016/j.electacta.2018.12.086.

[48] Roselló-Márquez, G., R. M. Fernández-Domene, R. Sánchez-Tovar, S. García-Carrión, B. Lucas-Granados and J. García-Antón. "Photoelectrocatalyzed degradation of a pesticides mixture solution (chlorfenvinphos and bromacil) by WO$_3$ nanosheets," *Science of the Total Environment*, vol. 674, pp. 88–95, July 2019. doi: 10.1016/j.scitotenv.2019.04.150.

[49] Sánchez-Montes, I., J. F. Pérez, C. Sáez, M. A. Rodrigo, P. Cañizares and J. M. Aquino. "Assessing the performance of electrochemical oxidation using DSA® and BDD anodes in the presence of UVC light," *Chemosphere*, vol. 238, 2020. doi: 10.1016/j.chemosphere.2019.124575.

[50] Chen, Y. et al. "Photoelectrocatalytic oxidation of metal-EDTA and recovery of metals by electrodeposition with a rotating cathode," *Chemical Engineering Journal*, vol. 324, pp. 74–82, 2017. doi: 10.1016/j.cej.2017.05.031.

[51] Wang, Q. et al. "In situ construction of semimetal Bi modified BiOI-Bi$_2$O$_3$ film with highly enhanced photoelectrocatalytic performance," *Separation and Purification Technology*, vol. 226, pp. 232–240, Nov. 2019. doi: 10.1016/j.seppur.2019.06.002.

[52] Liu, C., Y. Ding, W. Wu and Y. Teng. "A simple and effective strategy to fast remove chromium (VI) and organic pollutant in photoelectrocatalytic process at low voltage," *Chemical Engineering Journal*, vol. 306, pp. 22–30, Dec. 2016. doi: 10.1016/j.cej.2016.07.043.

[53] Pirsaheb, M., H. Hoseini and V. Abtin. "Photoelectrocatalytic degradation of humic acid and disinfection over Ni TiO$_2$-Ni/ AC-PTFE electrode under natural sunlight irradiation: Modeling, optimization and reaction pathway," *Journal of the Taiwan Institute of Chemical Engineers*, vol. 118, pp. 204–214, Jan. 2021. doi: 10.1016/j.jtice.2020.12.023.

[54] Rather, R. A. and I. M. C. Lo. "Photoelectrochemical sewage treatment by a multifunctional g-C$_3$N$_4$/Ag/AgCl/BiVO$_4$ photoanode for the simultaneous degradation of emerging pollutants and hydrogen production, and the disinfection of E. coli," *Water Research*, vol. 168, p. 115166, 2020. doi: 10.1016/j.watres.2019.115166.

[55] Montenegro-Ayo, R., J. C. Morales-Gomero, H. Alarcon, S. Cotillas, P. Westerhoff and S. Garcia-Segura. "Scaling up photoelectrocatalytic reactors: A TiO2 nanotube-coated disc compound reactor effectively degrades acetaminophen,"

Water (Switzerland), vol. 11, no. 12, pp. 1–14, 2019. doi: 10.3390/w11122522.

[56] Selcuk, H., J. J. Sene, H. Z. Sarikaya, M. Bekbolet and M. A. Anderson. "An innovative photocatalytic technology in the treatment of river water containing humic substances," *Water Science and Technology*, vol. 49, no. 4, pp. 153–158, 2004. doi: 10.2166/wst.2004.0248.

[57] Paschoal, F. M. M., M. A. Anderson and M. V. B. Zanoni. "The photoelectrocatalytic oxidative treatment of textile wastewater containing disperse dyes," *Desalination*, vol. 249, no. 3, pp. 1350–1355, 2009. doi: 10.1016/j.desal.2009.06.024.

[58] Xu, Y. L., D. J. Zhong, J. P. Jia, S. Chen and K. Li. "Enhanced dye wastewater degradation efficiency using a flowing aqueous film photoelectrocatalytic reactor," *Journal of Environmental Science and Health – Part A Toxic/Hazardous Substances and Environmental Engineering*, vol. 43, no. 10, pp. 1215–1222, 2008. doi: 10.1080/10934520802171790.

[59] Zhao, X. et al. "Photoelectrochemical treatment of landfill leachate in a continuous flow reactor," *Bioresource Technology*, vol. 101, no. 3, pp. 865–869, 2010. doi: 10.1016/j.biortech.2009.08.098.

[60] Cardoso, J. C., G. G. Bessegato and M. V. Boldrin Zanoni. "Efficiency comparison of ozonation, photolysis, photocatalysis and photoelectrocatalysis methods in real textile wastewater decolorization," *Water Research*, vol. 98, pp. 39–46, 2016. doi: 10.1016/j.watres.2016.04.004.

6

Advanced Oxidation Processes for Wastewater Treatment: Types and Mechanism

Mamta Chahar, Sarita Khaturia, Har Lal Singh, Anjali Bishnoi and CB Mahto

CONTENTS

6.1 Introduction

Due to global industrialization and population growth, water is contaminated by the discharge of domestic, agriculture, pharmaceutical and industrial waste [1, 2]. In 2017, the World Health Organization (WHO) reported that by 2025, 50% of the world's population will live in water-stressed areas [3]. Water containing unwanted substances, which cause adverse effects on its quality and thus make it unsuitable for use, is termed wastewater. Therefore, it is necessary to remove the pollutants from wastewater, treating it by using environmentally safe methods so that the water can be reused. Wastewater treatment plays an important role in controlling water pollution. Several researchers have found various wastewater treatment methods that reduce the cost for the reuse of the water [3, 4]. Household wastewater can be reused for floor and car washing, toilet flushing and landscape irrigation. Gray water contains nutrients, mainly P, N and K, and is an excellent source for irrigating landscaping and gardens [5]. For the removal of harmful substances from wastewater, which is essential for health and environmental protection, conventional methods such as reduction, precipitation, adsorption, oxidation and ion exchange are usually used. Advanced oxidation processes are efficient and eco-friendly methods for the degradation of toxic chemicals [6] such as pesticides, coloring matters, surfactants, pharmaceuticals and so on. In 1980, advanced oxidation processes were first proposed for the treatment of water. AOPs are based on the *in situ* generation of highly reactive strong oxidants like hydroxyl radicals and sulfate radicals by the oxidation of organic compounds.

For advanced oxidation processes, the following *oxidants can be used*: (i) ozone, (ii) hydrogen peroxide and (iii) oxygen. Ultraviolet light is used as an energy source, and titanium oxide can be used as a catalyst.

DOI: 10.1201/9781003165958-6

6.2 Advanced Oxidation Process Mechanism

6.2.1 Mechanism of the Process Involves Three Steps

1. Generation of hydroxyl radicals $\cdot$OH
2. Breakdown of target molecules into fragments by attack of this $\cdot$OH
3. This hydroxyl radical subsequently attacks until ultimate mineralization takes place

The mechanism of $\cdot$OH production (step 1) highly depends on the sort of AOP technique that is used. For example, ozonation, UV/H_2O_2 and photocatalytic oxidation rely on different mechanisms of $\cdot$OH generation:

- UV/H_2O_2:
  - $H_2O_2 + UV \rightarrow 2 \cdot OH$ *(homolytic bond cleavage of the O–O bond of H_2O_2 leads to formation of 2 $\cdot$OH radicals)*
- UV/HOCl:
  - $HOCl + UV \rightarrow \cdot OH + Cl\cdot$
- Ozone-based AOP:
  - $O_3 + HO^- \rightarrow HO_2^- + O_2$ *(reaction between O_3 and a hydroxyl ion leads to the formation of H_2O_2 (in charged form))*
  - $O_3 + HO_2^- \rightarrow HO_2\cdot + O_3^{-}\cdot$ *(a second O_3 molecule reacts with the HO_2^- to produce the ozonide radical)*
  - $O_3^{-}\cdot + H^+ \rightarrow HO_3\cdot$ *(this radical gives to $\cdot$OH upon protonation)*
  - $HO_3\cdot \rightarrow \cdot OH + O_2$
  - *The reaction steps presented here are just a part of the reaction sequence; see reference for more details.*
- Photocatalytic oxidation with TiO_2:
  - $TiO_2 + UV \rightarrow e^- + h^+$ *(irradiation of the photocatalytic surface leads to an excited electron (e^-) and electron gap (h^+))*
  - $Ti(IV) + H_2O \rightleftharpoons Ti(IV)\text{-}H_2O$ *(water adsorbs onto the catalyst surface)*
  - $Ti(IV)\text{-}H_2O + h^+ \rightleftharpoons Ti(IV)\text{-} \cdot OH + H^+$ *the highly reactive electron gap will react with water.*
  - *The reaction steps presented here are just a part of the reaction sequence; see reference for more details.*
  - A detailed mechanism for step 3 is not established yet, but researchers have cast light on the processes of initial attacks in step 2. In essence, $\cdot$OH is a radical species and should behave like a highly reactive electrophile. Thus, two type of initial attacks are supposed to be hydrogen abstraction and addition.

Primary oxidants such as ozone (O_3) and hydrogen peroxide (H_2O_2) break down compounds into intermediates, which then are converted into simple compounds such as water, carbon dioxide and salt by the mineralization process. Other components, like ultraviolet light (UV), are used as catalysts in the reaction to encourage the compounds to break down accordingly.

Ozone + UV: degradation of direct blue 86 dye in wastewater. Fenton process – fast removal of phenol/pesticides from wastewater.

6.2.2 Components in AOP

Advanced oxidation process comprises three components (Figure 6.1).

6.2.2.1 Ozone

Ozone interacts with hydrogen-containing compounds and, in an alkaline solution, decomposes them in a series to reduce to hydroxyl radicals. Ozone is a powerful oxidant and acts as a secondary oxidizer in the process. Because of its very short half-life, O_3 must be generated *in situ* during the process (on site) and used quickly as formed. Although it is a cost-effective method, proper care should be taken during its use – if bromide ions are present in the wastewater, they lead to the formation of bromates, which are poisonous.

6.2.2.2 Hydrogen Peroxide

For oxidation treatment, we cannot use hydrogen peroxide alone because it is not as strong an oxidizer as O_3 is. The important thing is its action; it can react with hydrogen and oxygen-containing compounds in a less complex manner. H_2O_2 also needs to be monitored for residuals left over after treatment.

6.2.2.3 Ultraviolet Light

Ultraviolet light is used for disinfection purposes due to its ability to kill or prohibit the growth of pathogens. Actually, UV is a wavelength of light and does not itself act as an oxidant, but it transfers photons to chemical compounds that are responsible for breaking their bonds quickly and easily. Presence of some contaminants such as suspended solids can reduce the efficiency of UV interaction by blocking it from the target compounds.

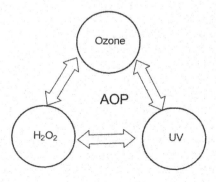

FIGURE 6.1 Components in AOP

AOPs for Wastewater Treatment   71

6.2.2.3.1 Combinations

To improve the efficiency of the overall AOP process, these treatments are used in some combination with one another; for example, O_3 is used with UV, O_3 is used with H_2O_2, H_2O_2 is used with UV, and O_3 is used with H_2O_2 and UV. These combinations are more effective than these individual processes. Each combination suffers from some drawbacks; therefore, the optimized process is chosen based on application.

6.3 Types of Advanced Oxidation Process

Different types of advanced oxidation processes like ozonation, photocatalysis and Fenton-based process were well studied to explain this technology. Their comparatives studies showed that among all of these processes, ozonation (Figure 6.2a) is found to be most effective (highest elimination is obtained when ozonation is combined with H_2O_2 and all the organic components are removed), and photocatalysis (Figure 6.2b) shows the least efficient results for TOC removal. The Fenton-based process is found to be moderately effective (Figure 6.3) for similar applications.

To remove biologically toxic or nondegradable materials like pesticides, aromatics, petroleum constituents and volatile organic compounds from wastewater, some other techniques were found to be more useful.

Classification of AOPs

Broadly, advanced oxidation process can be divided into the following types:

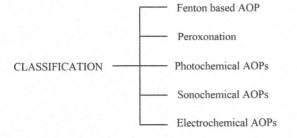

6.3.1 Fenton-Based AOP: It Uses Fenton's Reagent for Oxidation

When H_2O_2 and Fe^{2+} are used in combination, it is called the Fenton's reagent. It is used for the oxidation and destruction of tartaric acid [7]. Catalytic decomposition of H_2O_2 by Fe salts followed by a chain mechanism was reported by Haber and Weiss in the 1930s [8]. Because of its important development in water and soil treatment, the main research has been focused on Fenton's chemistry in the 20th century [9–14].

6.3.1.1 Mechanism

The Fenton process was initiated by the formation of a hydroxyl radical, and it is applied to the degradation of various organic pollutants [15–17].

$$Fe^{2+} + H_2O_2 \rightarrow Fe^{3+} + \cdot OH + OH^- \qquad (6.1)$$

In acidic medium, it can be also written as:

$$Fe^{2+} + H_2O + H^+ \rightarrow Fe^{3+} + H_2O + \cdot OH \qquad (6.2)$$

When the pH of the contaminated aqueous medium is in the range 2.8–3.0, the Fenton process can be efficiently applied. Catalytic behavior of the Fe^{3+}/Fe^{2+} couple leads to propagation in the Fenton's reaction:

$$Fe^{3+} + H_2O_2 \rightarrow Fe^{2+} + HO_2^{\cdot} + H^+ \qquad (6.3)$$

Reactions in the Fenton-like reaction category (i.e., reaction 6.3) are much slower than the Fenton's reaction itself (reaction 6.1) [18]. In fact, Fe^{2+} can be more rapidly regenerated by the reduction of Fe^{3+} with $HO_2^{\cdot}$ from reaction 6.4 [9], with an organic radical $R^{\cdot}$ from reaction 6.5 and/or with superoxide ion ($O_2^{\cdot -}$) from reaction 6.6 [19].

$$Fe^{3+} + HO_2^{\cdot} \rightarrow Fe^{2+} + O_2 + H^+ \qquad (6.4)$$

$$Fe^{3+} + R^{\cdot} \rightarrow Fe^{2+} + R^+ \qquad (6.5)$$

$$Fe^{3+} + O_2^{\cdot -} \rightarrow Fe^{2+} + O_2 \qquad (6.6)$$

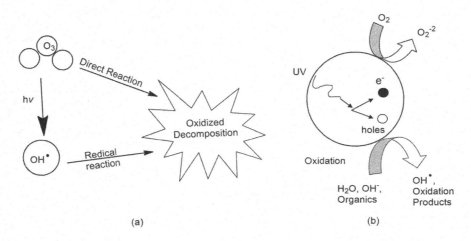

(a)   (b)

FIGURE 6.2 (a) Photo ozonation (b) photocatalysis.

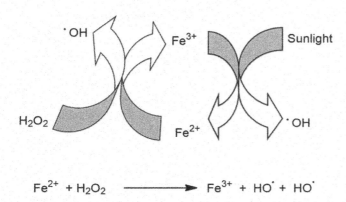

FIGURE 6.3 Fenton process.

6.3.1.2 Advantages of the Process

This process is advantageous in various ways, such as (i) it is a simple and flexible operation easily permitted for implementation in existing plants. (ii) Chemicals used in this process are inexpensive, and the process can be handled easily relative to others. (iii) The most important advantage associated with the process is that no energy input is needed.

6.3.1.3 Drawbacks of the Process

One of the main drawbacks is its high cost and risks generated due to the storage and transportation of H_2O_2. Before decontamination, effluents should be acidified at pH 2–4 and before disposal they should be neutralized; for this they are treated with chemicals, which are needed in higher amounts. In this process, Fe sludge also accumulates, which must be removed at the end of the treatment. Overall mineralization is not possible in this process due to the formation of Fe(III)-carboxylic acid complexes, which cannot be efficiently destroyed with bulk $^•OH$.

6.3.2 Peroxonation

A coupling between ozone and hydrogen peroxide is known as peroxonation. This process generates oxidizing radicals. Zaviska et al. found that this process works more efficiently than does ozonation alone [20]. Decomposition rate of O_3 in water increases by H_2O_2 due to production of reactive $^•OH$ radicals in a larger number. Paillard et al. explained the mechanism and application of peroxonation [21]. A very fast reaction between H_2O_2 under its ionized form (HO_2^-, pKa = 11.6) and ozone takes place, which leads to the formation of hydroxyl radicals:

$$O_3 + HO_2^- \rightarrow O_2 + {}^•OH + O_2^{-•} \qquad (6.7)$$

$HO_2^•$ radicals are also obtained by the reaction of $^•OH$ radicals with HO_2^-. Then, all these radicals can decompose H_2O_2 by other mechanisms occurring under optimum experimental conditions (pH = 7.7 and H_2O_2/O_3 ratio = 0.5) [21]. Several researchers used the peroxonation process for the elimination of micropollutants and toxic compounds (hydrocarbons,

pesticides etc.) in industrial waters, drinkable waters and groundwater [22]. O_3/H_2O_2 can be used in a filtration on sand and a filtration on active coal in a reactor through which water is running. By using this method, we can reduce the concentration of micropollutants on active coal.

6.3.2.1 Advantages

Due to its easier workup procedure and great bactericidal activity, this process can be effectively used for potable water treatment. It is applied for the removal of pesticides [22] like phenylureas, organochlorines and triazines from wastewater plants [23].

6.3.3 Photochemical AOPs

This is a simple, clean and relatively inexpensive technique but more efficient than chemical AOPs. It can disinfect waters and destroy pollutants. When UV radiation is coupled with powerful oxidants such as O_3 and H_2O_2 and catalysis with Fe^{3+} or TiO_2, the processes are called photochemical AOPs. These processes destroy pollutants by photodecomposition, UV radiation, excitation and degradation of pollutant molecules, oxidation by direct action of O_3 and H_2O_2, and oxidation by photocatalysis (with Fe^{3+}/TiO_2), inducing the formation of $^•OH$ radicals.

Here, we examined the methods and applications of these different AOPs, including H_2O_2 photolysis (H_2O_2/UV) [23], O_3 photolysis (O_3/UV) [24, 25], photo-Fenton process ($H_2O_2/Fe^{2+}/UV$) [26–28] and heterogeneous photocatalysis (TiO_2/UV) [29, 30]. In Table 6.1, experimental conditions, applications and performances of various photochemical AOPs have been given.

6.3.3.1 Photolysis of H_2O_2 (H_2O_2/UV)

The rate of formation of free radicals mainly depends on characteristics of UV lamps, pH, transmission of UV radiation, turbidity and so on [31]. The H_2O_2/UV AOP is used for the decontamination of groundwaters [32] and the removal of cyanides and organic pollutants such as benzene and trichloroethylene [33]. Hydrogen peroxide absorbs UV radiation at wavelengths ranging from 200 to 300 nm, and the homolytic fission of the O–O bond of the H_2O_2 molecule gives the hydroxyl ($^•OH$) radicals and contributes to the decomposition of H_2O_2 by secondary reactions [20, 34].

The steps of the process are as follows:

a. Initiation step: Photolysis of hydrogen peroxide is initiated by UV radiation at wavelengths ranging from 200 to 300 nm. The homolytic fission of the O–O bond of the H_2O_2 molecule gives the hydroxyl ($^•OH$) radicals.

$$H_2O_2 + h\nu \rightarrow 2\,{}^•OH \qquad (6.8)$$

b. Propagation steps: Reaction is propagated by the decomposition of H_2O_2 by secondary reactions.

$$^•OH + H_2O_2 \rightarrow H_2O + HO_2^• \qquad (6.9)$$

TABLE 6.1

Various Photochemical Advanced Oxidation Processes

S.No.	AOP	Contaminants	Removal	Ref.
1.	H_2O_2/UV	Chlorotriazine dyes	TOC	[23]
2.	O3/UV	Azo dyes	Color	[24]
1.		Pesticides: deltamethrin, lambda-cyhalothrin, triadimenol	Pesticides	[25]
3.	Photo-Fenton	PVA, Azo-dye orange	Color	[26–27]
1.		Abamectin pesticide	Pesticides	[28]
4. 4	TiO_2/UV	3-Chloropyridine pharmaceuticals	TOC	[29–30]

TOC = Total organic carbon

$$HO_2^{\bullet} + H_2O_2 \rightarrow {}^{\bullet}OH + H_2O + O_2 \qquad (6.10)$$

$${}^{\bullet}OH + HO_2^{-} \rightarrow HO_2^{\bullet} + OH^{-} \qquad (6.11)$$

c. Termination steps: Termination takes place in two steps. The disproportionation of the hydroperoxyl radical gives hydrogen peroxide and oxygen (6.12), and the subsequent combination of hydroxyl radical and hydroperoxyl radical produces water and oxygen (6.13). The hydroxyl radical recombines to form H_2O_2 (6.14).

$$2HO_2^{\bullet} \rightarrow H_2O_2 \qquad (6.12)$$

$${}^{\bullet}OH + HO_2^{\bullet} \rightarrow H_2O + O_2 \qquad (6.13)$$

$$2\,{}^{\bullet}OH \rightarrow H_2O_2 \qquad (6.14)$$

Generally, the reaction rate is larger in alkaline medium at pH > 10, which can be attributed to the fact that the HO_2^{-} anion (Equation 6.13), resulting from the ionization of H_2O_2, can strongly absorb UV radiation and produce free radicals ($HO_2^{\bullet}$ and ${}^{\bullet}OH$).

6.3.3.1.1 Drawback of this AOP

Due to the low molar absorption coefficient of H_2O_2, it is used in high concentration for the oxidation of organic pollutants. Ikehata and El Din studied the efficiency of H_2O_2/UV and photo-Fenton-type AOPs for the degradation in water of pesticides including aniline derivatives, organochlorinated derivatives, organophosphates, pyridine and pyrimidine derivatives and substituted ureas [35].

6.3.3.2 Photolysis of O_3 (O_3/UV)

The ozone photolysis process is more efficient than H_2O_2 photolysis, due to higher molar extinction coefficient (ε_{max}= 3600 L mol^{-1} cm^{-1}) of ozone [36]. Therefore, the O_3/UV process has been applied to the treatment of waters and wastewaters for the removal of persistent organic pollutants (POPs), including pesticides and phenolic compounds [37–41], according to the following successive and competitive steps [20, 36]:

a. Initiation: The photolysis of ozone in water leads to the formation of ${}^{\bullet}OH$ radicals, which are very reactive and efficient oxidizing species.

$$O_3 + H_2O + h\nu \rightarrow 2\,{}^{\bullet}OH + O_2 \qquad (6.15)$$

b. Propagation

$$O_3 + {}^{\bullet}OH \rightarrow HO_2^{\bullet} + O_2 \qquad (6.16)$$

$$O_3 + HO_2^{\bullet} \rightarrow {}^{\bullet}OH + 2O_2 \qquad (6.17)$$

c. Termination

$${}^{\bullet}OH + HO_2^{\bullet} \rightarrow H_2O + O_2 \qquad (6.18)$$

$$2\,{}^{\bullet}OH \rightarrow H_2O_2 \qquad (6.19)$$

Bhowmick and Semmens reported O_3/UV AOP for the removal of various volatile chlorinated organic compounds (VCOCs) such as $CHCl_3$, CCl_4, trichloroethylene (TCE), tetrachloroethylene and 1,1,2-trichloroethane (TCA) [42]. Irmak et al. found that bisphenol A (BPA) in aqueous medium oxidation rates are faster in the O_3/UV process than simple ozonation [43].

6.3.3.3 Photo-Fenton (H_2O_2/Fe^{2+}/UV)

Zepp et al. reported the photo-assisted Fenton reaction by using UV radiation to decompose H_2O_2 with catalyst ferrous or ferric ions [44].

In aqueous solutions, it stimulates the catalytic reduction of Fe^{3+}, which increases the formation of ${}^{\bullet}OH$ radicals (6.20):

$$Fe^{3+} + H_2O + h\nu \rightarrow Fe^{2+} + H^+ + {}^{\bullet}OH \qquad (6.20)$$

In the photo-Fenton process, different UV regions are used as a light energy source, such as UVA (λ = 315–400 nm), UVB (λ = 285–315 nm), and UVC (λ < 285 nm). It is noted that the intensity and wavelength of UV radiation have significant effect on the removal of organic pollutants. When sunlight (at wavelengths λ > 300 nm) is used as a renewable energy source, then this process is the called solar photo-Fenton process, which is applied for the photocatalytic decontamination of waters [45–47]. It is an economic and an environmentally safe method.

Primo et al. removed the organic matter in the treatment of landfill leachate by oxidations, and their removal efficiencies order: photo-Fenton > Fenton-like > Fenton > H_2O_2/UV > UV alone [48].

Macias-Sanchez et al. reported the performances of the photo-Fenton process for the degradation of the azo dye mixture [49]. They found that with the photo-Fenton AOP (λ = 365 nm), decolorization of the dye mixture sample within 70 min and complete mineralization were achieved, estimated by the reduction of TOC; whereas, using the Fenton reaction, only 75% of mineralization was obtained in the same time. These results show that the photo-Fenton process was much more efficient than the Fenton one for the decolorization and mineralization of an azo dye mixture. Vilar et al. investigated a solar photo-Fenton process combined with a biological nitrification and denitrification system for the decontamination of a landfill leachate by photocatalytic and biological devices [50]. In addition to this, solar collectors were used to degrade water contaminants by photocatalysis and to deactivate microorganisms present in waters [45].

6.3.3.4 Heterogeneous Photocatalysis (TiO₂/UV)

Fujishima and Honda discovered the photoexcited semiconductor titanium dioxide (TiO₂) to split water into hydrogen and oxygen in a photoelectrochemical solar cell [51]. Their fundamental work led to the development of a new AOP technology, based on semiconductor photocatalysis, for numerous environmental and energy applications [52–54]. In this method, catalyst TiO₂ irradiates with near-UV light in its rutile in front anatase form, which is easily photoexcited to form electron-donating and electron-accepting sites and persuade oxidation-reduction reactions. When the absorbed UV photons have an energy larger than the energy gap of the semiconductor, electron-hole pairs are formed, which can either recombine or migrate to the semiconductor surface and then react with chemical species adsorbed on the surface [20]. Titanium dioxide is selected for the water treatment because it is highly stable and biologically inert, very easy to produce, inexpensive and active from the photocatalysis, and it has an energy gap comparable to that of solar photons [20, 55]. Moreover, the photogenerated holes are strong oxidants, and the photogenerated electrons are reducing enough to yield superoxide from dioxygen. The energy band diagram for TiO₂ is presented in Figure 6.4. As can be seen, the redox potential for photogenerated holes is 2.53 V versus the standard electrode hydrogen (SHE).

In these potential conditions, the photogenerated holes are able to either directly oxidize the absorbed pollutants or oxidize the hydroxyl groups located at the TiO₂ surface to form ·OH radicals, whose redox potential is only slightly decreased [53]. Consequently, the degradation of pollutants contained in the contaminated waters can take place either directly at the semiconductor surface or indirectly through interactions with the ·OH radicals, the indirect oxidation by the radicals being the most favored degradation pathway. In addition, it is possible to again increase the number of ·OH radicals by adding into the photoreactor H_2O_2 or O_3 that can be photolyzed by UV irradiation [20]. During the heterogeneous photocatalytic process, the TiO₂ catalyst can be utilized either under dispersed form (powder, aqueous suspension) or in thin film form (fixed TiO₂ catalytic layer). The Fujishima group has widely participated in the preparation of TiO₂ films, by putting TiO₂ coatings on various types of support materials [20]. The dispersed TiO₂ catalyst presents several advantages: it is easy to use, it possesses an important specific surface and it can be aerated, which prevents the recombining of electron-hole pairs and increases the catalyst efficiency.

However, a drawback of the dispersed form is the progressive formation of dark catalytic sludge, which diminishes the efficiency of UV irradiation and reduces the photoreactor performances. In contrast, for TiO₂ films, there is no need to separate the catalytic particles at the end of the process, but the catalytic layer must be very stable and active. Also, the amount and type of catalyst to be used depend on the irradiation source, the nature and concentration of pollutant to be treated, and the photoreactor. Moreover, the pH value of the medium plays a crucial role in the efficiency of photocatalysis and must be optimized in a preliminary step, according to the type of pollutant under treatment. For example, in the case of weak acid pollutants, the photocatalysis efficiency increases when the pH diminishes, yielding a decrease of the polarity of the pollutant that is more easily adsorbed at the catalyst surface [20]. The heterogeneous TiO₂ photocatalysis has been widely applied in recent years, particularly in the case of organic pollutants refractory to oxidation by the other conventional AOPs

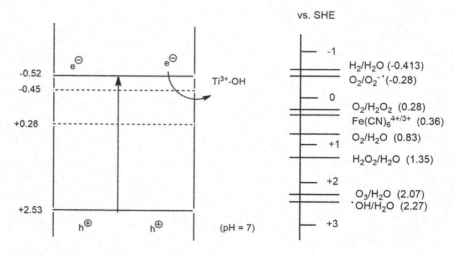

FIGURE 6.4 Band diagram of various redox processes occurring on the TiO₂ surface at pH 7.

[52]. Also, it is able to completely destroy pathogenic biologic pollutants, including viruses, bacteria and mold [20]. This technology is generally very efficient for treating a substantial range of inorganic as well as organic pollutants. Moreover, it is worthwhile to point out the recent development of various strategies to modify TiO_2 for the use of visible light, including nonmetal and/or metal doping, dye sensitization and coupling semiconductors [54]. A large number of applications of heterogeneous TiO_2 photocatalysis, particularly in the field of water purification, have been recently described [52–54]. For example, toxic, inorganic ions, such as cyanide, bromate, nitrite and sulfite, have been oxidized by this process into nontoxic or weakly toxic compounds (CO_2, bromide, nitrate, sulfate) [20, 52]. Heterogeneous TiO_2 photocatalytic degradation and/or mineralization were performed in the case of a number of organic pollutants, such as pesticides [56–61], pharmaceuticals [62], surfactants such as dodecylbenzenesulfonate [63], sulfur-containing organic compounds, dyes [64] and chloropyridine [29].

6.4 Sonochemical AOPs

In AOP technology, sonic waves or ultrasounds are used through indirect and direct mechanism. At higher frequency, an indirect mechanism is applied; for example, H_2O and O_2 molecules go through homolytic fragmentation and give $\cdot OH$, $HO_2 \cdot$, and $\cdot O$ radicals [65–67]. In a direct mechanism, there is the involvement of the sonication method, which includes the formation of cavitation bubbles by the ultrasounds which grow and then collapse, creating powerful breaking forces with very high temp (T = 2000–5000 K) and pressures. The sonolysis of water molecules generate highly reactive radicals and react with organic compounds present in the aqueous medium (Equations 6.22 and 6.23), and pyrolysis of degradation of organic compounds is taking place (Equation 6.24) [68–69]:

$$H_2O \xrightarrow{\text{ultrasound}} \cdot OH + \cdot H \qquad (6.22)$$

$$\cdot OH + X(O.C.) \rightarrow \text{Products} \qquad (6.23)$$

$$X(O.C.) + H \rightarrow \text{Products} \qquad (6.24)$$

where: X(O.C.) = organic compound.

Nowadays, ultrasounds have been successfully employed for the oxidation and degradation of organic pollutants in wastewaters and sewage sludges. The main drawback of this water treatment method done by ultrasounds is that the quantity of generated $\cdot OH$ radicals is generally insufficient. Ultrasounds have been used with oxidants including H_2O_2 and dioxygen, in combination with UV irradiation and with the Fenton's reagent [Fe(0), Fe(II), and Fe(III)] and Fenton-type reactions. This method is generally called sono-Fenton AOPs. This advanced hybrid technique is applied for the improvement of the degradation efficiency of organic pollutants in waters and of the reduction of the sonochemical treatment.

Ma reported that ultrasonic frequency (20–1700 kHz) and amplitude, the iron (Fe^0, Fe^{2+}, and Fe^{3+}) and dosage, and the solution pH must be optimized in order to improve the efficiency of the sonochemical AOP treatment of polluted waters [69]. Moreover, it has been also shown that the optimal values of ultrasonic frequency and amplitude depend on the characteristics of the effluents. In fact, the sonochemical AOPs have been broadly applied to the degradation of various pollutants such as pesticides [70–71], dyes [72–77], aromatic compounds [78–82], endocrine disrupters (bisphenol A) and pharmaceuticals [83], and disinfectant by-products [84] in wastewaters. The combination of ultrasounds with the Fenton-type reactions has resulted into the rapid and recent development of sonochemical methods for the removal of organic pollutants from waters for decontamination purposes [69]. Therefore, application of sonochemical AOPs at the industrial level in a real-time water (or wastewater) treatment plant would be needed to demonstrate the economic and commercial feasibility of these methods.

6.5 Electrochemical AOPs

This advanced oxidation technique is based on the transfer of electrons, which makes *in situ* hydroxyl radicals ($\cdot OH$), and is able to destroy a large variety of toxic and persistent organic pollutants. These $\cdot OH$ radicals can be electrochemically produced either directly or indirectly via *in situ* electrocatalytically generated Fenton's reagent (electro-Fenton process). The effectiveness of the process can be further increased by combining both electrochemical processes as the simultaneous AO with B-doped diamond anode and the classical EF process. These EAOPs using direct electrochemistry (AO) or indirect electrochemistry (EF) oxidations are applied for the detoxification of water. In order to enhance the EF process efficiency, its coupling has been recently proposed with other AOPs, such as PEF, solar photoelectro-Fenton, sonoelectro-Fenton, and peroxi-coagulation.

6.6 Conclusions

For the protection of surface and ground waters from wastewater, easy-to-handle and cost-effective water treatment technologies are required. In this regard, several researchers applied the AOP methods to water and wastewater treatments. For wastewater treatment purposes, different types of AOPs are successfully employed, such as chemical AOPs, Fenton reaction and peroxonation, photochemical AOPs, AOP via photolysis of H_2O_2 and O_3 by photo-Fenton process ($H_2O_2/Fe^{2+}/UV$) and heterogeneous photocatalysis (TiO_2/UV), sonochemical AOPs and electrochemical AOPs, such as AO, EF and related processes. Chemical AOPs, including the Fenton's reagent and peroxonation, are mainly used for treatment of wastewaters, in dye industries for discoloring of effluents, and for destruction of various toxic organic compounds such as aromatic compounds, haloalkenes and haloalkanes. Nowadays, because of the better performances of photochemical, sonochemical and electrochemical AOPs, they are widely applied in various processes. For example in photochemical technologies, the photochemical AOPs are extensively used because they are generally simple, clean, relatively inexpensive and more efficient than classical, chemical AOPs. In this category of photochemical AOPs, four main techniques such as H_2O_2 photolysis ($H_2O_2/$ UV), O_3 photolysis (O_3/UV), photo-Fenton process ($H_2O_2/Fe^{2+}/$ UV) and heterogeneous photocatalysis (TiO_2/UV) have been applied to degrade organic pollutants in waters. Photo-Fenton AOP, when compared to the H_2O_2/UV and O_3/UV processes, is found to be the best method for the rapidity of photodecomposition as well as for the mineralization. Similarly, solar photo-Fenton AOP is very efficient in reducing energy consumption and represents an alternative to the classical photo-Fenton process. Recently, this technique has been also utilized for the removal of various organic compounds present in natural or polluted waters, for degradation of herbicides and in treatment of wastewater effluents and landfill leachates. Moreover, it is important to stress that heterogeneous photocatalysis, an AOP mainly based on the use of the semiconductor titanium oxide (TiO_2/UV), has been the object of tremendous development in the last decade. Additionally, sonochemical AOPs, the combination of ultrasounds with Fenton-type reactions, have resulted in the rapid and recent development of sonochemical methods for the removal of organic pollutants from waters and lead to the decontamination process. The EAOPs seem to be very promising water and wastewater treatment processes and should be considered technologies of the future. These technologies now have become mature enough to be applied at the industrial stage.

REFERENCES

1. Theresa M., Pendergast M., Hoek E.M.V. (2011). A review of water treatment membrane nanotechnologies. *Energy Environ Sci*, 4, 1946–1971.
2. Huerta-Fontela M., Galceran M.T., Ventura F. (2010). Fast liquid chromatography quadrupole-linear ion trap mass spectrometry for the analysis of pharmaceuticals and hormones in water resources. *J Chromatogr*, A 1217(25), 4212e4222. https://doi.org/10.1016/j.chroma.2009.11.007
3. Ingole S., Chavhan A., Dhote J. (2013). Study of physico-chemical parameters of sewage water from some selected location on Amba Nalla, Amravati. *Indian J Appl*, 3(8), 377.
4. Dhote J., Chavhan A., Ingole S. (2013). Treatment of sewage using dual media filter & its sustainable reuse. *Indian Str Res J*, A7, 43–50.
5. Pangarkar B., Parjane S., Sane M. (2010). Design and economical performance of gray water treatment plant in rural region. *World Aca Sci, Eng Technol*, 4(1), 782–786.
6. Legrini O., Oliveiros E., Braun A.M. (1993). Photochemical processes for water treatment. *Chem Rev*, 93, 671–698.
7. Fenton H.J.H. (1894). Oxidation of tartaric acid in presence of iron. *J Chem Soc*, 65, 899–910.
8. (i) Haber F., Weiss J. (1932). Uber die katalyse des hydroperoxides. *Naturwissenschaften*, 20(51), 948–950. (ii) Haber F., Weiss, J. (1934). The catalytic decomposition of hydrogen peroxide by iron salts. *Proc R Soc London A*, 147(861), 332–351.
9. Merli C., Petrucci, E., Da Pozzo, A., Pernetti, M. (2003). Fenton-type treatment: State of the art. *Ann Chim*, 93(9–10), 761–770.
10. Neyens E., Baeyens J. (2003). A review of classic Fenton's peroxidation as an advanced oxidation technique. *J Hazard Mater*, 98(1), 33–50.
11. Ikehata K., El-Din M.G. (2006). Aqueous pesticide degradation by hydrogen peroxide/ultraviolet irradiation and Fenton-type advanced oxidation processes: A review. *J Environ Eng Sci*, 5(2), 81–135.
12. Ikehata K., Naghashkar N.J., El-Din M.G. (2006). Degradation of aqueous pharmaceuticals by ozonation and advanced oxidation processes: A review. *Ozone Sci Eng*, 28(6), 353–414.
13. Pignatello J.J., Oliveros E., MacKay A. (2006). Advanced oxidation processes for organic contaminant destruction based on the Fenton reaction and related chemistry. *Crit Rev Environ Sci Technol*, 36(1), 1–84.
14. Bautista P., Mohedano A.F., Casas J.A., Zazo J.A., Rodriguez J.J. (2008). An overview of the application of Fenton oxidation to industrial wastewaters treatment. *J Chem Technol Biotechnol*, 83(10), 1323–1338.
15. Metelitsa D.I. (1971). Mechanisms of the hydroxylation of aromatic compounds. *Russ Chem Rev*, 40(7), 563–580.
16. Sun Y., Pignatello J.J. (1993). Photochemical reactions involved in the total mineralization of 2,4-D by iron(3+)/hydrogen peroxide/UV. *Environ Sci Technol*, 27(2), 304–310.
17. Gallard H., de Laat J., Legube B. (1998). Influence du pH sur la vitesse d'oxydation de composes organiques par FeII/H_2O_2: Mecanismes reactionnelset modelisation. *New J Chem*, 22(3), 263–268.
18. Brillas E., Sires I., Oturan M.A. (2009). Electro-Fenton process and related electrochemical technologies based on Fenton's reaction chemistry. *Chem Rev*, 109(12), 6570–6631.
19. Rothschild W.G., Allen A.O. (1958). Studies in the radiolysis of ferrous sulfate solutions: III. Air-free solutions at higher pH. *Radiat Res*, 8(2), 101–110.
20. Zaviska F., Drogui P., Mercier G., Blais J.F. (2009). Proced' es d'oxydation avancee dansle traitement deseauxetdes effluents industriels: Applicationala degradation des polluants refractaires. *Rev Sci Eau*, 22(4), 535–564.
21. Paillard H. (1994). *Etude de la mineralisation de la matiere organique dissouteen milieu aqueux dilue par ozonation, oxydation avancee O_3/H_2O_2 et ozonation catalytique heterogene* (PhD thesis). University of Poitiers, Poitiers, France.
22. Galey C., Paslawski D. (1993). Elimination des micropolluants par lozone couple avec le peroxyde d'hydrog' ene dans le traitement de potabilisation deseaux. *LEau, LIndustrie, Les Nuisances*, 161(1), 46–49.
23. Alaton I.A., Balcioglu I.A., Bahnemann D.W. (2002). Advanced oxidation of a reactive dye bath effluent: Comparison of O_3, H_2O_2/UV-C and TiO_2/UV-A processes. *Water Res*, 36(5), 1143–1154.
24. Shu H.Y., Chang M.C. (2005). Decolorization effects of six azo dyes by O_3, UV/O_3 and UV/H_2O_2 processes. *Dyes Pigm*, 65(1), 25–31.
25. Lafi W.K., Al-Qodah Z. (2006). Combined advanced oxidation and biological treatment processes for the removal of pesticides from aqueous solutions. *J Hazard Mater*, 137(1), 489–497.
26. Kang S.F., Liao, C.H., Po S.T. (2000). Decolorization of textile wastewater by photo-Fenton oxidation technology. *Chemosphere*, 41(8), 1287–1294.

27. Maezono T., Tokumura M., Sekine M., Kawase Y. (2010). Hydroxyl radical concentration profile in photo-Fenton oxidation process: Generation and consumption of hydroxyl radicals during the discoloration of azo-dye Orange II. *Chemosphere*, 82(10), 1422–1430.

28. Kaichouh G., Oturan N., Oturan M.A., El Hourch A., El Kacemi K. (2008). Mineralization of herbicides imazapyr and imazaquin in aqueous medium by Fenton, photo-Fenton and electro-Fenton processes. *Environ Technol*, 29(5), 489–496.

29. Ortega-Liebana M.C., Sanchez-Lopez E., Hidalgo-Carrillo J., Marinas A., Marinas J. M., Urbano F.J. (2012). A comparative study of photocatalytic degradation of 3-chloropyridine under UV and solar light by homogeneous (photo-Fenton) and heterogeneous (TiO₂) photocatalysis. *Appl Catal B-Environ*, 127, 316–322.

30. Sakkas V.A., Calza P., Medana C., Villioti A.E., Baiocchi C., Pelizzetti E., Albanis T. (2007). Heterogeneous photocatalytic degradation of the pharmaceutical agent salbutamol in aqueous titanium dioxide suspensions. *Appl Catal B Environ*, 77(1), 135–144.

31. Crissot F. (1996). *Oxydation catalytique de composes organiques aqueux par le' peroxyde dehydrogene phase heterogene* (PhD thesis). University of Poitiers, Ecole Superieure d'Ingenieurs de Poitiers, Poitiers, France.

32. Eckenfelder W., Bowers A.R., Roth J.A. (1992). Chemical oxidation: Technologies for the nineties. In *Chemical oxidation: Technology for the nineties. Proceedings of the first international symposium*. Nashville: Vanderbilt University; Lancaster, PA: Technomic Pub. Co.

33. Dore M. (1989). *Chimie des oxydants et traitement des eaux*. Techn. Doc. Paris: Lavoisier.

34. Hernandez R., Zappi M., Colluci J., Jones R. (2002). Comparing the performance of various advanced oxidation process for treatment of acetone contaminated water. *J Hazard Mater*, 92(1), 33–50.

35. Ikehata K., Naghashkar N.J., El-Din M.G. (2006). Degradation of aqueous pharmaceuticals by ozonation and advanced oxidation processes: A review. *Ozone Sci Eng*, 28(6), 353–414.

36. Van Craeynest K., Van Langenhove H., Stuetz R.M. (2004). AOPs for VOCs and odour treatment. In S. Parsons (Ed.), *Advanced oxidation processes for water and wastewater treatment*. London: IWA Publishing, Alliance House.

37. Trapido M., Hirvonen A., Veressinina Y., Hentunen J., Munter R. (1997). Ozonation, ozone/UV and UV/H₂O₂ degradation of chlorophenols. *Ozone Sci Eng*, 19(1), 75–96.

38. Trapido M., Kallas, J. (2000). Advanced oxidation processes for the degradation and detoxification of 4-nitrophenol. *Environ Technol*, 21(7), 799–808.

39. Aaron J.J., Oturan M.A. (2001). New photochemical and electrochemical methods for the degradation of pesticides in aqueous media: Environmental applications. *Turk J Chem*, 25, 509–520.

40. Rosenfeldt E.J., Linden K.G., Canonica S., Von-Gunten U. (2006). Comparison of the efficiency of OH radical formation during ozonation and the advanced oxidation processes O₃/H₂O and UV/H₂O₂. *Water Res*, 40(20), 3695–3704.

41. Cernigoj U., Stangar U.L., Trebse P. (2007). Degradation of neonicotinoid insecticides by different advanced oxidation processes and studying the effect of ozone on TiO₂ photocatalysis. *Appl Catal B Environ*, 75(3), 229–238.

42. Bhowmick M., Semmens M.J. (1994). Ultraviolet photooxidation for the destruction of VOCs in air. *Water Res*, 28(11), 2407–2415.

43. Irmak S., Erbatur O., Akgerman A. (2005). Degradation of 17β-estradiol and bisphenol A in aqueous medium by using ozone and ozone/UV techniques. *J Hazard Mater*, 126(1–3), 54–62.

44. Zepp R.G., Faust B.C., Hoigne J. (1992). Hydroxyl radical formation in aqueous reactions (pH 3–8) of iron(II) with hydrogen peroxide: The photoFenton reaction. *Environ Sci Technol*, 26(2), 313–319.

45. Malato S., Blanco J., Alarcon D.C., Maldonado M.I., Fernandez-Ibanez P., Gernjak W. (2007). Photocatalytic decontamination and disinfection of water with solar collectors. *Catal Today*, 122(1–2), 137–149.

46. Oller I., Malato S., Sanchez-Perez J.A., Gernjak W., Maldonado M.I., Perez-Estrada L.A., Pulgarin C. (2007). A combined solar photocatalytic biological field system for the mineralization of an industrial pollutant at pilot scale. *Catal Today*, 122(1–2), 150–159.

47. Silva M.R.A., Trovo A.G., Nogueira R.F.P. (2007). Degradation of the herbicide tebuthiuron using solar photo-Fenton process and ferric citrate complex at circumneutral pH. *J Photochem Photobiol A*, 191(2–3), 187–192.

48. Primo O., Rivero M.J., Ortiz I. (2008). Photo-Fenton process as an efficient alternative to the treatment of landfill leachates. *J Hazard Mater*, 153(1–2), 834–842.

49. Macias-Sanchez J., Hinojosa-Reyes L., Guzman-Mar J.L., Peralta-Hernandez J.M., Hernandez-Ramırez A. (2011). Performance of the photo-Fenton process in the degradation of a model azo dye mixture. *Photochem Photobiol Sci*, 10(3), 332–337.

50. Vilar V.J.P., Rocha E.M.R., Mota F.S., Fonseca A., Saraiva I., Boaventura R.A.R. (2011). Treatment of a sanitary landfill leachate using combined solar photo-Fenton and biological immobilized biomass reactor at a pilot scale. *Water Res*, 45(8), 2647–2658.

51. Fujishima A., Honda K. (1972). Electrochemical photolysis of water at a semiconductor electrode. *Nature*, 238(5358), 37–38.

52. Mills A., Le Hunte S. (1997). An overview of semiconductor photocatalysis. *J Photochem Photobiol A Chem*, 108(1), 1–35.

53. Fujishima A., Rao T.N., Tryk D.A. (2000). Titanium dioxide photocatalysis. *J Photochem Photobiol C Photochem Rev*, 1(1), 1–21.

54. Pelaez M., Nolan N.T., Pillai S.C., Seery M.K., Falaras P., Kontos A.G., Dunlop P.S.M., Hamilton J.W.J., Byme, J.A., O'Shea, K., Entezari, M.H., Dionisiou D.D. (2012). A review on the visible light active titanium dioxide photocatalysts for environmental applications. *Appl Catal B Environ*, 125, 331–349.

55. Pignatello J.J. (1992). Dark and photoassisted iron(3+)-catalyzed degradation of chlorophenoxy herbicides by hydrogen peroxide. *Environ Sci Technol*, 26(5), 944–951.

56. Herrmann J.M., Guillard C., Arguello M., Aguera A., Tejedor A., Piedra L., Fernandez A.A. (1999). Photocatalytic degradation of pesticide pirimiphosmethyl: Determination of the reaction pathway and identification of intermediate products by various analytical methods. *Catal Today*, 54(2–3), 353–367.

57. Konstantinou K.I., Albanis A.T. (2003). Photocatalytic transformation of pesticides in aqueous titanium dioxide suspensions using artificial and solar light: Intermediates and degradation pathways. *Appl Catal B Environ*, 42(4), 319–335.

58. Cernigoj U., Stangar U.L., Trebse P. (2007). Degradation of neonicotinoid insecticides by different advanced oxidation processes and studying the effect of ozone on TiO2 photocatalysis. *Appl Catal B Environ*, 75(3), 229–238.

59. Hazime R., Ferronato C., Fine L., Salvador A., Jaber F., Chovelon J. M. (2012). Photocatalytic degradation of imazalil in an aqueous suspension of TiO2 and influence of alcohols on the degradation. *Appl Catal B Environ*, 126, 90–99.

60. Rivera-Utrilla J., Sanchez-Polo M., Abdel Daiem M.M., Ocampo-Perez R. (2012). Role of activated carbon in the photocatalytic degradation of 2,4- dichlorophenoxyacetic acid by the UV/TiO$_2$/activated carbon system. *Appl Catal Environ*, 126, 100–107.

61. Seck E.I., Dona-Rodriguez J.M., Fernandez-Rodrıguez C., Gonzalez-Dıaz O.M., Arana J., Perez-Pena J. (2012). Photocatalytic removal of 2,4- dichlorophenoxyacetic acid by using sol-gel synthesized nanocrystalline and commercial TiO$_2$: Operational parameters optimization and toxicity studies. *Appl Catal B Environ*, 125, 28–34.

62. Sakkas V.A., Calza P., Medana C., Villioti A.E., Baiocchi C., Pelizzetti E., Albanis, T. (2007). Heterogeneous photocatalytic degradation of the pharmaceutical agent salbutamol in aqueous titanium dioxide suspensions. *Appl Catal B Environ*, 77(1), 135–144.

63. Sanchez M., Rivero M.J., Ortiz I. (2011). Kinetics of dodecylbenzenesulphonate mineralisation by TiO2 photocatalysis. *Appl Catal B Environ*, 101(3–4), 515–521.

64. Lin F., Zhang Y., Wang L., Zhang Y., Wang D., Yang M., Yang J., Zhang B., Jiang Z., Li C., (2012). Highly efficient photocatalytic oxidation of sulfur-containing organic compounds and dyes on TiO2 with dual cocatalysts Pt and RuO$_2$. *Appl Catal B Environ*, 127, 363–370.

65. Riez P., Berdahl D., Christman C.L. (1985). Free radical generation by ultrasound in aqueous and non-aqueous solutions. *Environ Health Perspect*, 64, 233–252.

66. Lorimer J.P., Mason T.J. (1987). Sonochemistry. Part-1: The physical aspects. *Chem Soc Rev*, 16, 239–274.

67. Trabelsi S., Oturan N., Bellakhal N., Oturan M.A. (2012). Application of Doehlert matrix to determine the optimal conditions for landfill leachate treatment by electro-Fenton process. *J Mater Environ Sci*, 3(3), 426–433.

68. Hua I., Hoffmann M.R. (1997). Optimization of ultrasonic irradiation as an advanced oxidation technology. *Environ Sci Technol*, 31(8), 2237–2243.

69. Ma Y.S. (2012). Short review: Current trends and future challenges in the application of sono-Fenton oxidation for wastewater treatment. *Sustain Environ Res*, 22(5), 271–278.

70. Ma Y.S., Sung C.F. (2010). Investigation of carbofuran degradation by ultrasonic process. *Sustain Environ Res*, 20(4), 213–219.

71. Ma Y.S., Sung C.F., Lin J.G. (2010). Degradation of carbofuran in aqueous solution by ultrasonic and Fenton processes: Effect of system parameters and kinetic study. *J Hazard Mater*, 178(1–3), 320–325.

72. Jian-Hui S., Sheng-Peng S., Guo-Liang W., Li-Ping Q. (2007). Degradation of azo dye Amido black 10B in aqueous solution by Fenton oxidation process. *Dyes and Pigments*, 74(3), 647–652.

73. Ghodbane H., Hamdaoui O. (2009). Intensification of sonochemical decolorization of anthraquinonic dye acid blue 25 using carbon tetrachloride. *Ultrason Sonochem*, 16(4), 455–461.

74. Trabelsi F., At-Lyazidi H., Ratsimba B., Wilhem A.M., Delmas H., Fabre P.L., Berlan J. (1996). Oxidation of phenol in wastewater by sonoelectrochemistry. *Chem Eng Sci*, 51, 1857–1865.

75. Petrier C., Jiang Y., Francony A., Lamy M.F. (1999). Aromatics and chloroaromatics sonochemical degradation: Yields and by-products. *Ham. Ber. Siedlungswasserwirtschaft (Ultrasound Environ. Eng.)*, 25, 23–37.

76. Shah M.P. (2021). *Removal of emerging contaminants through microbial processes*. Singapore: Springer.

77. Shah M.P. (2020). *Advanced oxidation processes for effluent treatment plants*. Amsterdam: Elsevier.

78. Shah M.P. (2020). *Microbial bioremediation and biodegradation*. Singapore: Springer.

79. Dai Y., Li F., Ge F., Zhu F., Wu L., Yang X. (2006). Mechanism of the enhanced degradation of pentachlorophenol by ultrasound in the presence of elemental iron. *J Hazard Mater*, 137(3), 1424–1429.

80. Ku Y., Tu Y.H., Ma C.M. (2006). Decomposition of monochlorophenols by sonolysis in aqueous solution. *J Environ Eng Manage*, 16(4), 259–265.

81. Liang J., Komarov S., Hayashi N., Kasai E. (2007). Improvement in sonochemical degradation of 4-chlorophenol by combined use of Fenton-like reagents. *Ultrason Sonochem*, 14(2), 201–207.

82. Namkung K.C., Burgess A.E., Bremner D.H., Staines H. (2008). Advanced Fenton processing of aqueous phenol solutions: A continuous system study including sonication effects. *Ultrason Sonochem*, 15(3), 171–176.

83. Torres R.A., Abdelmalek F., Combet E., Petrier C., Pulgarin C. (2007). A comparative study of ultrasonic cavitation and Fenton's reagent for bisphenol A degradation in deionised and natural waters. *J Hazard Mater*, 146, 546–551.

84. Kim I.K., Huang C.P. (2007). Sonochemical process for the removal of DBPs and precursor in water. *J Environ Eng Manage*, 17, 39–48.

7

Nanotechnology for Advanced Oxidation Based Water Treatment Processes

Sougata Ghosh and Bishwarup Sarkar

CONTENTS

7.1 Introduction

Increasing industrialization has led to the accumulation of refractory pollutants in effluents, which can't be easily removed using conventional wastewater treatment processes. Hence, advanced oxidation processes (AOPs) are developed for effective degradation of biorefractory organics in industrial effluents. Currently used tertiary treatment processes employed for wastewater treatment are activated carbon adsorption, reverse osmosis, microfiltration, ultrafiltration and sand filters (Moreno et al. 2005). However, these treatment methods are often ineffective, and additional steps for purification are required to remove the most persistent pollutants, such as drugs, pesticides, solvents, oil and heavy metals (Mantzavinos and Psillakis 2004; Walid and Al-Qodah 2006). Given this background, advanced oxidation processes (AOPs) are employed for efficient removal of refractory pollutants from wastewater. Employing AOPs can completely mineralize the pollutants to CO_2, water and inorganic compounds. Also, the nonbiodegradable pollutants can be partially broken down to intermediates, which can be subjected to microbial biodegradation. Such combined processes where AOPs are used as pre-treatments, followed by biological processes, are economically viable and more efficient than an individual process (Cañizares et al. 2009).

As shown in Table 7.1, the AOPs use diverse mechanism and corresponding reagent systems such as photochemical degradation processes (UV/O_3, UV/H_2O_2), photocatalysis (TiO_2/UV, photo-Fenton reactives) and chemical oxidation processes (O_3, O_3/H_2O_2, H_2O_2/Fe^{2+}), that can generate hydroxyl radicals (•OH) radicals. The organic pollutants are then attacked by these highly reactive radicals (Skoumal et al. 2006; Rosenfeldt et al. 2007). As depicted in Figure 7.1, the AOPs can be classified as either homogeneous or heterogeneous. Homogeneous processes can be energy dependent or independent.

This chapter discusses nanotechnology-driven AOPs for wastewater treatment where several nanopolymers, carbon-based nanomaterials, nanoscale metal, metal oxide particles and zeolites are included as listed in Table 7.2. Further, the underlying mechanism of nanocomposite-based advanced oxidation systems are elaborated. Finally, the future prospects and scope of further modification in this area are presented.

7.2 Polymeric Dendrimers

The polymer-supported ultrafiltration (PSU) technique, introduced in 1968, is considered a promising alternative for the removal of heavy metal ions from industrial effluents. Numerous modifications and advancements have led to the fabrication of polymer-based nano- and microstructures for wastewater treatment using AOPs. Rether and Schuster (2003) described the synthesis of a water-soluble benzoylthiourea modified ethylenediamine (EDA) core-poly(amidoamine) (PAMAM) dendrimer that was used for selective removal of certain heavy metal ions that are toxic in nature. The surface of EDA core-PAMAM dendrimers were functionalized using a mixture of piperazine and diethylamine in a ratio of 1:9, which allows for the introduction of benzoylthiourea and the conversion of ester terminal ends to amide functional groups, respectively. Partially amine-terminated PAMAM derivatives were then reacted with an excess of benzoylisothiocyanate in the presence of acetonitrile leading to formation of benzoylthiourea groups. The PSU technique revealed the complexing properties of water-soluble benzoylthiourea modified PAMAM

DOI: 10.1201/9781003165958-7

TABLE 7.1

Major Mechanisms for Organics Removal During Wastewater Treatment by Different AOPs. (Reprinted with permission from Deng, Y., Zhao, R., 2015.) Advanced Oxidation Processes (AOPs) in Wastewater Treatment. (*Curr. Pollution Rep.* 1, 167–176. Copyright ©2015 Springer International Publishing AG.)

AOP Types	Oxidant for Advanced Oxidation	Other Occurring Mechanisms
O_3	OH·	Direct O_3 oxidation
O_3/H_2O_2	OH·	Direct O_3 oxidation H_2O_2 oxidation
O_3/UV	OH·	UV photolysis
UV/TiO_2	OH·	UV photolysis
UV/H_2O_2	OH·	UV photolysis H_2O_2 oxidation
Fenton reaction	OH·	Iron coagulation Iron sludge-induced adsorption
Photo-Fenton reaction	OH·	Iron coagulation Iron sludge-induced adsorption UV photolysis
Ultrasonic reaction	OH·	Acoustic cavitation generates transient high temperatures ($\geq$5000 K) and pressures ($\geq$1000 atm), and produce H· and HO_2·, besides OH·
Heat/persulfate	SO_4^-	Persulfate oxidation
UV/persulfate	SO_4^-	Persulfate oxidation UV photolysis
Fe(II)/persulfate	SO_4^-	Persulfate oxidation Iron coagulation Iron sludge-induced adsorption
OH^-/persulfate	SO_4^-/OH·	Persulfate oxidation

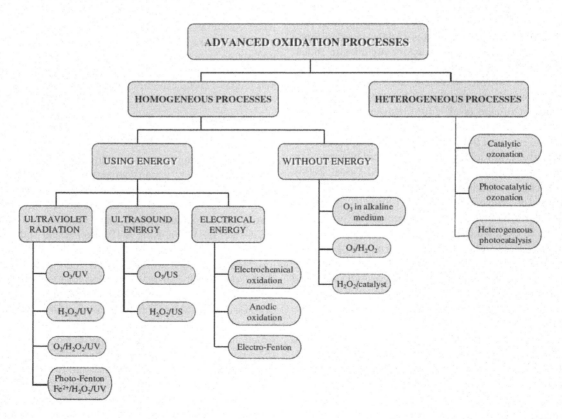

FIGURE 7.1 Advanced oxidation processes (AOPs) classification. Abbreviations used: O_3 ozonation; H_2O_2 hydrogen peroxide; UV ultraviolet radiation; US ultrasound energy; Fe^{2+} ferrous ion. (Reprinted with permission from Poyatos, J. M., Muñio, M. M., Almecija, M. C., Torres, J. C., Hontoria, E., Osorio, F., 2010. Advanced oxidation processes for wastewater treatment: State of the art. *Water Air Soil Pollut.* 205, 187–204. Copyright © 2009 Springer Science + Business Media B.V.)

TABLE 7.2

Nanomaterials Used for Advanced Oxidation Processes for Water Treatment

Nanomaterials	Contaminant Removal-	% Removal-	References-
Benzoylthiourea modified EDA core- PAMAM dendrimers	Co(II), Cu(II), Hg(II), Ni(II), Pb(II) and Zn(II)	90%	Rether and Schuster, 2003
Cotton fabrics grafted with PPI dendrimers	Direct red 80, disperse yellow 42 and basic blue 9	Around 40% (direct red 80)	Abkenar et al., 2015
EDA core- PAMAM dendrimers	Cu(II)	95%–100%	Diallo et al., 2005
Mesoporous alumina nanofibers- PAMAM-G1 dendrimer	Methyl orange dye	-	Shen et al., 2015
PAMAM dendrimer containing $Fe_3O_4@SiO_2$ nanoparticles	Naphthalene	89.17%	Aliannejadi et al., 2019
Al_2O_3-coated multiwall carbon nanotubes (MWCNTs)	Pb	100%	Gupta et al., 2011
Carbon nanotube/titanium dioxide (CNT/TiO_2) core–shell nanocomposites	Methylene blue dye	20% (TTIP), 30% (TEOTi) and 40% (TBT)	Li et al., 2011
Chitosan/carbon nanotubes (CS/CNT)	Congo red dye	-	Chatterjee et al., 2010
Chitosan-Fe^0 nanoparticles	Cr(VI)	92%	Geng et al., 2009
P-MWCNTs and H-MWCNTs	Sulfamethazine	68% (P-MWCNTs) and 87% (H-MWCNTs)	Yang et al., 2015
Ag@citrate nanoparticles (NPs)	Chlorpyrifos	100%	Bootharaju and Pradeep, 2012
Magnetic copper ferrite NPs $(CuFe_2O_4)$	2,4-dichlorophenoxyacetic acid	74.6%	Jaafarzadeh et al., 2017
Manganese functionalized silicate NPs	Methylene blue dye	-	Tušar et al., 2012
Polymer coated silicon NPs (SiNPs)	Methanol	2.4% (quantum yield)	Iqbal et al., 2016
Ag NP-modified mesoporous anatase	Rhodamine B dye, *E. coli*	-	Xiong et al., 2011
Cobalt-exchanged zeolites	Phenol	100%	Shukla et al., 2010
Cu-exchanged Y zeolite	Anionic dye	95%	Fathima et al., 2008
Fe(III)-exchanged Y zeolites	C.I. reactive yellow 84	96.9%	Neamțu et al., 2004
FeZSM5	Diclofenac	85%	Perisic et al., 2016
Zeolite Fe-beta	Chlorinated phenols	23%	Doocey et al., 2004

(BTUPAMAM) dendrimers with heavy metal ions. Selectivity of the polymer was observed against several heavy metal ions such as Co(II), Cu(II), Hg(II), Ni(II), Pb(II) and Zn(II). Mercury was observed to have a strong binding affinity even in strongly acidic media, whereas other metals depended on the pH for complexation. A dark-green water-soluble complex was formed when complexed with copper, indicating selective separation is possible at a pH value of less than 4, resulting in complete retention. The presence of sodium nitrate and other low-molecular competing agents such as ammonia, citric acid and triethanolamine did not have any effect on metal sorption. However, retention of zinc and nickel was reduced in the presence of tartarate at an acidic pH range. Interestingly, more than 90% of heavy metal ions could be potentially removed and recovered from industrial effluents or waste streams using these modified BTUPAMAM dendrimers even in the presence of high concentrations of sodium nitrate, ammonia and triethanolamine.

In another study, Abkenar et al. (2015) reported adsorption of dyes using cotton fabrics grafted with poly (propylene imine) (PPI) dendrimers. The initial concentrations of three model dyes, namely direct red 80 (DR80), disperse yellow 42 (DY42) and basic blue 9 (BB9), affected the removal efficacy. A decrease in dye removal percentage by the grafted sample with subsequent increase in dye concentration was reported. This was attributed to the decrease of available active sites present on PPI dendrimers. Presence of 0.01 mol/L of NaCl was found to decrease the adsorption of dye, which may be due to competition of salt anions with dye molecules. DR80 is an anionic dye with six sulfonic groups that could interact via electrostatic attraction with the positively charged primary and tertiary amine groups of PPI dendrimers along with the dye encapsulation process. DY42 is a nonpolar molecule that could be adsorbed through the expended cavities of PPI dendrimers. Optimum pH value of 3.0 was observed to be ideal for maximum adsorption of both DR80 and DY42 dyes. BB9 is a cationic dye and thus was able to adsorb on the grafted dendrimer at a high pH value of 11, which could be due to hydrogen bonding between the functional groups of the dye and PPI dendrimers. The Langmuir equation confirmed that all three model dyes were adsorbed onto the grafted samples with pseudo-second-order kinetics.

Dendrimer-enhanced ultrafiltration was reported by Diallo et al. (2005) for recovery of Cu(II) from aqueous solutions. Commercially available PAMAM dendrimers having an EDA core and terminal NH_2 groups were used in this study. Concentric shells of β-alanine units are produced around the EDTA initiator core present at the center using a two-step iterative reaction sequence. The retention of dendrimers and Cu(II) (Cu(II)-dendrimer complexes) were measured using ultrafiltration (UF) experiments. A stirred cell of 10 mL volume with an effective membrane area of 4.1 cm^{2}was used. Every UF experiment was carried out for 4.5 h with collection of permeate within every 30 min and measurement of flux every 10 min. Retention of G4-NH_2 PAMAM dendrimers was more than 90% in the presence of 10 kDa regenerated cellulose (RC) and polyethersulfone (PES) membranes. This particular dendrimer had a globular molecule with a molar mass of 14.2 kDa and a very low polydispersity. Retention of 95%–100% of Cu(II)-dendrimer complex with G4-NH_2 PAMAM dendrimers was observed by RC membranes at a pH value of 7.0 and for a concentration of 10 mg/L of Cu(II). The PES membranes also showed 92%–100% retention for the same parameters, thus suggesting that 100% of Cu(II) ions are bound to the G4-NH_2 PAMAM dendrimers at Cu(II) dendrimer terminal NH_2 groups molar ratio of 0.2 in the presence of neutral pH. The RC membranes showed negligible permeate fluxes of aqueous solutions of Cu(II)-dendrimer complex for both 5 and 10 kDa membrane sizes; however, significant decline of around 45%–63% in permeate fluxes were observed for 5 and 10 kDa PES membranes during filtration at acidic and neutral pH values. Hence, these results indicated that PES membranes are more prone towards membrane fouling by Cu(II)-dendrimer complex. Two models were proposed for membrane fouling, wherein pore blockage and decline in normalized permeate flux as a power-law function was expressed, respectively. Further characterization of RC and PES membranes exposed to dendrimer complexes were performed using atomic force microscopy (AFM). The roughness parameter (RMS) and mean roughness (R_a) were found to be significantly higher for the clean PES membranes than for the clean RC membranes. Cu(II)-dendrimer complex treated RC membranes were found to be smooth and uniform with a few "rough" spots in comparison to the dense, tightly packed and grainy structure characteristics of Cu(II)-dendrimer complex treated PES membranes with nodular skin morphology, which was proposed to be due to electrostatic attractions between the protonated terminal NH_2 groups of the G4-NH_2 PAMAM dendrimer and the negatively charged PES membranes resulting in sorption.

Shen et al. (2015) reported the synthesis of mesoporous alumina nanofibers using PAMAM-G1 dendrimer as the structure directing agent. Combination of sol-gel, electrospinning and calcination was used to prepare mesoporous structures on alumina nanofibers. Fourier transform infrared (FTIR) spectroscopy was performed to characterize the sample, which showed strong hydrophilicity along with removal of adsorbed water molecules upon increasing the temperature. The calcined samples revealed amorphous alumina with a phase type of γ-Al_2O_3,as evident from scanning electron microscopic (SEM) images (shown in Figure 7.2). Partial hydrolysis of aluminum nitrate to form hydroxyl aluminum ion in water could combine amino and imino groups of PAMAM-G1 molecules that may reduce the content of free hydroxyl aluminum ions. This led to the production of a higher quantity of $[Al(OH)_4]^-$that self-assembled along with PAMAM and polyvinylpyrrolidone (PVP) polymers forming a triple supramolecular composite. The nanofibers thus produced were used for adsorption against methyl orange dye. A series of adsorption experiments were performed using a wide range of methyl orange concentrations from 5 to 100 mg/L at a pH value of 5.5, and the saturated adsorption capacity was found to be 351.3 mg/g using the Langmuir equation. The adsorption process followed a pseudo-second-order model with a correlation coefficient of 0.9928 and an adsorption rate constant of 8.4×10^{-4} g/(mg.min). Heat treatment could reactivate the mesoporous alumina nanofibers indicating their reusability.

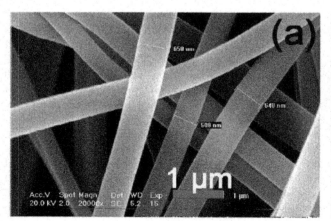
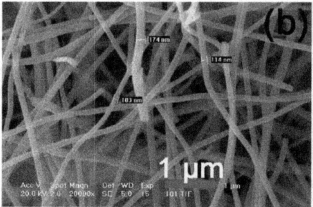

FIGURE 7.2 SEM images of the mesoporous alumina derived products: (a) before calcination and (b) after calcination at 450 °C. (Reprinted with permission from Shen, J., Li, Z., Wu, Y.N., Zhang, B., Li, F., 2015. Dendrimer-based preparation of mesoporous alumina nanofibers by electrospinning and their application in dye adsorption. *Chem. Eng. J.* 264, 48–55. Copyright © 2014 Elsevier B.V.)

A decrease of 8.7% in the ratio of adsorbed methyl orange was observed after three cycles. Hence, such nanofibers synthesized using PAMAM-G1 dendrimers shows great potential in dye removal from wastewater.

Unlike other nanostructures, magnetic nanoparticles (MNPs) can be recovered after reaction using a strong magnetic field. They can be multifunctionalized with various ligands in order to enhance their activity synergistically. Further, they are biocompatible and nontoxic, which is an added advantage to develop MNP-based AOPs (Ghosh et al. 2015a).

Modified MNPs using PAMAM dendrimers were reported by Aliannejadi et al. (2019) to be applicable in the removal of naphthalene from aqueous solutions. The coprecipitation method was used to synthesize magnetic Fe_3O_4 nanoparticles using a 1:2 mixture of ferrous chloride tetrahydrate ($FeCl_2.4H_2O$) and ferric chloride hexahydrate ($FeCl_3.6H_2O$). Further, core-shell $Fe_3O_4@SiO_2$ nanoparticles were synthesized by addition of tetraethylorthosilicate (TEOS). The $Fe_3O_4@SiO_2$ nanoparticles were then modified using 3-aminopropyl triethoxysilane (APTES) that was bound to PAMAM dendrimers and altered using benzaldehyde (BZ). FTIR spectroscopy indicated formation of Fe_3O_4 nanoparticles along with the presence of SiO_2 shells covering the magnetic nanoparticles. The nanoparticles were mostly spherical with an average particle size of 25–55 nm. The $Fe_3O_4@SiO_2$ nanoparticles displayed a saturation magnetization (Ms) of 43 emu/g suggesting the presence of paramagnetic properties. The nanomaterial was stable up to 250 °C, beyond which a 14% decrease in weight was observed due to decomposition of amine and organic-silane. The nanocomposite exhibited maximum adsorption efficiency of 89.17% was observed at pH 7 during 30 min contact time with 0.2 g/L of sorbent and 20 mg/L of naphthalene. A sorbent dosage of 0.4 g/L was obtained to be optimum for significant adsorption of naphthalene with a contact time of 15 min. Further, adsorption capacity was found to be increased up to 98.21 mg/g in the presence of 50 mg/L of naphthalene. An efficiency value of 89.31% was retained after four cycles of usage, suggesting effective reusability of the sorbent. Water samples from contaminated rivers and industrial effluents were also used to investigate the removal efficiency of naphthalene. High adsorption values were obtained with minimal deviation, suggesting that the method of naphthalene removal is efficient, precise and repetitive. Hence, a novel synthesized $Fe_3O_4@SiO_2/APTES@PAMAM(G10)/BZ$ nanosorbent has a high application potential in naphthalene removal from industrial effluents and contaminated water reservoirs.

7.3 Carbon-Based Nanomaterials

The large specific surface area and small, hollow and layered structures have made carbon nanotubes (CNTs) one of the most preferred nanomaterials for removal of several organic pollutants and heavy metals by adsorption. The surface of the carbon-based nanostructures can be easily modified by chemical treatment to enhance their adsorption capacity. Remarkable features of the CNTs, such as fibrous shape with high aspect ratio, large accessible external surface area and

well-developed mesopores, have attributed to their superior metal removal capacities.

Al_2O_3-coated multiwall carbon nanotubes (MWCNTs) were used by Gupta et al. (2011) to remove lead from aqueous solutions. The Al_2O_3/MWCNT composites were synthesized using aluminum nitrate and commercially available MWCNTs. The Al_2O_3 nanoparticles were successfully coated on the surface of MWCNTs to form multiwall carbon nanotube aluminum oxide composites. The alignment of Al_2O_3 on the surface of MWCNTs revealed close association of aluminum atom of alumina to the oxygen atom of the carbonyl group (C=O) of MWCNTs. This arrangement led to the enhancement of the affinity of the aluminum atom to interact with the oxygen electron pair, while the oxygen atom of alumina interacted closely with the hydrogen atom of the hydroxyl group (-OH) of MWCNTs through hydrogen bonding. Lead removal experiments showed a fast adsorption rate during the first hour of the process followed by a slower stage as the adsorbed amount of lead reached its equilibrium at 60 min. Adsorption of lead was also found to be increased when pH value was subsequently increased from 3 to 7, which could be due to electrostatic attraction between the negatively charged MWCNTs adsorbent surface and the positively charged cationic lead. However, it was highlighted that pH values higher than 7 could lead to precipitation of lead in the form $Pb(OH)_2$ and $Pb(OH)^+$ along with adsorption. Approximately 100% of 200 ppm of lead was adsorbed on Al_2O_3-coated MWCNTs when the composite concentration was increased to 50 mg. Three different mechanisms might be responsible for the adsorption of cationic lead on to the composite surface. It can involve van der Waals interactions between the hexagonally arrayed carbon atoms in the graphite sheet of MWCNTs and the positively charged lead ions. Similarly, electrostatic attraction may exist between the positive cationic lead and the negatively charged MWCNTs adsorbent surface. Likewise, electrostatic attraction between the pairs of electrons on the oxygen atoms of alumina and the positive cationic lead may also play a critical role in metal removal. An increase in agitation speed to 150 rpm resulted in a simultaneous increase in lead adsorption. Flow rate and thickness of the composite were also found to influence the rate of adsorption. Therefore, such Al_2O_3-coated MWCNTs could be useful in the removal of lead and thus have potential applications in industrial wastewater treatment.

CNTs combined with TiO_2 at a nanoscale were reported by Li et al. (2011) to form CNT/TiO_2 core–shell nanocomposites for photocatalytic applications. Nanocomposites were prepared using the surfactant wrapping sol-gel method by combining raw multiwall CNTs with three different precursors, namely titanium ethoxide (TEOTi), titanium isopropoxide (TTIP) and titanium butoxide (TBT). Sodium dodecylbenzesulfonate (NaDDBS) was used as a surface functionalizing agent in order to provide a well-defined, uniform, continuous, mesoporous TiO_2 layer over the CNTs. A continuous thin layer of TiO_2 coating covered the entire surface of CNTs, as shown in Figure 7.3. The diameter and thickness of raw CNTs and TiO_2 was estimated to be 10–30 nm and 5–15 nm, respectively. The composites had a lower surface area and pore volume as compared to pure CNTs, suggesting that the presence of TiO_2 may partially clog the nanotube

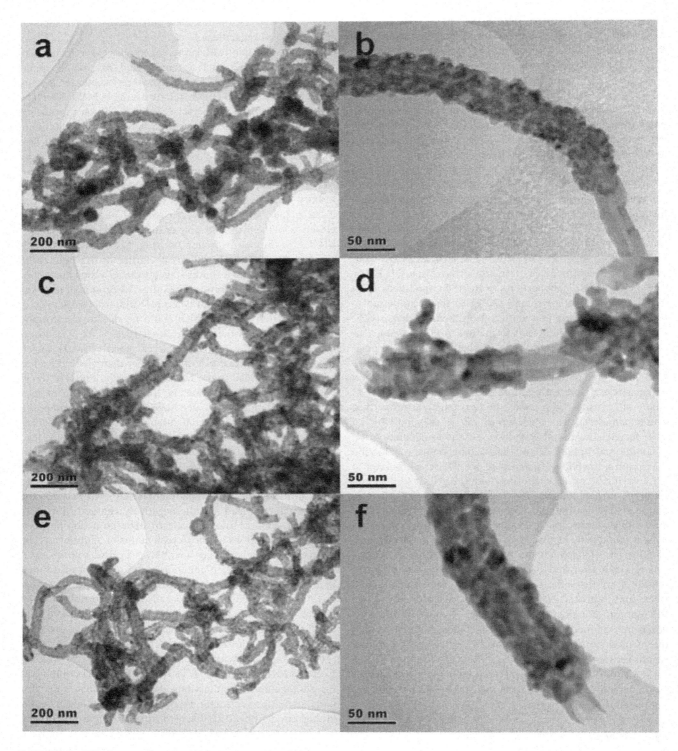

FIGURE 7.3 TEM images of calcined CNT/TiO$_2$ nanocomposites prepared from pristine CNTs with different TiO$_2$ precursors: (a and b) TTIP; (c and d) TBT; (e and f) TEOTi. Images on right-hand side show comparisons in thickness of TiO$_2$ layer with rare appearance of bare CNT. (Reprinted with permission from Li, Z., Gao, B., Chen, G.Z., Mokaya, R., Sotiropoulos, S., Puma, G.L., 2011. Carbon nanotube/titanium dioxide (CNT/TiO$_2$) core–shell nanocomposites with tailored shell thickness, CNT content and photocatalytic/photoelectrocatalytic properties. *Appl. Catal. B Environ.* 110, 50–57. Copyright © 2011 Elsevier B.V.)

openings. Thermal gravimetric analysis (TGA) showed significant weight loss between 550 and 750 °C, suggesting decomposition of CNTs leaving only TiO_2 and some C that has been incorporated into it. Nanocomposites made from TBT had lower levels of CNT and higher levels of C doping. Nanocomposites made from TTIP was found to convert around 20% of methylene blue dye, while TBT and TEOTi were observed to degrade about 40% and 30%, respectively. The enhanced activity of composites made from TBT could be due to TiO_2 film thickness that governs transfer of electrons and photocatalytic activity. Furthermore, the electronic conductivity increased with subsequent increase in CNT content that regulated the rate of electron removal from the photocatalyst when subjected to an external positive bias in an appropriate photoelectrochemical cell.

Congo red (CR) dye adsorption using MWCNT-impregnated chitosan hydrogel beads was reported by Chatterjee et al. (2010). Chitosan/carbon nanotube (CS/CNT) beads were dark blue in color with uniform distribution of MWCNTs throughout the beads. The porosity of CS/CNT beads was 95.15% with a diameter of around 2.66 mm. CNTs were uniformly distributed throughout the CS matrix along with some aggregates. TGA analysis indicated weight loss difference between the CS beads (90.6%) and the CS/CNT beads (73.0%) at a temperature of 1200 °C, suggesting that impregnation of CS beads within CNTs resulted in increased thermal stability. The adsorption of CR onto CS/CNT beads was via –OH and $-NH_2$ group of CS. The CNT concentration of 0.01 wt% was ideal for maximum adsorption. The addition of cetyltrimethylammonium bromide (CTAB) as a dispersant also resulted in increasing adsorption capacity, which may be due to hydrophobic interactions between the hydrophobic tail of CTAB and the hydrophobic moieties of CR dye. During adsorption, change in pH was observed at the initial stage and then leveled off to 6 from an initial value of 5, suggesting that protonation of amine groups is necessary for its interaction with negatively charged CR molecules. The adsorption of CR onto CS/CNT was highly dependent on pH, with maximum adsorption of 423.1 mg/g observed at a pH value of 4, which decreased with further increase in the value of pH. Adsorption isotherm studies revealed a better fit for Langmuir isotherm model with 0.998 value of correlation coefficient. Therefore, impregnation of CNT was highlighted to be a good strategy for enhancing capacity of adsorption of dyes such as CR.

Chitosan-Fe^0 nanoparticles were synthesized by Geng et al. (2009) and applied for hexavalent chromium removal. *In situ* reduction of Fe(II) with KBH_4 in the presence of chitosan as a stabilizer was performed to prepare chitosan-Fe^0 nanoparticles. A homocentric layered structure was observed with Fe^0 nanoparticles and chitosan polymers arranged on top of each other. A mean particle diameter of 82 nm was observed for chitosan-Fe^0. About 92% of chromium was present as Cr(III) while only 8% was in the form of Cr(VI), confirming the reduction of Cr(VI) by Fe^0 on the surface of the nanoparticles. Also, it was observed that Fe(III) was the predominant oxidation state, which could be due to synergistic redox of Fe^0 with Cr(VI) as well as reaction of Fe^0 with water and oxygen present in the experimental system. The rate constant of removal of

Cr(VI) gradually decreased with an increase in initial Cr(VI) concentrations. A two-step process was found during reduction of Cr(VI), wherein a decrease of Cr(VI) was higher during the first 2 min of the experiment with an average rate of about 10 mg/L/min, after which the reduction process followed a pseudo-first-order kinetics. Two kinds of sites were proposed to exist on the surface, that is, the active site where Fe^0 occupies the position for reduction of Cr(VI) to Cr(III) and other nonreactive sites where species other than Fe^0 are present. Hence, interaction of Cr(VI) with the nonreactive sites doesn't result in its reduction. Rate of reduction of Cr(VI) was found to increase upon an increase in iron loadings in the experimental system, suggesting more availability of Fe^0 active sites on the surface; thus, it was proposed that Fe(III)-Cr(III) hydroxides are co-precipitated on the surface of Fe^0 nanoparticles. However, introduction of chitosan as a support could hinder the precipitation due to presence of $-NH_2$ and –OH groups on the chitosan surface that can form coordinate bonds with the reduced metal ions. An increase in pH value from 4.0 to 8.0 was found to decrease the Cr(VI) reduction rate constant from 44×10^{-3} to 5.3×10^{-3}, respectively. This indicated that the acidic environment facilitated the reduction of Cr(VI) via electrostatic attraction between the positively charged nanoparticle surface and negatively charged Cr(VI) ions. However, the rate constant increased with the subsequent rise in temperature of the experimental system. Therefore, chitosan-Fe^0 nanoparticles can be promising nanomaterials for Cr(VI) reduction, which could be potentially applicable for wastewater treatment.

Yang et al. (2015) reported adsorption of sulfamethazine (SMZ) using pristine and hydroxylated multiwall carbon nanotubes (P-MWCNTs and H-MWCNTs). Both P-MWCNTs and H-MWCNTs had an outer diameter of 10–20 nm with 0.85% and 7.07% surface oxygen contents, respectively. A chemical vapor deposition method was used for synthesis of nanotubes using methane in hydrogen mixture at 700 °C in the presence of Ni nanoparticles as catalyst. An adsorption kinetics study indicated that SMZ adsorption increased significantly in the first 2 h and reduced thereafter. The reason for this could be due to the large number of vacant sites available on the surface of MWCNTs during the initial phase of adsorption. The pseudo-second-order model was observed to fit perfectly according to the kinetics result thus obtained. Also, it was observed that H-MWCNTs had comparatively slower and smaller SMZ adsorption than P-MWCNTs in spite of larger specific area and higher micro- and mesopore volumes of H-MWCNTs. The Langmuir model was observed to fit for the adsorption isotherms with correlation coefficient values of 0.995 and 0.998 for P-MWCNTs and H-MWCNTs, respectively. Maximum adsorption capacities of approximately 38.13 and 27.29 mg/g had been found for P-MWCNTs and H-MWCNTs, respectively. The lower adsorption capacities of H-MWCNTs was proposed due to the negative effects of functional groups present on H-MWCNTs by decreasing hydrophobicity in aqueous solutions, thus leading to a decrease in SMZ adsorption. Also, formation of a water-dense shell may lead to a decrease in surface-available sites for adsorption of SMZ in nanotubes. SMZ adsorption was maximum at a pH 7 that decreased with further increase in pH value. A reduction of 68% and 87% in adsorption capacity of P-MWCNTs and H-MWCNTs was observed, respectively upon increasing ionic strength from 0.02M to 0.20M. Increase in ionic strength could weaken the electrostatic attraction and enhance the competition between salt ions and SMZ for adsorption on surface sites thus, leading to reduction in adsorption capacity. Presence of metal ions such as Cu^{2+} at low pH values of 2.3 and 4.9 was observed to slightly decrease SMZ adsorption. This 10%–20% decrease could be a result of competition between metal ions and SMZ for attachment on the surface active sites. However, at higher pH values of 7.4 and 10.0, complex formation between SMZ and Cu^{2+} and Cd^{2+} facilitates easy adsorption on MWCNTs. Al^{3+} was observed to inhibit SMZ adsorption at both acidic and alkaline pH, which may be due to its larger hydrated radius that could form a water shell and block available adsorption sites. Similar results were obtained upon introduction of anions such as Cl^-, SO_4^{2-}, PO_4^{3-} and CO_3^{2-} that mainly led to blockage of sites. Further, it was speculated that the adsorption process in both P-MWCNTs and H-MWCNTs might involve π–π interactions.

7.4 Metal and Metal-Oxide Nanoparticles

Bootharaju and Pradeep (2012) reported degradation of chlorpyrifos (CP) pesticide using Ag-citrate nanoparticles (Ag@citrate NPs). Varying concentrations of CP, that is, 1 ppm, 10 ppm and 50 ppm, were treated with Ag@citrate NPs followed by monitoring absorption spectra after 24 and 48 h. The silver plasmon peak was observed to shift to 436 nm and 650 nm in the presence of 50 ppm of CP, which was speculated to be due to aggregation of NPs. In presence of 10 ppm of CP, a clear peak at 320 nm was observed after 48 h with a shift in silver plasmon peak to 457 and 700 nm. The elemental composition of NPs' surface in the aggregates was analyzed using TEM as illustrated in Figure 7.4. The 3,5,6-trichloro-2-pyridinol (TCP) and diethyl thiophosphate (DETP) were found to be the major degradation products. Complete degradation of CP was observed after 12 h of incubation at a reaction temperature of 35 °C. Furthermore, alumina-supported Ag@citrate NPs were observed to have an enhanced degradation activity. A mechanism of decomposition was thus proposed wherein a surface complex is formed when NPs bind with sulfur and

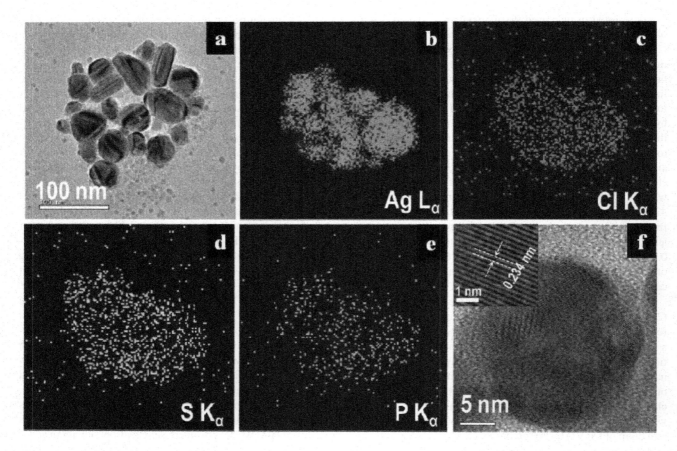

FIGURE 7.4 TEM image and elemental maps of Ag@citrate NPs after treating with 10 ppm CP solution (after 48 h). Part (a) is the TEM image of an aggregate. Parts (b), (c)(d) and (e) are elemental maps of Ag, Cl, S and P, respectively. Part (f) is the TEM image of a single NP, and the Ag(111) interplanar distance is marked in the inset. The elemental maps are rotated nearly 90° with respect to the TEM. (Reprinted with permission from Bootharaju, M.S., Pradeep, T., 2012. Understanding the degradation pathway of the pesticide, chlorpyrifos by noble metal nanoparticles. *Langmuir*. 28(5), 2671–2679. Copyright © 2012 American Chemical Society.)

nitrogen atoms. The side chain link to the pyrimidine ring in such surface complex may then be weakened due to electron polarization, leading to P-O bond cleavage. This can result in nucleophilic attack by water on the phosphorus site that may help in the formation of stable degradation products, that is, TCP and DETP, because of electron withdrawal from a coordination bond by nitrogen and sulfur, respectively.

Jaafarzadeh et al. (2017) reported efficient degradation of 2,4-dichlorophenoxyacetic acid (2,4-D) using magnetic copper ferrite nanoparticles ($CuFe_2O_4$, MCFNs). The coprecipitation method was used for synthesis of MCFNs followed by Fe-SEM analysis for particle size and morphology. Highly homogeneous, spherical-shaped particles were observed and were agglomerated, which may be due to their magnetic properties, with a size range of 22–27 nm. The Brunauer–Emmett–Teller (BET) specific surface area of MCFNs were found to be 32.3 m^2/g with an average pore diameter of 7.8 nm and a pore volume of 0.209 cm^3/g. MCFNs were speculated to be easily separated by a magnet due to an ideal magnetic saturation moment of 30.2 emu/g. An approximate ratio of 1:2 was observed for copper and iron with weight percentages of 20.37% and 40.4%, respectively. X-ray diffraction (XRD) and energy dispersive spectra (EDS) analyses revealed no impurity and the presence of Cu, Fe and O, respectively. Higher degradation of 20 mg/L of 2,4-D was observed within a 40 min reaction time at a pH range of 5–7 in the presence of 16 mg/L of ozone, 1 mM of peroxymonosulfate (PMS) and 0.10 g/L of MCFNs. Addition of 16 mg/L of ozone was found to increase 2,4-D removal from 33.6% to 74.6%, which could be due to more free radical generation with PMS or MCFNs in order to degrade 2,4-D present in the solution. Similarly, the addition of PMS may increase the possibility of production of sulfate radicals that further increase 2,4-D degradation. However, an increased concentration of PMS up to 4 mM was observed to create an inhibitory effect and reduce the removal efficiency. A remarkable enhancement in 2,4-D degradation was thus observed in the presence of solid heterogeneous catalyst in PMS/O_3 because of the tendency of PMS and ozone to be activated on the surface of the catalyst, along with more adsorption sites present with increasing loads of MCFNs. Reusability of MCFN catalyst was investigated, wherein no change in catalytic activity was observed after three consequent uses followed by only a slight reduction in activity after the fourth and fifth use. Also, the presence of anions such as bicarbonate, nitrite and phosphate ions was found to have a negative impact on degradation.

Manganese functionalized silicate nanoparticles that could act as Fenton-type nanocatalysts were reported by Tušar et al. (2012) to be useful in water purification for removal of organic pollutants. A two-step solvothermal synthesis was carried out to formulate the hybrid nanoparticle, wherein Mn was incorporated into an inorganic silicate matrix from the KIL-family (KIL-2) made from silicate nanoparticles with inter-particle porosity. The catalytic performance was investigated in the process of wet hydrogen peroxide catalytic oxidation (WHPCO), which is one of the most important industrial AOPs for organic pollutant decomposition in water samples. Various Mn concentrations (Mn/Si ratios from 0.005 to 0.05) were analyzed. The lowest activity for methylene blue

decomposition was observed for the 005MnKIL-2 catalyst, while the 01MnKIL-2, 02MnKIL-2 and 05MnKIL-2 solids exhibited the highest activities. A catalyst with a 0.01 Mn/Si ratio was considered to be optimal for catalysis due to comparable oxidation of methylene blue with higher ratios and a low tendency for H_2O_2 decomposition. Interestingly, the pH of the liquid was observed to have a slight influence on the initial removal rates of methylene blue, which could be due to electrostatic repulsion between the reactant and differently charged catalyst surface. Also, total organic carbon (TOC) removal of more than 80% was obtained using 01MnKIL-2 catalyst. Oxidation state of Mn was between +2 and +3 for all ratios of Mn/Si. Mn cations present in 01MnKIL-2 are like Lewis acid sites such that three framework oxygen atoms are coordinated in a distorted and coordinatively unsaturated 3-fold symmetry.

Surface-engineered amphiphilic polymer-coated silicon nanoparticles (SiNPs) were reported by Iqbal et al. (2016) to be useful as photocatalysts for degradation of methanol present in water samples. Thermal processing of commercial hydrogen silsesquioxane (HSQ) containing sub-10 ppb concentrations of metal impurities was performed to obtain SiNPs of three sizes, that is, 3 nm, 9 nm and 75 nm. SiNPs of 100 nm particle size was also obtained from commercially available oxide-embedded d = 100 nm SiNPs. Hydrofluoric acid (HF) was used to etch SiNPs in order to remove oxides as well as to provide a hydride surface useful for further modification. Thermally induced hydrosilylation with 1-dodecene was performed to modify SiNPs surface. An amphiphilic polymer based upon a poly(maleic anhydride) backbone was used for such alkyl-terminated SiNPs to make it water soluble. FTIR analyses of dodecyl-modified SiNPs showed presence of residual Si-H moieties and stretching of Si-O-Si, suggesting surface oxidation. Photocatalytic activity of SiNPs was investigated against a standard model containment, that is, methanol. A quasi-collimated beam with a medium-pressure UV lamp was used for degradation, and the SiNPs were isolated after 30 min of reaction and investigated using spectroscopic techniques. FTIR spectrum revealed the presence of characteristic alkyl group absorption along with attachment of amphiphilic polymer on the surface of SiNPs. The mechanism of SiNP-mediated photocatalytic AOPs was proposed that involves the migration of valence electrons to the surface states, creating a valence band hole that can be trapped on the surface of SiNPs and may induce the formation of –OH radicals, which could react with the methanol present as a pollutant to form α-hydroxy methyl radicals. These intermediate radicals may further react with dissolved oxygen present in water in order to produce formaldehyde. The Nash method was employed to quantitatively estimate the production of such formaldehyde, wherein acetylacetone and ammonium acetate react with formaldehyde to produce diacetyl dihydrolutidine (DDL), which could be spectrophotometrically determined by monitoring the absorption of DDL at 412 nm. All different sizes of SiNPs were used for photocatalytic studies at a concentration of 100 mg/L with an exposure time of 30 min. The highest quantum yield value for methanol decomposition of 2.4% as well as the highest yield factor for hydroxyl radical generation of 89% were obtained using SiNPs of 3 nm particle size.

Among various other nanoparticles, silver nanoparticles (AgNPs) with attractive physicochemical and optoelectronic properties are considered the most preferred nanostructures for catalytic degradation of toxic and hazardous pollutants (Ghosh et al. 2016a, 2016b; Shende et al. 2018, 2017). Several physical, chemical and biological routes are designed for the effective synthesis of AgNPs with exotic shape and size (Ghosh et al. 2016c; Shinde et al. 2018; Sant et al. 2013). In view of this background, silver-supplemented hybrid nanostructures have emerged as potential catalytic agents with multiple applications in environment, medicine and agriculture (Robkhob et al. 2020; Ghosh et al. 2015b; Salunke et al. 2014; Ranpariya et al. 2021).

AgNP-modified mesoporous anatase (TiO_2) was reported by Xiong et al. (2011) for purification of water samples. Mesoporous TiO_2 was prepared using titanium isopropoxide followed by modification with AgNPs using a photoreduction method. The BET surface area of mesoporous TiO_2 was approximately 166 m^2/g, which decreased upon modification with AgNPs along with a decrease in pore volume from 0.43 to 0.33 cm^{-3}/g. Also, a decrease in pore size was observed in the presence of 3.0% of Ag, thus indicating that AgNPs may have blocked some of the pores. TEM images confirmed the loading of Ag onto TiO_2 with AgNPs smaller than 1.5 nm in size. A strong absorption of mesoporous TiO_2 samples was seen at about 400 nm, and with increasing amount of Ag loading, surface plasmon absorption gradually shifted from 468 nm to 498 nm, thus suggesting more aggregation of AgNPs on the surface of anatase. Photocatalytic properties of mesoporous anatase TiO_2 was analyzed under UV light irradiation, wherein degradation of rhodamine B (RhB) dye degradation and inactivation of *E. coli* was observed. The degradation rate constant was observed to be 0.044 min^{-1}, which was enhanced to 0.069 min^{-1} in the presence of 0.25 wt % Ag. Similar results were obtained when *E. coli* bacterial suspension was irradiated with UV in the presence of mesoporous TiO_2 - 2 wt % Ag composites. Hence, it was speculated that the immobilization of AgNPs on the surface of TiO_2 may improve electron-hole separation through the formation of a Schottky barrier at the interface. Such Ag depositions may decrease pollutants adsorption; however, it greatly increases the adsorption of oxygen leading to production of reactive oxygen species (ROS) that could subsequently enhance photodegradation.

7.5 Zeolites

The inorganic microporous and microcrystalline materials called zeolites are highly capable of removing hazardous pollutants owing to their ability to form complex or adsorb small and medium-sized organic molecules. The efficiency of zeolites to purify water is dependent on its chemical properties, which are affected by synthesis (Si/Al ratio), post-synthesis treatments (ion exchange, temperature of calcinations, dealumination, etc.), composition, cations and framework structure. Various transition metals can be incorporated into zeolites, resulting in the fabrication of solid-phase-advanced oxidation catalysts.

Shukla et al. (2010) reported heterogeneous oxidation of phenol using cobalt-exchanged zeolites in the presence of peroxymonosulfate. A conventional ion-exchange technique was carried out for loading of cobalt ion onto zeolite samples (zeolites ZSM-5, A and X). UV-Vis diffuse reflectance spectroscopic (DRS) analysis of three samples was performed and the spectral minima were observed to appear around 530 nm in the visible region and 240 nm in the UV region, which suggests formation of octahedral $[Co(H_2O)_6]^{2+}$ complex. XRD patterns highlighted complete loss of crystalline structure of Co-zeolite-A, while Co-ZSM5 and Co-zeolite-X had 17% and 83% loss of crystalline integrity, respectively. Phenol oxidation was observed to be completed in less than 30 min with Co-zeolite-A/PMS and Co-zeolite-X/PMS, while it took 6 h in absence of PMS. It was speculated that such fast rates of oxidation could be due to leaching of cobalt ions into the solution due to homogeneous reaction between PMS and cobalt ions. The pH of the system was found to be around 3–3.5 during oxidation. Co-ZSM5 showed comparatively stable structural and catalytic properties. Si/Al ratio of Co-ZSM5 was found to be 25, suggesting approximately 50% of cobalt was present in ionic form and the rest existed in oxide form. A 40% loss in crystalline structure after reaction thus suggests that the damage may be due to exchangeable cations, whereas the remaining cobalt ions in the oxide form tend to remain in the structure. Apart from PMS, two other oxidants, namely H_2O_2 and persulfate, were also considered for phenol oxidation using Co-ZSM5. However, H_2O_2 and persulfate were only able to degrade less than 10% of phenol in 6 h, indicating the inability of Co-ZSM5 to effectively produce hydroxyl and sulfate radicals, respectively. Degradation of phenol was found to follow zero-order kinetics with activation energy of 69.7 kJ/mol showing 100% degradation within 4 h in the presence of 12.5 ppm of phenol.

Cu-exchanged Y zeolite was reported by Fathima et al. (2008) to be useful in catalytic wet hydrogen peroxide oxidation of an anionic dye. A flexible ligand method was used for preparation of the catalyst, wherein the metal is introduced in the pores of a solid via cation exchange. An atomic absorption spectrophotometer (AAS) was used to estimate the copper content in the catalyst and was found to be 7.47 wt%, indicating successful entrapment of Cu(II) nitrate inside the super cages of zeolite. BET surface area was observed to decrease upon modification, which could be due to the presence of compounds in the cages of the zeolite. Infrared (IR) spectra of ligand-encapsulated zeolite were observed to be weak due to their low concentration in zeolite matrix as compared to Cu-salen complex. Thermal analysis data showed zeolite catalysts may contain encapsulated Cu-salen complex that may be treated thermally up to a maximum of around 500 °C without any significant decomposition. Catalytic activity of the complex was studied towards decomposition of H_2O_2 at different reaction times. H_2O_2 was observed to enhance oxidation of the dye along with a catalyst with a residual concentration of 7.59 mg/L of dye for an initial concentration of 50 mg/L of dye. The addition of 0.175 M of H_2O_2 was observed to be the optimum concentration that was able to increase color removal up to 84% after 3 h of reaction. A pH value of 6.0 and above along with a higher temperature of around 60 °C was observed to increase the efficiency of dye removal. The efficiency of catalyst against commercial tannery dye effluents was also evaluated, wherein 95% of dye removal was observed along with 85% of chemical oxygen demand (COD) removal after treatment. The reusability of the catalyst was also highlighted in which no change in catalytic efficiency was observed even after 10 cycles of use. Finally, kinetic studies suggested that the decolorization of the dye followed pseudo-first-order kinetics with respect to the initial concentration of dye.

In another study, Fe(III)-exchanged Y zeolites were reported by Neamțu et al. (2004) to be applicable in the degradation of a reactive azo dye, C.I. reactive yellow 84 (RY84) using H_2O_2, as an oxidant, under very mild conditions. Commercially available ultra-stable Y (USY) zeolite was used in an ion-exchange method to synthesize the catalysts for dye degradation. Optimal concentration of H_2O_2 was observed to be 20 mmol/L, wherein 96.9% of color removal, 70.7% of COD reduction and 34.52% of TOC removal were achieved after 60 min of reaction time. The mechanism of degradation that was proposed involved the conversion of Fe(III) to Fe(II), which reacts with H_2O_2 to form a hydroxyl radical that can attack the aromatic ring present in the dye. With an increase in temperature, a simultaneous increase in dye degradation was observed, that is, at 50 °C, about 97.83% of color removal was observed. Likewise, the efficiency of degradation was found to be increased with higher concentrations of catalyst. A catalyst concentration of 1 g/L was suggested to be optimum to avoid leaching along with maximal color removal. Three different heterogeneous USY catalysts with varying degrees of dealumination were considered for investigation of wet catalytic oxidation of C.I. RY84. Maximum color removal of 99.96% was achieved with Fe-Y$_5$ catalyst along with most effective COD and TOC removal as well. Furthermore, high performance ion chromatography (HPICE) analysis revealed formation of acetate, nitrate, formate, malonate and oxalate as the main oxidation products. A toxicity assay was also carried out using bioluminescence in *V. fisheri*, in which an increase in toxicity was found after the initial 10 min of reaction with Fe-Y$_5$, highlighting the formation of toxic intermediates that were rapidly decreased after 60 min of reaction, indicating the disappearance of these toxic compounds. More than 62% toxicity was found to be reduced after 120 min of reaction.

Perisic et al. (2016) reported a process for zeolite-assisted removal of diclofenac (DCF) in water matrix. A solid-state ion-exchange method was carried out to prepare a heterogeneous Fenton-type catalyst (FeZSM5) using $FeSO_4.7H_2O$ and NH_4ZSM5. SEM analysis indicated that the shape and size of both NH_4ZSM5 and FeZSM5 were preserved after catalyst preparation. Inductively coupled plasma mass spectrometer (ICP-MS) analysis revealed that very trace amounts of iron are present in non-exchange zeolite, while significant iron content of around 3.76×10^{-2} g/g was found in FeZSM5 indicating its potential to be used as a heterogeneous Fenton-type catalyst. Removal of DCF by FeZSM5 under dark conditions were studied that showed adsorption of DCF with 85% and 73% TOC removal at pH value of 3.0 and 4.4, respectively. It was assumed that complexation of DCF occurs with surface Fe(III) via a deprotonated ethanoic functional group. A response surface methodology (RSM) approach was used to investigate the influence of UV-A irradiation in FeZSM5-mediated DCF removal. It was indicated that adsorption along with oxidative degradation mechanisms are involved in a concerted manner to remove DCF in the presence of UV-A rays. Almost complete DCF removal as well as higher mineralization was observed, wherein FeZSM5 was suggested to behave as an effective Fenton-type catalyst. After 2 h of treatment of DCF in water with an H_2O_2 concentration of 50 mM at a pH 4.0 and in the presence of 2.0 mM of [Fe]FeZSM5, a significant improvement in biodegradability and a decrease in toxicity was seen.

Doocey et al. (2004) demonstrated the use of zeolite Fe-beta in AOPs for removal of chlorinated phenols in aqueous waste. Zeolite Fe-beta was prepared using $FeSO_4.7H_2O$ along with 0.01 M ascorbic acid as a reducing agent. Two potential solid-phase Fenton oxidation catalysts were considered for further experiments after preliminary screening. Both zeolite Fe-beta and Fe-4A had an optimum pH value of 3.5 that was observed to be ideal for maximum degradation of hydrogen peroxide, which was used as an oxidant. A slightly better decomposition efficiency of 23% was observed for zeolite Fe-4A as compared to 19% efficiency of zeolite Fe-beta. AAS analysis was used to investigate the iron content. It was observed that zeolite Fe-4A was more stable with only 0.2% available iron being leached out of the zeolite, suggesting that degradation of hydrogen peroxide is due to solid-phase Fenton catalysis only. Moreover, Fe-beta zeolite was used for studying selective removal of 2,4-DCP from an aqueous waste stream in which a 25.03 mg/g of 2,4-DCP was observed to be adsorbed per unit of zeolite from an adsorption column packed with zeolite Fe-beta after the first cycle of experiment with a regeneration efficiency of 76.7% as measured by Cl$^-$ production.

7.6 Conclusions and Future Perspectives

Nanotechnology-based AOPs are ideal for removal of synthetic chemicals, dyes, organic matters, refractory organic waste and heavy metals from industrial effluents. Several polymeric, metal and metal oxide based nanoparticles, CNTs and zeolites can effectively reduce the concentration of the hazardous pollutants from water. However, since the efficiency of nanoparticles are mostly size and shape dependent, the parameters for synthesis of nanoparticles like time, temperature, pH, agitation, aeration, concentration of reagents and so on are needed to be optimized to get tunable morphology and surface properties. Similarly, special attention needs to be given for scaling up nanoparticle synthesis so that it can be implemented at community levels, where large volumes of wastewater need to be treated daily. Developing nanoparticles with a reusable and recyclable nature would reduce the cost of treatment. Likewise, rational surface functionalization of the nanoparticles or post-synthesis modification should be considered very carefully so that their properties can be further enhanced for AOPs.

Environmental aspects of the nanotechnology-driven solutions for wastewater treatment employing AOPs must be critically considered as, after treatment, retention of the nanoparticles in the environment can be deleterious. Hence, post-treatment removal of the nanoparticles and their disposal strategy should be designed. Further, during nanoparticle-mediated AOPs, the intermediates produced by degradation of the pollutants should be checked for toxicity. The intermediates and end products should be nontoxic. In case they are still hazardous, additional biological treatment should be employed for further detoxification of the intermediates and the end products. In view of this background, nanotechnology-based AOPs for wastewater treatment can serve as a powerful tool in the future to ensure clean water and environmental safety.

REFERENCES

Abkenar, S.S., Malek, R.M.A., Mazaheri, F., 2015. Dye adsorption of cotton fabric grafted with PPI dendrimers: Isotherm and kinetic studies. *J. Environ. Manage.* 163, 53–61.

Aliannejadi, S., Hassani, A.H., Panahi, H.A., Borghei, S.M., 2019. Fabrication and characterization of high-branched recyclable PAMAM dendrimer polymers on the modified magnetic nanoparticles for removing naphthalene from aqueous solutions. *Microchem. J.* 145, 767–777.

Bootharaju, M.S., Pradeep, T., 2012. Understanding the degradation pathway of the pesticide, chlorpyrifos by noble metal nanoparticles. *Langmuir.* 28(5), 2671–2679.

Cañizares, P., Paz, R., Saez, C., Rodrigo, M. A., 2009. Costs of the electrochemical oxidation of wastewaters: A comparison with ozonation and Fenton oxidation processes. *J. Environ. Manage.* 90(1), 410–420.

Chatterjee, S., Lee, M.W., Woo, S.H., 2010. Adsorption of congo red by chitosan hydrogel beads impregnated with carbon nanotubes. *Bioresour. Technol.* 101(6), 1800–1806.

Deng, Y., Zhao, R., 2015. Advanced oxidation processes (AOPs) in wastewater treatment. *Curr. Pollution Rep.* 1, 167–176.

Diallo, M.S., Christie, S., Swaminathan, P., Johnson, J.H., Goddard, W.A., 2005. Dendrimer enhanced ultrafiltration. 1. Recovery of Cu (II) from aqueous solutions using PAMAM dendrimers with ethylene diamine core and terminal NH_2 groups. *Environ. Sci. Technol.* 39(5), 1366–1377.

Doocey, D.J., Sharratt, P.N., Cundy, C.S., Plaisted, R.J., 2004. Zeolite-mediated advanced oxidation of model chlorinated phenolic aqueous waste: Part 2: Solid phase catalysis. *Process Saf. Environ. Prot.* 82(5), 359–364.

Fathima, N.N., Aravindhan, R., Rao, J.R., Nair, B.U., 2008. Dye house wastewater treatment through advanced oxidation process using Cu-exchanged Y zeolite: A heterogeneous catalytic approach. *Chemosphere.* 70(6), 1146–1151.

Geng, B., Jin, Z., Li, T., Qi, X., 2009. Kinetics of hexavalent chromium removal from water by chitosan-Fe^0 nanoparticles. *Chemosphere.* 75(6), 825–830.

Ghosh, S., Chacko, M.J., Harke, A.N., Gurav, S.P., Joshi, K.A., Dhepe, A., Kulkarni, A.S., Shinde, V.S., Parihar, V.S., Asok, A., Banerjee, K., Kamble, N., Bellare, J., Chopade, B.A., 2016a. *Barleria prionitis* leaf mediated synthesis of silver and gold nanocatalysts. *J. Nanomed. Nanotechnol.* 7, 4.

Ghosh, S., Gurav, S.P., Harke, A.N., Chacko, M.J., Joshi, K.A., Dhepe, A., Charolkar, C., Shinde, V.S., Kitture, R., Parihar, V.S., Banerjee, K., Kamble, N., Bellare, J., Chopade, B.A., 2016b. *Dioscorea oppositifolia* mediated synthesis of gold and silver nanoparticles with catalytic activity. *J. Nanomed. Nanotechnol.* 7, 5.

Ghosh, S., Harke, A.N., Chacko, M.J., Gurav, S.P., Joshi, K.A., Dhepe, A., Dewle, A., Tomar, G.B., Kitture, R., Parihar, V.S., Banerjee, K., Kamble, N., Bellare, J., Chopade, B.A., 2016c. *Gloriosa superba* mediated synthesis of silver and gold nanoparticles for anticancer applications. *J. Nanomed. Nanotechnol.* 7, 4.

Ghosh, S., More, P., Nitnavare, R., Jagtap, S., Chippalkatti, R., Derle, A., Kitture, R., Asok, A., Kale, S., Singh, S., Shaikh, M.L., Ramanamurthy, B., Bellare, J., Chopade, B.A., 2015a. Antidiabetic and antioxidant properties of copper nanoparticles synthesized by medicinal plant *Dioscorea bulbifera.* *J. Nanomed. Nanotechnol.* S6, 007.

Ghosh, S., Jagtap, S., More, P., Shete, U.J., Maheshwari, N. O., Rao, S. J., Kitture, R., Kale, S., Bellare, J., Patil, S., Pal, J. K., Chopade, B.A., 2015b. *Dioscorea bulbifera* mediated synthesis of novel $Au_{core}Ag_{shell}$ nanoparticles with potent antibiofilm and antileishmanial activity. *J. Nanomater.* 2015, Article ID 562938.

Ghosh, S., More, P., Derle, A., Kitture, R., Kale, T., Gorain, M., Avasthi, A., Markad, P., Kundu, G.C., Kale, S., Dhavale, D.D., Bellare, J., Chopade, B.A., 2015c. Diosgenin functionalized iron oxide nanoparticles as novel nanomaterial against breast cancer. *J. Nanosci. Nanotechnol.* 15(12), 9464–9472.

Gupta, V.K., Agarwal, S., Saleh, T.A., 2011. Synthesis and characterization of alumina-coated carbon nanotubes and their application for lead removal. *J. Hazard. Mater.* 185(1), 17–23.

Iqbal, M., Purkait, T.K., Goss, G.G., Bolton, J.R., Gamal El-Din, M., Veinot, J.G., 2016. Application of engineered Si nanoparticles in light-induced advanced oxidation remediation of a water-borne model contaminant. *ACS Nano.* 10(5), 5405–5412.

Jaafarzadeh, N., Ghanbari, F., Ahmadi, M., 2017. Efficient degradation of 2, 4-dichlorophenoxyacetic acid by peroxymonosulfate/magnetic copper ferrite nanoparticles/ozone: a novel combination of advanced oxidation processes. *Chem. Eng. J.* 320, 436–447.

Li, Z., Gao, B., Chen, G.Z., Mokaya, R., Sotiropoulos, S., Puma, G.L., 2011. Carbon nanotube/titanium dioxide (CNT/TiO2) core–shell nanocomposites with tailored shell thickness, CNT content and photocatalytic/photoelectrocatalytic properties. *Appl. Catal. B Environ.* 110, 50–57.

Mantzavinos, D., Psillakis, E., 2004. Enhancement of biodegradability of industrial wastewaters by chemical oxidation pretreatment. *J. Chem. Technol. Biotechnol.* 79(5), 431–454.

Moreno Escobar, B., Gomez Nieto, M. A., Hontoria García, E., 2005. Simple tertiary treatment systems. *Water Sci. Technol. Water Supply*, 5(3–4), 35–41.

Neamţu, M., Catrinescu, C., Kettrup, A., 2004. Effect of dealumination of iron (III) – exchanged Y zeolites on oxidation of reactive yellow 84 azo dye in the presence of hydrogen peroxide. *Appl. Catal. B Environ.* 51(3), 149–157.

Perisic, D.J., Gilja, V., Stankov, M.N., Katancic, Z., Kusic, H., Stangar, U.L., Dionysiou, D.D., Bozic, A.L., 2016. Removal of diclofenac from water by zeolite-assisted advanced oxidation processes. *J. Photochem. Photobiol. A Chem.* 321, 238–247.

Poyatos, J. M., Muñio, M. M., Almecija, M. C., Torres, J. C., Hontoria, E., Osorio, F., 2010. Advanced oxidation processes for wastewater treatment. *State of the Art Water Air Soil Pollut.* 205, 187–204.

Ranpariya, B., Salunke, G., Karmakar, S., Babiya, K., Sutar, S., Kadoo, N., Kumbhakar, P., Ghosh, S. 2021. Antimicrobial synergy of silver-platinum nanohybrids with antibiotics. *Front. Microbiol.* 11, 610968.

Rether, A., Schuster, M., 2003. Selective separation and recovery of heavy metal ions using water-soluble N-benzoylthiourea modified PAMAM polymers. *React. Funct. Polym.* 57(1), 13–21.

Robkhob, P., Ghosh, S., Bellare, J., Jamdade, D., Tang, I.M., Thongmee, S., 2020. Effect of silver doping on antidiabetic and antioxidant potential of ZnO nanorods. *J. Trace Elem. Med. Biol.* 58, 126448.

Rosenfeldt, E.J., Chen, P.J., Kullmanc, S., Linden, K.G., 2007. Destruction of estrogenic activity in water using UV advanced oxidation. *Sci. Total Environ.* 377, 105–113.

Salunke, G.R., Ghosh, S., Santosh, R.J., Khade, S., Vashisth, P., Kale, T., Chopade, S., Pruthi, V., Kundu, G., Bellare, J.R., Chopade, B.A., 2014. Rapid efficient synthesis and characterization of AgNPs, AuNPs and AgAuNPs from a medicinal plant, *Plumbago zeylanica* and their application in biofilm control. *Int. J. Nanomed.* 9, 2635–2653.

Sant, D.G., Gujarathi, T.R., Harne, S.R., Ghosh, S., Kitture, R., Kale, S., Chopade, B. A., Pardesi, K.R., 2013. *Adiantum philippense* L. frond assisted rapid green synthesis of gold and silver nanoparticles. *J. Nanoparticles,* 2013,1–9.

Shen, J., Li, Z., Wu, Y.N., Zhang, B., Li, F., 2015. Dendrimer-based preparation of mesoporous alumina nanofibers by electrospinning and their application in dye adsorption. *Chem. Eng. J.* 264, 48–55.

Shende, S., Joshi, K.A., Kulkarni, A.S., Charolkar, C., Shinde, V.S., Parihar, V.S., Kitture, R., Banerjee, K., Kamble, N., Bellare, J., Ghosh, S., 2018. *Platanus orientalis* leaf mediated rapid

synthesis of catalytic gold and silver nanoparticles. *J. Nanomed. Nanotechnol.* 9, 2.

Shende, S., Joshi, K.A., Kulkarni, A.S., Shinde, V.S., Parihar, V.S., Kitture, R., Banerjee, K., Kamble, N., Bellare, J., Ghosh, S., 2017. *Litchi chinensis* peel: A novel source for synthesis of gold and silver nanocatalysts. *Glob. J. Nanomed.* 3(1), 555603.

Shinde, S.S., Joshi, K.A., Patil, S., Singh, S., Kitture, R., Bellare, J., Ghosh, S., 2018. Green synthesis of silver nanoparticles using *Gnidia glauca* and computational evaluation of synergistic potential with antimicrobial drugs. *World J. Pharm. Res.* 7(4), 156–171.

Shukla, P., Wang, S., Singh, K., Ang, H.M., Tadé, M.O., 2010. Cobalt exchanged zeolites for heterogeneous catalytic oxidation of phenol in the presence of peroxymonosulphate. *Appl. Catal. B Environ.* 99(1–2), 163–169.

Skoumal, M., Cabot, P.L., Centellas, F., Arias, C., Rodríguez, R.M., Garrido, J.A., Brillas, E., 2006. Mineralization of paracetamol by ozonation catalyzed with Fe^{2+}, Cu^{2+} and UVA light. *Appl. Catal. B Environ.* 66, 228–240.

Tušar, N.N., Maučec, D., Rangus, M., Arčon, I., Mazaj, M., Cotman, M., Pintar, A., Kaučič, V., 2012. Manganese functionalized silicate nanoparticles as a Fenton-type catalyst for water purification by advanced oxidation processes (AOP). *Adv. Funct. Mater.* 22(4), 820–826.

Walid, K. L., Al-Qodah, Z., 2006. Combined advanced oxidation and biological treatment processes for the removal of pesticides from aqueous solutions. *J. Hazard. Mater.* B137, 489–497.

Xiong, Z., Ma, J., Ng, W.J., Waite, T.D., Zhao, X.S., 2011. Silver-modified mesoporous TiO_2 photocatalyst for water purification. *Water Res.* 45(5), 2095–2103.

Yang, Q., Chen, G., Zhang, J., Li, H., 2015. Adsorption of sulfamethazine by multi-walled carbon nanotubes: Effects of aqueous solution chemistry. *RSC Adv.* 5(32), 25541–25549.

8

Electron Beam Accelerators: Wastewater to Useable Water

Mrunal Deshpande, K. Sathish Kumar, R. Rengaraj, G.R. Venkatakrishnan and Hithu Anand

CONTENTS

8.1 Introduction

Around the world, the scarcity of water resources is increasing rapidly due to the imbalance caused between available fresh water and its rate of consumption. Though 71% of our world is covered with water, only 2.5% of the world's water is fresh water (Hossain et al., 2018). It is well known that fresh water is the most important substance on earth, without which there would be no life. Hence, accessing clean and safe water has become one of the major challenges in our modern society. The various reasons for this challenge are increase in population, rapid development in the industrial sector and climatic change, which leads to an increase in per capita use of fresh water. Further, the presence of pathogens and anthropogenic chemicals threaten the water quality of urban and rural water sources (Ta Wee Seow et al., 2016). In addition, various treatment plants that discharge their untreated wastes into water bodies is considered another major reason for water scarcity. Even in a highly industrialized country like China, most of their wastes are discharged into water bodies without any treatment, resulting in the contamination of fresh water (Hossain et al., 2018; Ta Wee Seow et al., 2016).

Any material that is dissolved or suspended in or on water may exist in the wastewater. In general, materials existing in wastewater are classified as organic, inorganic, particulate or pathogenic. The major concern about wastewater is

DOI: 10.1201/9781003165958-8

the presence of organic materials that decrease the dissolved oxygen in the fresh water. The different tests performed to measure the presence of organic materials in wastewater are biochemical oxygen demand (BOD), chemical oxygen demand (COD), total organic carbon (TOC) and total oxygen demand (TOD). Of the aforementioned tests, BOD is classified as a customary pollutant parameter, whereas the remaining three are considered non-customary pollutant parameters. Detailed definitions of the four tests are presented in Topare et al. (2011). The tests do not determine the exact organic material existing in the wastewater, but they estimate the presence of organic materials by measuring the oxygen level in it. Other organic evaluations are also important, such as those identifying oil and grease content, phenols, cyanide, surfactants and other organics containing toxic functional compounds that create problems for natural life. The various environmental problems associated with wastewater are that it depletes the dissolved oxygen in the fresh water, which is required for aquatic life; threatens human health as it contains many pathogenic, or disease-causing, microorganisms and toxic compounds; and leads to eutrophication in lakes and streams, as it stimulates the growth of aquatic plants and algal blooms, which leads to the production of stinking gases (Topare, et al., 2011).

For the urban world, the aforementioned problems associated with wastewater become an inevitable due to the rise in human population and urbanization. Wastewater-related issues are also significant in coastal areas, as they are occupied by over 60% of the human population (Maruthi et al., 2012a, 2012b; Satyanarayana et al., 2010). According to the Global Programme of Action 2011, the wastewater creates a significant threat to coastal environments worldwide. In Abbas et al. (2017), Danulat et al. (2002), WHO (2003), and Maruthi et al. (2012c, 2012d, 2012e), the adverse impacts of wastewater on the environment, human life, socioeconomics, quality of food and security are well illustrated. These problems are created by wastewater due to change in nutrient levels, diversity of organisms, bioaccumulation of organic and inorganic compounds and changes in interaction among species (Hossain et al., 2015, 2016; Singh et al., 2016; Hossain and Ismail, 2015). Hence, it becomes a necessary task to treat the wastewater before it enters natural freshwater sources.

Various physical and chemical treatment methods have been implemented for the treatment of wastewater such as biological degradation, ion exchange, chemical precipitation, adsorption, reverse osmosis, coagulation and flocculation. The performance of these methods is different and has different impacts on the environment. According to Rajasulochana and Preethy (2016), there exist different conventional methods for treating wastewater, but they are very costly. Also, it is necessary to develop an environmentally friendly method, as the conventional treatment method creates pollution in the environment. To overcome the difficulties and disadvantages faced by conventional methods, advanced new green methods have been developed to treat wastewater (Rajasulochana and Preethy, 2016). Though there exist very limited water treatment methods, with the help of an increase in human knowledge and technologies, many researchers have started developing methods that are environmentally friendly and effective for treating wastewater (Rajasulochana and Preethy, 2016)

Rajasulochana and Preethy (2016) have treated wastewater by removing arsenic using electrocoagulation and electrodialysis techniques. Wastewater containing lead (II) ions can be treated using a carbon nanotube (CNT)/magnesium oxide composite (Saleh et al.,2012). In the same way, alumina coated CNTs are used for removing lead from wastewater (Gupta and Saleh, 2011). A device that combines the process of adsorption and ultrasonic action is used by Ghaedi et al. (2018) to significantly improve the purification quality of wastewater.

For several decades, treating wastewater using accelerated electrons has attracted various researchers. But North America was the first to do large-scale studies on this topic in 1974, which led to the development of the first pilot plant system in Boston and later in Miami, Florida, in 1976 (Trump et al., 1984). Though many small pilot plants were constructed under the Electron Beam Research Facility (EBRF), the largest commercial plant for removing dye from wastewater is constructed in Daegu, Korea, in 2005 (Han et al., 2012). During the time at which industrial pilot plants were constructed, researchers concentrated on the electron beam (EB) treatment on a laboratory scale, which revealed that the treatment could remove bacteria, viruses and parasites (Capodaglio, 2017; Engohang-Ndong et al., 2015; Rawat and Sarma, 2013). In addition, solubility of carbohydrates, proteins and lipids (Engohang-Ndong et al., 2015; Lim et al., 2016) and the removal of color and odor (Engohang-Ndong et al., 2015) could be performed using EB. One of the critical issues is the effect caused on BOD and COD, which increases when EB is used for treating wastewater (Lim et al., 2016). The reason for the increase in BOD and COD is the conversion of nonbiodegradable compounds into biodegradable form. Also, some researchers claim that due to breakdown of biodegradable components into water, carbon oxide and salts, BOD and COD decrease after EB irradiation (He et al., 2016; Son, 2017). Hence, it is necessary to do a literature survey on the applications of electron beam radiation in treating the wastewater and the complications existing in it.

8.2 Sources of Wastewater

Concerning the origins of wastewater, they are categorized into domestic/municipal, industrial/commercial and agricultural sources, which is shown in Figure 8.1 (Arash, 2016).

8.2.1 Domestic/Municipal Wastewater

Wastewater produced from activities related to residential sources like food preparation, laundry, cleaning and personal hygiene (bathrooms and sanitary appliances such as toilets, sinks, washing machines) comes under the category of domestic/municipal wastewater. Major volumes of wastewater are produced in this category. It is to be noted that the water produced due to storm, rain, snow and melting of ice also comes under this category.

More contamination occurs in this type of wastewater than in sewage. In addition, wastewater produced by shops, stores, repair shops, workshops and other similar establishments comes under this category. The different sources of domestic wastewater are depicted in Figure8.2 (Arash, 2016).

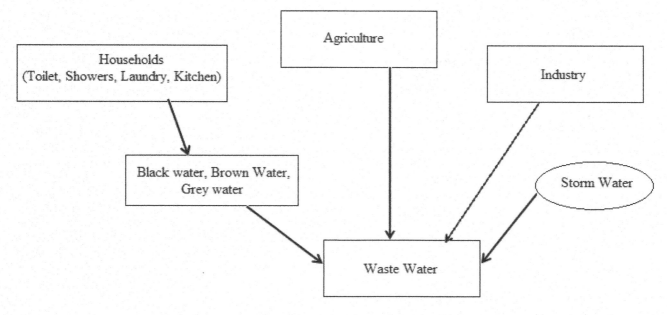

FIGURE8.1 Sources of wastewater.

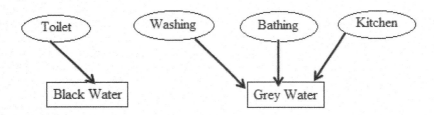

FIGURE8.2 Sources of domestic/municipal wastewater.

8.2.2 Industrial/Commercial Wastewater

Wastewater produced from manufacturing and commercial industries like printing, food, pharmaceuticals and beverage processing and production constitutes industrial/commercial wastewater. The wastewater produced from these industries and their effluents depends on the product that the industry manufactures. The most common foreign particles that exist in effluent of mining, steel and chemical factories is mineral, and in other factories like food products and starch manufacturing factories it is organic matter. The major differences between industrial and domestic wastewater are (i) the presence of more chemical substances, (ii) the existence of corrosive substances, (iii) the alkalinity or acidic property is high and (iv) the presence of fewer living organisms. The different industries that constitute the industrial/commercial wastewater are textile, glass, catering, printing, semiconductor, pharmaceutical, automobile, refinery, metal and so on (Arash, 2016).

8.2.3 Agricultural Wastewater

Wastewater produced by agricultural facilities using water for cleaning animal farms, washing harvested products and cleaning equipment comes under this category. At present day, chemical fertilizers and pesticides that contain nutrients and potassium constitute the main sources of wastewater (Arash, 2016).

8.2.4 Specification of Wastewater

There are some specifications that indicate the presence of wastewater. They are as follows (Arash, 2016):

i.	Color	ii.	Temperature
iii.	Smell	iv.	Foreign particles
v.	Degree of acidity	vi.	Living organisms

8.2.4.1Color

The parameter that helps in indicating the age of wastewater is color. The color of fresh wastewater is gray, which turns into dark and black after some duration of time when the elements present in it rots.

8.2.4.2Smell

When the organic matter existing in the wastewater decomposes, gases will be released that can be sensed by their smell. Compared to decayed wastewater, the smell produced by fresh wastewater can be tolerated. The reason behind this is the existence of hydrogen sulfide and carbon dioxide, which is produced by anaerobic bacteria. Anaerobic bacteria will stop producing hydrogen sulfide only when the wastewater has enough air and oxygen. To make the wastewater odorless, a sufficient amount of oxygen is supplied to wastewater.

8.2.4.3 Degree of Acidity

The degree of acidity in wastewater is determined by its pH level. The pH level of fresh wastewater will be neutral or slightly alkaline. The pH level of wastewater changes only when the wastewater starts decaying. During the decaying process, acidic gases are produced that make the wastewater acidic. The amount of time it takes to start the decaying process depends directly on the temperature of the environment.

8.2.4.4 Temperature

Considering the same environmental conditions, the temperature of wastewater is higher than fresh water due to the presence of biological activities.

8.2.4.5 Foreign Particles

Foreign particles, which may be dissolved or undissolved and suspended, constitutes about 0.1% of wastewater. Most foreign particles are organic in nature, and the remaining are mineral materials. The particles that settle at the bottom are called settleable organic materials, and the others are called no settleable organic materials.

8.2.4.6 Living Organisms

In addition, large amounts of pathogenic microscopic living organisms like viruses, bacteria and so on might exist in the wastewater.

8.3 Treatment Technologies of Industrial Wastewater

Wastewater from industrial outlets is the major contaminant of land and aquatic environments, which in turn affects the world's living beings. Hence, it is necessary to treat wastewater to provide and protect a balanced ecosystem (Volesky and Holan, 1995). Different treatment technologies are categorized into physical, chemical and biological methods, and combinations of these three could be carried out more effectively nowadays (Van Hoof et al., 1999). The treatment method could be based on the choices of the sources of waste, nature of waste and its concentration, rate of production and constituents.

8.3.1 Physical Treatment

The physical mode of treatment technology involves separation using equalization tanks, settlers, sedimentation tanks,

clarifiers or filters. Also, methods like filtration, flocculation, mixing, screening and gas transfer also use the physical forces of separation. The treatment involves just a physical separation; therefore, the use of chemicals or any other microorganisms is not required. Filtration involves special kinds of filters to pass the sewage through, which is very difficult to remove otherwise. The most commonly used filter is a sand filter, and some wastewater containing grease or oil on its surface could be removed from the surface easily. To make the sedimentation treatment effective, sufficient time should be provided for the settlement of solids. Another effective mode that can be used for physical water treatment is aeration. In this process, air is circulated through the water to provide oxygen. Chemicals used in this method ensure that the separation process has taken part with maximum effectiveness, also to some extent the chemicals act as disinfectant and cleaning agents. To name a few, chlorine, hydrogen peroxide and sodium chloride act as agents that disinfect, sanitize and assist in purification (Yonar et al., 2005).

8.3.2 Chemical Treatment

Chemical treatment involves processes such as adsorption using different adsorbents, disinfection and precipitation. Adsorbents vary widely and range from activated carbon from different natural resources to recent techniques involving nanoparticles using biopolymers. Membrane filtration, ultrafiltration, nanofiltration, reverse osmosis, nanoparticles with immobilized enzymes, hydrogels and their nanoparticles all fall under the category of chemical treatment. Low-cost adsorbents like baggase (Putra et al., 2014), sand, starch, seeds of fruits and vegetables (Ozcimen and Ersoy-Mericboyu, 2009), barks of different trees in powder form and polymers like chitosan, chitin, polyvinylalcohol (PVA) (Anitha et al., 2015) and polyacrylonitrile (PAN) could be used effectively. Biological wastewater treatment is an extremely cost-effective and energy-efficient system for the removal of BOD, since only microorganisms are involved, not chemicals. Biological wastewater treatment is designed to degrade pollutants dissolved in effluents by the action of microorganisms. The microorganisms utilize these organic substances to live and reproduce, leaving a pollutant-free waterbody.

8.3.3 Biological Treatment

Biological processes provide the advantage of recovery of resources and utilization of the trace elements present in wastewater, thereby eliminating the maximum contaminants present in water bodies. The organic content of the biological matter is degraded and removed using microorganisms and their activity. The different methods include aerobic and anaerobic digestion, activated sludge process (Ayoub, 2010), upflow mode anaerobic sludge blanket, suspended growth process and attached growth process with the involvement of microorganisms. Different parameters such as dissolved oxygen concentration, biological oxygen demand, solid retention time, sludge volume index, organic loading rate and food-to-microorganisms ratio could be studied to optimize the reactor run. Recently, reactors with immobilized enzymes or microorganisms are handled to provide the maximum removal of contaminants involving nanoparticles (Ahammad et al., 2013). The steps involved in biological treatment are preliminary or pre-treatment, primary, secondary and tertiary treatment. The preliminary treatment involves the removal of bulk solids and heavy matters using a screening system or grid system. The primary treatment is again a physical separation of settling solids by means of settlers or sedimentation tanks. Secondary treatment involves the decomposition of organic compounds, and the final tertiary treatment step deals with the output of higher effluent quality involving ozonation, ultraviolet mode of treatment and the complete removal of residual microorganisms.

8.3.3.1 Aerobic Digestion Process

The treatment of wastewater utilizes microorganisms in the presence of molecular oxygen. Aeration can be continuous or discontinuous depending on the microorganism involved. The main disadvantage of the process is the production of bulk biomass, which leads to the problem of sludge disposal. The main gases vented out of this reaction are mostly carbon dioxide and water vapor (Juteau et al., 2004).

8.3.3.2 Anaerobic Digestion Process

This treatment involves microbes that are anaerobic (absent of oxygen) in nature. The production of sludge is minimum in this case, where the reaction leads to the production of a byproduct called methane, which can be further converted into biofuel (Forster-Carneiro, 2008).

8.3.3.3 Suspended Growth Process

Organic biological content can be dispersed in liquid phase to be made available for the growth and utilization of microbes. Some microbes may depend on the supply of oxygen and some may not. That the biological matter is completely utilized by the microorganism when it is present in a liquid medium is the main advantage of the suspended growth process. Activated sludge treatment and aerated lagoons are two modes of aerobic treatment. Suspended growth could be carried out as a batch mode or a continuous mode of operation (Zia et al., 2013).

8.3.3.4 Activated Sludge Process

Aerobic suspended cell systems could utilize a consortium of microorganisms for the decomposition of the organic compounds in wastewater. The two parts of the design process system are divided as aeration and secondary clarifier. Aeration is provided through a surface sparger, where the microorganism uses the dissolved oxygen supplied for its growth utilizing the organic matter. Then, the left out solid residue is taken to the clarifier where the sediments are separated out and the clear supernatant is given further treatment. The next step will be disinfection and the use of effluent for the receiving bodies of water. The settled and separated out solids can then be sent to the digestion tank of solid disposal.

8.3.3.5 Aerated Lagoons

Aerated lagoons are structurally very simple, where the treatment allows the system to be exposed to the open surface atmosphere. The open surface makes oxygen available for the growth of microbes utilizing the complex organic matter and breaking them down into simple molecules. Also, mechanical dispersers and diffusers could be added, which enhances the oxygen supply rate. Diffusers supply oxygen in the form of bubbles that could be easily broken down and consumed by the microbes. This method facilitates and speeds up the process of aerobic treatment (Zia et al., 2013).

Anaerobic treatment includes both completely mixed type digesters and upward plug flow digesters. In this mode of anaerobic treatment, a complete system of mixing makes it possible that the microbes grow faster upon consumption of organic matter. The proper residence time of the mixture for the microbes makes it easier for the faster digestion of organic loads; therefore, the reactor would be effective. Agitation could be continuous or intermittent based on the requirement of digestion (Berni et al., 2014).

8.3.3.6 Plug Flow Digesters

The flow rate is tuned in such a way that the upward velocity of the fluid doesn't disturb the settlement of solids present. A uniform contact time is provided so that the digestion of organics takes place. Also, the plug flow type of digesters could make use of immobilized cells or enzymes for the digestion of matter.

8.3.3.7 Other Types of Reactors

Different types of reactors, particularly packed bed reactors and fluidized bed reactors (Mao et al., 2015), make use of immobilized cells. Also, hybrid reactors (Jabeen et al., 2015) are used nowadays for the treatment of both industrial and domestic wastewater. Anaerobic rotating drums, biological contacting drums, upflow anaerobic sludge blanket (Arimi 2015) and airlift reactors are also applied in the treatment of wastewater.

8.4 Electron Beam Wastewater Treatment

With growing water demands, reusing treated water is a good option to partially meet the shortage. As seen previously, many chemical, biological and physical processes are being implemented to treat water. They are all found to be reliable to great extent, though each has its drawbacks. Some conventional methods use a lot of chemicals to treat water and also produce toxic sludge. River osmosis (RO) and ultraviolet (UV) radiation are also commonly used to treat groundwater. But it is noted that these methods are not able to remove a few chemicals, specifically bromate, from contaminated water. So, the trend is now to use electron beam (e-beam) for treating wastewater. It produces high energy accelerating electrons, which are said to effectively take care of the toxicities existing even after other treatment methods have been used. So, e-beam now finds usage in treating water for potable purpose and in many industries, specifically the textile industry, to treat wastewater produced during dyeing and so on. It removes color by destroying double bonds. The different merits of e-beam technology are listed as follows:

i. Removes odors.
ii. Disinfects microorganisms by destroying their DNA.
iii. Destroys residual chemicals such as POPs, endocrine disruptors, pesticides and pharmaceutical residues.
iv. Allows recycling of the water for irrigation, impoundment and individual use.
v. Removes toxic chemicals.
vi. Reduces pathogens.
vii. Does not give toxic residuals.
viii. Produces no air emissions.

8.4.1 Principle of Electron Beam

The water to be irradiated is exposed to a stream of high-energy electrons. These electrons interact with water and eventually produce a large number of low-energy electrons. The electrons further interact with molecules to produce an excited state of molecules, ions and electrons. Water is highly dielectric, and the ions and electrons react with the impurities in water.

E-beam technology consists of high-energy electron accelerators. Depending on the thickness of water to be penetrated, the acceleration of electrons is controlled. High-energy electrons thus generated irradiate a thick layer of water. While penetrating some of the beam energy is lost, but it can be neglected as the electrons that penetrate are capable of ionizing water molecules and thus neutralizing the pollutants.

The e-beam when used for treating contaminated water completely decomposes the organic waste by releasing carbon dioxide from it. It also removes odor and also makes water colorless. The toxic contents in the water are also treated. The efficiency of the e-beam can be improved by optimizing irradiation.

8.4.2 Working of Electron Beam

Water containing organic-based chemicals, biologically resistant water and water with pathogens and other impurities, which generally cannot be treated by conventional methods, can be treated for reuse by electron beam. It is one of the most effective oxidation methods used for inducing oxidizing species through water radiolysis to remove unwanted H^+ and e^-_{aq} species.

$$H_2O \rightarrow \text{electron beam} \rightarrow OH\text{-}, e^-_{aq}, H^+, H_3O, H_2O_2, H_2$$

It is known that pure water under ionizing radiation generates powerful oxidizing and reducing species and molecular products. So, electron beam irradiation is a good means to inactivate microorganisms and remove organic impurities from water.

The e-beam is similar to a cathode ray tube. It has an electron gun with an accelerator. The accelerator is designed to control the speed of the electrons. It generally emits electrons at 25,000V. There can be more than one electron beam. A beam scanner is used to scan the electron beam, and the beam reflectors direct the electrons to irradiate wastewater. The high-density and accelerated electrons penetrate through water, forming electrically excited states and free radicals. This is the unique feature of e-beam, where oxidizing and reducing chemistry are simultaneously generated. The results obtained with e-beam are appreciated, as it effectively mineralizes the water and prevents reformation. Due to high-energy electrons, there is no sludge formation and lethal impurities are effectively removed. The accelerated beam gives rise to rapid chemical reactions and destroys all organics non-selectively.

The methods used for wastewater treatment in the textile industry mostly reduce the toxicity and degrade the complex organic compounds. But e-beam is now preferred, as it gives results without adding any chemicals. It uses radiation to degrade complex organic compounds. The literature shows, however, that the addition of hydrogen peroxide to wastewater before treatment gives better results as it improves degradation efficiency and radiation. This water is then irradiated by e-beam and then checked for color, pH value and dye concentration.

Electron beaming is in application under various industrial domains. Its working, applications and advantages are discussed. It is identified to have many benefits such as its operation without casing heat onto the material. Further, its energy can be precisely controlled to make it applicable for precision processes. Its control possibility will also allow us to set the amount of energy required for a particular process. Only an optimal energy setting can provide the highest efficiency. An increase or decrease from optimal value will cause a detrimental effect.

8.5 Conclusion

In this chapter, the necessity or requirement for fresh water is addressed. The different sources of wastewater and conventional methods for treating it are discussed. Also discussed is the principle of operation and various applications of electron beaming, which has been identified to have versatile industrial applications. It is also identified based on requirement or application, design and varying energy levels of electron beam. Electron beaming is economical due to advent of knowledge and technologies; however, certain specific applications prefer other methods, for example laser, water jet or plasma arc in cutting.

REFERENCES

Abbas SZ, Rafatullah M, Hossain K, Ismail N, Abdul Tajarudin HA, Khali HPS 2017, A review on mechanism and future perspectives of cadmium-resistant bacteria. *Int J Environ Sci Technol.* https://doi.org/10.1007/s13762-017-1400-5

Ahammad SZ, Graham DW, Dolfing J 2013, Wastewater treatment: Biological. *Encyclopedia of Environmental Management.* http://doi.org/10.1201/9781003045045-61

Anitha, T, Kumar, PS, Kumar, KS 2015, Binding of Zn(II) ions to chitosan – PVA blend in aqueous environment: Adsorption kinetics and equilibrium studies. *Environmental Progress and Sustainable Energy*, 34, 1, 15–22.

Arash, Jebrail 2016, *Waste water treatment process.* Construction Engineering, Visamäki Campus, Häme University of Applied Sciences.

Arimi MM, Knodel J, Kiprop A, Namango SS, Zhang Y, Geißen S-U 2015, Strategies for improvement of biohydrogen production from organic-rich wastewater: A review. *Biomass Bioenerg* 75, 101–118. https://doi.org/10.1016/j.biombioe.2015.02.011

Ayoub K, van Hullebusch ED, Cassir M, Bermond A 2010, Application of advanced oxidation processes for TNT removal: A review. *J. Hazard. Mater.* 178, 10–28.

Berni M etal.2014 Anaerobic digestion and biogas production: Combine efuent treatment with energy generation in UASB reactor as biorefinery annex. *Int J Chem Eng* 2014, 8. https://doi.org/10.1155/2014/543529

Capodaglio AG 2017, High-energy oxidation process: An efficient alternative for wastewater organic contaminants removal. *Clean Techn Environ Policy* 19, 1995–2006. https://doi.org/10.1007/s10098-017-1410-5

Danulat E, Muniz P, García-Alonso J, Yannicelli B 2002, Fist assessment of the highly contaminated harbour of Montevideo, Uruguay. *Mar Pollut Bull* 44, 554–565.

Engohang-Ndong J, Uribe RM, Gregory R, Gangoda M, Nickelsen MG, Loar P (2015) Effect of electron beam irradiation on bacterial and Ascaris ova loads and volatile organic compounds in municipal sewage sludge. *Radiat Phys Chem* 112, 6–12. https://doi.org/10.1016/j.radphyschem.2015.02.013

Forster-Carneiro T, Pérez M, Romero LI 2008, Anaerobic digestion of municipal solid wastes: Dry thermophilic performance. *Biores Technol* 99:8180–8184. https://doi.org/10.1016/j.biort ech.2008.03.021

Ghaedi M, Roosta M, Ghaedi AM 2018, Removal of methylene blue by silver nanoparticles loaded on activated carbon by an ultrasound-assisted device: Optimization by experimental design methodology. *Res Chem Intermed* 44:2929–2950. https://doi.org/10.1007/s11164-015-2285-x

Gupta V, Saleh TA 2011, Syntheses of carbon nanotube-metal oxides composites; Adsorption and Photo-degradation, *Carbon Nanotubes - From Research to Applications*. Stefano Bianco, IntechOpen, doi: 10.5772/18009

He S, Sun W, Wang J, Chen L, Zhang Y, Yu J 2016, Enhancement of biodegradability of real textile and dyeing wastewater by electron beam irradiation. *Radiat Phys Chem* 124, 203–207. https://doi.org/10.1016/j.radphyschem.2015.11.033

Hossain K, Ismail N 2015, Bioremediation and detoxification of pulp and paper mill efuent: A review. *Res J Environ Toxicol* 9(3), 113–134.

Hossain K, Ismail N, Rafatullah M, Quaik S, Nasir M, Maruthi AY, Shaik R 2015, Bioremediation of textile efuent with membrane bioreactor using the white-rot fungus Coriolus versicolor. *J Pure Appl Microbiol* 9(3), 1979–1986.

Hossain, K, Maruthi, YA, Das, NL et al. 2018, Irradiation of wastewater with electron beam is a key to sustainable smart/green cities: A review. *Appl Water Sci* 8, 6. https://doi.org/10.1007/s13201-018-0645-6

Hossain K, Quaik S, Ismail N, Rafatullah M, Maruthi A, Rameeja S 2016, Bioremediation of textile wastewater with membrane bioreactor using the white-rot fungus and reuse of wastewater. *Iran J Biotechnol*14.

Jabeen M, Yousaf S, Haider MR, Malik RN 2015, High-solids anaerobic co-digestion of food waste and rice husk at different organic loading rates. *Int Biodeterior Biodegrad* 102, 149–153. https://doi. org/10.1016/j.ibiod.2015.03.023

Juteau P, Tremblay D, Ould-Moulaye C, Bisaillon J, Beaudet R 2004, Swine waste treatment by self-heating aerobic thermophilic bioreactors. *Water Research* 38, 539–546.

Lim SJ, Kim TH, Kim JY, Shin IH, Kwak HS 2016, Enhanced treatment of swine wastewater by electron beam irradiation and ion-exchange biological reactor. *Sep Purif Technol* 157, 72–79. https://doi.org/10.1016/j.seppur.2015.11.023

Mao C, Feng Y, Wang X, Ren G 2015, Review on research achievements of biogas from anaerobic digestion. *Renew Sustain Energy Rev* 45, 540–555. https://doi.org/10.1016/j.rser.2015.02.032

Maruthi YA, Hossain K, Apta Chaitanya D 2012a, Incidence of dermatophytes school soils of Visakhapatnam: A case study. *Asian J Plant Sci Res* 2(4), 534–538.

Maruthi YA, Hossain K, Goswami A 2012b, Assessment of drinking water quality in some selected primary schools in Visakhapatnam. *Der Chemia* 3(5), 1071–1074.

Maruthi YA, Hossain K, Priya DH, Tejaswi B 2012c, Prevalence of keratinophilic fungi from sewage sludge at some wastewater out lets along the coast of Visakhapatnam: A case study. *Adv Appl Sci Res* 3(1), 605–610.

Maruthi YA, Hossain K, Apta Chaitanya D 2012e, Incidence of dermatophytes school soils of Visakhapatnam: A case study. *Asian J Plant Sci Res* 2(4), 534–538.

Maruthi YA, Hossain K, Sultana M 2012d, Optimization studies on pollution abatement: Biodegradation of nitroso dye efuents by two fungi (Phanerochaetechrysosporium & Trameteshirsuta) under static conditions. *Int J Pharm Sci* 4(5), 262–267.

Ozcimen, D, Ersoy-Mericboyu, A 2009, Removal of copper from aqueous solutions by adsorption onto chestnut shell and grapeseed activated carbons. *Journal of Hazardous Materials*, 168(2), 118–1125.

Putra, WP, Kamari, A, Yusoff, SNM, Ishak, CF, Mohamed, A, Hashim, N, Isa, IM 2014, Biosorption of Cu(II), Pb(II) and Zn (II) ions from aqueous solutions using selected waste materials: Adsorption and characterization studies. *Journal of Encapsulation and Adsorption Sciences* 4, 25–35.

Rajasulochana, P, Preethy V 2016, Comparison on efficiency of various techniques in treatment of waste and sewage water – a comprehensive review. *Resource-Efficient Technologies* 2(4), 175–184.

Rawat KP, Sarma KSS 2013, Enhanced biodegradation of wastewater with electron beam pretreatment. *Appl Radiat Isot* 74, 6–8. https://doi.org/10.1016/j.apradiso.2012.12.013

Satyanarayana CH, Ramakrishna Rao S, Hossain K 2010, Assessment of water quality along the coast of Andhra Pradesh. *Nat Environ Pollut Technol* 9(1), 19–23.

Singh M, Pant G, Hossain K, Bhatia AK 2016, Green remediation – tool for safe and sustainable environment: A review. *Appl Water Sci.* https://doi.org/10.1007/s13201-016-0461-9; Iran J Biotechnol 14(3), e124–e126. https://doi.org/10.15171/ijb.1216

Son Y-S 2017, Decomposition of VOCs and odorous compounds by radiolysis: A critical review. *Chem Eng J* 316, 609–622. https://doi. org/10.1016/j.cej.2017.01.063

Ta Wee Seow, Chi Kim Lim, Muhamad Hanif Md Nor, Mohd Fahmi Muhammad Mubarak, Chi Yong Lam, Adibah Yahya, Zaharah Ibrahim 2016, Review on wastewater treatment technologies. *International Journal of Applied Environmental Sciences* 11(1), 111–126.

Topare NS, Attar SJ, Manfe MM 2011, Sewage/wastewater treatment technologies: A review. *Sci Revs Chem Commun* 1, 18–24.

Trump JG, Merrill EW, Wright KA 1984, Disinfection of sewage wastewater and sludge by electron treatment. *Radiat Phys Chem* 24, 55–66.

Van Hoof SCJM, Hashim A, Kordes AJ 1999, The effect of ultrafiltration as pretreatment to reverse osmosis in wastewater reuse and seawater desalination applications. *Desalination* 124(1–3), 231–242.

Volesky B, Holan ZR 1995, Biosorption of heavy metals. *Biotechnology Progress* 11, 235–250.

WHO (World Health Organization) 2003, *Looking back, looking ahead: Five decades of challenges and achievements in environmental sanitation and health.* World Health Organisation.

Yonar T, Yonar GK, Kestioglu K, Azbar N 2005, Decolorization of textile effluent using homogeneous photochemical oxidation processes. *Color. Technol.* 121(5), 258–264.

Zia S, Graham DW, Dolfing J 2013, Waste water treatment: Biological. *Encyclopedia of Environmental Management.* http:/doi.org/ 10.1081/E-EEM-120046063

9

Insight into Advanced Oxidation Processes for Wastewater Treatment

Surbhi Sinha, Sonal Nigam and Muskan Syed

CONTENTS

9.1 Introduction

A safe and clean water supply is a requirement for human health, the environment and a flourishing economy. Currently, decline in the quality of water and constant reduction of fresh water are issues of concern globally, especially in developing nations like India and other sparsely populated regions. A number of industries release into water reservoirs a variety of toxic organic contaminants, such as synthetic dyes, heavy metals, pesticides and herbicides, that deteriorate water quality and make it deadly, causing severe damage to human health and aquatic environments. These contaminated waters are treated mainly by conventional methods such as adsorption, oxidation, flocculation, coagulation, chlorination and reverse osmosis (Litter and Quici, 2010; Shah, 2020, 2021). These conventional methods, however, are not always sufficient to adequately treat the contaminated water and achieve the purity grade vital for local or international regulations. In these circumstances, advanced oxidation processes (AOPs) prove to be an attractive option because of benefits like easy implementation, high efficiency, environmentally sound nature, less use of chemicals, minimal production of sludge and ability to oxidize a broad range of pollutants (Huang and Zhang, 2019).

9.2 Advanced Oxidation Processes

AOPs were first suggested in the 1980s for the treatment of potable water by using strong oxidizing agents (Glaze, 1987; Glaze et al., 1987; Shah, 2021). Subsequently, these processes were widely utilized for the treatment of various types of industrial wastewaters. Generally, AOPs are defined as the water treatment processes at ambient temperature and pressure that involve the production of oxidants. Typically, AOPs are based upon the *in situ* production of highly reactive free radicals that have the ability to inactivate microorganisms and to oxidize toxic, nondegradable organic pollutants completely or convert these to harmless products, delivering an absolute solution to wastewater remediation (Garrido-Cardenas et al., 2020). Mainly, AOPs include two stages in the degradation of contaminants: (1) *in situ* development of reactive oxygen species and (2) the interaction of these oxidative species with the particular contaminant (Cuerda-Correa et al., 2020). These AOPs utilize a wide variety of photocatalysts or the amalgamation of oxidizing agents like hydrogen peroxide, persulfate or peroxymonosulfate with UV or visible light radiation and metal catalysts (Hodges et al., 2018). Out of all of these, the processes that involve the production of hydroxyl radicals are

DOI: 10.1201/9781003165958-9

the most discussed for wastewater treatment. The hydroxyl radicals are efficient in the degradation or removal of complex organic contaminants because these are highly reactive electrophiles that interact not just rapidly, but also unselectively with organic pollutants (Bokare and Choi, 2014). Lately, other free radicals, namely sulfate radicals and superoxide radicals, have also garnered a lot of attention in water treatment, due to their high oxidation potential.

9.3 Classification of AOPs

9.3.1 Hydroxyl Radical–Based AOPs

Hydroxyl radical (OH˙) is considered an extremely reactive oxidant (redox potential – 2.8 V) in the treatment of wastewater (Tchobanoglous and Eddy, 2013). The OH• radicals are unselective in nature and react rapidly with a variety of toxic contaminants at a rate constant of 10^6–10^9 $M^{-1}s^{-1}$. These OH• radicals interact with the pollutants via the following mechanisms: (1) addition of radical, (2) hydrogen abstraction, (3) transfer of electrons and (4) radical combination (Buxton et al., 1988). Generally, their interaction with pollutants generates carbon-centered radicals. In the presence of oxygen (O_2), these carbon-centered radicals are converted to peroxyl radicals. These radicals react further, leading to the generation of H_2O_2 and O_2^-. The reactive species, H_2O_2 and O_2^-, cause the complete breakdown of the organic/inorganic pollutant by forming OH• radicals.

9.3.1.1 Ozone-Based Processes

Ozone (O_3) is a powerful oxidizing agent, often utilized to treat wastewater and degrade harmful organic pollutants, due to its high oxidizing potential of 2.07 V (Rekhate and Srivastava, 2020). O_3 interacts with the pollutants via two mechanisms: a direct mechanism that involves the selective electrophilic reaction of O_3 with functional groups, namely amines, double bonds and aromatic rings of organic contaminant, and an indirect mechanism that leads to the breakdown of O_3 in water, resulting in the formation of OH• radicals. This method causes an unselective and faster oxidation of the contaminant. The overall process involving the generation of hydroxyl radicals is stated as (Gottschalk et al., 2009):

$$3O_3 + H_2O \rightarrow 2OH\bullet + 4O_2 \qquad (9.1)$$

The generation of OH• radicals can be significantly enhanced in ozone-based AOPs by the addition of other oxidizing agents or radiation. For example, in a peroxone process, the generation of OH• radicals can be increased by the decomposition of hydrogen peroxide (H_2O_2) via the formation of hydroperoxide (HO_2^-).

$$H_2O_2 \rightarrow HO_2^- + H^+ \qquad (9.2)$$
$$HO_2^- + O_3 \rightarrow OH\bullet + O_2^- + O_2 \qquad (9.3)$$

In another process, O_3 can be combined with UV irradiation to generate H_2O_2, which then undergoes photolysis to produce OH• radicals.

$$O_3 + H_2O + h\nu \rightarrow H_2O_2 + O_2 \qquad (9.4)$$
$$H_2O_2 + h\nu \rightarrow 2OH\bullet \qquad (9.5)$$

9.4 UV-Based Processes

UV-light-based AOPs encompass techniques that focus on UV irradiation (especially UV-C) or combining UV irradiation with various radical enhancers (Miklos et al., 2018). The most commonly utilized light sources during AOPs include low- or medium-pressure mercury lamps along with mono- or polychromatic emission spectra, respectively. Lately, ultraviolet light emitting diode (LED) technology in AOP has been utilized for wastewater treatment due to numerous benefits such as exceptional peak emission wavelength, small size, versatile application and eradication of mercury (Song et al., 2016). The most common oxidant utilized to produce OH• radicals in the present method is H_2O_2. Usually, H_2O_2 in the presence of UV light produces two hydroxyl radicals.

$$H_2O_2 + h\nu \rightarrow 2OH\bullet \qquad (9.6)$$

Additionally, OH• radicals can also be generated via photolysis of water in the presence of UV radiation at a wavelength less than 242 nm.

$$H_2O + h\nu \rightarrow OH\bullet + H\bullet \qquad (9.7)$$

Titanium dioxide (TiO_2) is the most frequently used catalyst in the UV-light-based AOPs. In the presence of UV light, TiO_2 particulates get excited to generate positive holes in the valence band and the electrons at the conduction band. On reaction with OH^-, H_2O and O_2^-, these electrons and holes generate hydroxyl radicals (Tang, 2003). Additionally, persulfate, chlorine and nitrate have also been successfully applied in UV-based AOPs as oxidants to treat wastewater.

9.5 Fenton-Based Processes

The traditional Fenton method includes the amalgamation of peroxide (usually H_2O_2) with metal salt, mostly iron (Mohajeri et al., 2010). Generally, in the Fenton process, H_2O_2 interacts with Fe^{2+} or Fe^{3+} to produce OH• radicals and to some extent peroxyl radicals ($RO_2\bullet$). These extremely reactive radicals, under ambient working conditions, can result in complete degradation of the pollutant (Ali et al., 2014). The conventional Fenton process involves the sequence of following reactions (Umar et al., 2010):

$$Fe^{2+} + H_2O_2 \rightarrow Fe^{3+} + OH^- + HO\bullet \qquad (9.8)$$
$$Fe^{3+} + H_2O_2 \rightarrow Fe^{2+} + H^+ + HO\bullet_2 \qquad (9.9)$$
$$OH\bullet + H_2O_2 \rightarrow HO\bullet_2 + H_2O \qquad (9.10)$$
$$OH\bullet + Fe^{2+} \rightarrow Fe^{3+} + OH^- \qquad (9.11)$$
$$Fe^{3+} + HO\bullet_2 \rightarrow Fe^{2+} + O_2H^+ \qquad (9.12)$$
$$Fe^{2+} + HO\bullet_2 + H^+ \rightarrow Fe^{3+} + H_2O_2 \qquad (9.13)$$
$$2HO\bullet_2 \rightarrow H_2O_2 + O_2 \qquad (9.14)$$

Based on the traditional Fenton treatment system, three different processes have been identified. (a) A Fenton-like process, in which transition metal ions react with H_2O_2 to oxidize organic/inorganic substances by generating reactive OH• radical. (b) Photo-Fenton process where UV radiation, H_2O_2 and Fe^{2+} or Fe^{3+} interact to produce hydroxyl radicals. In this process, H_2O_2 acts as an oxidant, iron ions act as a catalyst and UV radiation enhances the production of hydroxyl radicals and also recycles Fe ions. (c) Electro-Fenton process that involves the *in situ* electrogeneration of Fenton's reagent.

9.6 Electrochemical Advanced Oxidation Processes

Electrochemical advanced oxidation method is considered a major advanced oxidation process for the degradation of toxic contaminants in an aqueous solution (Sirés et al., 2014). The prime motive of this process is to reduce the concentration of the toxic contaminants by generating oxidants right in the medium by means of electrochemistry. The process has numerous benefits like ease of operation, high efficiency, eco-friendly nature, safe, selective and low cost (Oturan, 2014; Feng et al., 2013). Among the current electrochemical advanced oxidation processes, anodic oxidation process (direct) and electro-Fenton process (indirect) are widely utilized for the breakdown of contaminants.

9.6.1 Anodic Oxidation

Anodic oxidation, also known as electrooxidation is a process extensively utilized for the treatment of industrial wastewater. Generally, the process consists of two electrodes, anode and cathode, linked to a power supply. When sufficient power input and electrolytes are given to the system, strong oxidizing agents, particularly hydroxyl radicals, are formed. The OH• radicals then cause the degradation of the contaminants (Anglada et al., 2009). The mechanism of degradation in anodic oxidation is represented as follows:

$$S + H_2O \rightarrow S(OH\bullet) + H\bullet + e^- \quad (9.15)$$

Where S represents the adsorption site on the surface of the electrode. Various electrodes such as Pt, boron-doped diamond (BDD), Ti/SnO_2 and PbO_2 with high oxygen voltage are commonly utilized in the present method (Fernandes et al., 2014; Chaplin et al., 2013). Also, the type of electrode and the conditions for electrolysis play a crucial role in wastewater treatment.

9.6.2 Electro-Fenton

Another major electrochemical AOP that has emerged as a powerful technology for the complete degradation of toxic contaminants is electro-Fenton. The process has been widely utilized by researchers for the treatment of various industrial effluents. In the process of electro-Fenton, the Fe^{2+} ions interact with hydrogen peroxide (H_2O_2) to electrochemically produce OH• radicals (Nidheesh and Gandhimathi, 2012). Moreover, electro-Fenton, in combination with other techniques, releases OH• radicals more efficiently, causing degradation of toxic contaminants (Teymori et al., 2020).

9.7 Photocatalysis

Another widely used AOP for the degradation of contaminants is heterogeneous photocatalysis. In this process, different types of photocatalysts (present in the solid phase) are utilized for the breakdown of pollutants (present in the liquid phase). The photocatalyst leads to the stimulation of photo reactions, creating electron hole pairs, which produces highly reactive hydroxyl radicals (Bello and Raman, 2017). A good photocatalyst must be biologically or chemically inert, photostable, innocuous and cheap (Chong et al., 2010). Various photocatalysts, namely ZnO, Si, TiO_2, ZnS, Fe_2O_3, $SrTiO_3$, CdS and WSe_2 have been widely employed to accelerate the reduction of numerous organic/inorganic pollutants in wastewater (Karunakaran and Anilkumar, 2007; Elkacmi and Bennajah, 2019). Among all, the most preferred photocatalyst for the oxidation of contaminants is titanium dioxide (TiO_2). When the TiO_2 is exposed to UV light, OH• radicals are produced on the surface of the TiO_2, resulting in the oxidization of the pollutant. The general mechanism of photocatalytic degradation using TiO_2 is represented as follows:

$$TiO_2 + h\nu \rightarrow e^-_{CB} + h^-_{VB} \quad (9.16)$$

$$h^-_{VB} + R \rightarrow CO_2 + H_2O \quad (9.17)$$

$$H_2O + h_{VB} \rightarrow OH\bullet + H^+ \quad (9.18)$$

$$OH\bullet + R \rightarrow CO_2 + H_2O \quad (9.19)$$

9.8 Sonochemical Advanced Oxidation Process

Sonochemical AOP is one of the latest techniques developed by researchers for the removal of a variety of contaminants from wastewater. The technique uses ultrasound radiation for degrading the pollutants. Basically, the process involves the application of ultrasound frequencies of 20 kHz and 1 MHz in water to generate OH• radicals. High-energy sound wave frequencies create microbubble cavitations in water, which produce OH• radicals capable of degrading pollutants in wastewater (Elkacmi, 2019).

$$H_2O + irradiation \rightarrow OH\bullet + H\bullet \quad (9.20)$$

9.8.1 Sulfate-Based AOPs

Recently, sulfate radical–based AOPs have also gained a huge amount of attention. These radicals also play a significant role in the breakdown of pollutants due to their high oxidation potential (2.5–3.1 V). Moreover, sulfate radicals interact proficiently with organic substances over a broad pH range varying from 2 to 8 (Zhang et al., 2015). Additionally, sulfate radicals interact more selectively and easily through electron transfer with organic compounds that comprise unsaturated bonds or aromatic Π electrons (Zhao et al., 2017; Liang and Bruell, 2008).

Generally, peroxymonosulfate (PMS) or persulfate (PS) are employed to produce sulfate radicals, owing to their high oxidation potential (1.82 and 2.1 V, respectively). Direct reaction with pollutant occurs at a very low rate, however, so they should be activated to produce sulfate radicals (Guerra-Rodríguez et al., 2018). PMS and PS activation can take place by different methods such as heat, UV, ultrasound, transition metal activation or alkaline activation (Wang and Wang, 2018; Wacławek et al., 2017). The general mechanism for the activation of peroxymonosulfate (HSO_5^-) or persulfate ($S_2O_8^{-2}$) for the production of sulfate radicals are represented as follows:

$$S_2O_8^{2-} + h\nu \rightarrow 2SO_4^- \tag{9.21}$$

$$HSO_5^- + h\nu \rightarrow + SO_4^- + OH\bullet \tag{9.22}$$

$$S_2O_8^{2-} + heat \rightarrow 2SO^{4-} \tag{9.23}$$

$$HSO_5^- + heat \rightarrow SO_4^- + OH\bullet \tag{9.24}$$

$$S_2O_8^{2-} + M^{n+} \rightarrow M^{n+1} + SO_4^{2-} + SO_4^- \tag{9.25}$$

$$HSO_5^- + M^{n+} \rightarrow M^{n+1} + OH\text{-} + SO_4^- \tag{9.26}$$

$$S_2O_8^{2-} + 2H_2O \rightarrow HO_2^- + 2SO_4^{2-} + 3H^+ \tag{9.27}$$

$$S_2O_8^{2-} + 2\,HO_2^- \rightarrow SO_4^- + SO_4^{2-} + H^+ + O_2^- \tag{9.28}$$

Additionally, hydroxyl radicals can also be generated from sulfate radicals via Equations 9.29 and 9.30:

$$SO_4^- + H_2O \rightarrow OH\bullet + SO_4^{2-} + H^+ \tag{9.29}$$

$$SO_4^- + OH\text{-} \rightarrow OH\bullet + SO_4^{2-} \tag{9.30}$$

Equation 9.30 demonstrates that the production of hydroxyl radicals can be enhanced under alkaline conditions.

9.9 Applications of AOPs

The utilization of AOPs for the removal of pollutants from wastewater is an interesting and simple method. Several AOPs are being applied for the removal of pollutants from wastewater. Table 9.1 lists the different types of AOPs used for the various pollutants' removal from wastewater.

9.10 Conclusion and Future Recommendations

Increasing urbanization and industrialization has caused quite a rise in the generation of contaminants in water. In this respect, advanced oxidation processes have appeared as a powerful and appropriate water treatment technique that efficiently and rapidly removes contaminants from wastewater. Advanced oxidation processes involve the production of strong oxidizing agents such as hydroxyl or sulfate radicals for the degradation of toxic recalcitrant pollutants, which are otherwise not oxidized by traditional physical-chemical methods. Due to the rapid and non-selective nature of advanced oxidation processes, these have been widely applied in the degradation of different pollutants present in water. Future study should focus on the (i) identification of the toxicological effects of degraded compounds on human health and the

TABLE 9.1

Different Types of AOPs Used for the Removal of Pollutants from Wastewater

Pollutant	AOPs	Result	Reference
Direct Blue 86	Ozone and ozone combined with UV	62% reduction of COD	Hassaan et al. (2017)
Real washing machine effluent	Electro-Fenton	99.5% reduction of COD	Ghanbari and Martínez-Huitle (2019)
Blue BR dye	Electrochemical AOP	100% color removal	Alcocer et al. (2018)
Ciprofloxacin	Combination of UV, H_2O_2, n(ZVI)	100% removal	Mondal et al. (2018)
Atrazine	Photoelectro-Fenton	99% removal	Komtchou et al. (2017)
Carmoisine dye	Fenton process	92.7% removal	Sohrabi et al. (2017)
Ethyl paraben	UV radiation-mediated persulfate activation	98.1% removal	Dhaka et al.(2018)
Thyroxin, Diatrizoate	UV radiation-mediated persulfate activation	100% removal	Duan et al.(2017)
Ciprofloxacin	UV/peroxymonosulfate	97% removal	Mahdi Ahmed and Chiron (2014)
Arsenic (III)	TiO_2-based AOP	100% removal	Bissen et al. (2001)
Chromium (VI)	TiO_2-based AOP	100% removal	Marinho et al.(2017)
Ofloxacin	Fenton process	100% removal	Carbajao et al. (2015)
Palm oil mill effluent	Photocatalysis with ZnO	95% color removal	Ng and Cheng (2016)
Landfill leachate	Electro-Fenton	96% removal of dissolved organic carbon	El Kateb et al. (2019)
Atachlor	Ozonation	75% TOC removal	Li et al. (2013)
N-diethyl metatolu-amide	Photocatalytic ozonation	60% mineralization	Mena et al. (2017)
Acetaminophen, caffeine	UV-based ozonation	99% removal	Sanchez Montez et al. (2020)
4-chloro-3-methyl phenol	Anodic oxidation	100% removal; 49% TOC removal	Song et al. (2010)

environment, (ii) establishment of control measures for releasing contaminants in the water, (iii) determination of scale-up factors and economic sustainability and (iv) examination of the detailed degradation mechanism of pollutants.

REFERENCES

Alcocer, S., Picos, A., Uribe, A.R., Pérez, T. and Peralta-Hernández, J.M., 2018. Comparative study for degradation of industrial dyes by electrochemical advanced oxidation processes with BDD anode in a laboratory stirred tank reactor. *Chemosphere*, 205, pp. 682–689.

Ali, M.E., Gad-Allah, T.A., Elmolla, E.S. and Badawy, M.I., 2014. Heterogeneous Fenton process using iron-containing waste (ICW) for methyl orange degradation: process performance and modeling. *Desalination and Water Treatment*, 52(22–24), pp. 4538–4546.

Anglada, A., Urtiaga, A. and Ortiz, I., 2009. Contributions of electrochemical oxidation to waste-water treatment: fundamentals and review of applications. *Journal of Chemical Technology & Biotechnology*, 84(12), pp. 1747–1755.

Bello, M.M. and Raman, A.A.A., 2017. Trend and current practices of palm oil mill effluent polishing: application of advanced oxidation processes and their future perspectives. *Journal of Environmental Management*, 198, pp. 170–182.

Bissen, M., Vieillard-Baron, M.M., Schindelin, A.J. and Frimmel, F.H., 2001. TiO$_2$-catalyzed photooxidation of arsenite to arsenate in aqueous samples. *Chemosphere*, 44(4), pp. 751–757.

Bokare, A.D. and Choi, W., 2014. Review of iron-free Fenton-like systems for activating H$_2$O$_2$ in advanced oxidation processes. *Journal of Hazardous Materials*, 275, pp. 121–135.

Buxton, G.V., Greenstock, C.L., Helman, W.P. and Ross, A.B., 1988. Critical review of rate constants for reactions of hydrated electrons, hydrogen atoms and hydroxyl radicals (·OH/·O– in aqueous solution. *Journal of Physical and Chemical Reference Data*, 17(2), pp. 513–886.

Carbajo, J.B., Petre, A.L., Rosal, R., Herrera, S., Letón, P., García-Calvo, E., Fernández-Alba, A.R. and Perdigón-Melón, J.A., 2015. Continuous ozonation treatment of ofloxacin: transformation products, water matrix effect and aquatic toxicity. *Journal of Hazardous Materials*, 292, pp. 34–43.

Chaplin, B.P., Hubler, D.K. and Farrell, J., 2013. Understanding anodic wear at boron doped diamond film electrodes. *Electrochimica Acta*, 89, pp. 122–131.

Chong, M.N., Jin, B., Chow, C.W. and Saint, C., 2010. Recent developments in photocatalytic water treatment technology: a review. *Water Research*, 44(10), pp. 2997–3027.

Cuerda-Correa, E.M., Alexandre-Franco, M.F. and Fernández-González, C., 2020. Advanced oxidation processes for the removal of antibiotics from water: an overview. *Water*, 12(1), p. 102.

Dhaka, S., Kumar, R., Lee, S.H., Kurade, M.B. and Jeon, B.H., 2018. Degradation of ethyl paraben in aqueous medium using advanced oxidation processes: efficiency evaluation of UV-C supported oxidants. *Journal of Cleaner Production*, 180, pp. 505–513.

Duan, X., He, X., Wang, D., Mezyk, S.P., Otto, S.C., Marfil-Vega, R., Mills, M.A. and Dionysiou, D.D., 2017. Decomposition of iodinated pharmaceuticals by UV-254 nm-assisted advanced oxidation processes. *Journal of Hazardous Materials*, 323, pp. 489–499.

El Kateb, M., Trellu, C., Darwich, A., Rivallin, M., Bechelany, M., Nagarajan, S., Lacour, S., Bellakhal, N., Lesage, G., Héran, M. and Cretin, M., 2019. Electrochemical advanced oxidation processes using novel electrode materials for mineralization and biodegradability enhancement of nanofiltration concentrate of landfill leachates. *Water Research*, 162, pp. 446–455.

Elkacmi, R. and Bennajah, M., 2019. Advanced oxidation technologies for the treatment and detoxification of olive mill wastewater: a general review. *Journal of Water Reuse and Desalination*, 9(4), pp. 463–505.

Feng, L., van Hullebusch, E.D., Rodrigo, M.A., Esposito, G. and Oturan, M.A., 2013. Removal of residual anti-inflammatory and analgesic pharmaceuticals from aqueous systems by electrochemical advanced oxidation processes: A review. *Chemical Engineering Journal*, 228, pp. 944–964.

Fernandes, A., Santos, D., Pacheco, M.J., Ciríaco, L. and Lopes, A., 2014. Nitrogen and organic load removal from sanitary landfill leachates by anodic oxidation at Ti/Pt/PbO$_2$, Ti/Pt/SnO$_2$-Sb2O$_4$ and Si/BDD. *Applied Catalysis B: Environmental*, 148, pp. 288–294.

Tchobanoglous, G., 2013. *Wastewater engineering: Treatment and resource recovery*, 4th Ed. Metcalf & Eddy.

Garrido-Cardenas, J.A., Esteban-García, B., Agüera, A., Sánchez-Pérez, J.A. and Manzano-Agugliaro, F., 2020. Wastewater treatment by advanced oxidation process and their worldwide research trends. *International Journal of Environmental Research and Public Health*, 17(1), p. 170.

Ghanbari, F. and Martínez-Huitle, C.A., 2019. Electrochemical advanced oxidation processes coupled with peroxymonosulfate for the treatment of real washing machine effluent: a comparative study. *Journal of Electroanalytical Chemistry*, 847, p. 113182.

Glaze, W.H., 1987. Drinking-water treatment with ozone. *Environmental Science & Technology*, 21(3), pp. 224–230.

Glaze, W.H., Kang, J.W. and Chapin, D.H., 1987. The chemistry of water treatment processes involving ozone, hydrogen peroxide and ultraviolet radiation. *Ozone: Science & Engineering*, 9(4), pp. 335–352. https://doi.org/10.1080/01919518708552148

Gottschalk, C., Libra, J.A. and Saupe, A., 2009. *Ozonation of water and waste water: A practical guide to understanding ozone and its applications*. John Wiley & Sons.

Guerra-Rodríguez, S., Rodríguez, E., Singh, D.N. and Rodríguez-Chueca, J., 2018. Assessment of sulfate radical-based advanced oxidation processes for water and wastewater treatment: a review. *Water*, 10(12), p. 1828.

Hassaan, M.A., El Nemr, A. and Madkour, F.F., 2017. Testing the advanced oxidation processes on the degradation of direct blue 86 dye in wastewater. *The Egyptian Journal of Aquatic Research*, 43(1), pp. 11–19.

Hodges, B.C., Cates, E.L. and Kim, J.H., 2018. Challenges and prospects of advanced oxidation water treatment processes using catalytic nanomaterials. *Nature Nanotechnology*, 13(8), pp. 642–650.

Huang, J. and Zhang, H., 2019. Mn-based catalysts for sulfate radical-based advanced oxidation processes: a review. *Environment International*, 133, p. 105141.

Litter, M.I. and Quici, N., 2010. Photochemical advanced oxidation processes for water and wastewater treatment. *Recent Patents on Engineering*, 4(3), pp. 217–241.

Karunakaran, C. and Anilkumar, P., 2007. Semiconductor-catalyzed solar photooxidation of iodide ion. *Journal of Molecular Catalysis A: Chemical*, *265*(1–2), pp. 153–158.

Komtchou, S., Dirany, A., Drogui, P., Robert, D. and Lafrance, P., 2017. Removal of atrazine and its by-products from water using electrochemical advanced oxidation processes. *Water Research*, *125*, pp. 91–103.

Li, H., Huang, Y. and Cui, S., 2013. Removal of alachlor from water by catalyzed ozonation on Cu/Al 2 O 3 honeycomb. *Chemistry Central Journal*, *7*(1), pp. 1–6.

Liang, C. and Bruell, C.J., 2008. Thermally activated persulfate oxidation of trichloroethylene: experimental investigation of reaction orders. *Industrial & Engineering Chemistry Research*, *47*(9), pp. 2912–2918.

Mahdi-Ahmed, M. and Chiron, S., 2014. Ciprofloxacin oxidation by UV-C activated peroxymonosulfate in wastewater. *Journal of Hazardous Materials*, *265*, pp. 41–46.

Marinho, B.A., Djellabi, R., Cristóvão, R.O., Loureiro, J.M., Boaventura, R.A., Dias, M.M., Lopes, J.C.B. and Vilar, V.J., 2017. Intensification of heterogeneous TiO2 photocatalysis using an innovative micro – meso-structured-reactor for Cr (VI) reduction under simulated solar light. *Chemical Engineering Journal*, *318*, pp. 76–88.

Mena, E., Rey, A., Rodríguez, E.M. and Beltrán, F.J., 2017. Reaction mechanism and kinetics of DEET visible light assisted photocatalytic ozonation with WO3 catalyst. *Applied Catalysis B: Environmental*, *202*, pp. 460–472.

Miklos, D.B., Remy, C., Jekel, M., Linden, K.G., Drewes, J.E. and Hübner, U., 2018. Evaluation of advanced oxidation processes for water and wastewater treatment – a critical review. *Water Research*, *139*, pp. 118–131.

Mohajeri, S., Aziz, H.A., Isa, M.H., Bashir, M.J., Mohajeri, L. and Adlan, M.N., 2010. Influence of Fenton reagent oxidation on mineralization and decolorization of municipal landfill leachate. *Journal of Environmental Science and Health Part A*, *45*(6), pp. 692–698.

Mondal, S.K., Saha, A.K. and Sinha, A., 2018. Removal of ciprofloxacin using modified advanced oxidation processes: kinetics, pathways and process optimization. *Journal of Cleaner Production*, *171*, pp. 1203–1214.

Ng, K.H. and Cheng, C.K., 2016. Photo-polishing of POME into CH4-lean biogas over the UV-responsive ZnO photocatalyst. *Chemical Engineering Journal*, *300*, pp. 127–138.

Nidheesh, P.V. and Gandhimathi, R., 2012. Trends in electro-Fenton process for water and wastewater treatment: an overview. *Desalination*, *299*, pp. 1–15.

Oturan, M.A., 2014. Electrochemical advanced oxidation technologies for removal of organic pollutants from water. In *Environmental science and pollution research* (Vol. 21, Issue 14, pp. 8333–8335). Springer Verlag. https://doi.org/10.1007/s11356-014-2841-8

Rekhate, C.V. and Srivastava, J.K., 2020. Recent advances in ozone-based advanced oxidation processes for treatment of wastewater-A review. *Chemical Engineering Journal Advances*, p. 100031.

Sánchez-Montes, I., García, I.S., Ibañez, G.R., Aquino, J.M., Polo-López, M.I., Malato, S. and Oller, I., 2020. UVC-based advanced oxidation processes for simultaneous removal of microcontaminants and pathogens from simulated municipal wastewater at pilot plant scale. *Environmental Science: Water Research & Technology*, *6*(9), pp. 2553–2566.

Shah, M.P., 2021. *Removal of emerging contaminants through microbial processes*. Springer.

Shah, M.P., 2020. *Advanced oxidation processes for effluent treatment plants*. Elsevier.

Shah, M.P., 2020. *Microbial Bioremediation and Biodegradation*. Springer.

Sirés, I., Brillas, E., Oturan, M.A., Rodrigo, M.A. and Panizza, M., 2014. Electrochemical advanced oxidation processes: today and tomorrow: a review. *Environmental Science and Pollution Research*, *21*(14), pp. 8336–8367.

Sohrabi, M.R., Khavaran, A., Shariati, S. and Shariati, S., 2017. Removal of carmoisine edible dye by Fenton and photo Fenton processes using Taguchi orthogonal array design. *Arabian Journal of Chemistry*, *10*, pp. S3523–S3531.

Song, K., Mohseni, M. and Taghipour, F., 2016. Application of ultraviolet light-emitting diodes (UV-LEDs) for water disinfection: A review. *Water Research*, *94*, pp. 341–349.

Song, S., Zhan, L., He, Z., Lin, L., Tu, J., Zhang, Z., Chen, J. and Xu, L., 2010. Mechanism of the anodic oxidation of 4-chloro-3-methyl phenol in aqueous solution using Ti/SnO2 – Sb/PbO2 electrodes. *Journal of Hazardous Materials*, *175*(1–3), pp. 614–621.

Solarchem Environmental Systems, 1994. *The UV/oxidation handbook*. Solarchem Environmental Systems.

Tang, W.Z., 2003. *Physicochemical treatment of hazardous wastes*. CRC Press.

Teymori, M., Khorsandi, H., Aghapour, A.A., Jafari, S.J. and Maleki, R., 2020. Electro-Fenton method for the removal of Malachite Green: effect of operational parameters. *Applied Water Science*, *10*(1), pp. 1–14.

Umar, M., Aziz, H.A. and Yusoff, M.S., 2010. Trends in the use of Fenton, electro-Fenton and photo-Fenton for the treatment of landfill leachate. *Waste Management*, *30*(11), pp. 2113–2121.

Wacławek, S., Lutze, H.V., Grübel, K., Padil, V.V., Černík, M. and Dionysiou, D.D., 2017. Chemistry of persulfates in water and wastewater treatment: a review. *Chemical Engineering Journal*, *330*, pp. 44–62.

Wang, J. and Wang, S., 2018. Activation of persulfate (PS) and peroxymonosulfate (PMS) and application for the degradation of emerging contaminants. *Chemical Engineering Journal*, *334*, pp. 1502–1517.

Zhang, B.T., Zhang, Y., Teng, Y. and Fan, M., 2015. Sulfate radical and its application in decontamination technologies. *Critical Reviews in Environmental Science and Technology*, *45*(16), pp. 1756–1800.

Zhao, Q., Mao, Q., Zhou, Y., Wei, J., Liu, X., Yang, J., Luo, L., Zhang, J., Chen, H., Chen, H. and Tang, L., 2017. Metal-free carbon materials-catalyzed sulfate radical-based advanced oxidation processes: a review on heterogeneous catalysts and applications. *Chemosphere*, *189*, pp. 224–238.

10

Implementation of Progressive and Advanced Oxidation Techniques for the Efficient Treatment of Cytotoxic Effluents

Ishani Joardar and Subhasish Dutta

CONTENTS

10.1 Introduction

Water is an essential component for all living beings. Although it constitutes almost 70% of the earth's geography, not every resource is suitable for glaciers, salty water, rivers, seas and ocean and so on. Besides, due to industrialization, the amount of detrimental emissions in such water bodies is skyrocketing. Henceforth, there will be an increased demand for these wastes' incineration using efficient advanced oxidation processes (AOPs). First discovered in 1987 by Glaze et al. (1987), they defined it as the series of water treatment methods that aid in the decontamination of wastewater with hydroxyl radicals at average temperatures and pressure [1]. The mechanism of AOPs depends entirely on the *in situ* generation of these hydroxyl radicals, which readily react with most recalcitrant organic compounds. There are two kinds of the initial attack of this oxidant. It abstracts a hydrogen atom from water that involves alcohols and alkanes or it might attach itself to the pollutant that occurs in aromatic compound olefins. Thus, it works by either hydroxylation or dehydrogenation. Suppose the attack occurs in the presence of oxygen. In that case, it initiates an oxidative reaction, which finally leads to the mineralization of this organic compound into water, carbon dioxide and inorganic ions [2]. This generation of hydroxyl radicals or the types of oxidation processes is classified in Figure 10.1. The types of AOPs or advanced oxidation techniques (AOTs) that have been discussed are ozonation at increased pH conditions, UV rays, ozone along with UV radiation (O_3/UV), ozone and hydrogen peroxide, or peroxonation (O_3/H_2O_2) and visible solar irradiation. Photochemical processes like ozone combined with H_2O_2 and UV, photo-Fenton techniques, are general conventional methods for removing toxic components from the water. Other progressive and evolving techniques that have been proven to be much more cost-effective and efficient include the usage of ultrasound (sonolysis), microwave radiation, zero-valent iron and pulsed plasma technologies. Most of them include a combination of conventional processes like photochemical, non-photochemical, sonochemical, electrochemical, physical and photocatalytic methods (transition metal ions). [3]. Fenton's integrated technologies are the oldest known method for treating various persistent organic pollutants and effluents from contaminated waters. Other efficient ways eventually came into view, namely photo-assisted Fenton technologies and solar photo-Fenton methods, among others [4].

10.2 Types of AOP

10.2.1 Non-Photochemical AOP

10.2.1.1 Fenton's Reagent Integrated Technologies

Discovered by H.J.H. Fenton in the late 1800s for the oxidation of maleic acid, Fenton's reagent is now used in the oxidation process of organic compounds for wastewater treatment technology. Fenton reported this reaction, but later this method was proposed by Haber and Weiss in 1934 [5]. It is a mixture of hydrogen peroxide (H_2O_2) and ferrous iron (Fe (II)) as a catalyst under acidic conditions of optimum pH 2.8–3. The reaction pathway is given as follows [6]:

$$Fe^{2+} + H_2O_2 \rightarrow Fe^{3+} + OH\text{-} + \bullet OH \tag{10.1}$$
$$OH\bullet + Fe^{2+} \rightarrow OH^- + Fe^{3+} \tag{10.2}$$

Equations 10.1 and 10.2 are chain initiation and chain termination reactions, respectively. In Equations 10.3 to 10.7, we will see how the ferric ions catalyze hydrogen peroxide and decompose it to form oxygen and water [7].

$$Fe^{3+} + H_2O_2 \leftarrow\rightarrow Fe - OOH^{2+} + H^+ \tag{10.3}$$

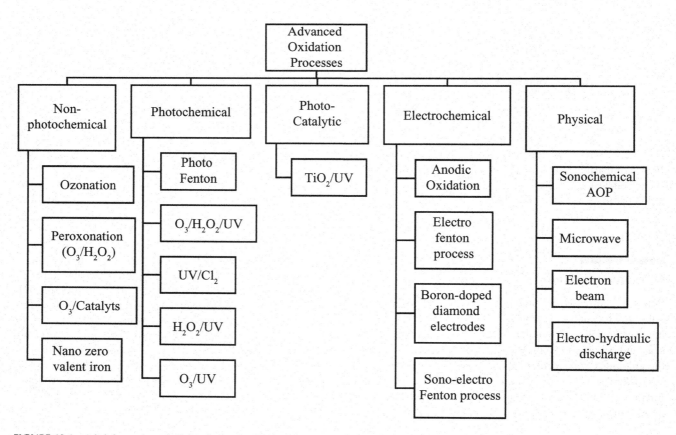

FIGURE 10.1 A brief overview of all the advanced oxidation processes discussed in this chapter.

$$Fe^- OOH^{2+} \rightarrow HO_2\bullet + Fe^{2+} \tag{10.4}$$

$$Fe^{2+} + HO_2\bullet \rightarrow Fe^{3+} + HO \tag{10.5}$$

$$Fe^{3+} + HO_2\bullet \rightarrow Fe^{2+} + O_2 + H^+ \tag{10.6}$$

$$OH\bullet + H_2O_2 \rightarrow H_2O + HO_2\bullet \tag{10.7}$$

$$HO\bullet + RH \rightarrow H2O + R\bullet \tag{10.8}$$

$$R\bullet + Fe^{3+} \rightarrow R^+ + Fe^{2+} \tag{10.9}$$

In alkaline conditions, iron precipitation occurs, which further lowers the concentration of Fe^{3+} in the solution. Moreover, UV radiation and adding H_2O_2 in excess will promote the reduction of Fe^{3+} to Fe^{2+} [8]. Fenton's reagent is simple and easy to operate and implement across industries; it is a manageable, cheap and inexpensive alternative. Moreover, the residual iron can be magnetically separated [9]. Some of the notable disadvantages include that the bulk OH cannot completely replenish ferric carboxylic acid complexes. Also, the deposited iron sludge residues must be removed at the end of the treatment. Storing and transporting H_2O_2 also involves high risks and cost [5].

10.2.1.1.1 Ozonation

With the increase in pH, ozone decomposes more readily in water so that the half-life of ozone decreases to less than a minute at a pH of 10. Thus, if the water to be treated has higher alkalinity or high pH, then ozonation might be favorable. Degradation of organic components takes place because of molecular ozone reactions with hydroxyl radicals. Ozone and hydroxide ions react to form O_2^- and $HO_2\bullet$, which further reacts with the superoxide anion radical to form O^{3-} and gives OH [10]. The reaction is given as follows:

$$3O^{3-} + OH- + H^+ \rightarrow 2OH + 4O_2 \tag{10.10}$$

The energy requirements for the synthesis of ozone varies from 22 kWh/kg O_3 to 34 kWh/kg O_3 using air as a feed gas and from pure oxygen (excluding oxygen costs) is 12 to 18 kWh per kg of O_3 [2].

10.2.1.2 Peroxonation Process (O_3/H_2O_2)

This process involves the formation of hydroxyl radicals by the coupling reaction of H_2O_2 and ozone (O_3). The reactions are shown below [2].

$$H_2O_2 \rightarrow HO_2^- + H^+ \tag{10.11}$$

$$HO_2^- + O_3 \rightarrow HO_2 + O_3^- \tag{10.12}$$

Optimum ozone measures are 1–20 mg/l at an optimum pH of 7.7. The optimum molar ratio of H_2O_2 to O_3 is considered to be 0.5 mol/mol and should be kept constant. The reaction takes place readily between ozone and H_2O_2 in its ionized form, leading to OH radicals' formation, as shown next [11].

$$O_3 + HO_2^- \rightarrow O_2 + OH + O^{2-} \tag{10.13}$$

It was also observed that radicals produced by the OH and HO_2 reaction further decompose H_2O_2 under optimum conditions. This process is used mainly to remove micropollutants

and harmful compounds like pesticides (phenylurea, triazines, organochlorines) from wastewaters. Paillard et al. (1988) conducted a study in the Seine River for atrazine eradication, which proved that the peroxone process was more efficient than ozonation alone [12]. Optimum H_2O_2/O_3 ratio was said to be around 0.3 to 0.45. Wastewaters said to have a pesticide concentration of 0.1 µg L^{-1} were reduced by 80% to 90% by this method. One of the most significant advantages is that it is easy to handle and has prophylactic activity, making it suitable for drinking water, groundwater and more [13].

10.2.1.3 O3/Catalysts AOP

Catalytic oxidation can be of two types, homogeneous and heterogeneous, dependent on the solubility of the catalyst in water. Metal oxides like Fe_2O_3, MnO_2, Ru/CeO_2, Al_2O_3-Me and MnO_2 and metal ions like Fe^{3+}, Fe^{2+} and Mn^{2+} have been used for the efficient degradation of target effluents. In a study [14] in which Ru/CeO_2 was taken as a catalyst for the oxidation process of succinic acid, the chemical oxygen demand (COD) and total organic carbon (TOC) were efficiently reduced with this process's help, which failed to show significant results in the case of ozonation at elevated pH.

Moreover, a comparative study was done to examine the efficiency of O_3/TiO_2 over simple ozonation and O_3/H_2O_2 coupled with oxalic acid as a model [15]. Results proved that the O_3/TiO_2 model reduced the overall TOC better than any other process. Other catalysts include granulated activated carbon, among others, for the breakdown of biorefractory compounds; however, more research needs to be done regarding this.

10.2.1.4 Nano Zero-Valent Iron (nZVI)-Based AOP

Nanotechnology is emerging in today's world, and the production and development of nanomaterials are opening up an array of research routes. Nanoscale zero-valent iron is one such nanomaterial that is highly effective in bioremediation. It is inexpensive, possesses adequate absorption capacity, is a potent reducing agent and is highly reactive. Bare nZVI is 10–1000 times more readily reactive than granular zero-valent iron. It is covered with a thin oxide layer when it comes in contact with air or water and is mixed with zero-valent iron/ ferrous or ferric ions in water [16]. It is generally produced either physically or chemically by reducing the bulk iron size to the nanoscale (also called top-down approach) or constructing nano iron from atoms that are made from iron-containing molecules of dissolved iron salts (by bottom-up approach) [17]. Moreover, this method is more efficient than Fenton's reagent and can overcome its limitations like radical decomposition, low pH environment and so on. Due to its good reducing properties, it can transform harmful compounds into lesser toxic components; however, complete mineralization is not possible in an oxygen-free environment. It is used to remove halogenated organic compounds such as perfluorooctanoic acid, perfluorooctane sulfonate, perfluorodecanoic acid, with Mg-amino clay coating [18]. Other compounds include halonitromethanes, chloramphenicol, lindane, florfenicol, chlorinated hydrocarbons, iopromide [19, 20, 21] and more. Bautitz et al. (2012) implemented a heterogeneous Fenton system for the mineralization of diazepam, a common drug found in pharmaceutical wastewater [22]. Li et al. (2007) proposed nZVI coated with silica and polydopamine to eliminate anthracene and phenanthrene [23]. It has also been used to remove heavy metal residues like arsenic, mercury, nickel, lead, copper, selenium, uranium and zinc.

Thus, nZVI has many benefits, mainly due to its smaller size, which makes it highly advantageous. More research needs to be done to unfurl other biological effects and increase its applications.

10.2.2 Photochemical AOPs

10.2.2.1 Photo-Fenton Process

The formation of OH radicals under low pH conditions by coupling the photolysis of hydrogen peroxide and Fenton's reagent is called photo-Fenton reaction. Low pH conditions (pH = 3) lead to the formation of $Fe(OH)^{2+}$ ions, which, when exposed to UV radiation, leads to Fe^{2+} and OH decomposition. [2]

$$Fe^{3+} + H_2O \rightarrow Fe(OH)^{2+} + H^+ \quad (10.14)$$

$$Fe(OH)^{2+} \leftrightarrow Fe^{3+} + OH^- \quad (10.15)$$

$$FeOH^{2+} \rightarrow Fe^{2+} + OH \quad (10.16)$$

Hydrogen peroxide has a lower extinction coefficient, and its absorption spectra only limits to 300 nm, unlike ferric ion which stretches to near UV/visible spectrum and has a large extinction coefficient. Photo-Fenton processes have been used to remove hydrocarbons from saline wastewaters, diesel contamination [23] and dyes from an olive mill, which each showed more than 90% removal [24].

10.2.2.2 O3/H₂O₂/UV Process

Hydrogen peroxide can be added to the mixture of O_3/UV to help in faster decomposition of ozone and accelerate the hydroxyl ion generation from the overall reactions. In case the contaminants weakly absorb UV radiation, adding hydrogen peroxide will be economical as compared to photooxidation [2].

10.2.2.3 Chlorine Photolysis (UV/Cl₂) Process

In this method, UV-activated chlorine reacts to form $\bullet Cl^{2-}$, $\bullet Cl$ and $\bullet OH$ radicals, which further helps to oxidize pollutants from the wastewater. It has shown similar efficiency compared to other AOP methods like UV/hydrogen peroxide. for the disruption of the same [25]. Since chlorine radicals are more powerful oxidants, they react more readily than hydroxyl radicals do. This method is advantageous for wastewater with lower pH values [26]. Other two primary oxidants that are used in this process are chlorine dioxide and hypochlorite. However, one its limitations is that it generates disinfection by-products (DBPs) like perchlorate, chlorite, chlorate and bromate for a reaction time of less than one minute [26]. Many pilot-scale studies have been done to date with only two full-scale implementations in a drinking water supply system in Canada and Los Angeles.

10.2.2.4 H₂O₂/UV Process

Ultraviolet radiation at a 200–280 nm wavelength causes a photolytic cleavage of two hydroxyl radicals. The chemical reactions describing the same are as follows [27]:

$$\text{Rate limiting reaction: } H_2O_2 + h\nu \rightarrow 2\ HO\bullet \quad (10.17)$$

Reaction 10.17 is called a rate-limiting reaction because, unlike other reactions, this reaction's rate is less. It has been proved that taking a higher initial concentration of hydrogen peroxide will yield higher hydroxyl radicals resulting in more of the desired compound in the end (because H_2O_2 has a weak molar absorption coefficient in the UV region). However, a higher than optimum level of H_2O_2 will result in the formation of $HO_2\bullet$, as shown in Equation 10.19.

$$H_2O_2 + HO\bullet \rightarrow HO_2\bullet + H_2O \quad (10.18)$$

$$H_2O_2 + HO_2\bullet \rightarrow HO\bullet + H_2O + O_2 \quad (10.19)$$

$$2\ HO\bullet \rightarrow H_2O_2 \quad (10.20)$$

$$2\ HO_2\bullet \rightarrow H_2O_2 + O_2 \quad (10.21)$$

$$HO\bullet + HO_2\bullet \rightarrow H_2O + O_2 \quad (10.22)$$

The reaction occurs more readily at a pH greater than 10, in an alkaline medium because HO_2 anions can absorb strong UV radiation and further produce free radicals. One of its limitations is that hydrogen peroxide has a weak absorption coefficient in the UV spectrum [2]. This process can efficiently decontaminate organic substances from the water like benzene, ethylene derivatives like tetrachloroethylene, trichloroethylene and so on. It is used to remove pesticides, carbamates, pyrimidine derivatives, triazines, organophosphate derivatives, chlorophenoxy acids, aniline derivatives, ureas and more [28].

10.2.2.5 O3/UV Process

In this process, UV radiation causes a cleavage of the ozone in the aqueous solution. Atomic oxygen and water react to form hydrogen peroxide. Further, highly reactive OH radicals are formed by the photolysis of ozone in water. The reactions describing the same are provided next [29]:

Initiation step

$$O_3 + H_2O + h\nu \rightarrow 2 \bullet OH + O_2 \quad (10.23)$$

Propagation steps

$$O_3 + \bullet OH \rightarrow HO_2\bullet + O_2 \quad (10.24)$$

$$O_3 + HO_2\bullet \rightarrow \bullet OH + 2O_2 \quad (10.25)$$

Termination reactions

$$\bullet OH + HO_2\bullet \rightarrow H_2O + O_2 \quad (10.26)$$

$$\bullet OH \rightarrow H_2O$$

Ozone is said to absorb UV radiation from a wavelength of 200 to 360 nm in an aqueous solution with 253.7 nm being the maximum wavelength. The photolysis process is more efficient than hydrogen peroxide under low-pressure Hg lamp at a wavelength of 253.7 nm because the εmax value of O_3 at 18.6 $mol^{-1}cm^{-1}$ is much higher than hydrogen peroxide [14]. It eliminates volatile chlorinated organic compounds like CCl_4, $CHCl_3$, trichloroethylene (TCE), tetrachloroethylene and 1,1,2-trichloroethane (TCA). Moreover, it was found that direct ozonation promoted the oxidation of $CHCl_3$, whereas hydroxyl radicals helped in the oxidation of $CHCl_3$, TCE and TCA compounds. CCl_4 could not be broken down by either of them [30]. Other effluents degraded by this process include TCE (with the help of a hybrid pilot reactor coupled with air pollution control system), surfactants, antibiotics, nitrobenzene derivatives, medical components, and endocrine agitators or disrupters (namely bisphenol A (BPA) and 17β-estradiol in an aqueous medium)[31].

10.2.3 Photocatalytic AOPS

10.2.3.1 Titanium Dioxide/UV Processes

In this process, hydroxyl radicals are generated with a photo-excited semiconductor that leads to UV light absorption. This kind of AOP technology is based on heterogeneous photocatalysis, preferably with a semiconducting catalyst like TiO_2. It was discovered by Fujishima and Honda (1972) for splitting water into H_2 and O_2 in a photoelectrochemical solar-powered cell. It works by photoexcitation of the semiconductor with UV radiation, which forms electron-accepting and -donating sites to enable oxidation-reduction reactions. The reactions are given next [32]. In Equation 10.27, the band electrons reacted with the molecular oxygen for yielding superoxide anions as shown in Equation 10.28, whereas the band holes reacted with water molecules for producing hydroxyl radicals (Equation 10.29).

$$TiO_2 + h\nu \rightarrow e_{cb} + h^+ \qquad (10.27)$$

$$e_{cb} + O_2 \rightarrow O_2^- \qquad (10.28)$$

$$h^+ + H_2O \rightarrow H^+ + HO\bullet \qquad (10.29)$$

This method is relatively cheap, available in many crystalline forms, abundant and capable of oxidizing many organic compounds. Optimum temperatures of 20–80 °C have to be maintained, as an elevation of temperatures beyond 80 °C might cause a decrease in the reduction rates. Moreover, there is an increase of reaction rate at a lower or an acidic pH for pollutants that are slightly acidic in nature and vice versa [3]. This method's application includes decontamination of olive mill wastewaters, phenols, azoreactive dyes, cyanide and formate.

10.2.3.2 Electrochemical AOPs

This is an efficient and practical technique for eliminating recalcitrants from wastewater in more environmentally friendly and straightforward ways. Electrochemical AOPs mineralize the organic species completely without their further recurrence, don't need any chemical reagent (unless coupled with other processes) and are economical and energy efficient. The hydroxyl radical is electrochemically generated either directly (anodic oxidation) or indirectly (*in situ* generated methods like electrochemical-Fenton's reagent) [14]. Apart from anodic oxidation and EF processes, boron-doped diamond electrodes will also be reviewed, which have been preferred over other electrodes due to their effectiveness, lower costs and high stability under the diamond's anodic polarization [33].

10.2.3.3 Anodic Oxidation

Anodic oxidation is the most abundantly used electrochemical method for the treatment of effluents from wastewater. In this method, a high O_2-overvoltage anode like PbO_2, SnO_2, RuO_2 or boron-doped diamond is used to generate hydroxyl radicals. However, metallic anodes do not promote the generation of a considerable number of oxidants, so metal oxides and dimensionally stable anodes (DSA) are used. Anodic oxidation facilitates hypochlorous acid formation from aqueous chloride ions, which is a relatively good oxidant [13].

10.2.3.4 BDD Electrodes

Boron-doped diamond (BDD) is the most efficient electrode material available, compared to the other conventional anodes like Pt, Au, glassy carbon, PbO_2, SnO_2, IrO_2 and other metallic anodes studied. BDD is electrochemically stable, has high oxidizing power and has higher O_2 overvoltage than other anodes do [34]. The reaction looks like the following:

$$M(\bullet OH) + R \rightarrow M + mCO_2 + nH_2O + pX$$

(M = the anode material; M($\bullet$OH) = heterogeneous radicals absorbed on the anode; R = organic compound; X = inorganic compounds). The reaction produces a large number of OH radicals on the anodes' surface, and these BDD hydroxyl radicals are said to be more highly reactive than other electrodes are. The process can also produce many heterogeneous hydroxyl radicals from the wastewater and other oxidant radicals like C_2O_{62}, P_2O_{82}, HClO [35]. Palma-Goyes et al. (2010) conducted a study to remove hexamethyl pararosaniline chloride or crystal violet using BDD anodes. They observed that 100% of contaminants were completely removed within 35 minutes, including all the by-products [36]. BDD anodes have also been used to treat and remove chlorophenoxy herbicides, parathion, amitrole, propham, methamidophos, atrazine, pharmaceutical effluents, phenol derivatives, dyes, surfactants, landfill leachates, tannery wastewater and more.

10.2.3.5 Electro-Fenton Processes

In this method, the hydroxyl radicals are generated indirectly via the Fenton's reagent. There are two different kinds of electro-Fenton techniques. Primarily, Fenton's reagents are applied from the outside with the anode having a high catalytic activity. The second one uses sacrificial cast iron anodes for the supply of Fe(II), and the hydrogen peroxide is added from outside [37]. Although hydrogen peroxide's oxidizing powder is weaker than any other chemical oxidant, the Fe^{2+} ions from Fenton's reagent help to enhance its oxidizing power.

Fe^{3+} ions are electrochemically reduced at the cathode for the generation of Fe^{2+} ions and form OH radicals that react with the wastewater's organic contaminants. Mercury was used in hydrogen peroxide; however, it has been discarded due to mercury's potential toxicity. Carbon cathodes have less toxicity, possess a low catalytic potential for the decomposition of hydrogen peroxide, are stable, have good conductivity and are energy efficient [37,38,13]. To date, electro-Fenton technologies have been used in treating different kinds of contaminants from wastewater, whether hazardous recalcitrant, pesticides, synthetic dyes like azo dyes or textile effluents, industrial effluents or harmful emissions from hospital wastes [37].

This method has been used to treat pentachlorophenol, which is the most cytotoxic component among all the chlorophenols found in wastewater. Z. Heidari et al. (2015) has discussed how electro-Fenton techniques are used to remove this potential carcinogen from wastewater. Furthermore, this method has also been used to treat landfill leachate, degrade phenols like p-nitrophenol, remove dyes from landfill leachate, remove cyanides, treat formaldehyde-polluted aquatic water and remove insecticides [39, 40].

10.2.3.6 Sonoelectro-Fenton AOP

Sonolysis, when coupled to the electro-Fenton process, is called sonoelectro-Fenton. The wastewaters' effluents are then destroyed with hydroxyl radicals' combined action generated from Fenton's reaction and other methods with sonolysis [13, 40]. To date, sonoelectro-Fenton processes have been used to remove various synthetic dyes, industrial wastewaters like wine, alkydic resins, pharmaceuticals, and pulp mills agrochemicals, crude oil and more. A three-electrode divided cell is used similar to the one used in the case of electro-Fenton, and the ceramic that produces the ultrasonic waves is directly placed under the base of the cell. The solutions that were to be degraded are 2,4-D along with 4,6-dinitro-o-cresol. The radiation at frequencies of 460 and 465 kHz and low frequencies around 28 kHz are transmitted to the solution via a transductor. There is a clear increase of the degradation rate compared to simple electrolysis or EF process at lower frequencies with low energy (20 and 60 watts). However, at higher frequencies, a loss in its mineralizing activities was noted, possibly due to oxygen depletion. This increase infers only to ultrasonic radiation's mechanistic effects instead of the chemical effects [41, 42].

10.2.4 Physical AOPs

10.2.4.1 Electron Beam Induced AOP

Ionizing radiation from an electron beam source of the range (0.01–10 MeV) is used to treat wastewater. The free electrons in the water surface result in excited species ultimately exhibiting high oxidizing power to break down the contaminants ($E_{EO} < 3$ kWh/m^3 *order). The excited species' maximum infiltration depth is directly proportional to the energy exhibited by the incident electrons [43]. Duarte et al. (2002) used an electron beam accelerator with 1.5 MeV of energy and 37 kW power to remove industrial

wastes. He reported that the optimum dose for eliminating the contaminants was 20 kGy. The process efficiently removed the impurities composed of phenol, xylene, methyl isobutyl ketone, fluoxetine, chloroform, dichloroethane, crude oil and synthetic dyes. This method, however, is not deemed profitable due to its high costs (> 1 million USD) and risk associated with X-rays [44].

10.2.4.2 Microwave-Based AOP

Microwave irradiation of the range 300 MHz to 300 GHz (3×10^8 cycles to 3×10^{11} cycles) is used to treat wastewater and other oxidants or catalysts TiO_2 or coupled with other AOP techniques like Fenton, photolysis, catalysis or photocatalytic methods [45]. Homogeneous microwave processes are avoided due to their high operating cost. That is why it is coupled with other oxidants or catalysts (Cu, Pt, TiO2, granular activated carbon) to decrease reaction time, promote faster degradation and minimize entire treatment costs [46]. The microwave advanced oxidation process works by the dielectric heating mechanism, in which heat is generated because of the dielectric materials' interaction with electromagnetic irradiation. Besides organic compounds, pathogens can also be disrupted by this method known as athermal effects [46]. Microwave-associated AOPs use hydrogen peroxide along with microwave radiation for the generation of hydroxyl radicals. This process is believed to be similar to wet air oxidation methods. Lo and Liao (2011) used microwave-enhanced advanced oxidation to treat dairy manure. MW/H_2O_2 methods were used for the pre-treatment of solids from daily wastes. The optimum microwave temperatures reported from this process were 120–160 °C with a low H_2O_2 concentration and a treatment period of 15–20 min [47]. This process is suitable for operating continuously, and hence it shows better outcomes than a batch process. One of the limitations of this process is that most of the energy irradiated is converted into heat, so the water body gets immensely heated up to mitigate this limitation. Cooling devices should be set up to prevent the water body from being overheated.

10.2.4.3 Sonolysis/Sonication or Sonochemical AOP

The AOPs that involve the use of high-energy ultrasounds with low to moderate levels of frequency (in the range of 20–1000 kHz) for destructing harmful contaminants in water are called sonochemical AOPs, and the process is called sonolysis, the chemical effects of which initiate acoustic cavitation, triggering the formation and implosion of microbubbles [48]. Ultrasound works by two kinds of mechanisms. The physical process is called a direct mechanism, and the chemical process is called an indirect mechanism. Sonication or the direct method involves the initiation of acoustic cavitation to form and destroy microbubbles and then reach an unstable critical resonance size. It violently implodes, generating high temperatures (2000 to 5000 K) and high pressure (> 1000 bar). The transient cavitation results from intensive compression and rarefaction of water, following which hydroxyl radicals needed for the oxidation of the pollutants are formed

[49]. However, there have to be two critical components for this process to work. Acoustic cavitation can only occur in a liquid medium and a source of increased energy vibrations. Sonochemical AOPs have been used to decontaminate many organic and inorganic components, pesticides, pharmaceutical components, endocrine disrupters, disinfectants and more [13, 48–49]. However, combined processes that involve ultrasound with other oxidation methods like ozonation, UV radiation and electro-Fenton have shown to be more effective and faster to eliminate effluents than either of them applied alone.

10.2.4.4 Electrohydraulic Discharge (Plasma)

Electrical or electrohydraulic discharge reactors have been rigorously investigated in the present time for the decontamination of pollutants from wastewater. A strong electrical field or high voltage electrical discharge is applied directly to the water body for initiating both physical and chemical processes. This leads to the generation of hydrogen peroxide, hydrogen, molecular oxygen, hydroxyl and hydroperoxyl, among other radicals. There are two kinds of electrohydraulic discharge: the corona-like system works with a discharge energy of approximately 1 J/pulse and the pulsed arc discharge works with a discharge energy of more than 1 kJ/pulse [50]. The operating frequency of pulsed corona like system is 10^2–10^3 Hz with a peak current less of than 100A. A moderate amount of bubbles and UV rays are generated with the radicals emitted near the electrodes. The pulsed arc works at a frequency of 10^{-2}–10^{-3} Hz with a peak current level greater than 1 kA. Strong waves are emitted around the zone of cavitation, and the water reaches transient supercritical conditions. However, strong UV irradiation, plasma bubbles and radicals are short-lived in that cavitation area [51].

Depending upon the magnitude and other factors, UV light or shock waves might also be generated. This efficiently removes various recalcitrant organic compounds like benzene derivatives, phenol derivatives, toluene derivatives, aniline derivatives, methyl tert-butyl ether (MTBE), anthraquinone, acetophenone, trichloroethylene, perchloroethylene and organic dyes. It also eradicates pharmaceutical wastes, inorganic ions, pathogens and harmful microorganisms (virus, yeasts, bacteria) from the water body [52].

10.3 Comparing AOPs and Understanding Their Applications

Glaze et al. (1987) had determined the oxidation of nitrobenzene's concentrated solutions using various methods of AOP, that is, ozonation, peroxonation, O_3/UV and H_2O_2/UV [1]. Some of the results included the oxidation rate being on par with peroxidation or unassisted ozonation alone. Primo et al. (2008) arranged the efficiencies of AOP based on UV< H_2O_2/ O_3 < Fenton < Fenton-like < photo-Fenton for the removal of organic matter in the treatment of landfill leachates [53]. EAOPs, and more prominently BDD among the electrochemical methods of AOP, are said to be the most efficient over other electrode choices [35]. Peyton et al. (1988) assessed the O_3/UV process's kinetic model describing the radiation flux,

the ozone concentration in its liquid phase and reaction constants [54]. In the purification of groundwater contaminated by TCE and PCE, both O_3/H_2O_2 and UV/H_2O_2 were proven to be efficient in removing chlorinated ethylenes. However, the UV/H_2O_2 model was efficient in the partial stripping of PCE and TCE; thus, it was chosen for further studies. Trapido et al. (1997) also conducted the degradation of nitrophenols with ozone. Ozone combined with H_2O_2/UV was found to be the most efficient for the oxidation of nitrophenols at low pH levels [55].

The order of efficiency of AOP for the efficient removal of polyaromatic hydrocarbons (PAHs) and anthracene was given as UV < O3/H_2O_2/UV < O_3/UV = O_3/H_2O_2 [56]. Beltran et al. (1998) mathematically modeled the advanced oxidation of PAHs, chlorophenols and nitroaromatic organic compounds [57–60]. It proves that the efficiency of many AOPs is positively compound specific.

10.4 Conclusion

For decades, AOP has been used to remediate and completely diminish wastewater. Despite extensive and intensive research throughout the years, it has the potential for comprehensive study but not much to reveal the costs and implementation expenses. Its overall economic efficiency and awareness cannot be adequately estimated. It will also evince that AOP is a more economical alternative than all other established conventional technologies available rather than more efficient. Neither have many papers revealed the kinetic and mechanistic models for representing the substrate degradation. Coupled techniques of the combined AOPs and their efficiencies should be rigorously studied due to the same cost competitiveness. More emphasis should be given to the electrochemical and physical methods, utilizing solar radiation efficiently and conducting more pilot-plant scale studies using existing industrial wastewater.

In the present chapter, principles, limitations, advantages and applications of various AOPs with Fenton's and peroxonation processes are discussed. A thorough discussion was also carried out on the most frequently applied techniques like removing dyes, toxic organic compounds, and aromatic and synthetic dyes. Photochemical methods are simple, clean, economical and more efficient than the classical methods. Among these, heterogeneous photocatalysis, photo-Fenton, is more promising than ozonation and peroxonation alone. Solar photo-Fenton is preferred for its reduced energy consumption compared to the classical photo-Fenton method. Concerning the photocatalytic methods, TiO_2 is an ideal catalyst because of its stability, inertness, easy production and economic feasibility. Sonochemical AOPs and their combination with existing methods like Fenton and electro-Fenton are gaining more prominence; however, not much research has been conducted at the industrial level, without which the economic feasibility cannot be understood. Among the electrochemical processes, BDD anodes have been said to be the most efficient of all due to their improved stability, reactivity and mineralizing power compared to other electrodes. The hydroxyl ion formed is physisorbed on the electrode's surface and is found to be very reactive.

On the other hand, electro-Fenton overcame all the existing limitations of conventional Fenton processes like cost of reagent, iron decomposition and sludge formation. Coupling these with other AOPs, like sonoelectro-Fenton, photoelectro-Fenton and sonophotoelectro-Fenton, is gaining more prominence because of its increased efficiency. These methods have been widely applied in treating phenol derivatives, pesticides, textile effluents, synthetic dyes, landfill leachates, antibiotics and other pharmaceutical wastes. Besides EAOPs, other physical methods like electrohydraulic discharge, microwave and sonolysis are cost-effective and environmentally friendly for their low sludge formation. Thus, they are extremely promising over existing conventional approaches. To assess further, more industrial applications need to be done in this area.

REFERENCES

1. Glaze, W.H., Kang, J.W. and Chapin, D.H., 1987. The chemistry of water treatment processes involving ozone, hydrogen peroxide and ultraviolet radiation. *Ozone: Science and Engineering*, 9(4).
2. Munter, R., 2001. Advanced oxidation processes – current status and prospects. *Proc. Estonian Acad. Sci. Chem*, 50(2), pp. 59–80.
3. Herrmann, J.M., Guillard, C., Arguello, M., Agüera, A., Tejedor, A., Piedra, L. and Fernandez-Alba, A., 1999. Photocatalytic degradation of pesticide pirimiphos-methyl: Determination of the reaction pathway and identification of intermediate products by various analytical methods. *Catalysis Today*, 54(2–3), pp. 353–367.
4. Andreozzi, R., Caprio, V., Insola, A. and Marotta, R., 1999. Advanced oxidation processes (AOP) for water purification and recovery. *Catalysis Today*, 53(1), pp. 51–59.
5. Hermosilla, D., Cortijo, M. and Huang, C.P., 2009. Optimizing the treatment of landfill leachate by conventional Fenton and photo-Fenton processes. *Science of the Total Environment*, 407(11), pp. 3473–3481.
6. Kitis, M., Adams, C.D. and Daigger, G.T., 1999. The effects of Fenton's reagent pretreatment on the biodegradability of nonionic surfactants. *Water Research*, 33(11), pp. 2561–2568.
7. Neyens, E. and Baeyens, J., 2003. A review of classic Fenton's peroxidation as an advanced oxidation technique. *Journal of Hazardous Materials*, 98(1–3), pp. 33–50.
8. Safarzadeh-Amiri, A., Bolton, J.R. and Cater, S.R., 1996. The use of iron in advanced oxidation processes. *Journal of Advanced Oxidation Technologies*, 1(1), pp. 18–26.
9. Bautista, P., Mohedano, A.F., Casas, J.A., Zazo, J.A. and Rodriguez, J.J., 2008. An overview of the application of Fenton oxidation to industrial wastewaters treatment. *Journal of Chemical Technology & Biotechnology: International Research in Process, Environmental & Clean Technology*, 83(10), pp. 1323–1338.
10. Gottschalk, C., Libra, J.A. and Saupe, A., 2009. *Ozonation of water and waste water: A practical guide to understanding ozone and its applications*. John Wiley & Sons.
11. Pisarenko, A.N., Stanford, B.D., Yan, D., Gerrity, D. and Snyder, S.A., 2012. Effects of ozone and ozone/peroxide on trace organic contaminants and NDMA in drinking water and water reuse applications. *Water Research*, 46(2), pp. 316–326.
12. Paillard, H., 1988. Optimal conditions for applying an ozone-hydrogen peroxide oxidizing system. *Water Research*, 22, pp. 91–103.
13. Oturan, M.A. and Aaron, J.J., 2014. Advanced oxidation processes in water/wastewater treatment: principles and applications: A review. *Critical Reviews in Environmental Science and Technology*, 44(23), pp. 2577–2641.
14. Kaptijn, Jan P., 1997. The Ecoclear® process. *Results from Full-Scale Installations*, pp. 297–305.
15. Paillard, H., Dore, M. and Bourbigot, M.M., 1991, March. Prospects concerning applications of catalytic ozonation in drinking water treatment. In *Proc. 10th Ozone World Congress* (Vol. 1, pp. 313–329). Intl. Ozone Assoc., European-African Group.
16. Ling, L., Pan, B. and Zhang, W.X., 2015. Removal of selenium from water with nanoscale zero-valent iron: Mechanisms of intraparticle reduction of Se (IV). *Water Research*, 71, pp. 274–281.
17. Mackenzie, K. and Georgi, A., 2019. NZVI synthesis and characterization. In *Nanoscale zerovalent iron particles for environmental restoration* (pp. 45–95). Springer.
18. Arvaniti, O.S., Hwang, Y., Andersen, H.R., Stasinakis, A.S., Thomaidis, N.S. and Aloupi, M., 2015. Reductive degradation of perfluorinated compounds in water using Mg-aminoclay coated nanoscale zero valent iron. *Chemical Engineering Journal*, 262, pp. 133–139.
19. Chen, H., Cao, Y., Wei, E., Gong, T. and Xian, Q., 2016. Facile synthesis of graphene nano zero-valent iron composites and their efficient removal of trichloronitromethane from drinking water. *Chemosphere*, 146, pp. 32–39.
20. Xia, S., Gu, Z., Zhang, Z., Zhang, J. and Hermanowicz, S.W., 2014. Removal of chloramphenicol from aqueous solution by nanoscale zero-valent iron particles. *Chemical Engineering Journal*, 257, pp. 98–104.
21. Pasinszki, T. and Krebsz, M., 2020. Synthesis and application of zero-valent iron nanoparticles in water treatment, environmental remediation, catalysis, and their biological effects. *Nanomaterials*, 10(5), p. 917.
22. Bautitz, I.R., Velosa, A.C. and Nogueira, R.F.P., 2012. Zero valent iron mediated degradation of the pharmaceutical diazepam. *Chemosphere*, 88(6), pp. 688–692.
23. Li, J., Zhou, Q., Liu, Y. and Lei, M., 2017. Recyclable nanoscale zero-valent iron-based magnetic polydopamine coated nanomaterials for the adsorption and removal of phenanthrene and anthracene. *Science and Technology of advanced Materials*, 18(1), pp. 3–16.
24. Gonçalves, C., Lopes, M., Ferreira, J.P. and Belo, I., 2009. Biological treatment of olive mill wastewater by non-conventional yeasts. *Bioresource Technology*, 100(15), pp. 3759–3763.
25. Watts, M.J., Rosenfeldt, E.J. and Linden, K.G., 2007. Comparative OH radical oxidation using UV-Cl2 and UV-H₂O₂ processes. *Journal of Water Supply: Research and Technology – Aqua*, 56(8), pp. 469–477.
26. Wang, D., Bolton, J.R., Andrews, S.A. and Hofmann, R., 2015. Formation of disinfection by-products in the ultraviolet/chlorine advanced oxidation process. *Science of the Total Environment*, 518, pp. 49–57.

27. Buxton, G.V., Greenstock, C.L., Helman, W.P. and Ross, A.B., 1988. Critical review of rate constants for reactions of hydrated electrons, hydrogen atoms and hydroxyl radicals (OH/O− in aqueous solution. *Journal of Physical and Chemical Reference Data*, 17(2), pp. 513–886.

28. Roth, J.A., 1993. *Chemical oxidation: technology for the nineties* (Vol. 2). CRC Press.

29. Parsons, S. ed., 2004. *Advanced oxidation processes for water and wastewater treatment.* IWA Publishing.

30. Bhowmick, M. and Semmens, M.J., 1994. Ultraviolet photooxidation for the destruction of VOCs in air. *Water Research*, 28(11), pp. 2407–2415.

31. Irmak, S., Erbatur, O. and Akgerman, A., 2005. Degradation of 17β-estradiol and bisphenol A in aqueous medium by using ozone and ozone/UV techniques. *Journal of Hazardous Materials*, 126(1–3), pp. 54–62.

32. Crittenden, J.C., Trussell, R.R., Hand, D.W., Howe, K.J. and Tchobanoglous, G., 2012. *MWH's water treatment: Principles and design.* John Wiley & Sons.

33. Chaplin, B.P., 2014. Critical review of electrochemical advanced oxidation processes for water treatment applications. *Environmental Science: Processes & Impacts*, 16(6), pp. 1182–1203.

34. Comninellis, C. and De Battisti, A., 1996. Electrocatalysis in anodic oxidation of organics with simultaneous oxygen evolution. *Journal de Chimie Physique*, 93, pp. 673–679.

35. Panizza, M. and Cerisola, G., 2003. Electrochemical oxidation of 2-naphthol with in situ electrogenerated active chlorine. *Electrochimica Acta*, 48(11), pp. 1515–1519.

36. Palma-Goyes, R.E., Guzmán-Duque, F.L., Peñuela, G., González, I., Nava, J.L. and Torres-Palma, R.A., 2010. Electrochemical degradation of crystal violet with BDD electrodes: Effect of electrochemical parameters and identification of organic by-products. *Chemosphere*, 81(1), pp. 26–32.

37. Nidheesh, P.V. and Gandhimathi, R., 2012. Trends in electro-Fenton process for water and wastewater treatment: An overview. *Desalination*, 299, pp. 1–15.

38. Sudoh, M., Kodera, T., Sakai, K., Zhang, J. Q., and Koide, K., 1986. Oxidative degradation of aqueous phenol effluent with electrogenerated Fenton's reagent. *J. Chem. Eng. Jpn.*, 19(6), 513–518.

39. Heidari, Z., Motevasel, M. and Jaafarzadeh, N.A., 2015. Application of Electro-Fenton (EF) process to the removal of pentachlorophenol from aqueous solutions. *Iranian Journal of Oil & Gas Science and Technology*, 4(4), pp. 76–87.

40. Babuponnusami, A. and Muthukumar, K., 2012. Advanced oxidation of phenol: A comparison between Fenton, electro-Fenton, sono-electro-Fenton and photoelectro-Fenton processes. *Chem. Eng. J.*, 183, 1–9.

41. Yasman, Y., Bulatov, V., Gridin, V.V., Agur, S., Galil, N., Armon, R. and Schechter, I., 2004. A new sono-electrochemical method for enhanced detoxification of hydrophilic chloroorganic pollutants in water. *Ultrasonics Sonochemistry*, 11(6), pp. 365–372.

42. Oturan, M.A., Sirés, I., Oturan, N., Pérocheau, S., Laborde, J.L. and Trévin, S., 2008. Sonoelectro-Fenton process: A novel hybrid technique for the destruction of organic pollutants in water. *Journal of Electroanalytical Chemistry*, 624(1–2), pp. 329–332.

43. Nickelsen, M.G., Cooper, W.J., Lin, K., Kurucz, C.N. and Waite, T.D., 1994. High energy electron beam generation of oxidants for the treatment of benzene and toluene in the presence of radical scavengers. *Water Research*, 28(5), pp. 1227–1237.

44. Duarte, C.L., Sampa, M.H.O., Rela, P.R., Oikawa, H., Silveira, C.G. and Azevedo, A.L., 2002. Advanced oxidation process by electron-beam-irradiation-induced decomposition of pollutants in industrial effluents. *Radiation Physics and Chemistry*, 63(3–6), pp. 647–651.

45. Banik, S., Bandyopadhyay, S. and Ganguly, S., 2003. Bioeffects of microwave – a brief review. *Bioresource Technology*, 87(2), pp. 155–159.

46. Hong, S.M., Park, J.K. and Lee, Y.O., 2004. Mechanisms of microwave irradiation involved in the destruction of fecal coliforms from biosolids. *Water Research*, 38(6), pp. 1615–1625.

47. Lo, K.V. and Liao, P.H., 2011. Microwave enhanced advanced oxidation in the treatment of dairy manure. *Microwave Heating*, pp. 91–106.

48. Mason, T.J., 2002. *Lorimer JP-applied sonochemistry: The uses of power ultrasound in chemistry and processing.* Wiley.

49. Sathishkumar, P., Mangalaraja, R.V. and Anandan, S., 2016. Review on the recent improvements in sonochemical and combined sonochemical oxidation processes – a powerful tool for destruction of environmental contaminants. *Renewable and Sustainable Energy Reviews*, 55, pp. 426–454.

50. Zastawny, H., Romat, H., Karpel, N. and Chang, J.S., 2003. *Pulse arc discharges for water treatment and disinfection.* Electrostatics.

51. Šunka, P., 2001. Pulse electrical discharges in water and their applications. *Physics of Plasmas*, 8(5), pp. 2587–2594.

52. Locke, B.R., Sato, M., Sunka, P., Hoffmann, M.R. and Chang, J.S., 2006. Electrohydraulic discharge and nonthermal plasma for water treatment. *Industrial & Engineering Chemistry Research*, 45(3), pp. 882–905.

53. Primo, O., Rivero, M.J. and Ortiz, I., 2008. Photo-Fenton process as an efficient alternative to the treatment of landfill leachates. *Journal of Hazardous Materials*, 153(1–2), pp. 834–842.

54. Peyton, G.R. and Glaze, W.H., 1988. Destruction of pollutants in water with ozone in combination with ultraviolet radiation. 3. Photolysis of aqueous ozone. *Environmental Science & Technology*, 22(7), pp. 761–767.

55. Trapido, M., Hirvonen, A., Veressinina, Y., Hentunen, J. and Munter, R., 1997. Ozonation, ozone/UV and UV/H$_2$O$_2$ degradation of chlorophenols. *Ozone Science Engineering*, 19(1), pp. 75–96.

56. Hirvonen, A., Tuhkanen, T. and Kalliokoski, P., 1996. Treatment of TCE-and PCE contaminated groundwater using UV/H$_2$O$_2$ and O$_3$/H$_2$O$_2$ oxidation processes. *Water Science and Technology*, 33(6), pp. 67–73.

57. Beltran, F.J., Encinar, J.M. and Alonso, M.A., 1998. Nitroaromatic hydrocarbon ozonation in water. 2. Combined ozonation with hydrogen peroxide or UV

radiation. *Industrial & Engineering Chemistry Research,* *37*(1), pp. 32–40.

58. Beltran, F.J., Ovejero, G., Garcia-Araya, J.F. and Rivas, J., 1995. Oxidation of polynuclear aromatic hydrocarbons in water. 2. UV radiation and ozonation in the presence of UV radiation. *Industrial & Engineering Chemistry Research,* *34*(5), pp. 1607–1615.

59. Beltrán, F.J., Rivas, F.J. and Montero-de-Espinosa, R., 2005. Iron type catalysts for the ozonation of oxalic acid in water. *Water Research,* *39*(15), pp. 3553–3564.

60. Gimeno, O., Carbajo, M., Beltrán, F.J. and Rivas, F.J., 2005. Phenol and substituted phenols AOPs remediation. *Journal of Hazardous Materials,* *119*(1–3), pp. 99–108.

11

Advanced Oxidation Processes and Bioremediation Techniques for Treatment of Recalcitrant Compounds Present in Wastewater

Apoorva Sharma and Praveen Dahiya

CONTENTS

DOI: 10.1201/9781003165958-11

11.1 Introduction

The scarcity of clear drinking water has become a severe issue globally, resulting in devastating effects to the entire environment. As per the World Health Organization (WHO), water-related diseases are proving to be fatal for around 1 million people each year. In times like these, before wastewater reuse, proper treatment strategies to remove the harmful contaminants is the present requirement. The world of wastewater engineering involves the process of continuous development and rejuvenation of strategies for efficient treatment of physical and chemical contaminants in the wastewater. Research and development in environmental engineering and technical developments are currently the answers to the problem of domestic and industrial effluent contamination in drinking water, existing in many developed and developing countries.

Immense scientific intricacies and extensive scientific challenges are being faced by the bioprocess engineering and environmental engineering sciences. For the past several years, one of the main concerns in maintaining the quality of water is linked to the analysis of minute chemical pollutants and their removal from municipal and industrial wastewater. Depending upon water solubility, these contaminants, both naturally occurring and organic chemicals, enter the marine system, are easily transported and are distributed in the water cycle (Oller et al. 2011). The health afflictions and harmful complications associated with these contaminants include impaired and anomalous physiological processes such as the development of malignant tumors in bodies of aquatic and terrestrial species, reproductive impairment, and antagonistic or synergistic effects of the toxic pollutants resulting in increasing wastewater toxicity post-treatment.

The treatment system deals with industrial effluents that are partly degradable or totally nonbiodegradable with conventional biotreatment processes. These pollutants accumulate in the environment and lead to the interruption of biological activity, hampering the development of terrestrial and aquatic organisms (Kasprzyk-Hordern et al. 2008, 2009). With the continuous release of harmful effluents, immense degradation of water resources has been observed, making improvement in quality of water an immediate step that needs to be taken to decrease the associated complications and smooth functioning of the environment (Yang 2011).

This chapter elucidates the processes required to eradicate recalcitrant organics from industrial wastewaters. Along with the extremely advantageous separation processes required for water reuse, biodegradative techniques are needed for breaking down the harmful organic substances that can contaminate the environment if released. Usually the economically beneficial conventional biological treatments are sufficient for the degradation of organic and inorganic wastewater pollutants, but the refractory and recalcitrant compounds are still not removed efficiently as the biorecalcitrant might result in microbial death, reducing the efficiency of the process (Jeworski and Heinzle 2000). The treatment of these recalcitrant compounds through processes like ionic exchange resins, activated carbon and membrane filtration become very strenuous because the quantity of these polluting substances is on the lower side in large volumes of the wastewater (Sandhya et al. 2005). Contaminants are treated using nonbiological techniques, which can be broadly divided into oxidative and reductive techniques. These technologies can be used for improving the waste biodegradability potential by converting it into lower chain or less harmful compounds that are easily degraded biologically. They include chemical oxidative treatment technology, which uses chemical oxidizing agents such as O_3, K_2MnO_4 and H_2O_2 to transform the chemical waste into smaller biodegradable by-products. The chemical oxidative wastewater treatment comprises the classic chemical oxidation and AOP that involves wastewater incineration and wet air oxidation for wastewaters rich in organic pollutants. AOP is an efficient method that can oxidize contaminants into mineral end products, forming CO_2 and inorganic ions (Estrada et al. 2012). The mineralization of recalcitrant substances in the wastewater is not cost-effective but is useful for the removal of the organic recalcitrant that the biodegradation treatment failed to remove (Oller et al. 2011). Therefore, this method is used for the pre- or post-treatment of the recalcitrant using specialized microbial cultures needed for successful removal of the recalcitrant organic contaminants present in wastewater (Sandhya et al. 2005).

Thus, further research is required on the organic pollutants, treatment options and advancements in the scientific endeavors associated with wastewater treatments. This chapter gives a deeper understanding of the treatment processes for biodegradation and removal of recalcitrant compounds from wastewater.

11.2 Wastewater Pollutants

Water quality is affected by anthropogenic activities; water containing an increased level of contaminants is known as wastewater. Various sources like domestic, agricultural and industrial sectors produce wastewater. These sources of wastewater are not just diversified based on their location but also based on the pollutants they release into the water bodies. These wastewater contaminants can be divided into four major categories: organic, inorganic, macroscopic pollutants and pathogens. Other than some of the organic and inorganic pollutants, the rest of the contaminants are easily treatable with conventional biotreatment procedures. However, some organic matter is more refractory than others, resisting biodegradation and becoming a challenge to treat. The presence of organic compounds in water bodies leads to enhanced oxygen demand, hence lowering the level of dissolved oxygen available for other marine organisms. This results in an adverse effect on the metabolism and physiology of marine life and cause their death. These pollutants are composed of recalcitrant organic compounds, which are perilous to animal as well as human health.

This chapter primarily focuses on the recalcitrant organic chemical compounds, which are slowly degrading or are nonbiodegradable because of chemical factors, as these compounds are toxic even to the microbes they are being treated with (Mara and Horen 2003). These compounds range from halogenated hydrocarbons to complex polymers, which usually comprises fungicides and phenols that are chloro- and nitro- derivatives. The compound concentration influences the fate of these recalcitrant substances. If the concentration of these contaminants is too high, the chances of toxicity may increase. Some of the recalcitrant effluents include pharmaceuticals, chlorinated congeners, personal care products, pesticides, plasticizers and endocrine disrupters like benzyl chloride, 2,6-dihydroxybenzoic acid 3TMS, benzyl alcohol, butyl octyl phthalate, 4-chloro-3-methyl phenol, 2'6'-dihydroxyacetophenone, benzyl butyl phthalate and 4-biphenyltrimethylsiloxane that can still be present in the water treated by biological mode (Bharagava et al. 2018; Shah 2021). This shows that wastewater discharged into water bodies even after the secondary treatment has a very high percentage of pollutants. Upon exposure, these contaminants have detrimental effects on human and environmental health, even when present at trace levels (Lai et al. 2009). Therefore, this chapter explains in detail all the possible prospects that can be undertaken in order to properly treat and manage the recalcitrant compounds in wastewater. Wastewater pollutants can be categorized into non-refractory and refractory organic compounds. Some of these organic wastewater pollutants are listed in Table 11.1.

11.3 Conventional Treatment Processes

To protect public health and safeguard the environment, it is essential to properly treat wastewater effluents before discharging them into water resources. This wastewater is collected and remediated by several processes at centralized facilities. A typical treatment plant comprises primary, secondary and tertiary treatments. These treatment processes can be further categorized into physical remediation, phytoremediation process, chemical remediation and microbial remediation.

TABLE 11.1

Classification of the Organic Compounds

Category	Compounds
Non-Refractory Compounds	Benzyl alcohol propane, 1,2-dichloro- 3- (2- chloro -1- (chloromethyl)ethoxy)-phallic acid, isobutyl nonyl ester ethane, 2-chloro 1,1- dimethoxy propane, 2,2' – oxy bis (1,3 dichloro- phenol, 4- propyl-
Refractory Compounds	3,3-dichloro-propene Dibutyl phthalate Benzene, 1,1'-(1,2-ethanediyl) bis(4-methoxy- Benzene, 1-methoxy-4-propyl- 2,6 dichlorobenzaldoxime 1,3, benzenediamine, 4-methyl- 4-(But-2-oxy)benzaldehyde Phenyl acetic acid, 4-methoxy-, methyl ester Bis((4-methoxyphenyl)methyl)disulfide Hydrazonomethyl1)-phenyl ester 3-Furan-2-yl-acrylicacid 4-((2-chloro-bezoyl)-

11.3.1 Primary Treatment

11.3.1.1 Physical Remediation

This includes physical processes such as screening, aeration, sedimentation, flotation, filtration, skimming and precipitation (Hamby 1996). Physical remediation techniques are generally used in combination with other treatment processes. This process includes separating the solid waste contaminants and the insoluble particles from the wastewater. This generally involves screening, comminution and sedimentation processes where the floating, nonbiodegradable solids are screened out for the smooth functioning of the treatment plant, preventing any possible operational or maintenance issues and damages. The waste is then shredded into smaller particles that are suspended in the water and, because of being heavier than the water, can easily settle down in the sedimentation process. Physical treatment involves procedures that help in making further treatment processes easier and smoother to carry out. Flow equalization is also a part of physical remediation but is not exactly considered a treatment process as it is a technique that helps to enhance efficiency of the secondary and advanced treatment techniques. Presence of extensive variation with respect to flow and time can degrade the efficiency; therefore, flow equalization is used to minimize the variation in rate of flow and composition of the water and wastewater.

11.3.2 Secondary Treatment

11.3.2.1 Phytoremediation Process

Phytoremediation involves plants to stabilize or remove the pollutants present in the water. During this process, the contaminants are transformed into simpler, harmless and valuable products or are destroyed. Certain phytoremediation techniques involved in pollutant removal from wastewater include phytodegradation, phytoimmobilization and phytoextraction (Todd and Josephson 1996). The application of phytoremediation can be classified based on mechanisms involved in the process and the fate of the contaminants. It is an environmentally friendly, clean, cost-effective treatment, which is generally utilized to treat contaminants like pesticides, metals, fuels, semi-volatile organic compounds, explosives and volatile organic compounds. For the treatment of radioactive substances, chelating agents are sometimes used so that the contaminants can be made more amenable to plant uptake.

11.3.2.2 Microbial Remediation

This process uses microorganisms like fungi, bacteria, algae, protozoa and rotifers for breaking the organic pollutants into smaller, treatable compounds (Wilson 2017). The environmental conditions are the deciding factors for microbial remediation processes because the treatment greatly depends upon factors like molecular oxygen, pH and nutrient conditions. This is an emerging technology that transforms or eliminates the contaminants by using living organisms to degrade polluted wastewater. The remediation of a specific type of waste needs a chain of different environmental conditions for variety of microorganisms to carry out the reaction (Bellandi 1988). The procedure could be a non-specific or an aspecific process. A non-specific process involves a chain of microbial events resulting in the degradation of the pollutants, and an aspecific process uses a microbe that targets a particular site of a molecule marking the degradation process (Jegatheesan et al. 2004; Davies 2005).

11.3.2.3 Chemical Remediation

Chemical remediation is a process that uses chemical compounds to treat wastewater pollutants. This method prevents the bacteria from multiplying in the water, in turn making the water pure. The process removes persistent organic and inorganic compounds and includes various treatment techniques that vary based on the pollutants being dealt with. Some of the processes involved in wastewater remediation include advanced oxidation processes, chemical extraction, radiocolloid treatment and wet air oxidization (Hamby 1996; Rubalcaba et al. 2007).

This work reviews in detail the most studied processes, fundamentals and recent innovations of the AOP. It also discusses the combination of AOP and bioremediation technique to effectively treat industry effluent.

11.3.3 Tertiary Treatment

Tertiary treatment technique includes biological, physical and chemical treatment strategies utilized for the removal of contaminants, left out once the traditional technique is performed. It takes place using mechanical and photochemical processes in one single step. To proceed, microfiltration or synthetic membranes usually are used after the conventional sewage treatment sequence. The process is required for irradiating the viruses or bacteria or for eliminating infectious capabilities. This process involves treatment of sanitary sewage wastewater, which contains microorganisms that require disinfecting.

11.4 Wastewater Remediation Using Advanced Oxidation Processes

Advanced oxidation processes (AOPs), being a part of chemical treatment procedures, are designed to remove the organic or inorganic persistent pollutants present in the wastewater. This method is carried out by utilizing the hydroxyl ($^{\bullet}OH$) and sulfate ($SO_4^{\bullet-}$) radicals as the oxidizing agents. Advanced oxidation processes first came into action in the 1980s when they were utilized for potable water treatment (Glaze 1987; Glaze et al. 1987). They are defined as oxidation processes that involve the formation of hydroxyl radicals and some other reactive oxygen species such as H_2O_2, singlet oxygen and superoxide anion radicals, which are sufficient in quantity for the treatment of wastewater contaminants. These strong oxidants produced can degrade the organic recalcitrant and certain inorganic compounds and transform them into nontoxic or less harmful products, thereby imparting a complete treatment of the wastewater (Huang et al. 1993). AOPs can be used as either a pre-treatment or a post-treatment method, where in

the former case they improve the biotreatability of the contaminants by breaking them into smaller compounds (Karrer et al. 1997; Bila et al. 2005). Subsequently, wastewater already treated with microorganisms is exposed to a chemical oxidation step to completely degrade the available contaminants (Poole 2004). However, this chapter has expounded the pre-treatment step of the AOP involved in the treatment of recalcitrant compounds. AOPs are usually considered better and far more important than other processes because the compounds, rather than being transformed or concentrated into the diffused phase, are degraded totally, avoiding the generation and treatment of secondary wastewater pollutants.

11.4.1 Hydroxyl Radical–Based AOPs

This AOP depends on the generation of very reactive hydroxyl ($^{\bullet}OH$) radicals that, when applied in the wastewater, can degrade any compound. $^{\bullet}OH$ readily reacts with multiple species at a constant rate and has an oxidation potential ranging from 2.8V (pH 0) and 1.95V (pH 14) vs. SCE (most common reference electrode – saturated calomel electrode). The radical is very non-selective in its behavior and the pollutants nearby are fragmented and converted into small inorganic molecules. Hydroxyl radical production takes place using primary oxidants like ozone or oxygen, catalysts like titanium oxide and energy sources like UV light or ultrasound. These hydroxyl radicals react with organic contaminants through four basic pathways, naming electron transfer, hydrogen abstraction, radical addition and radical combination (System 1994). When these pollutants react with organic compounds, production of carbon-centered radicals takes place ($R^{\bullet}$ or R-OH), which then react with molecular oxygen in order to create organic peroxyl radicals ($ROO^{\bullet}$), which undergo further reactions forming oxidation products like alcohols, ketones or aldehydes. Hydroxyl radicals can form a radical cation with the process of abstraction of an electron from an electron substrate, which can easily give an oxidized product by hydrolyzing in aqueous media. All the radicals further react and form stronger reactive species, which leads to the chemical degradation of these refractory organic compounds, which in turn produce inorganic ions, CO_2 and H_2O. The process is clearly illustrated with the help of the diagram in Figure 11.1.

The major AOPs using hydroxyl radical generation mechanism for the remediation of the wastewater are summarized in Table 11.2.

11.4.2 Fenton and Photo-Fenton Process

The process employed for wastewater treatments are strong, energy efficient and highly effective techniques to remove toxic and recalcitrant pollutants, when used singularly or coupled with other biological or chemical procedures.

11.4.2.1 Fenton Process

The process was first discovered by H. J. Fenton and is defined as the increased oxidative potential of H_2O_2, using Fe as a catalyst under acidic conditions. Iron is most recurrently used among all the metals activating H_2O_2 and forms hydroxyl

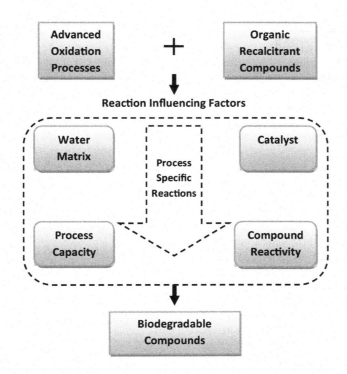

FIGURE 11.1 Advanced oxidation mechanism.

TABLE 11.2

Advanced Oxidation Processes Using Hydroxyl Radical Generation Mechanism

Homogeneous AOPs	Heterogeneous AOPs
Fenton AOP (Fe^{2+} + H$_2$O$_2$)	Heterogeneous Photocatalysis (Light + Catalyst)
Photo-Fenton (Solar Light + Fenton)	Photoelectrocatalysis
Ozone (O$_3$)-based AOPs	
UV-based AOPs	
Sonolysis (Ultrasound)	

radicals in the water. It is because hydrogen peroxide reacts with Fe^{2+} to generate strong reactive species that makes the process very powerful. The reaction leads to highly reactive ·OH radicals and ferryl ions at pH 3.0. The reactions involved in the classical Fenton mechanism are as follows:

$$Fe^{2+} + H_2O_2 \rightarrow Fe^{3+} + OH\cdot + OH^- \qquad (11.1)$$

$$Fe^{3+} + H_2O_2 \rightarrow Fe^{2+} + HO_2 + H^+ \qquad (11.2)$$

$$OH\cdot + H_2O_2 \rightarrow HO_2 + H_2O \qquad (11.3)$$

$$OH\cdot + Fe^{2+} \rightarrow Fe^{3+} + OH^- \qquad (11.4)$$

$$Fe^{3+} + HO_2 \rightarrow Fe^{2+} + O_2H \qquad (11.5)$$

$$Fe^{2+} + HO_2 + H^+ \rightarrow Fe^{3+} + H_2O_2 \qquad (11.6)$$

$$2HO_2 \rightarrow H_2O_2 + O_2 \qquad (11.7)$$

$$\text{Organic Pollutant} + OH\cdot \rightarrow \text{Degraded Products} \qquad (11.8)$$

The previous equations show that ferrous ion (Fe^{2+}) was readily oxidized and transformed to form ferric ions (Fe^{3+}). Ferrous ions were then regenerated by a Fenton-like reaction at a very slow rate. The reaction generates OH· in Equation 11.1 by the process of electron transfer. Equation 11.1 is the core of the Fenton reaction.

11.4.2.2 Photo-Fenton Process

The photo-Fenton process is marked by the formation of hydroxyl ions produced by the combination of UV radiation and hydrogen peroxide along with Fe^{2+} and Fe^{3+} ions, to decrease the degradation rate of the recalcitrant contaminants (Kim and Vogelpohl 1998). The Fenton reaction cumulates Fe^{3+} ions and the reaction does not advance, once all Fe^{2+} ions are consumed. The following reaction takes place during photo-Fenton process:

$$Fe^{3+} + H_2O + h\nu \rightarrow HO\cdot + Fe^{2+} + H^+ \qquad (11.9)$$

$$Fe^{3+} + H_2O + h\nu \rightarrow HO_2\cdot + Fe^{2+} + H^+ \qquad (11.10)$$

In photo-Fenton reaction Equation 11.9, photoreduction of the ferric ions leads to photochemical regeneration of the ferrous ions. The ferrous ions generated in this process react with hydrogen peroxide producing ferric ions and hydroxyl radicals and the process continues in the same way until the organic compound is completely degraded.

11.4.3 Ozone-Based Processes

Ozone (O_3) is considered to be a very selective electrophile and is used for organic and inorganic wastewater pollutant degradation (Von 2003). Ozone is categorized as a powerful oxidant and a strong disinfectant, which has an oxidizing potential of 2.07 V vs. SCE (Glaze et al. 1987). The ozone-based processes leading to the production of hydroxyl ions can be categorized into three pathways, namely ozonation, peroxone process and hydrogen peroxide photolysis.

11.4.3.1 Ozonation

Ozonation is a treatment process that can be utilized for decolorization, elimination of taste and odor, removal of micropollutants, treatment of non-protonated amines and remediation of microorganisms in wastewater (Konsowa 2003; Hoigné 1998; Bourgin et al. 2017; Gottschalk et al. 2009; Camel and Bermond 1998). The complex HO· formation has been explained through various detailed mechanisms, and the overall reaction equation involved in hydroxyl ion generation is depicted as follows (Gottschalk et al. 2009).

$$3O_3 + H_2O \rightarrow 2OH\cdot + 4O_2 \qquad (11.11)$$

In this oxidation reaction equation, the O_3 selectively reacts with the segregated and ionized form of organic compounds, instead of the neutral form that results in hydroxyl ion (HO·) formation, initiating the oxidation indirect mechanisms.

11.4.3.2 Peroxone Process

Peroxone process is an effective treatment procedure to deal with the contaminants present in the wastewater system. In this process, amalgamation of hydrogen peroxide and ozone takes place, resulting in the generation of persistent hydroxyl radicals, which are capable of oxidizing pollutants (organic, inorganic) more efficiently than the standalone process. Research shows that the use of H_2O_2 in the process of ozonation enhances the oxidation of the organic compounds. The incorporation of H_2O_2 increases the ozone transfer rate, which is in turn useful for the degradation and reclamation of wastewater. The peroxone system reaction can be expressed as follows:

$$H_2O_2 \rightarrow HO_2^- + H^+ \qquad (11.12)$$
$$HO_2 + O_3 \rightarrow OH\cdot + O_2^- + O_2 \qquad (11.13)$$

In the peroxone (O_3/H_2O_2) reaction, the process of decomposing the hydrogen peroxide (H_2O_2) molecule leads to the formation of hydroperoxide (HO_2^-) ions. These hydroperoxide ions initiate the decomposition cycle of ozone, resulting in the formation of highly degradative hydroxyl ions.

11.4.3.3 Hydrogen Peroxide Photolysis

This process involves the generation of hydroxyl radicals via photolysis of H_2O_2. Although these hydroxyl ions are very powerful oxidizing agents, they cannot be prepared and used as a ready-made disinfectant because of their very short half-life in water (approximately 10^{-9} s) and are instead prepared by the irradiation of 3% hydrogen peroxide with 365–405 nm light, and the process is known as photolysis (Ikai et al. 2010; Sies et al. 1992). In O_3/UV irradiation, generation of H_2O_2 takes place as an additional oxidant, through O_3 photolysis:

$$O_3 + H_2O + h\nu \rightarrow H_2O_2 + O_2 \qquad (11.14)$$

This H_2O_2 then further undergoes a photolysis reaction, and as a consequence, the formation of hydroxyl ions (OH·) takes place.

$$H_2O_2 + h\nu \rightarrow 2OH\cdot \qquad (11.15)$$

11.4.4 UV-Based Remediation (Heterogeneous Photocatalysis)

UV-based remediation process is a tertiary treatment utilized for the degradation of organic compounds and minute microorganisms. After the surface absorption of the UV light, the contaminant electrons jump from ground to excited state and are transferred to an oxygen molecule. This molecule converts both pollutant molecule and the O_2 molecule into a radical. The radical formed is a highly reactive species that oxidizes other molecules to acquire stability. The reaction equations involved in the generation of degradative hydroxyl radicals through the process of UV-based AOP are as follows:

$$TiO_2 + h\nu \rightarrow e^-cb + h\nu\ ^+vb \qquad (11.16)$$

The reaction shows that hydroxyl radicals can be formed via action of photons in the presence of oxidants or catalysts. The catalyst most commonly used is a RO-type semiconductor known as titanium dioxide (TiO_2), which is common due to its capability of achieving total mineralization of organic pollutants under ambient pressure and temperature.

The oxidation and reduction processes are used simultaneously, utilizing photocatalyst and light irradiation, and are known as heterogeneous photocatalysis. The TiO_2 particles reach an excitation state and develop positive holes in the valance band with oxidative capacity and negative electrons at the conduction band with reductive capacity.

Along with the reactions of H_2O, OH^- and O_2^-, these electrons and holes further lead to the formation of hydroxyl radicals at the surface of TiO_2 (Tang 2003).

$$h\nu\ ^+vb + OH^-\ (\text{surface}) \rightarrow OH\cdot \qquad (11.17)$$

$$hv \, ^+vb + H_2O \text{ (absorbed)} \rightarrow OH\cdot + H^+ \qquad (11.18)$$

$$e^-cb + O_2 \text{ (absorbed)} \rightarrow O_2^- \qquad (11.19)$$

In addition, hydroxyl radical can be formed through the photolysis of H_2O when carried out at a wavelength of 242 nm.

$$H_2O + hv \rightarrow OH\cdot + H\cdot \qquad (11.20)$$

11.4.5 Sonolysis

Sonolysis includes ultrasonic irradiation to produce hydroxyl radicals, without a catalyst. Wastewater containing organic contaminants are successfully removed using this method (Ollis and Al-Ekabi 1993; Asim et al. 2012). The dissolved bubbles interact with the sound energy in the liquid, which results in the generation of bubbles near adiabatic collapse; this process is known as acoustic cavitation (Pilli et al. 2011; Shah 2020). In the process of ultrasound irradiation, refraction and alternate compression cycles of the sound waves can be the reason for the formation of three stages of cavities, namely nucleation, growth and implosive collapse. These cavities are made up of gas-filled microbubbles and vapor. The process of cavitation generates conditions of higher temperature and pressure within the collapsing bubbles (Mason et al. 2011; Margulis and Margulis 2004; Entezari et al. 2005). Such extreme conditions lead to water molecule fragmentation in the microbubbles, and this induces the generation of hydroxyl radicals (as shown in Equation 11.21), which further helps in the process of degradation of organic pollutants (Henglein and Gutierrez 1988; Mason et al. 1994; Mason and Bernal 2012).

$$H_2O \rightarrow OH\cdot + H\cdot \qquad (11.21)$$

11.4.6 Photoelectrocatalysis

This process combines electrocatalysis with the solar cell to direct as much solar energy as possible towards the target reaction. This reaction requires the photoelectrocatalysis (PEC) device to be working at a voltage between 1.6 V and 1.8 V. Various compounds are used in this process that can utilize light irradiation in order to undergo catalytic photolysis, to run the redox reactions. These compounds consist of a filled valance and an empty conduction band (Fenton 1894). Upon light irradiation, the electrons in the excited state shift from valance to the conduction band, generating positive holes in the semiconductor photocatalyst. In this process, the number of recombinations is diminished with the help of the applied differential potential. An external positive bias is applied to the photocatalytic system, through which the rates of reactions can be controlled. This positive bias can increase the photocatalytic reaction rates by moving the photo-produced electron-hole pairs in opposite directions. This bias extracts the

photogenerated electrons and brings them up to the cathode of the electrolytic cell. The organic compounds are oxidized, and the hydroxyl radicals are formed from the water oxidation with holes.

11.5 Latest Innovations in Advanced Oxidation Processes

Advanced oxidation processes serve as a bridge between the daily increasing requirements of the environment and the treatability acquired by conventional biological and physico-chemical processes. AOPs are widely utilized for wastewater treatment, and because they can degrade persistent toxic contaminants from wastewater, they have been highly investigated for years. Some of the recent research is briefly discussed in this chapter, including the nano-enabled technique for wastewater remediation, advancements in the UV/persulfate process and green Pickering HIPEs method for the adsorption of contaminants from the wastewater. Moreover, the combination of AOP and bioremediation techniques for managing the effluent is also discussed in detail. These techniques will lead to the decomposition of anthropogenic contaminants and are particularly used for the breakdown and removal of recalcitrant compounds present in the wastewater.

11.5.1 Nano-Enabled Technique for Wastewater Remediation

Recent research works show that graphene and its derivatives can result in the remediation of the environment as novel metal-free compounds. Nitrogen-doped graphene nanomaterials can be synthesized by controlled pyrolysis of a combination of urea, glucose and ferric chloride. In this technique, lower levels of oxygen and nitrogen doping level can be acquired at specific temperature. The nitrogen-doped graphene that is obtained this way is utilized as a metal-free catalyst to degrade phenol effectively by the activation of peroxymonosulfate (PMS). To unveil the phenol degradation and PS activation mechanisms on nanocarbons, electron paramagnetic resonance (EPR) is used to detect the radical generation. The results show that $\cdot OH$ and $SO_4^{\cdot -}$ radicals are generated during the process of catalytic oxidation, proving significant for phenol oxidation (Wang et al. 2016).

11.5.2 Advancements in the UV/Persulfate (UV/PS) Process

As per the latest studies, the UV/persulfate (UV/PS) process can effectively degrade microorganic contaminants and algal cells present in the wastewater. The process can be used for the degradation of 2,4,6-trichlorophenol (TCP) and *Microcystis aeruginosa* (*M. aeruginosa*) simultaneously (Wang et al. 2020). The degradation of TCP and algal cells is carried out by $\cdot OH$ and $SO_4^{\cdot -}$ radicals, and increased intensity UV radiation is used for enhancing the wastewater remediation performance of this procedure. The extracellular materials that are released outside the algal cells because of cell lysis can be easily degraded by the UV/PS technique, which inhibits the presence of TCP. The study proved that it can simultaneously help to remove the microorganic pollutants and algal cells.

11.5.3 Green Pickering HIPEs Method for Adsorption of Contaminants

For the past few years, Pickering high internal phase emulsions (Pickering HIPEs) are being used to fabricate macroporous materials. However, this technique is harmful for the environment and is expensive; therefore, recently, a novel monolithic macroporous material made up of carboxymethyl cellulose-*g*-poly(acrylamide)/montmorillonite (CMC-*g*-PAM/ MMT) has been prepared by free radical polymerization through oil-in-water Pickering medium internal phase emulsions (Pickering MIPEs) (Wang et al. 2019). This process utilizes environmentally friendly flaxseed oil as continuous phase, Tween-20 as stabilizer and MMT. The pore structure of the resultant macroporous material is adjustable, as it can be easily tuned by rearranging the content of co-surfactant T-20, MMT and oil-phase volume fraction. The macroporous monolith exhibits high adsorption capacities for Cd^{2+} and Pb^{2+}, and unparalleled reusability was observed when used for five consecutive cycles of adsorption-desorption. This technique is an eco-friendly alternative that can be utilized to prepare multiporous adsorption material for adsorption of heavy metals and purification of wastewater.

11.6 Bioremediation Techniques Focused on Wastewater Treatment

Bioremediation involves the removal of contaminants present in wastewater by using microbes. This process uses natural organisms for degradation, detoxification, mineralization or transformation of hazardous materials, breaking them down into less harmful or nontoxic materials. The microbes carrying out bioremediation are called bioremediators. However, some contaminants like organic recalcitrant and certain inorganic compounds are difficult or impossible to decompose with the help of these bioremediators. This process focuses on the degradation, detoxification and immobilization of diverse physical and chemical pollutants present in the wastewater. This process can only be effective in favorable environmental conditions for the procedural requirements and allow microbial growth and activity. When brought into application, this treatment involves manipulation of the environmental parameters and changes them to conditions that are suitable for the growth of microorganisms and degradation at a faster pace (Vidali 2001). The bioremediation technologies are extensively being used today for the remediation of the pollutants present in water sources, and researchers are making sure that these techniques undergo further advanced upgradation and developments that can successfully restore the polluted environments (Shah 2020b). The conventional process of bioremediation is depicted with the help of the diagram in Figure 11.2.

In the past few decades, many degradation processes are utilized for treatment of recalcitrant pollutants. Organic wastewater is difficult to treat as it contains a plethora of contaminants such as hormones, steroids, antiepileptics, diuretics,

analgesics, poly-hydroalkanoates, polycyclic aromatic hydrocarbons (PAHs), atrazine, chlorpyrifos, biosurfactants, TNT, PCB and dibenzothiophene (DBT). These hazardous materials discharged into water bodies must be removed or treated and converted into harmless substances. Therefore, bioremediation is the most eco-friendly and cheapest method for wastewater treatment. It involves the use of various microbes for the process of degradation of complex pollutants to simpler compounds. The major microorganisms involved in the process of bioremediation of wastewater to degrade specific compounds are listed in Table 11.3.

11.6.1 Aerobic Treatment

Aerobic treatment bioremediation is the process in which the microorganisms involved require oxygen for breaking down the organic pollutants and other contaminants like phosphorus and nitrogen. Oxygen is required as it is considered the preferred electron acceptor due to the high energy yield and its requirement by some of the important enzymes to initiate the degradation. Here, oxygen is continuously mixed with the sewage or wastewater with the help of mechanical aeration devices, such as compressors or air blowers. Aeration is the process of aerobic treatment that is used to eliminate the trace volatile organic compounds (VOCs) from wastewater. It is also utilized for the transfer of substances like oxygen, from the gas phase to the water phase, through the process known as oxidation or gas adsorption. This process is helpful in oxidizing materials like manganese or iron and allows the escape of dissolved gases like H_2S and CO_2. This treatment is considered helpful in reducing the harmful gaseous emissions of greenhouse gases such as ammonia, CH_4 and N_2O.

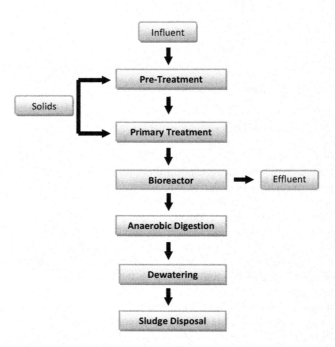

FIGURE 11.2 Bioremediation process.

TABLE 11.3

Microorganisms Used in Wastewater Treatment Bioremediation

Microorganisms Used	Compounds Degraded	References
Maclura pomifera	Pharmaceutics (bioactive peptides)	Corrons et al. 2012
Propionibacterium shermanii, Propionibacterium acidipropionici	Agricultural waste and pharmaceutics	Anderson et al. 1986 Woskow and Glatz 1991
Brevibacterium casei SRKP2	Plastics (polyhydroalkanoates)	Pandian et al. 2009
Bacillus licheniformis M104, Candida bombicola	Pharmaceutics (biosurfactants)	Gomaa 2013
Pleurotus ostreatus, Daedalea elegans, Coriolus versicolor	Polycyclic aromatic hydrocarbons (PAHs)	Arun et al. 2008
Enterobacter strain B-12	Chlorpyrifos	Singh et al. 2004
Pseudomonas sp.	Atrazine	Newcombe and Crowley 1999
Rhizobium meliloti	Dibenzothiophene (DBT)	Setti et al. 1999

11.6.2 Anaerobic Treatment

This is a treatment procedure in which the microorganisms carry out the process of bioremediation in the absence of oxygen. The presence of oxygen will hinder the activity or the enzyme generation; therefore, not even trace oxygen is allowed in the procedure. This process usually includes wastewater treatment containing biodegradable organic matter that is used further for the treatment of sedimentation sludge. Complete bioremediation by aerobic treatment is difficult because it involves fulfilling the elevated oxygen demand to preserve the aerobic conditions. However, treatment of wastewater through anaerobic bioremediation method provides an ease of treatment and a multitude of advantages such as raised levels of purification; generation of small amounts of sludge, which is very stable; the ability to deal with a huge quantity of organic load; and methane generation as an end product. This is a multistep technique involving the degradation of organic materials into smaller constituents, which then finally are converted into methane gas. This process includes the addition of an electron donor, so that the background electron acceptors like oxygen, oxidized iron, nitrates, sulfate and manganese can be depleted and the stimulation of biological, chemical reduction of the oxidized contaminants takes place. This method can be employed to treat oxidized pollutants like chlorinated ethanes, chlorinated ethenes, chloromethanes, chlorinated cyclic hydrocarbons, nitrates and many energetics (perchlorate, RDX, etc.).

11.6.3 Activated Sludge

This is an efficient bioremediation process that involves a mixture of a growing bacterial population suspended in an aqueous medium under aerobic conditions. If an unlimited and continuous flow of nutrients and oxygen is provided to the tank, then high rates of respiration and bacterial biomass are obtained. This bacterial population can lead to the consumption or degradation of the organic matter to oxidized end products or can cause biosynthesis of new microbes. This process is based on five elements, namely bioreactor, activated sludge, mixing system and aeration, sedimentation tank and returned sludge. The sludge settles down in the tank because downstream sedimentation is continuously recycled to the aeration tank. The microorganisms that return to the main reactor from this sludge are activated when they receive fresh food and oxygen in the tank (Pal 2017). Therefore, through this process, the reactor always remains activated, which results in conversion of immense contaminant loads. This method involves the removal of pollutants present in wastewater, and the cost of running the treatment is low.

11.6.4 Bioreactor

A bioreactor is an engineered system or an apparatus that controls the biologically active environment. The bioreactor is in the form of a vessel that carries out biochemical processes and involves biochemically active substances like enzymes and microorganisms such as algae, bacteria and fungi. These bioreactors are usually made up of stainless steel, range in size from liters to cubic meters and are cylindrical in shape. The bioreactors can be classified into plug, batch and continuous flow reactors. They can be used for wastewater treatment utilizing aerobic and anaerobic processes. This system can be used for growing cells or tissues in the cell culture process (Popovic and Portner 2012), which are utilized for bioprocess engineering or tissue engineering.

11.6.5 Microbial Fuel Cells (MFCs)

An MFC is a bioelectrochemical device that uses the respiring microbes' power to convert organic contaminants directly into electrical energy. An MFC is a fuel cell at the core, and it uses oxidation-reduction reactions for the transformation of chemical energy into electrical energy. In this process, the bacteria are allowed by the microbial fuel cells to grow on the anode via oxidation of the organic material, resulting in electron release. As an effect of this, the cathode sparks with air and gives rise to dissolved oxygen. This happens for proton reaction, electrons and oxygen on the cathode, resulting in completion of the electrical circuit and production of electrical energy.

11.7 Latest Innovations in Bioremediation Technology

For the sake of environmental sustainability and development, advanced and modified bioremediation technologies are emerging for improved treatment and wastewater management. Advanced bioremediation techniques such as modified activated sludge process, phytotechnologies and treatment with modified organisms can remove pollutants from wastewater. These emerging technologies would help improve the degradation rate and remediation efficiency.

11.7.1 Modified Activated Sludge Process

As per the latest studies, this is an innovative approach for the aerobic activated sludge wastewater remediation plant's design, resulting in a process operating with more efficient treatment results, decreased sludge accumulation, comparatively low energy needs and decreased amount of excess sludge produced. The approach used here is known as the complete solids retention activated sludge (CRAS) process, which involves slow degradation of unbiodegradable particulate organic matter, which can include cell debris via cellular lysis (Lubello et al. 2009; Amanatidou et al. 2016). The CRAS process provides the opportunity to have a treatment system under the conditions like high mixed liquor suspended solids (MLSS) concentration and nutrient limiting conditions in absence of filamentous bacterial growth, which is controlled by a microbial selector (Samiotis et al. 2018). Although the process has an increased organic and volumetric load, it still offers increased stability, minimization of sludge accumulation and excellent bioremediation efficiency.

11.7.2 Usage of Aquatic Weed in Phytoremediation of Wastewater

Phytoremediation is a process that employs the use of plants for the removal of pollutants from wastewater. The most used plant is *Pistia stratiotes, or water lettuce, which has the tendency to exhibit weedy behavior and* is utilized in the treatment of domestic, agricultural and industrial wastewater (Mishima et al. 2008). Research recently conducted on the mechanism of this plant shows that the plant improves the quality of wastewater by reducing the turbidity, BOD, sulfate, nitrate, chlorine, dissolved oxygen (DO) and conductivity up to a significant percentage. Moreover, a reduction in the putrid smell produced by the pollutants in the wastewater and in the pH level is also observed. Decreased total suspended solids (TSS) was observed when fecal coliform removal is taken into consideration, demonstrating the ability of these plants to supply toxins for pathogen disinfection and nutrient uptake (Yasar et al. 2018). *P. stratiotes* plants are also considered to be hyperaccumulators in the phytoremediation of sugar effluents and heavy metals like Fe and Pb present in wastewater (Haidara et al. 2019). Current works also show that these plants are utilized as an agent of phytoremediation for the post-treatment of sewage and domestic wastewater (Schwantes et al. 2019).

11.7.3 Genetically Engineered Microorganisms (GEMs)

Recent research analysis presents the enhanced bioremediation capability using genetically engineered microorganisms

(GEMs). Recent advancements in the genetics of bioremediation and the knowledge-based methods of rational protein modification have helped in the development of designer biocatalysts by the integration of efficient metabolic and novel pathways. This is done by enhancing the stability of the catabolic activity and broadening the substrate range of current pathways (Paul et al. 2005). Therefore, the derivative pathway of GEMs with a target pollutant could increase the efficiency of bioremediation when using the biological approach. Also, research in the multiplication of genetically engineered microbes and the parallel gene transfer indicates that these are encouraging approaches for environmental restoration. Recent articles show that the recalcitrant and resistant chemical dyes that are harmful for animals, plants and humans can be degraded completely through the microbial remediation procedure using engineered microorganisms.

11.8 Combination of AOP and Bioremediation Technique for Industry Effluent Management

The need to develop alternative wastewater treatment and management technologies is increasing continually. Considering this, AOPs are developing as a highly competitive wastewater remediation technique to treat and eliminate persistent organic contaminants. Although chemical oxidation for absolute mineralization can turn out to be pricey, it is proven that organic recalcitrant compounds are not effectually removed from the wastewater through conventional biological treatments because of low biodegradability or chemical stability (Kumar et al. 2005). Therefore, a helpful alternative would be to integrate AOP and bioremediation for the treatment of wastewater. Advanced oxidation process is used as a pretreatment to convert the initially resistant organic recalcitrant into simpler and more biodegradable compounds, which can then be further degraded with a biological oxidation process with a lower cost.

11.8.1 Utilization of Combination Technique (AOP and Bioremediation) for Degradation of Phenolic Wastewater

Increased contamination of water resources by phenol has become an issue of growing concern. Phenol is a very harmful contaminant as it has detrimental effects on human health, causing digestive problems, necrosis, liver problems and kidney damage. It is a known carcinogen and is of great concern even when present at low concentrations. Therefore, the treatment of phenolic wastewater becomes a necessity. Phenolic compounds can be found in several industrial chemicals, petroleum refineries, coke gasifiers, metals, medicinal plants, pesticides, plastics and organic chemical plants (Mohan et al. 2005; Shah 2020a).

Remediation of phenolic wastewater is very important to protect the environment and avoid any ecological issues. Currently, chemical and physical methods and additional biological techniques are utilized for the remediation of phenolic wastewater. An improved version of the conventional treatment process known as activated sludge, which uses a sequencing batch reactor (SBR) to treat the phenolic effluents, has been prepared in the past few decades.

The process of SBR has more advantageous results when compared with other reactors. In this process, the oxygen is supplied to the wastewater to reduce the chemical and biological oxygen demand (COD, BOD), making the water suitable for its further use or for dispersing it into the sewers. The initial treatment cost is also comparatively lower than other methods, but in some cases the low biodegradability of phenolic pollutants may affect the working efficiency and productivity of the SBR method. To overcome this drawback, a pre-treatment step can be included in the SBR treatment process. This step involves the method of AOP; when used as the initial step, AOPs tend to enhance the biodegradability and biotreatability of a wide range of pollutants. It is done using processes like sonochemical remediation, ozonation and UV-based AOPs, which are considered to be effective for several industrial effluents, as the potential of these processes in producing highly reactive degradative hydroxyl radicals has proven to be successful in treating organic recalcitrant compounds.

The combination technique utilizes ultrasound, ozonation and ultraviolet processes to obtain more degradation, and SBR is used for the bioremediation process (Wang and Loh 1999). The SBR process helps in obtaining an enhanced rate of decomposition and wider area. In the pre-treatment step, phenol is treated with UV radiation, ultrasound decomposition and ozonation that increase the biodegradability of the pollutants for further bioprocessing. The second step includes the use of a sequencing batch reactor, carrying out *fill, react, settle* and *draw* activated sludge processes for the remediation of the pollutants. In this process, the wastewater is poured into an acrylic column reactor to degrade the phenolic contaminants, removing the unwanted compounds present in the water and then discharging it (Vijayaraghavan et al. 2014). This reactor can help carrying out processes like aeration, equalization and clarification using time-controlled reactions. In the phenolic degradation, after the *fill* step is the *react* period, where effluent concentration is fixated. Proper aeration is carried out in both the *fill* and *react* periods. During each SBR cycle, certain amount of feed solution is introduced into the SBR reactor, which is filled with activated sludge. The process is repeated, and the next cycle is started, with an increased phenol concentration and higher *fill* period timings. It is observed that with each cycle, the amount of the degraded phenol increased even with variable pre-treatment. However, the ozonation method needed a shorter *react* time than did the other two procedures, showing that the process of ozonation was more advantageous for providing the remediation at a higher organic loading rate. Among all the three pre-treatment methods, ozonation shows a higher percentage of phenolic degradation. Therefore, the combination technique using ozonation as pre-treatment and SBR activated sludge process for bioremediation can be useful for the complete degradation of highly recalcitrant phenolic contaminants and prove to be an economical method for removing phenol from wastewater.

11.9 Conclusion

The chapter mainly focuses on the great depth and scientific ingenuity in applications of advanced oxidation processes and bioremediation techniques in the treatment of recalcitrant compounds present in wastewater. The work discusses in detail the scientific perspectives of conventional and non-conventional environmental engineering processes, which mainly include bioremediation and AOP for the treatment of persistent recalcitrant pollutants. The principles, mechanisms, subprocesses and degradation efficiency of both individual methods and the combined process of AOPs and bioremediation techniques are discussed. The chapter also delineates new scientific advancements in the field of treatment technologies involved in the degradation of recalcitrant compounds. The previous investigations listed in the chapter show that these processes have the capability to increase separation efficiency, enhance water quality and decrease the amount of energy consumed in the process. Thus, the various processes discussed have high potential applications in the treatment of industrial wastewater and can solve numerous existing problems with wastewater treatment, in turn preventing environmental biota.

REFERENCES

Amanatidou E, Samiotis G, Trikoilidou E, Tsikritzis L. Particulate organics degradation and sludge minimization in aerobic, complete SRT bioreactors. *Water Research*. 2016 May 1;94:288–295.

Anderson TM, Bodie EA, GoodmaN N, Schwartz RD. Inhibitory effect of autoclaving whey-based medium on propionic acid production by Propionibacterium shermanii. *Applied and Environmental Microbiology*. 1986 Feb 1;51(2):427–428.

Arun A, Raja PP, ARthi R, Ananthi M, Kumar KS, Eyini M. Polycyclic aromatic hydrocarbons (PAHs) biodegradation by basidiomycetes fungi, Pseudomonas isolate, and their cocultures: comparative in vivo and in silico approach. *Applied Biochemistry and Biotechnology*. 2008 Dec;151(2):132–142.

Asim N, Badeiei M, Ghoreishi BK, Ludin NA, Reza M. *Advances in fluid mechanics and heat & mass transfer*. Proceedings of the 10th WSEAS International Conference on Heat Transfer, Thermal Engineering and Environment (HTE'12), Istanbul, Turkey. 2012 Aug;21–23.

Bellandi R, editor. *Hazardous waste site remediation: The engineer's perspective*. Van Nostrand Reinhold Company. 1988.

Bharagava RN, Saxena G, Mulla SI, Patel DK. Characterization and identification of recalcitrant organic pollutants (ROPs) in tannery wastewater and its phytotoxicity evaluation for environmental safety. *Archives of Environmental Contamination and Toxicology*. 2018 Aug;75(2):259–272.

Bila DM, Montalvao AF, Silva AC, Dezotti M. Ozonation of a landfill leachate: Evaluation of toxicity removal and biodegradability improvement. *Journal of Hazardous Materials*. 2005 Jan 31;117(2–3):235–242.

Bourgin M, Borowska E, Helbing J, Hollender J, Kaiser HP, Kienle C, McArdell CS, Simon E, Von Gunten U. Effect of operational and water quality parameters on conventional ozonation and the advanced oxidation process O_3/H_2O_2: kinetics of micropollutant abatement, transformation product and bromate formation in a surface water. *Water Research.* 2017 Oct 1;122:234–245.

Camel V, Bermond A. The use of ozone and associated oxidation processes in drinking water treatment. *Water Research.* 1998 Nov 1;32(11):3208–3222.

Corrons MA, Bertucci JI, Liggieri CS, López LM, Bruno MA. Milk clotting activity and production of bioactive peptides from whey using Maclura pomifera proteases. *LWT-Food Science and Technology.* 2012 June 1;47(1):103–109.

Davies PS. *The biological basis of wastewater treatment.* Strathkelvin Instruments Ltd. 2005;3.

Entezari MH, Heshmati A, Sarafraz-Yazdi A. A combination of ultrasound and inorganic catalyst: Removal of 2-chlorophenol from aqueous solution. *Ultrasonics Sonochemistry.* 2005 Jan 1;12(1–2):137–141.

Estrada AL, Li YY, Wang A. Biodegradability enhancement of wastewater containing cefalexin by means of the electro-Fenton oxidation process. *Journal of Hazardous Materials.* 2012 Aug 15;227:41–48.

Fenton HJ. LXXIII. – Oxidation of tartaric acid in presence of iron. *Journal of the Chemical Society, Transactions.* 1894;65:899–910.

Glaze WH. Drinking-water treatment with ozone. *Environmental Science & Technology.* 1987 Mar;21(3):224–230.

Glaze WH, Kang JW, Chapin DH. The chemistry of water treatment processes involving ozone, hydrogen peroxide and ultraviolet radiation. *Ozone: Science & Engineering.* 1987 June; 9(4):335–352.

Gomaa EZ. Antimicrobial activity of a biosurfactant produced by Bacillus licheniformis strain M104 grown on whey. *Brazilian Archives of Biology and Technology.* 2013 Apr;56(2):259–268.

Gottschalk C, Libra JA, Saupe A. *Ozonation of water and wastewater: A practical guide to understanding ozone and its applications.* John Wiley & Sons; 2009 Dec 9.

Haidara AM, Magami IM, Sanda A. Bioremediation of aquacultural effluents using hydrophytes. *Bioprocess Engineering.* 2019 Mar 13;2(4):33.

Hamby DM. Site remediation techniques supporting environmental restoration activities – a review. *Science of the Total Environment.* 1996 Nov 22;191(3):203–224.

Henglein A, Gutierrez M. Sonolysis of polymers in aqueous solution: New observations on pyrolysis and mechanical degradation. *The Journal of Physical Chemistry.* 1988 Jun;92(13):3705–3707.

Hoigné J. Chemistry of aqueous ozone and transformation of pollutants by ozonation and advanced oxidation processes. In: *Quality and treatment of drinking water II.* Springer. 1998;83–141.

Huang CP, Dong C, Tang Z. Advanced chemical oxidation: Its present role and potential future in hazardous waste treatment. *Waste Management.* 1993 Jan 1;13(5–7):361–377.

Ikai H, Nakamura K, Shirato M, Kanno T, IwAsawa A, SasaKi K, Niwano Y, Kohno M. Photolysis of hydrogen peroxide, an effective disinfection system via hydroxyl radical formation. *Antimicrobial Agents and Chemotherapy.* 2010 Dec 1;54(12):5086–5091.

Jegatheesan V, Visvanathan C, Ben Aim R. *Advances in biological wastewater treatment. Concise encyclopaedia of bioresource technology.* Haworth Press, Inc. 2004 Jan 1.

Jeworski M, Heinzle E. Combined chemical-biological treatment of wastewater containing refractory pollutants. *Biotech Annual Rev.* 2000;6:163–196.

Karrer NJ, Ryhiner G, Heinzle E. Applicability test for combined biological-chemical treatment of wastewaters containing biorefractory compounds. *Water Research.* 1997 May 1;31(5):1013–1020.

Kasprzyk-Hordern B, Dinsdale RM, Guwy AJ. The occurrence of pharmaceuticals, personal care products, endocrine disruptors and illicit drugs in surface water in South Wales, UK. *Water Research.* 2008 July 1;42(13):3498–3518.

Kasprzyk-Hordern B, Dinsdale RM, Guwy AJ. The removal of pharmaceuticals, personal care products, endocrine disruptors and illicit drugs during wastewater treatment and its impact on the quality of receiving waters. *Water Research.* 2009 Feb 1;43(2):363–380.

Kim SM, Vogelpohl A. Degradation of organic pollutants by the photo-Fenton-process. *Chemical Engineering & Technology: Industrial Chemistry-Plant Equipment-Process Engineering-Biotechnology.* 1998 Feb;21(2):187–191.

Konsowa AH. Decolorization of wastewater containing direct dye by ozonation in a batch bubble column reactor. *Desalination.* 2003 Aug 1;158(1–3):233–240.

Kumar A, Kumar S, Kumar S. Biodegradation kinetics of phenol and catechol using Pseudomonas putida MTCC 1194. *Biochemical Engineering Journal.* 2005 Jan 1;22(2):151–159.

Lai HT, Hou JH, Su CI, Chen CL. Effects of chloramphenicol, florfenicol, and thiamphenicol on growth of algae Chlorella pyrenoidosa, Isochrysis galbana, and Tetraselmis chui. *Ecotoxicology and Environmental Safety.* 2009 Feb 1;72(2):329–334.

Lubello C, Caffaz S, GoRi R, Munz G. A modified activated sludge model to estimate solids production at low and high solids retention time. *Water Research.* 2009 Oct 1;43(18):4539–4548.

Mara D, Horan NJ, editors. *Handbook of water and wastewater microbiology.* Elsevier. 2003 Aug 7.

Margulis MA, Margulis IM. Mechanism of sonochemical reactions and sonoluminescence. *High Energy Chemistry.* 2004 Sep;38(5):285–294.

Mason TJ, Bernal VS. An introduction to sonoelectrochemistry. *Power Ultrasound in Electrochemistry.* 2012:21–44.

Mason TJ, Cobley AJ, Graves JE, Morgan D. New evidence for the inverse dependence of mechanical and chemical effects on the frequency of ultrasound. *Ultrasonics Sonochemistry.* 2011 Jan 1;18(1):226–230.

Mason TJ, Lorimer JP, Bates DM, Zhao Y. Dosimetry in sonochemistry: The use of aqueous terephthalate ion as a fluorescence monitor. *Ultrasonics Sonochemistry.* 1994 Jan 1;1(2):S91–S95.

Mishima D, Kuniki M, Sei K, Soda S, Ike M, Fujita M. Ethanol production from candidate energy crops: Water hyacinth (Eichhornia crassipes) and water lettuce (Pistia stratiotes L.). *Bioresource Technology.* 2008 May 1;99(7):2495–2500.

Mohan SV, Rao NC, Prasad KK, Madhavi BT, Sharma PN. Treatment of complex chemical wastewater in a sequencing batch reactor (SBR) with an aerobic suspended growth configuration. *Process Biochemistry.* 2005 Apr 1;40(5):1501–1508.

Newcombe DA, Crowley DE. Bioremediation of atrazine-contaminated soil by repeated applications of atrazine-degrading bacteria. *Applied Microbiology and Biotechnology.* 1999 Jun;51(6):877–882.

Oller I, Malato S, Sánchez-Pérez J. Combination of advanced oxidation processes and biological treatments for wastewater decontamination-a review. *Science of the Total Environment.* 2011 Sep 15;409(20):4141–4166.

Ollis DF, Al-Ekabi H, editors. *Photocatalytic purification and treatment of water and air: proceedings of the 1st international conference on TiO2 photocatalytic purification and treatment of water and air. 1992 Nov 8–13.* Elsevier Science Limited. 1993.

Pal P. *Industrial water treatment process technology.* Butterworth-Heinemann; 2017 Mar 31.

Pandian SR, Deepak V, Kalishwaralal K, Muniyandi J, Rameshkumar N, Gurunathan S. Synthesis of PHB nanoparticles from optimized medium utilizing dairy industrial waste using Brevibacterium casei SRKP2: A green chemistry approach. *Colloids and Surfaces B: Biointerfaces.* 2009 Nov 1;74(1):266–273.

Paul D, Pandey G, Pandey J, Jain RK. Accessing microbial diversity for bioremediation and environmental restoration. *Trends in Biotechnology.* 2005 Mar 1;23(3):135–142.

Pilli S, Bhunia P, Yan S, LeBlanc RJ, Tyagi RD, Surampalli RY. Ultrasonic pretreatment of sludge: A review. *Ultrasonics Sonochemistry.* 2011 Jan 1;18(1):1–8.

Poole AJ. Treatment of biorefractory organic compounds in wool scour effluent by hydroxyl radical oxidation. *Water Research.* 2004 Aug 1;38(14–15):3458–3464.

Popović MK, Pörtner R. *Bioreactors and cultivation systems for cell and tissue culture.* Encyclopedia of life support systems (EOLSS), Developed under the Auspices of the UNESCO. Eolss Publishers. www. eolss. net. 2012.

Rubalcaba A, Suárez-Ojeda ME, Stüber F, Fortuny A, Bengoa C, Metcalfe I, Font J, Carrera J, Fabregat A. Phenol wastewater remediation: advanced oxidation processes coupled to a biological treatment. *Water Science and Technology.* 2007 Jun;55(12):221–227.

Samiotis G, Tzelios D, Trikoilidou E, Koutelias A, Amanatidou E. Innovative approach on aerobic activated sludge process towards more sustainable wastewater treatment. *Multidisciplinary Digital Publishing Institute Proceedings.* 2018; 2(11):645.

Sandhya S, Padmavathy S, Swaminathan K, Subrahmanyam YV, Kaul SN. Microaerophilic – aerobic sequential batch reactor for treatment of azo dyes containing simulated wastewater. *Process Biochemistry.* 2005 Feb 1;40(2):885–890.

Schwantes D, Gonçalves Jr AC, Schiller AD, Manfrin J, Campagnolo MA, Somavilla E. Pistia stratiotes in the phytoremediation and post-treatment of domestic sewage. *International Journal of Phytoremediation.* 2019 Jun 7;21(7):714–723.

Setti L, Farinelli P, Di Martino S, FraSsinetti S, Lanzarini G, Pifferi PG. Developments in destructive and non-destructive pathways for selective desulfurizations in oil-biorefining processes. *Applied Microbiology and Biotechnology.* 1999 Jul;52(1):111–117.

Shah MP. *Removal of emerging contaminants through microbial processes.* Springer. 2021.

Shah MP. *Advanced oxidation processes for effluent treatment plants.* Elsevier. 2020a.

Shah MP. *Microbial bioremediation and biodegradation.* Springer. 2020b.

Sies H, Stahl W, Sundquist AR. Antioxidant functions of vitamins: Vitamins E and C, Beta-Carotene, and other carotenoids a. *Annals of the New York Academy of Sciences.* 1992 Sep;669(1):7–20.

Singh BK, Walker A, Morgan JA, Wright DJ. Biodegradation of chlorpyrifos by Enterobacter strain B-14 and its use in bioremediation of contaminated soils. *Applied and Environmental Microbiology.* 2004 Aug 1;70(8):4855–4863.

System SE. *The UV/oxidation handbook.* System. 1994.

Tang WZ. *Physicochemical treatment of hazardous wastes.* CRC Press; 2003 Dec 29.

Todd J, Josephson B. The design of living technologies for waste treatment. *Ecological Engineering.* 1996 May 1;6(1–3):109–136.

Vidali M. Bioremediation: An overview. *Pure and Applied Chemistry.* 2001 Jul 1;73(7):1163–1172.

Vijayaraghavan G, Rajasekaran R, Shantha Kumar S. Effectiveness of hybrid techniques (Aop & Bio): Simulation of degradation of phenolic wastewater. *Environ. Sci. Pollut. Res. Int.* 2014;4:102–108.

Von GU. Ozonation of drinking water: Part I. Oxidation kinetics and product formation. *Water Research.* 2003 Apr 1;37(7):1443–1467.

Wang C, Kang J, Sun H, Ang HM, Tadé MO, Wang S. One-pot synthesis of N-doped graphene for metal-free advanced oxidation processes. *Carbon.* 2016 Jun 1;102:279–287.

Wang F, Zhu Y, Xu H, Wang A. Preparation of carboxymethyl cellulose-based macroporous adsorbent by eco-friendly Pickering-MIPEs template for fast removal of Pb2+ and Cd2+. *Frontiers in Chemistry.* 2019 Sep 10;7:603.

Wang J, Wan Y, Yue S, Ding J, Xie P, Wang Z. Simultaneous removal of microcystis aeruginosa and 2, 4, 6-trichlorophenol by UV/persulfate process. *Frontiers in Chemistry.* 2020 Nov 4;8:926.

Wang SJ, Loh KC. Modeling the role of metabolic intermediates in kinetics of phenol biodegradation. *Enzyme and Microbial Technology.* 1999 Aug 1;25(3–5):177–184.

Wilson DJ. *Hazardous waste site soil remediation: Theory and application of innovative technologies.* Routledge; 2017 Nov 22.

Woskow SA, Glatz BA. Propionic acid production by a propionic acid-tolerant strain of Propionibacterium acidipropionici in batch and semicontinuous fermentation. *Applied and Environmental Microbiology.* 1991 Oct 1;57(10):2821–2828.

Yang M. A current global view of environmental and occupational cancers. *Journal of Environmental Science and Health, Part C.* 2011 Jul 1;29(3):223–249.

Yasar A, Zaheer A, TABinda AB, Khan M, Mahfooz Y, Rani S, Siddiqua A. Comparison of reed and water lettuce in constructed wetlands for wastewater treatment. *Water Environment Research.* 2018 Feb;90(2):129–135.

12

Innovative Advanced Oxidation Processes for Micropollutants in Wastewater

Sevde Üstün Odabaşi and Hanife Büyükgüngör

CONTENTS

12.1 Introduction

It is seen that aquatic environment sustainability is at risk due to the micropollutants that are formed as a result of meeting the needs of more and more people in the world. Especially the aquatic environment, which is one of the physical elements of the natural environment, human beings pollute without realizing it. As a result of this situation, the living beings in the aquatic environment whose natural cycle is disrupted cannot fully perform their functions. In recent years, pharmaceuticals, endocrine-disrupting chemicals and personal care products are classified as emerging pollutants due to their persistent nature in the aquatic environment. These pollutants have negative impacts on aquatic environments due to their persistence and toxicity (Kanakaraju et al., 2018). However, these emerging pollutants are also referred to as micropollutants due to their low concentrations, such as µg/L and ng/L, and can be found in surface waters, groundwater and even drinking water. Micropollutants are mixed into the aquatic environment from many sources such as domestic wastewater, hospital wastewater discharges, agriculture runoffs, industries and wastewater treatment plants (Mompelat et al., 2009). Figure 12.1 shows the distribution paths of micropollutants into the environment.

Agriculture and livestock runoffs are composed of pesticides used to increase production and hormones and antibiotics used in animal treatment, and these are important sources of micropollutants (Barbosa et al., 2016). Other sources of micropollutants are leaks from landfills and industrial waste systems and septic tanks. Wastewater treatment plants have been identified as the main source of micropollutants as they collect discharges from households, hospitals and industries. At the same time, wastewater treatment plants act as the primary barrier to the spread of micropollutants into the aquatic environment. Micropollutants can preserve their original structures and concentrations as a result of their discharge to the aquatic environment or they can transform into other active or inactive compounds by undergoing changes (Kim & Zoh, 2016; Luo et al., 2014). Their low biodegradability makes the discharge of treated wastewater from wastewater treatment plants an important pathway for micropollutants into the receiving water bodies (Lado Ribeiro et al., 2019). However, conventional wastewater treatment plants are not specifically designed for micropollutant removal and therefore are insufficient for removing micropollutants. In this context, micropollutants that are poorly removed in wastewater treatment plants are likely to be found in surface waters, groundwater, sediments and even drinking water (Feng et al., 2013).

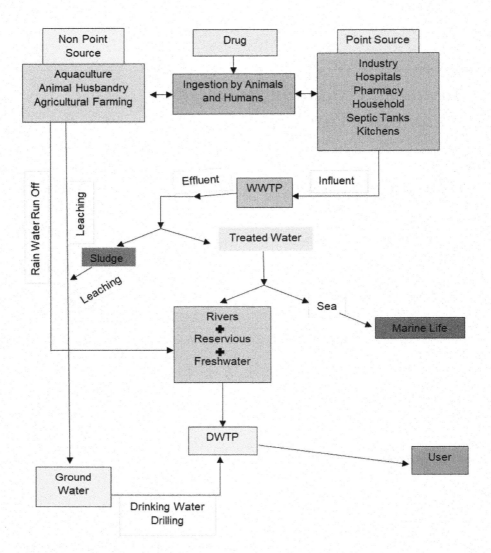

FIGURE 12.1 Distribution pathways of micropollutants into the aquatic environment.

Among the treatment techniques used up to now, advanced oxidation processes (AOPs) are among the most effective processes in removing micropollutants. AOP is a process based on the production of non-selective and highly reactive hydroxyl radicals for the degradation of organic substances (Kanakaraju et al., 2018). The most common processes among AOPs are processes with free hydroxyl radicals such as ozone and hydrogen peroxide based on photolysis of water (Klavarioti et al., 2009). At the same time, there are many innovative AOPs (visible light heterogeneous photocatalysis, electrochemical AOPs [electro-Fenton etc.], ultrasound-based AOPs [sono-Fenton etc.], plasma and electron beam) for removing micropollutants from water, and they are being reported on continuously in new studies (Miklos et al., 2018; Salimi et al., 2017). This chapter provides an overview of recent studies using AOPs to remove micropollutants from wastewater and also focuses on examining innovative AOP technologies.

12.2 Micropollutants

Micropollutants, which have been present in trace amounts in the aquatic environment in recent years, cause an increase in environmental concern and prevent the progress of environmental sustainability. For this reason, the fate of micropollutants in the aquatic environment has become a worldwide issue. Micropollutants, also known as emerging pollutants, are the name given to organic or mineral substances that are toxic, durable and bioaccumulative and harm the environment and ecosystem, originating from human activities. The reason these substances are called micropollutants is because their concentration in the aquatic environment is quite low (at the level of ng/L or µg/L) (Jiang et al., 2013; Luo et al., 2014). It can be caused by a variety of substances such as micropollutants, pharmaceuticals, hormones, cosmetics, toothpastes, perfumes, disinfectants, pesticides, waterproofing agents, plasticizers and flame retardants (Kim & Zoh, 2016). Despite their low concentrations, the effects of these pollutants on the ecosystem are enormous, and it is difficult to evaluate these effects. The flexibility in the regulations, the use of more chemicals in parallel with developing technology and the fact that the removal methods are not fully known have enabled micropollutants to spread rapidly to the environment (Ebele et al., 2017; Kim & Zoh, 2016). These micropollutants must be taken under control in order to ensure sustainability in the environment and aquatic ecosystems. Otherwise, they cause irreversible undesirable effects on beings living in the aquatic environment. Some of these effects are short-term and long-term toxicity, feminization of fish, reduction of the population, destruction of the nervous system of aquatic organisms, effect on photosynthesis of algae, endocrine-disrupting effects and many negative effects that increase the antibiotic resistance of microorganisms (Antakyali et al., 2015; Schwarzenbach et al., 2006).

Micropollutants can enter the aquatic environment in a variety of ways. Wastewater treatment plants are at the beginning of these ways and are an important pathway for micropollutants to enter surface water; they also act as a primary barrier to the spread of micropollutants. However, conventional wastewater treatment plants are designed to remove suspended solids, degradable organic substances and nutrients and are insufficient in removing micropollutants (Lado Ribeiro et al., 2019; Luo et al., 2014). For this reason, most of the micropollutants cannot be removed in wastewater treatment plants and are discharged into the receiving environment (Yang et al., 2017). Discharged micropollutants enter surface waters, groundwater and even drinking water. When the studies conducted are examined, it has been determined that there is an accumulation of micropollutants in soils and sediments (Kalyva, 2017; Luo et al., 2014; Sui et al., 2015).

Emerging pollutants are compounds without legal status. However, in 2000, the EU watchlist (Directive 2008/105/EC) was added to the list of 33 priority substances to be removed from surface water. Pharmaceutically active compounds (diclofenac, ibuprofen, iopamidol, musks, carbamazepine, clofiric acid, triclosan and phthalates) were added to this list in 2007 (Hena et al., 2021). Finally, in the first quarter of 2015, it was revised to include three macrolide antibiotics (azithromycin, clarithromycin and erythromycin), along with other natural hormones (estron E1), some pesticides, a UV filter and a widely used antioxidant, along with the aforementioned substances (Barbosa et al., 2016; Ebele et al., 2017).

12.2.1 Types of Emerging Micropollutants

12.2.1.1 Pharmaceuticals

Pharmaceuticals are generally produced from natural or synthesized organic compounds and are used for the prevention and treatment of diseases to increase the quality of life of humans and animals (Maletz et al., 2013; Salimi et al., 2017). Types of pharmaceuticals can be classified according to their applications. Analgesics/anti-inflammatories, cytostatics, antibiotics, X-ray contrast, anticonvulsants/tranquilizers, lipid regulators, beta blockers and more are pharmaceuticals detected in the ecosystem through various studies (Jiang et al., 2013). The pharmaceuticals commonly detected in wastewater are given in Table 12.1. After

pharmaceuticals are taken into the body, some of them are metabolized in the body, while others are not metabolized or converted into metabolites and excreted from the body. These compounds, which reach the wastewater treatment plant via sewage after being discharged from the body, are persistent compounds that can remain without degradation (Maryam et al., 2020). Hospital wastewater, pharmaceutical industries and domestic discharges are the main sources of these pollutants (Salimi et al., 2017).

12.2.1.2 Endocrine-Disrupting Chemicals

Endocrine-disrupting chemicals can be defined as compounds that interfere with the functioning of hormones and have a negative impact on environmental health (Wang et al., 2020). When endocrine-disrupting chemicals enter the human body, these synthesized chemicals can replace the hormones produced by the body (Salimi et al., 2017). The effects of these endocrine-disrupting chemicals are cumulative and occur in future generations. Moreover, these subsequent effects are irreversible and affect the sustainable development of living beings (Jiang et al., 2013). Many effects have been reported in aquatic wildlife such as endocrine system disorders and reproductive abnormalities/population declines. In humans, effects such as development disorders, female/male reproductive disorders, cardiovascular disorders, disorders of the immune system, cancer of reproductive tissues, adrenal dysfunction and metabolic disorders – obesity, diabetes and thyroid dysfunction, and neurological disorders (behavioral) have been observed (Darbre, 2019). The endocrine-disrupting chemicals commonly detected in wastewater are given in Table 12.1.

12.2.1.3 Personal Care Products

Personal care products consist of a large group of organic compounds. They are accepted as "pseudo persistent" due to their frequent use and release in human life (Meng et al., 2021). Nowadays, many personal care products such as perfumes, deodorants, toothpastes, disinfectants, detergents, shaving lotions and shampoos are used continuously (Abedi et al., 2018). While some of these substances and their by-products discharged to wastewater treatment plants are metabolized, some of them are attached to solid materials as sludge or they are originally discharged into the receiving environment without any change. The substances with antiseptic properties among personal care products disrupt the functioning of the wastewater treatment plant as it provides resistance to bacteria in the biological treatment part of the wastewater treatment plant. In addition, these substances cause the cells of living things in the aquatic environment to change (Jiang et al., 2013). The personal care products commonly detected in wastewater are given in Table 12.1.

12.3 Background of Advanced Oxidation Processes for Micropollutant Removal in Wastewater

AOPs are a good alternative among the water treatment technologies used until now to remove emerging pollutants. AOPs use the high reactivity of hydroxyl radicals to gradually

TABLE 12.1

Types of Micropollutants Commonly Detected in Wastewater (Barbosa et al., 2016; Ebele et al., 2017; Jiang et al., 2013; Sui et al., 2015; Vymazal et al., 2017; Wang et al., 2020)

Micropollutants	
Type of Pharmaceuticals	**Specific Compounds**
Analgesic/anti-inflammatory	Diclofenac
	Naproxen
	Ibuprofen
	Paracetamol
	Acetylsalicylic acid
Antibiotics	Amaoxilin
	Ciprofloxacin
	Clarithromycin
	Erythromycin
	Sulfamethoxazole
	Azithromycin
	Penicillin
Anticonvulsants/tranquilizers	Primidone
	Carbamazepine
	Diazepam
Cytostatics	Cyclosphosphamide
	Ifosfamide
X-ray contrast	Diatrizoate
	Iopamidol
	Iopromide
	Diatrizoic acid
Beta blockers	Atenolol
	Metaprolol
	Propranolol
Lipid regulators	Bezafibrate
	Clofibric acid
	Fenofibric acid
	Gemfibrozil
Stimulants	Caffeine
Endocrine-Disrupting Chemicals	**Specific Compounds**
Hormones and steroids	17α-ethinylestradiol
	Estrone (E1)
	Estrodiol (E2)
	Estriol (E3)
	17 β estrodiol
	Progesterone
	Tamoxifen
	Phytosterols
Pesticides and herbicides	Atrazine
	2,4-Dichlorophenoxy acid
	Dichlorodiphenytrichloroethane (DDT)
	Lindane
Plasticizers	Bisphenol A (BPA)
	Dimethyl phthalate (DMP)
	Diethyl phthalate (DEP)
	Dibutyl phthalate (DBP)
Other organic pollutants	Dioxin
	4-t-octylphenol
	Nonylphenol
	Phenanthene
	Polychlorinated biphenyls (PCBs)
Personal Care Products	**Specific Compounds**
Disinfectants	Triclosan
	Triclocarban

Micropollutants	
Type of Pharmaceuticals	**Specific Compounds**
Musks	Galoxolide
	Tonalide
Sunscreen agents	Octocrylene
	Ethylhexyl methoxycinnamate
Others	Benzophenone-3
	Ethyl paraben
	Methyl paraben

oxidize organic compounds to simple and nontoxic molecules (Kanakaraju et al., 2018). AOPs consist of two stages: the *in situ* formation of strong oxidants (e.g., hydroxyl radicals) and the reaction of these oxidants with target contaminants in water. Radical formation depends on parameters such as water quality and system design and is affected by their changes (Miklos et al., 2018). Unlike methods such as adsorption, ion exchange and stripping, AOPs do not present secondary waste that needs to be removed after treatment, so operating costs are low (Fast et al., 2017). Figure 12.2 shows the types of AOPs used in wastewater treatment.

12.3.1 Ozone-Based AOPs

Ozone is used as an oxidant and disinfectant (taste and odor control, decoloration, elimination of emerging pollutants, etc.) in water treatment (Kim & Zoh, 2016). Especially in drinking water treatment plants, ozonation has been a method that has been used for many years, and recently it has been used after pre-treatment and/or secondary treatment in the removal of micropollutants in wastewater treatment plants (Klavarioti et al., 2009; Salimi et al., 2017). It has been reported that over 90% of micropollutants have been removed through the ozone process (Fast et al., 2017; Huber et al., 2003). Ozone is a very selective oxidant. It is classified as an AOP as it generally forms OH in aqueous solutions (Miklos et al., 2018). Various parameters such as pH, ozone dose, water matrix and temperature affect the treatment efficiency and mineralization of micropollutants (Kanakaraju et al., 2018). Natural organic matter (NOM) and carbonate/bicarbonate ions affect the formation of •OH by interfering with the ozone decomposition rate (Lado Ribeiro et al., 2019). Ozone production must be done on site as it cannot be stored, and this has serious effects on operating costs. Again, ozone's short lifetime and high energy cost are among the cost-increasing factors. Therefore, it creates a disadvantage for real applications. Moreover, ozone's production of dangerous and carcinogenic by-products (such as formaldehyde, glycosal, methyl glycosal and bromate) poses a significant disadvantage (Fast et al., 2017; Kanakaraju et al., 2018). In studies, it has been determined that micropollutants have a low mineralization capacity despite their high removal efficiency due to the by-products formed as a result of ozonation. These results emphasize that toxicity should be evaluated before and after ozonation. Gomes et al. (2018) examined the removal of two different drugs (diclofenac and sulfamethaxazole), which are frequently detected in two different types of wastewater, in

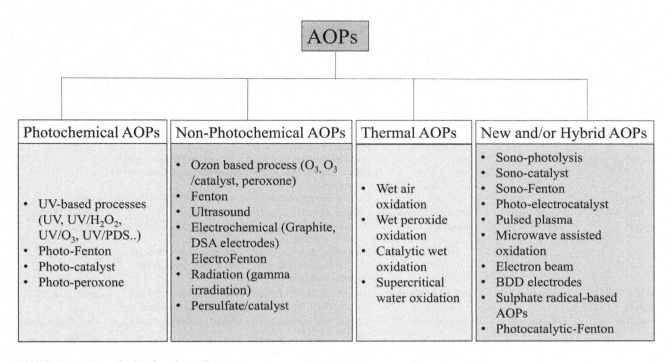

FIGURE 12.2 Types of AOPs for micropollutants wastewater treatment.

two different water matrices (secondary treatment wastewater and distilled water). As a result of the study, it was observed that the drugs completely degraded at the end of 45 and 60 minutes, but their COD removal efficiency was lower. In another study, Zhao et al. (2017) examined the degradation of the anti-inflammatory drug indomethacin by the ozonation process. It was reported that the drug (35 mg dose of O_3) completely disintegrated in 7 minutes, but the TOC yield was measured as 50% after 30 minutes. In the ozonation process, methods such as ozonation at evaluated pH, peroxane (O_3/H_2O_2), $O_3/$ UV and O_3/catalyst, O_3/photocatalyst and O_3/ultrasound are frequently used to increase the formation of hydroxyl radicals. The pH of the wastewater is a factor that directly affects ozonation, and if the pH of the water to be treated is higher than 8, the increase in •OH production as a result of the increase in hydroxide ions and therefore ozonation is affected positively (Miklos et al., 2018). The peroxane process, which is a system that combines ozone and H_2O_2, accelerates the •OH conversion of O_3, reducing the reaction time required to degrade micropollutants (Lado Ribeiro et al., 2019). For the peroxane process, the optimum molar ratio $H_2O_2/O_3 = 0.5$ mol/mol or slightly higher (Takeuchi & Mizoguchi, 2017). The most important problem for the peroxane process is residual H_2O_2. Before the treated water is discharged into the receiving environment, the residual H_2O_2 must be destroyed (Miklos et al., 2018). Catalysts are used to accelerate the decomposition of ozone in catalytic ozonation. In addition, the addition of catalyst helps degrade nonbiodegradable compounds (Rekhate & Srivastava, 2020). Generally, ozonation, when used alone, cannot provide full mineralization of micropollutants. Organic compounds can produce by-products that are more toxic when they are not fully degraded. However, the addition of catalysts

in catalytic ozonation increases the complete oxidation process and the conversion of by-products to CO_2 (Lado Ribeiro et al., 2019). Catalytic ozonation is divided into homogeneous and heterogeneous, depending on the solubility of the catalyst in water (Miklos et al., 2018). In homogeneous catalytic ozonation, transition metal ions (Fe^{2+}, Mn^{2+}, Ni^{2+}, Co^{2+}, Cd^{2+}, Cu^{2+}, Ag^+, Cr^{2+}, Zn^{2+}) are generally used to degrade micropollutants in wastewater. Metal ions affect the reaction rate, ozone selectivity and efficiency (Guo et al., 2012). The homogeneous photocatalytic ozonation process consists of two stages. In the first stage, free radicals are produced by separating ozone by metal ions. In the second stage, a complex formation takes place between the catalyst and the organic molecule (Rekhate & Srivastava, 2020). The most important disadvantage of this system is that the ions added to the water cause a second pollution and create extra treatment costs (Guo et al., 2012). Heterogeneous catalytic ozonation, with the presence of a solid catalyst, is a promising system to improve ozonation of micropollutants. Among the most important advantages of this system are that it has high stability and low loss, increases ozone decomposition efficiency and is recyclable/reusable. Among the catalysts used in heterogeneous catalytic ozonation, metal oxides (Al_2O_3, TiO_2, ZnO, MnO_2, $FeOOH$) are widely used in the degradation of micropollutants (Guo et al., 2012; Rekhate & Srivastava, 2020). Table 12.2 summarizes some of the studies on AOPs.

12.3.2 UV-Based AOPs

UV-based AOPs are generally used as an effective system in wastewater treatment of micropollutants. UV energy helps degrade micropollutants that cannot convert without UV light

TABLE 12.2

Summary of Some of the Studies on AOPs

AOPs	Micropollutants	Water Matrix	Research Findings	References
UV and UV/H_2O_2 process	23 pharmaceutical compounds	Wastewater (secondary effluent)	Capacity of 25 L photochemical reactor and irradiance of 15,47W/m^2. Optimal H_2O_2 dosage was 20 mg/L. UV adsorbance 254 nm. Removal rate of 93%. Residence time was 75 min. TOC remained unchanged after removal of pharmaceuticals. Metamizole, diclofenac, ketoprofen were mostly removed by photolysis. Caffeine and parxanthine were resistant to hydroxyl radicals.	(Afonso-Olivares et al., 2016)
UV and UV/H_2O_2 process	12 antibiotics, 10 analgesics	Wastewater (secondary effluent)	UV dose of 2768 mS/cm^2 (only UV process). UV dose of 923 mS/cm^2 (UV/H_2O_2 process) and 90% removal efficiencies. Ketoprofen, diclofenac and antipyrine had good removal. Clarithromycin, erythromycin and azithromycin had low removal with only UV process.	(Kim & Tanaka, 2009)
UV and UV/H_2O_2 process	Amoxicillin	Synthetic wastewater	UV dose: 3.8×10^{-3} Einstein/L, UV adsorbance 254 nm, 10 mM H_2O_2, 50% TOC removal, reaction time 80 min.	(Jung et al., 2012)
Ozonation	Salicylic acid	Deionized water	Salicylic acid removal was rapid pH 4 compared to pH 8 and pH 10. Ozone dose of 1 mg/L at pH=4. Salicylic acid removal was 95%.	(Hu, Zhang, & Hu, 2016)
O_3/H_2O_2	Diclofenac Sulfamethoxazole	Wastewater (secondary effluent)	Ozone dose of 20 g N/m^3, H_2O_2 dose of 5 mM, COD removal was 91%. Reaction time 120 min. Sulfamethoxazole was 45 min totally removed. Diclofenac was 60 min totally removed.	(Jung et al., 2012)
Fenton process	Carbamazepine	Synthetic wastewater	Carbamazepine is H_2O_2 and $FeSO_4$ under changing pH (3, 3.5, 4) and contact time (15, 20, 30 min). Results were 75.86% COD removal and 45.69% TOC removal and 53.73% carbamazepine removal.	(Üstün Odabaşı, Maryam, & Büyükgüngör, 2018)
Fenton-like process (Heterogeneous)	Carbamazepine Ibuprofen	Synthetic wastewater	Nano-magnetite (Fe_3O_4) heterogeneous Fenton-like reactions. CBZ/IBU solution concentration was 15 mg/L. Initial pH CBZ/IBU solution (%5.32–8.68).	(Sun et al., 2013)
Fenton-like process (Electrochemical assisted Fenton Fe^{2+}/HOCl)	1,4-dioxone	Synthetic wastewater	Single cell with a ruthenium dioxide-coated anode and a stainless steel cathode, 6.92 mA/cm^2 current density. 1.0 mmol/L Fe^{2+} concentration, removal rate was 90%.	(Kishimoto & Sugimura, 2010)
Heterogeneous Fenton/persulfate	Sulfadiazine	Synthetic wastewater	Two solid Fenton catalysis (α-FeO_3 and Fe_3O_4). The highest degradation of sulfadiazine (81% removal rate). Initial concentration of sulfadiazine = 20 ppm, pH = 4, inlet pressure = 10 atm, $Na_2S_2O_8$ = 348.5 mg/L, H_2O_2 = 0.95 ml/L, α-FeO_3 (181.8 mg/L) $[Fe^{2+}]/[H_2O_2]$ = 4.	(Roy & Moholkar, 2020)
Sonolytic Photocatalytic Sonophotocatalytic	Ibuprofen	Synthetic wastewater	Homogeneous (Fe^{3+}), heterogeneous photocatalyst (TiO_2). Ultrasound frequency of 213 kHz. Neutral pH (5.65), temperature (20 ± 2 °C). TiO_2 and Fe^{3+} were a slight synergistic enhancement.	(Madhavan, Grieser, & Ashokkumar, 2010)
Photocatalytic activity (TiO_2/AC)	Amoxicillin Ampicillin Paracetamol Diclofenac	Synthetic wastewater	TiO_2/AC composite was prepared by a temperature impregnation method. Amoxicillin (89%), ampicillin (83%), diclofenac (68%) and paracetamol (57%) were found with bare TiO_2. Amoxicillin and ampicillin were completely degraded by TiO_2/AC. Diclofenac (68%) and paracetamol (57%) were with TiO_2/AC process. TiO_2/AC dosage 0.4–1.6 g/L, TiO_2 dosage 0.2–0.8 g/L. The reactor volume 4 L, pharmaceutical solution concentration 50 mg/L, reaction time 210 min. First 30 min only adsorption process, 180 min photocatalytic process. pH range 3–10.	(Alalm, Tawfik, & Ookawara, 2016)

AOPs	Micropollutants	Water Matrix	Research Findings	References
Photocatalytic degradation (TiO$_2$ and ZnO)	Chloramphenicol	Synthetic wastewater	365 nm, 9 W lamp, TiO$_2$ (P-25) 4 h photocatalytic degradation. Chloramphenicol concentration was 50 mg/L. TiO$_2$ (P-25) was 1 g/L. Removal rate was 90%.	(Chatzitakis et al., 2008)
Homogeneous sonophotocatalytic (H$_2$O$_2$/UV/Fe/US)	Antipyrine	Synthetic wastewater	Operation conditions [H$_2$O$_2$] = 1500 ppm, pH = 2.7, amplitude 100%, λ = 190–280, 24 kHz, 200 W sonicator, Fe 10–20 ppm, pulse length (cycles) = 0.3, time 15 min, TOC removal was 92%.	(Durán et al., 2013)
Sonophotoreactor (US/UV/H$_2$O$_2$)	Chloramphenicol Diclofenac Salicylic acid Paracetamol Nitrobenzene Benzoic acid Phenol	Synthetic wastewater	Initial pH 3.9±0.1, TOC concentration was 12 mg/L, H$_2$O$_2$ concentration was 1200 mg/L, ultrasound power 140 W (20 kHz), TOC removal concentration was 98%. Flow rate 2 L/min, UV adsorbance 253.7 nm, total reactor volume 7 L.	(Ghafoori, Mowla, Jahani, Mehrvar, & Chan, 2015)
Radiolysis	Carbamazepine	Synthetic wastewater	Radiolytic degradation of CBZ with and without H$_2$O$_2$. Without H$_2$O$_2$, 75 mg/L CBZ removal efficiency of 100% within 20 kGy^{-1} adsorbed dose. Best degradation efficiency of CBZ with 10 mM/H$_2$O$_2$ 2.363 kGy^{-1}. Removal rate was 41% with 50 mM H$_2$O$_2$. 1.8 MeV, variable current (0–10 mA) electron beam.	(Liu et al., 2016)
Radiolysis	Carbamazepine	Synthetic wastewater	Radiation-induced activation of peroxymonosulfate (PMS), single ration system and Fe(II) activated PMS system. TOC removal was 38.3% with gamma radiation induced. Adsorbed dose of 100–800 Gy. The maximum decomposition efficiency was found with the molar ratio of PMS to CBZ of 20:1 at adsorbed dose of 300 Gy. Hydroxyl radicals, sulfate radicals and PMS contributed to degradation of CBZ. However, the main radical was hydroxyl radical.	(Wang et al., 2018)

(Sharma et al., 2018). In these systems, most of the UV irradiation takes place in the UV-C (usually 254 nm wavelength) region and is based on combinations of UV light with different systems (UV/H$_2$O$_2$, UV/O$_3$, UV/catalyst, UV/SO$_4$, UV/Cl$_2$, etc.) (Miklos et al., 2018) Photolysis leads to the degradation of pollutants through light exposure and photon absorption. Photons cause pollutants to degrade by destabilizing external electrons in compounds. In this process, UV lamps can be used as a light source and rarely can be used in sunlight (Fast et al., 2017). Low- and medium-pressure lamps are used in UV systems. Since the density of medium-pressure lamps is greater than low-pressure lamps, they require fewer lamps (Fast et al., 2017; Klavarioti et al., 2009; Miklos et al., 2018). One of the main problems in UV systems is the need to replace the lamp as a result the UV lamp's surface being covered by the contaminant. Another problem is that the system has high energy consumption (Fast et al., 2017). In addition, the performance of the system is affected by the pollutant adsorption spectrum, the water matrix, the H$_2$O$_2$ concentration used and the photolysis quantum efficiency. Variations in the molecular structure of the target pollutant directly affect photolysis and radical reactions, causing changes in the removal efficiency of the UV system (Kanakaraju et al., 2018; Klavarioti et al.,

2009). With the combination of UV combined with H$_2$O$_2$, more •OH radicals are produced. In this way, a better removal efficiency is provided for micropollutants with generally poor UV absorption (Kanakaraju et al., 2018). UV/H$_2$O$_2$ is an AOP that combines the instantaneous UV photocatalytic effect (directly or indirectly) with the •OH radical resulting from the homolytic degradation of H$_2$O$_2$ (Lado Ribeiro et al., 2019). That is, between UV/H$_2$O$_2$ oxidation and UV catalysis, direct photolysis and indirect oxidation (radical type oxidation, e.g., •OH, h$^+$, and e$^-$) are involved (Wang et al., 2018). The success of UV/H$_2$O$_2$ processes depends on factors such as H$_2$O$_2$ concentration, •OH radical formation rate, UV light intensity, water components, chemical structure of the micropollutant and pH of the solution (Kanakaraju et al., 2018). Water quality parameters such as NOM and alkalinity can adversely affect the degradation of micropollutants. The formation of scavengers in the water matrix can inhibit the removal of dissolved organic matter (DOM) that constitutes the main fraction of organic waste matter (Lado Ribeiro et al., 2019). The most important issue to be considered in UV/H$_2$O$_2$ process is the amount of H$_2$O$_2$ dose. High doses of H$_2$O$_2$ may be required to treat some micropollutants from wastewater. Overdosing of H$_2$O$_2$ also scavenges high concentrations of •OH radicals and prevents oxidation of target micropollutants. For this reason, in order to maximize the removal efficiency of micropollutants, the amount of H_2O_2 dose should be paid attention to (Lado Ribeiro et al., 2019). In addition, excessive use of H_2O_2 causes residual H_2O_2 at the wastewater outlet, and this residue must be treated before the wastewater is discharged into the receiving environment. The use of the UV/H_2O_2 process is more advantageous than the use of ozone because there is no potential bromate production (Fast et al., 2017). The UV/H_2O_2 process was not created for AOP due to low UV transmittance and high scavenging capacity of secondary or tertiary treated wastewater. This process is mostly used in integrated membrane systems (ultrafiltration/reverse osmosis) (Miklos et al., 2018). Afonso-Olivares et al. (2016) compared the UV and UV/H_2O_2 processes and removal efficiencies of samples taken from the biological treatment part of 23 pharmaceutical compounds from urban wastewater treatment plants. At the end of the study, they determined that only UV-C light was not effective in removal, and some micropollutants were completely mineralized with the addition of H_2O_2. Kim et al. (2009) found that UV/H_2O_2 also reduces the UV energy required to degrade micropollutants. Jung et al. (2012) investigated the degradation of amoxicillin and found that the UV/H_2O_2 (10 mM) process increased the degradation rate compared to the UV process alone. Combined with UV radiation, ozonation (UV/O_3) is an effective method for removing micropollutants in wastewater. The process starts with the photolysis of ozone and then continues with the reaction of •O water with the production of •OH radicals (Rekhate & Srivastava, 2020) because O_3 undergoes photolysis at 254 nm wavelength. Thus, micropollutants that cannot be transformed without UV light are broken down (Sharma et al., 2018). The most important disadvantage of this process is that UV lamps and ozone generators require large amounts of electrical energy and therefore increase the cost (Miklos et al., 2018).

As a different and new alternative to OH-based AOPs, $SO_4{}^{•-}$ radicals have also been used recently to remove micropollutants from water. Although $SO_4{}^{•-}$ radicals have more selective oxidizing power than •OH, low quantum yield and high cost are among their disadvantages. Peroxidesulfate (PDS, $S_2O_8{}^{2-}$) is cleaved homolytically by UV-C activation. Compared to H_2O_2, it has a higher potential to form oxidation by-products (Ahn et al., 2017; Miklos et al., 2018).

UV/Cl_2 is among other promising AOPs that oxidize target compounds with radical species of UV-activated chlorine forms (•Cl, •$Cl_2{}^-$). Mainly used are hypochlorite and chlorine dioxide. Although •Cl is a more selective oxidant than •OH, •Cl^--based reactions are not suitable for wastewater treatment as they produce chlorate, perchlorate and halogenated oxidative by-products. At the same time, NOM and alkalinity affect the reactivity of radical species, negatively affecting the degradation of micropollutants (Miklos et al., 2018; Sharma et al., 2018).

Heterogeneous photocatalysis has been studied extensively in water and wastewater treatment for the last few decades (Lado Ribeiro et al., 2019; Miklos et al., 2018). The heterogeneous photocatalysis method is based on the use of wide band gap semiconductors that produce conduction band electrons and valence band holes with a luminous energy (hv) equal to or higher than the semiconductor band gap energy. The light-formed conduction band electrons and valence band holes can recombine to dissipate heat, pass onto the semiconductor and have a mechanism to react with the species adsorbed on the catalyst surface (Lado Ribeiro et al., 2019). The basis for the photocatalytic oxidation degradation of micropollutants occurs directly or indirectly through photo-created holes and •OH radical formation (Salimi et al., 2017). Although there are many photocatalysts (e.g., ZnO, WO_3, CdS, etc.), TiO_2 is widely used because it is cheap, nontoxic, not prone to photocorrosion, photochemically stable and available in various crystal forms (Klavarioti et al., 2009; Miklos et al., 2018; Salimi et al., 2017). In TiO_2-based photocatalysis, the semiconductor material is irradiated with UV light ($\lambda = < 400$ nm) (Miklos et al., 2018). However, the disadvantage of this photocatalysis is that it cannot actively utilize the limited absorption of sunlight (due to its ability to absorb 3%–5% of the solar radiation reaching the earth's surface) since it requires near-UV radiation (Dewil et al., 2017).

Since the use of renewable energy sources in terms of economy and environment will provide great benefits, it is of great importance that the micropollutants in wastewater can be photocatalyzed by solar radiation. Wastewater treatment studies of heterogeneous photocatalytics are continuing, and new approaches are in development. Recently, researchers have tried to improve the apparent activity of TiO_2 by surface modification, noble metal addition (e.g., Au, Ag, Pt) (Lado Ribeiro et al., 2019), metal ion addition (e.g., V, Mn, Cu, Fe, Cr), nonmetal ion addition (N, S, B) (Salimi et al., 2017) and the addition of materials such as graphene-based materials (Ribeiro et al., 2017), activated carbon (Alalm et al., 2016) and carbon nanotubes (Santos et al., 2016). Again, as an alternative to TiO_2 (3.2 eV), ZnO semiconductor is used as a photocatalyst due to its very close band gap energy of 3.37 eV. Chatzitakis et al. (2008) compared the degradation efficiencies of the chloramphenicol antibiotic with TiO_2, P-25 and ZnO photocatalysis, and at the end of the study, they reported that ZnO was slightly more effective than TiO_2. However, the biggest disadvantage of ZnO compared to TiO_2 is photo corrosion (Lado Ribeiro et al., 2019). Factors affecting photocatalysis performance are catalysis concentration, UV light wavelength and intensity, solution pH and water matrix (presence of humic substances, bicarbonates, dissolved gases) (Klavarioti et al., 2009). Organic and inorganic substances in wastewater affect the performance of the system. Suspended particles and dissolved substances weaken the effectiveness of UV light. NOM and other oxidizing species prevent the degradation of micropollutants by partially depleting the •OH radical. •OH radicals scavenging ions such as bicarbonate, sulfate and chloride are among the factors that prevent the mineralization of micropollutants. Again, the organic salts (NaCI, $FeCl_3$, $FeCl_2$, $AlCl_3$, $CaCl_2$) in the water matrix have a significant deactivation effect at higher salt concentrations, causing a negative effect on the activity of the photocatalyst (Lado Ribeiro et al., 2019). Photocatalytic reactions generally occur in accordance with the Langmuir-Hinshelwood kinetic adsorption model according to the so-called first- or zero-order kinetics depending on operating conditions (Klavarioti et al., 2009). Table 12.2 summarizes some of the studies on AOPs.

12.3.3 Fenton-Based AOPs

Fenton process associated with •OH formation in wastewater treatment is divided into homogeneous and heterogeneous. The homogeneous Fenton process is also called the classical Fenton process. The Fenton process is a complex series of reactions that form hydroxyl radicals (•OH) under acidic conditions (pH = 2–3) using hydrogen peroxide (H_2O_2) as oxidant and Fe^{2+} ion as catalyst (Bokare & Choi, 2014; Salimi et al., 2017). The produced •OH radical reacts with organic pollutants, converting them into less toxic or nontoxic products (Sharma et al., 2018). The process is a metal-catalyzed oxidation reaction in which iron (usually at pH = 3) acts as a catalyst. Hydroxyl radicals (•OH) are powerful structures that can remove electrons from any substance in solution to form hydroxide. The efficiency of the process is related to the COD/H_2O_2 ratio in the feeding, which has an optimum pH value of 2–4 (Klavarioti et al., 2009). The process is limited in acidic conditions to prevent the precipitation of iron. While scavenging of hydroxyl radicals may increase in cases where pH is too low, oxidation potential and degradation rates will decrease in cases where pH is too high. Therefore, the pH of the solution is very important for the Fenton process (Fast et al., 2017). The advantages of the Fenton process, such as low cost, simple process, easy separation of residual iron and negligible activation energy, make this process attractive (Lado Ribeiro et al., 2019; Miklos et al., 2018). In addition, due to the high mineralization efficiency, organic pollutants are transformed into nontoxic CO_2 and H_2O (Lado Ribeiro et al., 2019). Due to its simple structure, it can be easily integrated into hybrid systems. Another advantage of the Fenton system is that oxidizing radicals can be easily produced at ambient temperature and pressure without the need for a complex reaction system (Bokare & Choi, 2014). For this reason, Fenton systems are widely used in the treatment of industrial wastewater (hazardous hospital wastes and pharmaceutical wastewater), as well as the removal of micropollutants in surface water (Mirzaei et al., 2017). However, Fenton processes also have disadvantages, such as the need to work at narrow pH in order to prevent the formation and precipitation of iron oxyhydroxides, Fe^{2+} rapid depletion and slow regeneration rate, the amount of sludge that needs additional treatment, and scavenging effect or loss of oxidant by self-degradation (Klavarioti et al., 2009). Moreover, the need to neutralize wastewater with bases after secondary treatment results in increased salt concentration and subsequent restriction of reuse of treated water (Lado Ribeiro et al., 2019). Since the hydroxyl radicals are not selective, undesirable by-products may occur in radical-induced reactions. For this reason, complete mineralization is targeted in wastewater treatment plants rather than removing micropollutants completely because the generated by-product can be more dangerous than the main pollutant. For this reason, TOC measurements must be made in removal studies and full mineralization (CO_2 and H_2O) must be controlled (Mirzaei et al., 2017).

In addition to the Fenton reaction, the process can be reinforced with UV radiation, which is called the photo-Fenton process. The photo-Fenton process has advantages over the Fenton process. Among the advantages are the faster Fe^{2+} regeneration rate of Fe^{3+} complexes, the use of less catalyst to generate the hydroxyl radical, less sludge production and more hydroxyl radical production (Sharma et al., 2018). In addition, the use of UV or sunlight in the photo-Fenton process is important in the disinfection of water treated by inactivation of microorganisms. For this reason, the removal efficiency of micropollutants with the photo-Fenton process is higher than the Fenton process (Mirzaei et al., 2017). The most important disadvantage of the photo-Fenton process is that the system is more costly (Lado Ribeiro et al., 2019).

Generally, additional treatment is required to reduce the concentration of Fe^{n+} ions in both Fenton and photo-Fenton processes, and a large number of chemicals are required to treat iron-containing sludge. All of these mean extra costs (Mirzaei et al., 2017). Due to these disadvantages and limitations, heterogeneous Fenton processes as an alternative to the Fenton process have recently been of great interest in the removal of micropollutants. As mentioned before, while the Fenton process proceeds under acidic conditions, there is no such requirement in the heterogeneous Fenton process (Miklos et al., 2018). With heterogeneous Fenton reactions, the applicable working range (no need for solution pH adjustment, no neutralization at the end of the process) is wide and is aimed to reduce such problems as separation of high-dose iron ions remaining after treatment (Mirzaei et al., 2017). In addition, the formation of by-products resulting from the complexation of iron ions in real wastewater can be prevented. Iron oxide minerals (magnetite [Fe_3O_4], hematite [Fe_2O_3], goethite [α-FeOOH], pyrite [FeS_2] and lepidocrocite [γ-FeOOH]) are used as catalysts in heterogeneous Fenton processes (Mirzaei et al., 2017). When magnetite is used as an example of solid catalyst in a simple Fenton-like process, it can be easily removed from solution thanks to its magnetism feature (Sun et al., 2013). It is used in support materials in the heterogeneous Fenton process. Supported metal (mostly iron) compounds and iron-coated particles show the potential to overcome the problems associated with homogeneous Fenton reactions such as limited pH operating range, iron release into wastewater and removal of iron sludge (Hu et al., 2011). The support materials used in the heterogeneous Fenton process are pillared clays, activated carbon, alumina and zeolite, and they mostly aim to improve working conditions (Mirzaei et al., 2017).

The electro-Fenton process consists of the process of producing Fe^{2+} by anodic oxidation of zero valence iron or the process of producing Fe^{2+} by cathodic reduction of Fe^{3+} ion and/or by cathodic reduction of oxygen gas. H_2O_2 is produced *in situ* in the electro-Fenton process, but has a theoretical limitation. The competition between Fe^{3+} and Fe^{2+} poses a handicap with the cathodic reduction of O_2 to H_2O_2. Electro-Fenton is a promising method for removing micropollutants from wastewater (Kishimoto & Sugimura, 2010). Table 12.2 summarizes some of the studies on AOPs.

12.3.4 Electrochemical-Based AOPs

Electrochemical advanced oxidation processes (EAOPs) have recently been used as promising technologies in the treatment of micropollutants, as they treat electrically without the need for chemicals and therefore without creating secondary waste

(Kanakaraju et al., 2018). They have been recognized as safe processes that can remove contaminants even at low concentrations. Electrochemical techniques offer the advantage of providing electrodes that are clean, versatile and effectively reactive for contaminant removal. They are easy to use and do not produce any toxic compounds (Chanikya et al., 2021). Two oxidation mechanisms, direct and indirect, occur in the treatment of micropollutants with EAOPs. First, in direct oxidation, direct charge transfer takes place between the micropollutants and the anode surface, and direct oxidation occurs with the anode. In indirect oxidation, on the other hand, it occurs through on-site production of reactive oxygen species by oxidants on the electrode surface (Kanakaraju et al., 2018). Recently, new electrode materials are produced based on technology development and can effectively destroy organic compounds. The main electrodes commonly used in water and wastewater treatment processes are carbon-based electrodes (boron-doped diamond [BDD], graphite), metal (platinum, titanium, stainless steel), stable DSA electrode (PbO_2, SnO_2, RuO_2) and sub-stoichiometric and doped TiO_2 (Kanakaraju et al., 2018; Moradi et al., 2020). BDD electrodes are the most preferred electrode for EAOPs due to the high stability of the diamond layer compared to other electrodes. BDD electrodes are effective in the degradation of micropollutants due to their corrosion stability, high oxygen overpotentials and inertness (Kanakaraju et al., 2018; Miklos et al., 2018). The use of BDD electrodes draws attention as an eco-friendly and efficient method as radicals are produced without adding chemicals. Good degradation of diclofenac, sulfamethoxazole and 17-α-estrylestradiol was determined using BDD electrodes (Kanakaraju et al., 2018). In water and wastewater treatment, BDD electrodes are used for the oxidation of microcontaminants as well as for disinfection or COD removal (Miklos et al., 2018). The results of the laboratory-scale EAOP studies are promising. However, the implementation of these studies on a pilot scale and the comparison of various techniques for the treatment of wastewater in the field are important for EAOPs. Table 12.2 summarizes some of the studies on AOPs.

12.3.5 Sonolysis

In recent years, AOP-based ultrasound, also known as sonolysis, has become popular in removing micropollutants. This technique is based on the production of •OH radicals by water pyrolysis of high-density acoustic void bubbles of fluids at cavitation-forming frequencies (20–1000 kHz) (Kanakaraju et al., 2018; Klavarioti et al., 2009). Ultrasound-induced removal of micropollutants occurs by two mechanisms: thermal decomposition and reaction with •OH radicals (Salimi et al., 2017). There are three regions in sonochemical reactions: (i) the cavitation bubble itself, (ii) the interface region between the bubble and (iii) the surrounding liquid and solution bulk (Klavarioti et al., 2009). With AOP-based ultrasound, the degradation efficiency of micropollutants, the applied ultrasound power and frequency are important (Kanakaraju et al., 2018). It has been determined that as the ultrasonic power increases, as a result of the increase of active cavitation bubbles, the removal of micropollutants also increases and there is a linear ratio between them (Madhavan et al., 2010). High-volatility

and/or low-solubility organics tend to accumulate at or around the gas-liquid interface and are well suited for the treatment of micropollutants as they are subject to rapid sonochemical degradation (Klavarioti et al., 2009). Parameters affecting AOP-based ultrasound treatment are ultrasound frequency and density, reactor geometry, type and nature of the contaminant, bulk temperature and water matrix (Klavarioti et al., 2009). Energy efficiency in ultrasound application is very low compared to other AOPs. Therefore, when this process is combined with UV irradiation (sonophotolysis-UV/US), oxidants (O_3,H_2O_2), catalysts (TiO_2) or their combination (e.g., sonophotocatalysis), both removal efficiency increases and additional advantages (energy efficiency) can be provided (Miklos et al., 2018). In addition, H_2O_2 is released during water sonolysis, and in the presence of iron ions (sono) the Fenton reaction takes place, mimicking the Fenton reaction (Klavarioti et al., 2009). Durán et al. (2013) found that 92% TOC removal efficiency was achieved in the removal of antipyrine by the H_2O_2/UV/Fe/US method. Therefore, although sonolysis does not require additional chemicals, it should be supplemented with hybrid AOPs due to disadvantages such as not providing complete mineralization and low energy efficiency. In this way, it will contribute more to the removal of micropollutants (Kanakaraju et al., 2018). Table 12.2 summarizes some of the studies on AOPs.

12.3.6 Radiation

In recent studies, gamma irradiation and electron beam radiation stand out as effective methods for the degradation of micropollutants. Compared to UV-based AOPs, gamma irradiation has shown that micropollutants facilitate the degradation, resulting in better degradation. As a result, it increases the formation of hydroxyl radicals and hydrated electrons (e^-_{aq}) (Kanakaraju et al., 2018). In gamma irradiation, the rate of degradation generally increases with the addition of an oxidizer. In their study, Wang and Wang (2018) stated that although there is only degradation in gamma radiation in the removal of carbamazepine, complete degradation is achieved with the presence of hydroxy radicals, perhydroxyl radicals and superoxide radical anions formed by the addition of peroxymonosulfate. Another radiation method in water and wastewater treatment is electron beam radiation. The accelerated electrons contact at the water surface to form electronically excited species in water, such as various ionic species and free radicals. This process exhibits a high oxidizing power, causing micropollutants to degrade (Miklos et al., 2018). The efficiency of electron beam radiation increases in the presence of an oxidizer. Liu et al. (2016) compared the removal efficiency of carbamazepine by adding hydrogen peroxide (H_2O_2) to the electron beam radiation. As a result of the study, they stated that there was an increase in the degradation of carbamazepine with the addition of H_2O_2, but the amount of H_2O_2 added was significant, otherwise it would cause a decrease in the rate of degradation. Although electron beam radiation and gamma irradiation do not require additional chemicals to initiate the reaction, the energy cost is minimal, and this method is expressed as a clean process when the process is performed at various temperatures, but the environmental concentrations of such potential ionizing radiation must be investigated absolutely (Kanakaraju et al., 2018). Table 12.2 summarizes some of the studies on AOPs.

12.3.7 Combined AOPs

In conventional wastewater treatment plants, the removal of permanent pollutants is insufficient and trace amounts of micropollutants are introduced to surface water. AOPs can often be included as add-ons to wastewater treatment plants, as they have better efficiencies in removing micropollutants. Combinations of classical AOP and biological processes can be used for the treatment of wastewater containing bioresistant and biodegradable fractions. They are applied both as a pre-treatment in the removal of toxins and as inhibiting compounds in wastewater treatment plants and as a post-treatment polishing step. Moreover, AOPs can be used together to remove the fraction of organic pollutants in the biological treatment step (Dewil et al., 2017). Instead of installing a new system, integrating AOPs into existing conventional wastewater treatment plants has been preferred in recent years due to both cost reduction and environmental friendliness. Although a single AOP is generally considered in the treatment of micropollutants, recently combined AOPs have been preferred due to better mineralization, increased removal of micropollutants and synergistic effects. AOP integration may be required to maximize treatment performance, especially in the treatment of wastewater treatment plant and industrial effluents (Klavarioti et al., 2009). Combined AOPs have benefits compared to single AOPs due to their cumulative effect (the increased production of reactive oxygen species [ROS]) and synergistic effects (positive effects between the same processes). Although not very common, combining AOPs in a few cases can also lead to antagonistic effects. An example of these effects is the self-scavenging effect observed as a result of excessive ROS production (Dewil et al., 2017). The most important point to be considered in combined AOPs is that although combined AOPs provide better mineralization, it must be determined that they have less environmental risk than single AOPs. It should be noted that in some cases the products of combined AOPs might be more toxic than the main products. For this reason, knowing the toxicity levels of the by-products is important in preventing greater damage to the environment (Kanakaraju et al., 2018).

Depending on the combinations of AOPs, type of micropollutants and water matrix, treatment sequences vary from study to study. Micropollutant removal studies with combined AOPs are given in Table 12.2. Ghafoori et al. (2015) measured TOC levels by using sonolysis and photolysis (UV/H_2O_2) systems together to degrade pharmaceutical wastewater containing chloramphenicol, diclofenac, paracetamol and salicylic acid. In the study, removal efficiencies were compared by applying only sonolysis, only photolysis and a sonophotolysis combination to pharmaceutical wastewater. TOC removal efficiency of 3% was obtained in sonolysis only, 8% in photolysis only and 91% in a sonophotolysis ($US/UV/H_2O_2$) system. As a result of the study, they decided that using sonolysis as a hybrid rather than using sonolysis alone is more advantageous in removing pharmaceutical wastewaters. Roy and Moholkar

(2020) examined the removal of sulfadiazine by heterogeneous Fenton/persulfate hybrid AOP in their study. Successful removal efficiency had been achieved compared to other AOPs. In another study, Antonio da Silva et al. (2021) carried out COD removal and toxicity studies with an AOP/GAC/AOP system in the polishing of secondary treatment effluent. Three different combined AOPs ($O_3/GAC/O_3$, $O_3/H_2O_2/GAC/O_3/H_2O_2$ and $UV/H_2O_2/GAC/UV/H_2O_2$) were compared. At the end of the study, it was determined that combining GAC with O_3/H_2O_2 had the highest removal efficiency. Since the combined AOP techniques outperform single AOPs, more studies should be conducted in this research area. It is noteworthy that although AOPs are effective in removing micropollutants, it is an expensive method. It is considered more appropriate to combine AOPs with existing conventional water treatment methods to overcome this disadvantage. In this way, both removal efficiency will be increased and cost will be reduced. For this reason, conventional wastewater treatment plants should be supported with AOPs in order to make the discharge of wastewater into the environment safer.

12.4 Conclusions and Future Outlook

Micropollutants and their metabolites pose a risk to the aquatic environment because they are nonbiodegradable, cumulative, toxic and persistent. Increasing environmental and health concerns reveal that micropollutants in wastewater are persistent pollutants. Many studies show that micropollutants in conventional wastewater treatment plants cannot be completely removed and that trace amounts of micropollutants are mixed into surface water. For this reason, AOPs appear to be a good option to increase removal efficiencies from wastewater treatment plants, which are the main input step of micropollutants into the aquatic environment. With AOP, many factors such as pollutant chemistry, water matrix and the presence of organic and inorganic substances are effective in the degradation of micropollutants in the aquatic environment. The most suitable and reliable AOPs should be selected to ensure full mineralization of micropollutants. Otherwise, numerous biologically active and more dangerous transformation products are formed from the parent compound as a result of micropollutant degradation, causing greater damage to the environment. Therefore, when experimenting with AOPs, studies should be done with real wastewater samples instead of synthetic water samples. For example, chlorine in water can cause greater damage by forming chlorinated by-products as a result of degradation of micropollutants. When comparing AOPs and conventional wastewater treatment methods, it is known that the most important challenge we encounter is cost. For this reason, adding AOPs to existing systems instead of completely changing the existing treatment techniques appears to be the most cost-effective solution. At this stage, a sustainable treatment method can be realized with the proper design, smart thinking and process optimization. AOPs must have a well-defined treatment target. Within the scope of this target, a degradation strategy should be established and focus on specific treatment instead of all treatment. In the removal of micropollutants with AOPs, new solutions such as renewable energy sources, for example solar energy or catalysts produced from waste, can be used to ensure sustainability. Moreover, treatment performance and cost effectiveness can be increased with hybrid systems created by adding systems such as ultrasound and radiation to conventional photocatalytic and ozonation systems. For the continuation of environmental sustainability, priority may be given to electrochemical AOPs (BDD, platinum, graphite and carbon-based electrodes) due to their "Green Label" (chemical-free, no waste production, operation at room temperature and atmospheric pressure). In addition, sulfate radical–based AOPs are popular recently because they can be treated without pH limit. Because while pH adjustment is needed after treatment with OH-based AOPs, pH adjustment is not needed in sulfate radical–based AOPs and unnecessary chemical substance use can be prevented.

As a result, many methods are available for the removal of micropollutants from wastewater treatment plants. However, the aim is not only to increase the performance of wastewater treatment plants with AOP, but also to obtain safe and clean wastewater by minimizing the formation of toxic by-products. Therefore, when conducting future studies, the most appropriate AOP protocol should be selected according to the water matrix, taking into account the performance, environmental impact and costs of treatment technologies. Fully operational (real wastewater) studies rather than academic (laboratory-scale) studies constitute the most important step to be applied next. In addition to all these, the most important step to take is to raise public awareness. Because a "zero-cost" treatment technology is not possible, it should be instilled in the public that a sustainable approach can be achieved by choosing low-cost useful systems.

REFERENCES

Abedi, G., Talebpour, Z., & Jamechenarboo, F. (2018). The survey of analytical methods for sample preparation and analysis of fragrances in cosmetics and personal care products. *TrAC – Trends in Analytical Chemistry*, *102*, 41–59. https://doi.org/10.1016/j.trac.2018.01.006

Afonso-Olivares, C., Fernández-Rodríguez, C., Ojeda-González, R. J., Sosa-Ferrera, Z., Santana-Rodríguez, J. J., & Rodríguez, J. M. D. (2016). Estimation of kinetic parameters and UV doses necessary to remove twenty-three pharmaceuticals from pre-treated urban wastewater by UV/H_2O_2. *Journal of Photochemistry and Photobiology A: Chemistry*, *329*, 130–138. https://doi.org/10.1016/J.JPHOTOCHEM.2016.06.018

Ahn, Y., Lee, D., Kwon, M., Choi, I. hwan, Nam, S. N., & Kang, J. W. (2017). Characteristics and fate of natural organic matter during UV oxidation processes. *Chemosphere*, *184*, 960–968. https://doi.org/10.1016/j.chemosphere.2017.06.079

Alalm, M. G., Tawfik, A., & Ookawara, S. (2016). Enhancement of photocatalytic activity of TiO2 by immobilization on activated carbon for degradation of pharmaceuticals. *Journal of Environmental Chemical Engineering*, *4*(2), 1929–1937. https://doi.org/10.1016/j.jece.2016.03.023

Antakyali, D., Morgenschweis, C., Kort, D., Sasse, R., Schulz, J., & Herbst, H. (2015, October). Micropollutants in the aquatic environment and their removal in wastewater treatment works. *Micropollutant Removal Processes*. https://www.researchgate.net/publication/312879251_micropollutants_in_the_aquatic_environment_and_their_removal_in_wastewater_treatment_works

Antonio da Silva, D., Pereira Cavalcante, R., Batista Barbosa, E., Machulek Junior, A., César de Oliveira, S., & Falcao Dantas, R. (2021). Combined AOP/GAC/AOP systems for secondary effluent polishing: Optimization, toxicity and disinfection. *Separation and Purification Technology*, *263*, 118415. https://doi.org/https://doi.org/10.1016/j.seppur.2021.118415

Barbosa, M. O., Moreira, N. F. F., Ribeiro, A. R., Pereira, M. F. R., & Silva, A. M. T. (2016). Occurrence and removal of organic micropollutants: An overview of the watch list of EU decision 2015/495. *Water Research*, *94*, 257–279. https://doi.org/10.1016/J.WATRES.2016.02.047

Bokare, A. D., & Choi, W. (2014). Review of iron-free Fenton-like systems for activating H_2O_2 in advanced oxidation processes. *Journal of Hazardous Materials*, *275*, 121–135. https://doi.org/10.1016/j.jhazmat.2014.04.054

Chanikya, P., Nidheesh, P. V., Syam Babu, D., Gopinath, A., & Suresh Kumar, M. (2021). Treatment of dyeing wastewater by combined sulfate radical based electrochemical advanced oxidation and electrocoagulation processes. *Separation and Purification Technology*, *254*(August 2020), 117570. https://doi.org/10.1016/j.seppur.2020.117570

Chatzitakis, A., Berberidou, C., Paspaltsis, I., Kyriakou, G., Sklaviadis, T., & Poulios, I. (2008). Photocatalytic degradation and drug activity reduction of Chloramphenicol. *Water Research*, *42*(1–2), 386–394. https://doi.org/10.1016/j.watres.2007.07.030

Darbre, P. D. (2019). The history of endocrine-disrupting chemicals. *Current Opinion in Endocrine and Metabolic Research*, *7*, 26–33. https://doi.org/10.1016/j.coemr.2019.06.007

Dewil, R., Mantzavinos, D., Poulios, I., & Rodrigo, M. A. (2017). New perspectives for advanced oxidation processes. *Journal of Environmental Management*, *195*, 93–99. https://doi.org/10.1016/j.jenvman.2017.04.010

Durán, A., Monteagudo, J. M., Sanmartín, I., & García-Díaz, A. (2013). Sonophotocatalytic mineralization of antipyrine in aqueous solution. *Applied Catalysis B: Environmental*, *138–139*, 318–325. https://doi.org/10.1016/j.apcatb.2013.03.013

Ebele, A. J., Abou-Elwafa Abdallah, M., & Harrad, S. (2017). Pharmaceuticals and personal care products (PPCPs) in the freshwater aquatic environment. *Emerging Contaminants*, *3*(1), 1–16. https://doi.org/10.1016/j.emcon.2016.12.004

Fast, S. A., Gude, V. G., Truax, D. D., Martin, J., & Magbanua, B. S. (2017). A critical evaluation of advanced oxidation processes for emerging contaminants removal. *Environmental Processes*, *4*(1), 283–302. https://doi.org/10.1007/s40710-017-0207-1

Feng, L., van Hullebusch, E. D., Rodrigo, M. A., Esposito, G., & Oturan, M. A. (2013). Removal of residual anti-inflammatory and analgesic pharmaceuticals from aqueous systems by electrochemical advanced oxidation processes. A review. *Chemical Engineering Journal*, *228*, 944–964. https://doi.org/10.1016/j.cej.2013.05.061

Ghafoori, S., Mowla, A., Jahani, R., Mehrvar, M., & Chan, P. K. (2015). Sonophotolytic degradation of synthetic pharmaceutical wastewater: Statistical experimental design and modeling. *Journal of Environmental Management, 150,* 128–137. https://doi.org/10.1016/j.jenvman.2014.11.011

Gomes, D. S., Gando-Ferreira, L. M., Quinta-Ferreira, R. M., & Martins, R. C. (2018). Removal of sulfamethoxazole and diclofenac from water: Strategies involving O_3 and H_2O_2. *Environmental Technology (United Kingdom), 39*(13), 1658–1669. https://doi.org/10.1080/09593330.2017.133 5351

Guo, Y., Yang, L., & Wang, X. (2012). The application and reaction mechanism of catalytic ozonation in water treatment. *Journal of Environmental & Analytical Toxicology, 2*(6). https://doi.org/10.4172/2161-0525.1000150

Hena, S., Gutierrez, L., & Croué, J. P. (2021). Removal of pharmaceutical and personal care products (PPCPs) from wastewater using microalgae: A review. *Journal of Hazardous Materials, 403*(August 2020). https://doi.org/10.1016/j.jhazmat.2020.124041

Hu, R., Zhang, L., & Hu, J. (2016). Study on the kinetics and transformation products of salicylic acid in water via ozonation. *Chemosphere, 153,* 394–404. https://doi.org/10.1016/j.chemosphere.2016.03.074

Hu, X., Liu, B., Deng, Y., Chen, H., Luo, S., Sun, C., . . . Yang, S. (2011). Adsorption and heterogeneous Fenton degradation of 17α-methyltestosterone on nano Fe3O4/MWCNTs in aqueous solution. *Applied Catalysis B: Environmental, 107*(3–4), 274–283. https://doi.org/10.1016/j.apcatb.2011.07.025

Huber, M. M., Canonica, S., Park, G., & Gunten, U. R. S. V. O. N. (2003). *Huber_2003_EST.pdf. 37*(5), 1016–1024.

Jiang, J. Q., Zhou, Z., & Sharma, V. K. (2013). Occurrence, transportation, monitoring and treatment of emerging micro-pollutants in waste water – A review from global views. *Microchemical Journal, 110,* 292–300. https://doi.org/10.1016/j.microc.2013.04.014

Jung, Y. J., Kim, W. G., Yoon, Y., Kang, J. W., Hong, Y. M., & Kim, H. W. (2012). Removal of amoxicillin by UV and UV/H_2O_2 processes. *Science of the Total Environment, 420,* 160–167. https://doi.org/10.1016/j.scitotenv.2011.12.011

Kalyva, M. (2017). *Fate of pharmaceuticals in the environment -A review.* https://doi.org/10.1016/j.ica.2010.12.018

Kanakaraju, D., Glass, B. D., & Oelgemöller, M. (2018). Advanced oxidation process-mediated removal of pharmaceuticals from water: A review. *Journal of Environmental Management, 219,* 189–207. https://doi.org/10.1016/j.jenvman.2018.04.103

Kim, I., Yamashita, N., & Tanaka, H. (2009). Performance of UV and UV/H_2O_2 processes for the removal of pharmaceuticals detected in secondary effluent of a sewage treatment plant in Japan. *Journal of Hazardous Materials, 166*(2–3), 1134–1140. https://doi.org/10.1016/J.JHAZMAT.2008.12.020

Kim, I., & Tanaka, H. (2009). Photodegradation characteristics of PPCPs in water with UV treatment. *Environment International, 35*(5), 793–802. https://doi.org/10.1016/J.ENVINT.2009.01.003

Kim, M. K., & Zoh, K. D. (2016). Occurrence and removals of micropollutants in water environment. *Environmental Engineering Research, 21*(4), 319–332. https://doi.org/10.4491/eer.2016.115

Kishimoto, N., & Sugimura, E. (2010). Feasibility of an electrochemically assisted Fenton method using Fe 2+/HOCl system as an advanced oxidation process. *Water Science and Technology, 62*(10), 2321–2329. https://doi.org/10.2166/wst.2010.203

Klavarioti, M., Mantzavinos, D., & Kassinos, D. (2009). Removal of residual pharmaceuticals from aqueous systems by advanced oxidation processes. *Environment International, 35*(2), 402–417. https://doi.org/10.1016/J.ENVINT.2008.07.009

Lado Ribeiro, A. R., Moreira, N. F. F., Li Puma, G., & Silva, A. M. T. (2019). Impact of water matrix on the removal of micropollutants by advanced oxidation technologies. *Chemical Engineering Journal, 363*(October 2018), 155–173. https://doi.org/10.1016/j.cej.2019.01.080

Liu, N., Lei, Z.-D., Wang, T., Wang, J.-J., Zhang, X.-D., Xu, G., & Tang, L. (2016). Radiolysis of carbamazepine aqueous solution using electron beam irradiation combining with hydrogen peroxide: Efficiency and mechanism. *Chemical Engineering Journal, 295,* 484–493. https://doi.org/https://doi.org/10.1016/j.cej.2016.03.040

Luo, Y., Guo, W., Ngo, H. H., Nghiem, L. D., Hai, F. I., Zhang, J., . . . Wang, X. C. (2014). A review on the occurrence of micropollutants in the aquatic environment and their fate and removal during wastewater treatment. *Science of The Total Environment, 473–474,* 619–641. https://doi.org/10.1016/J.SCITOTENV.2013.12.065

Madhavan, J., Grieser, F., & Ashokkumar, M. (2010). Combined advanced oxidation processes for the synergistic degradation of ibuprofen in aqueous environments. *Journal of Hazardous Materials, 178*(1), 202–208. https://doi.org/10.1016/j.jhazmat.2010.01.064

Maletz, S., Floehr, T., Beier, S., Klümper, C., Brouwer, A., Behnisch, P., . . . Hollert, H. (2013). In vitro characterization of the effectiveness of enhanced sewage treatment processes to eliminate endocrine activity of hospital effluents. *Water Research, 47*(4), 1545–1557. https://doi.org/10.1016/j.watres.2012.12.008

Maryam, B., Buscio, V., Odabasi, S. U., & Buyukgungor, H. (2020). A study on behavior, interaction and rejection of Paracetamol, Diclofenac and Ibuprofen (PhACs) from wastewater by nanofiltration membranes. *Environmental Technology and Innovation, 18,* 100641. https://doi.org/10.1016/j.eti.2020.100641

Meng, Y., Liu, W., Liu, X., Zhang, J., Peng, M., & Zhang, T. (2021). A review on analytical methods for pharmaceutical and personal care products and their transformation products. *Journal of Environmental Sciences (China), 101,* 260–281. https://doi.org/10.1016/j.jes.2020.08.025

Miklos, D. B., Remy, C., Jekel, M., Linden, K. G., Drewes, J. E., & Hübner, U. (2018). Evaluation of advanced oxidation processes for water and wastewater treatment – A critical review. *Water Research, 139,* 118–131. https://doi.org/10.1016/j.watres.2018.03.042

Mirzaei, A., Chen, Z., Haghighat, F., & Yerushalmi, L. (2017). Removal of pharmaceuticals from water by homo/heterogonous Fenton-type processes – A review. *Chemosphere, 174,* 665–688. https://doi.org/10.1016/j.chemosphere.2017.02.019

Mompelat, S., Le Bot, B., & Thomas, O. (2009). Occurrence and fate of pharmaceutical products and by-products, from resource to drinking water. *Environment International, 35*(5), 803–814. https://doi.org/10.1016/j.envint.2008.10.008

Moradi, M., Vasseghian, Y., Khataee, A., Kobya, M., Arabzade, H., & Dragoi, E. N. (2020). Service life and stability of electrodes applied in electrochemical advanced oxidation processes: A comprehensive review. *Journal of Industrial and Engineering Chemistry*, *87*, 18–39. https://doi.org/10.1016/j.jiec.2020.03.038

Rekhate, C. V., & Srivastava, J. K. (2020, June). Recent advances in ozone-based advanced oxidation processes for treatment of wastewater- A review. *Chemical Engineering Journal Advances*, *3*, 100031. https://doi.org/10.1016/j.ceja.2020.100031

Ribeiro, R. S., Rodrigues, R. O., Silva, A. M. T., Tavares, P. B., Carvalho, A. M. C., Figueiredo, J. L., . . . Gomes, H. T. (2017). Hybrid magnetic graphitic nanocomposites towards catalytic wet peroxide oxidation of the liquid effluent from a mechanical biological treatment plant for municipal solid waste. *Applied Catalysis B: Environmental*, *219*, 645–657. https://doi.org/10.1016/j.apcatb.2017.08.013

Roy, K., & Moholkar, V. S. (2020). Sulfadiazine degradation using hybrid AOP of heterogeneous Fenton/persulfate system coupled with hydrodynamic cavitation. *Chemical Engineering Journal*, *386*, 121294. https://doi.org/10.1016/j.cej.2019.03.170

Salimi, M., Esrafili, A., Gholami, M., Jonidi Jafari, A., Rezaei Kalantary, R., Farzadkia, M., . . . Sobhi, H. R. (2017). Contaminants of emerging concern: a review of new approach in AOP technologies. *Environmental Monitoring and Assessment*, *189*(8). https://doi.org/10.1007/s10661-017-6097-x

Santos, D. F. M., Soares, O. S. G. P., Silva, A. M. T., Figueiredo, J. L., & Pereira, M. F. R. (2016). Catalytic wet oxidation of organic compounds over N-doped carbon nanotubes in batch and continuous operation. *Applied Catalysis B: Environmental*, *199*, 361–371. https://doi.org/10.1016/j.apcatb.2016.06.041

Schwarzenbach, R. P., Escher, B. I., Fenner, K., Hofstetter, T. B., Johnson, C. A., Von Gunten, U., & Wehrli, B. (2006). The challenge of micropollutants in aquatic systems. *Science*, *313*(5790), 1072–1077. https://doi.org/10.1126/science.1127291

Sharma, A., Ahmad, J., & Flora, S. J. S. (2018, June). Application of advanced oxidation processes and toxity assessment of transformation products. *Environmental Research*, *167*, 223–233. https://doi.org/10.1016/j.envres.2018.07.010

Sui, Q., Cao, X., Lu, S., Zhao, W., Qiu, Z., & Yu, G. (2015). Occurrence, sources and fate of pharmaceuticals and personal care products in the groundwater: A review. *Emerging Contaminants*, *1*(1), 14–24. https://doi.org/10.1016/j.emcon.2015.07.001

Sun, S. P., Zeng, X., & Lemley, A. T. (2013). Nano-magnetite catalyzed heterogeneous Fenton-like degradation of emerging contaminants carbamazepine and ibuprofen in aqueous suspensions and montmorillonite clay slurries at neutral pH. *Journal of Molecular Catalysis A: Chemical*, *371*, 94–103. https://doi.org/10.1016/j.molcata.2013.01.027

Takeuchi, N., & Mizoguchi, H. (2017). Study of optimal parameters of the H_2O_2/O_3 method for the decomposition of acetic acid. *Chemical Engineering Journal*, *313*, 309–316. https://doi.org/10.1016/j.cej.2016.12.040

Üstün Odabaşı, S., Maryam, B., & Büyükgüngör, H. (2018). Fenton oxidation of carbamazepine in wastewater with fewer reagents. *Sigma Journal of Engineering and Natural Sciences*, *36*(1), 289–298.

Vymazal, J., Dvořáková Březinová, T., Koželuh, M., & Kule, L. (2017). Occurrence and removal of pharmaceuticals in four full-scale constructed wetlands in the Czech Republic – the first year of monitoring. *Ecological Engineering*, *98*, 354–364. https://doi.org/10.1016/J.ECOLENG.2016.08.010

Wang, R., Ma, X., Liu, T., Li, Y., Song, L., Tjong, S. C., . . . Wang, Z. (2020, March). Degradation aspects of endocrine disrupting chemicals: A review on photocatalytic processes and photocatalysts. *Applied Catalysis A: General*, *597*, 117547. https://doi.org/10.1016/j.apcata.2020.117547

Wang, S., & Wang, J. (2018). Degradation of carbamazepine by radiation-induced activation of peroxymonosulfate. *Chemical Engineering Journal*, *336*, 595–601. https://doi.org/10.1016/j.cej.2017.12.068

Wang, W. L., Wu, Q. Y., Huang, N., Xu, Z. Bin, Lee, M. Y., & Hu, H. Y. (2018). Potential risks from UV/H_2O_2 oxidation and UV photocatalysis: A review of toxic, assimilable, and sensory-unpleasant transformation products. *Water Research*, *141*, 109–125. https://doi.org/10.1016/j.watres.2018.05.005

Yang, Y., Ok, Y. S., Kim, K. H., Kwon, E. E., & Tsang, Y. F. (2017). Occurrences and removal of pharmaceuticals and personal care products (PPCPs) in drinking water and water/sewage treatment plants: A review. *Science of the Total Environment*, *596–597*, 303–320. https://doi.org/10.1016/j.scitotenv.2017.04.102

Zhao, Y., Kuang, J., Zhang, S., Li, X., Wang, B., Huang, J., . . . Yu, G. (2017). Ozonation of indomethacin: Kinetics, mechanisms and toxicity. *Journal of Hazardous Materials*, *323*, 460–470. https://doi.org/10.1016/j.jhazmat.2016.05.023

13

Application of Advanced Oxidation Processes in Combined Systems for Wastewater Reuse

Feryal Akbal, Burcu Özkaraova and Ayşe Kuleyin

CONTENTS

DOI: 10.1201/9781003165958-13

13.1 Introduction

Water consumption has increased as a result of population growth, rapid industrial development and differentiation in lifestyle. According to a projection by the Organisation for Economic Co-Operation and Development (OECD) (2012), an increase in global water demand of 55% is expected between 2000 and 2050 from 3500 to 5425 km³, based on demands by the manufacturing sector (+400%), thermal power production (+140%) and domestic consumption (+130%). Water withdrawal for agricultural irrigation is estimated to increase about 5.5% from 2008 to 2050 (FAO, 2011). Thus, projections represent that non-agricultural demand will rise more than agricultural demand, which will continue to remain the highest share (UN, 2018).

Higher water withdrawals have resulted in recognizable deterioration in the quantity of water resources. Generally, the recharge of water resources relies on the precipitation regime and climatic and geological conditions. Thus, the renewability of water resources is highly variable and therefore questionable. When renewable water potentially does not compensate for water withdrawal, water scarcity (lack of water) appears as a problem to be solved. According to the United Nations Development Programme (UNDP, 2016), more than 1.7 billion people are living in river basins where the water withdrawal is higher than the recharge. The business-as-usual climate scenario foresees a global water deficiency of 40% by 2030 (2030 WRG, 2009). According to the United Nations World Water Developing Report (UN, 2018), more than half of the world population will experience water scarcity to some degree. Water stress, on the other hand, may also occur due to water quality deterioration and reduced accessibility to water resources in addition to reduced availability/quantity (Schulte and Morrison, 2014). Water management generally covers all three aspects, namely availability, quality and quantity. Water management can be carried out following two different perspectives – water demand management and water supply management. Water demand management generally involves policies and strategies that enable more efficient water use and thus a reduction in water use. Reuse of wastewater (e.g., gray water) or treated wastewater, usage of water-efficient equipment/machinery and reduction in water losses (e.g., leakage) are among the main water demand management practices. Water supply management, on the other hand, covers engineering incentives for finding new resources and/or developing new water supply systems (e.g., reservoirs). Here, water resources management becomes an important issue especially for water-scarce countries, when such engineering incentives become costly. The protection of water resources, regarding both its quantity and its quality, becomes more and more important.

Numerous effects of global warming and climate change are expressed in conferences and presented in scientific reports. Impact on water resources is expected to be the most critical one. Extreme weather conditions such as more prolonged droughts or heavy rainfall and flooding regimes highlight the theme of water security. While threats from heavy rainfall and flooding require different water management scenarios involving water infrastructure, management scenarios for prolonged droughts involve enhanced efficiency in water use and water reuse issues. As existence relies on water, development of new strategies and infrastructure opportunities on a sectoral basis (e.g., agriculture, industry and domestic) for water demand management becomes important for the future. Agricultural irrigation, with the highest share (69%) of freshwater requirement, has already started a change towards efficient water irrigation systems (e.g., drip irrigation). The reuse of treated wastewater is regarded as another solution in future planning.

Wastewater is regarded as an unconventional water resource and an alternative to conventional water resources. Because domestic wastewater daily produced is more than 680–960 million m³, its potential as an alternative water resource can be better understood when appropriately treated. Unfortunately, only very little (~2 million m³) is treated beyond secondary treatment for reuse (Lautze et al., 2014; Shah, 2021). On the other hand, the reuse of 50% of the total treated wastewater theoretically available for irrigation in the EU is assumed to avoid more than 5% of direct abstraction from water resources, consequently reducing more than 5% in water stress overall (European Commission, 2018). Other alternative water resources are rain/stormwater and gray water. Once collected/harvested, these alternative water resources can be managed similarly to wastewater. Captured stormwater may be reused for restoration of surface water resources (e.g., stream flow augmentation), aquifer recharge, management of saltwater intrusion and fire protection. Individual on-site treatment of gray water in residential buildings, schools and campuses, hotels and shopping centers may enable its indoor (e.g., toilet flushing) and/or outdoor reuse (e.g., irrigation). Developing new strategies for the inclusion of alternative water resources, reduction in wasted water streams and wastewater will create an integrated water management system.

13.2 Reuse of Treated Wastewater

Conventional water resources management may not always ensure total water security. Especially when water resources are not recharged due to prolonged droughts, general precautions like water withdrawal restriction and interbasin water transfer may not address severe water scarcity conditions. Reuse of untreated and treated wastewater has been accepted as an alternative source of water (Yıldız Töre and Ata, 2019). Actually, reuse of wastewater has been experienced in water-scarce countries for years, delivering case-specific data and outcomes (Hamilton et al., 2007). Untreated or partially reclaimed water has been used for irrigating approximately 10% of irrigated land area globally (Ungureanu et al., 2020, Shah, 2021). However, the use of untreated or partially reclaimed wastewater for irrigation may have some short- and long-term consequences on both human and environmental health.

Most wastewaters from combined systems (e.g., gray water and black water, domestic and industrial wastewater) contain high numbers of various pathogenic microorganisms from toilet and urinal flushing. Past experiences have shown that especially excreta-related infections such as hepatitis, polio, schistosomiasis, cholera and typhoid may occur because of a lack of or improper treatment of wastewater. Thus, any end use with an intended or unintended exposure pathway to an adult (e.g., farmer, worker, consumer, etc.) or child will create risks to human and public health. Thus, as WHO suggests, a risk-based approach may help to determine the best precautions, level of water treatment (secondary/tertiary/advanced, with/without disinfection), and infrastructures (e.g., reservoir, distribution and/or irrigation system) to protect humankind and the environment (soil, ground and surface water). Quantitative microbial risk analysis (QMRA) is the most widely used analysis of health risks from pathogens.

Actually, the best solution would be to separate the black water from all other streams of domestic, municipal and/or industrial wastewater. This would enable the selection of another treatment scheme capable of an effective control on pathogenic microorganisms in wastewater and wastewater sludge. This separation would also reduce the volume of wastewater and sludge to be treated regarding pathogens, reduce related treatment costs and simplify final disposal options. Especially because of the presence of Covid-19 as the current pathogen found in both wastewater and sludge (Lahrich et al., 2021; Kocamemi et al., 2020), the amount and area of exposure to pathogens and related risks should be decreased as well. Various publications represent the presence of pathogens in soil and groundwater that have been exposed to untreated and treated wastewater without disinfection and the possible risks to human health (Moazeni et al., 2017; Adegoke et al., 2018). Additionally, ultraviolet (UV) disinfection is preferred over chlorine disinfection due to reduced flow rates and related costs. Chlorine disinfection should be minimized as hazardous chlorinated organic compound formation is generally expected after chlorine disinfection.

Thus, appropriate management strategies need to be developed and implemented regarding either the wastewater composition and/or the end use of treated wastewater. The secret is to detect the most appropriate treatment scheme and end use of treated wastewater. With the recognition of "fit for purpose," the most environmental and cost-effective integrated water management system can be established. End use of treated wastewater can be in agricultural, industrial (including energy), urban (municipal) and environmental sectors (Figure 13.1). Regarding the status of wastewater reuse in Europe, the highest shares were found as agricultural irrigation (32.0%), landscape irrigation (20.0%) and industrial use (19.3%), followed by lower shares as nonpotable urban uses (8.3%), environmental enhancement (8.0%), recreational use (6.4%), indirect potable use (2.3%) and groundwater recharge (2.2%) (Lazarova, 2017). The share of reuse in different sectors changes from country to country – some countries with limited reuse as recreational irrigation, other countries with additional reuse as agricultural irrigation and industrial reuse and minor countries with reuse as potable water.

13.3 Agricultural Reuse

As has been shown, the highest share of freshwater use is in the agricultural sector, mainly as irrigation water. For this reason, the reuse of treated wastewaters is generally within the agricultural sector. During agricultural production and related activities, wastewater such as tailwater, vehicle and equipment washing water, plant-washing water and food processing water may have been generated. The management of this wastewater may similarly enable feasible wastewater reclamation for its recycling and reuse in the very same sector. Sometimes, a simple pre-treatment may be enough for its recycling and reuse (e.g., plant washing). Similarly, the intention of municipal and industrial wastewater treatment is generally for recreational and agricultural reuse. Reuse of treated wastewater for the irrigation of food crops, processed food crops and non-food crops may require different water quality standards in order to assure reasonable levels of human and environmental protection. Especially, standards for water used for the irrigation of crops for raw consumption intend to eliminate risk from microbial contamination. Other crop-specific water quality requirements may also be implemented for effective crop production.

13.4 Urban and Environmental Reuse

Urban reuse of treated wastewater is one among the major sectors, after the agricultural sector. Treated municipal wastewaters are often reused for the irrigation of recreational areas and landscape areas. As golf courses require a lot of water, reuse of treated wastewater for irrigation is often preferred. The reuse of treated wastewaters can be for irrigation of restricted (no or controlled public access, e.g., highway medians) or unrestricted (free public access, e.g., parks) recreational areas. The level of wastewater treatment generally determines this restriction, generally in relation with the number of pathogens in the treated wastewater. The irrigation of free public areas requires a lower number of pathogens. Other public reuses include fire protection, toilet flushing, vehicle washing, dust control and construction.

Environmental reuse of treated wastewater, on the other hand, aims to restore wetlands, supplement stream and river flow and/or aquifer recharge (potable aquifer recharge, saltwater intrusion control, storage) that are under impact of high water withdrawal. The treated wastewater can also be used for the creation of impounds.

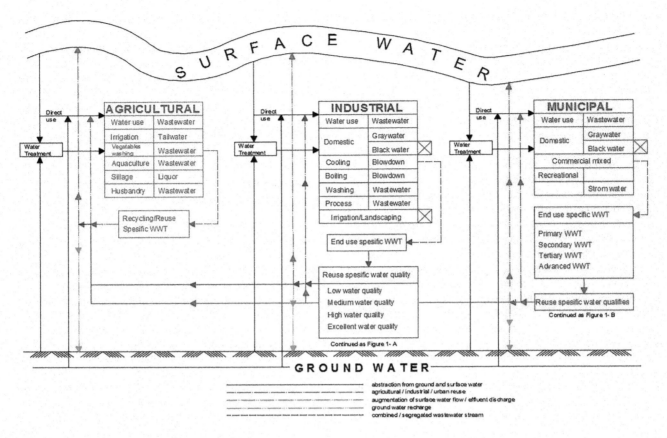

FIGURE 13.1 Wastewater reuse options in different sectors.

13.5 Industrial Reuse

Reuse of treated wastewater in the industrial sector can be for various purposes like daily facility unity cleaning water, reactor flushing water, process water, boiler feed water and cooling water. While process water and boiler feed water may require treated water of higher quality, water used for cleaning purposes may not require extensive treatment depending on the manufacturing unit processes. Thus, wastewater treatment will be end-use specific within the industrial sector. Knowing the category of water quality (e.g., excellent, high, medium and low water quality) required by the unit process and the relevant parameters of concern (e.g., hardness, alkalinity, suspended solids, etc.) is important for determining the sequence of wastewater treatment units. As can be seen from Figure 13.1a, various sequences of water treatment unit operations are possible for reaching excellent, high, medium and low water quality categories. Especially, water-intensive industries such as food and beverage production, pulp and paper production and textile production try to adopt new initiatives for the reduction of water consumption, for recycling water and/or reusing treated wastewater for different unit processes during manufacturing. Here, the water quality criteria may be specifically developed by the industrial sector with regard to the requirements of the manufacturing process.

The wastewater generally considered for reuse does not cover wastewater from all sectors but is rather limited to domestic and/or municipal wastewater with limited contributions from the industrial sector (generally the food manufacturing industry, e.g., milk processing, potato processing, soft drink and beverage processing, breweries and malt houses, fish and meat processing industries). Depending on the end use of reclaimed wastewater, all wastewaters have the potential to be appropriately treated to meet reuse water quality criteria. Wastewater treatment generally requires more treatment sequences going beyond primary and even secondary treatments (e.g., tertiary or advanced treatment) to meet excellent and high water quality (Figure 13.1b). The main issue here is the cost-benefit analyses of such attempts, the economic status of the country and the need/reason for treated wastewater reuse.

Industries generally are composed of different units with regard to processes and/or activities, each potentially generating wastewater streams of different quality and quantity. In conventional systems, these streams are combined and treated in one industrial wastewater treatment plant. This wastewater generally contains toilet and urinal flushing water and thus high numbers of various pathogenic microorganisms, making its reuse more difficult. The level of treatment totally depends on the water characteristics of combined industrial wastewater and the water of concern for specific end use. Another possible solution would be a separation of wastewater streams within the sector according to the wastewater composition or, in other words, the strength of wastewater (low, medium, high and very high strength). Thus, instead of treating a mixed wastewater

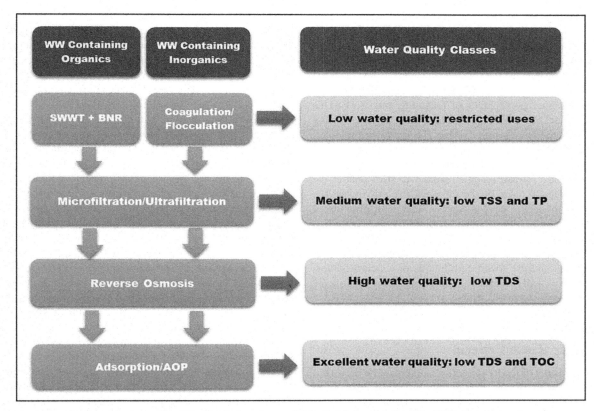

TSS: Total suspended solids; TP: Total phosphorus; TDS: Total dissolved solids; TOC: Total organic carbon
SWWT: Secondary wastewater treatment; BNR: Biological nurtient removal; AOP: Advanced oxidation process

FIGURE 13.1A   Reuse-specific water qualities.

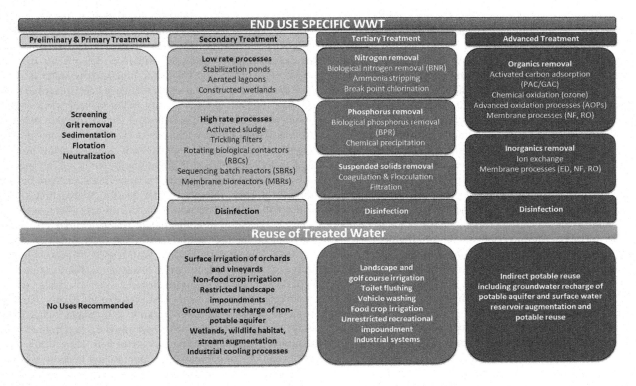

FIGURE 13.1B   End use–specific wastewater treatments.

at high cost, segregated different-strength wastewater streams can be treated with optimum cost, aiming to achieve a water quality status that meets the exact area of end use. As each unit and/or activity within the industry may require different water quality, the treatment of the most appropriate stream with the most cost-effective conventional and/or innovative technology will enable higher percentages of recycling and reuse. This strategy may act like a matchmaker accomplishing the cycling of water sources among units within the industry, industrial park (industrial symbiosis) and other sectors (sectoral symbiosis with, e.g., agriculture, municipal, etc.). Additionally, the segregation of wastewater streams will also enable the accomplishment of reuse of the most feasible streams in the other industries (e.g., heavy industry). The determination of the quality and quantity of water required by each manufacturing unit process per time, the quality and quantity of wastewater generated after the process, the variation in wastewater characteristics and the specific pollutants of concern is most important. This will enable an appropriate match within the facility or among nearby facilities.

13.6 Water Reuse Criteria

Strategic planning of the treatment of municipal and industrial wastewater is primarily carried out for reuse in the agricultural sector, since this sector is and will be the leading sector in water consumption in the future. Additionally, the agricultural sector is speculated to be the most influenced sector by climate change, meaning that higher amounts of irrigation water will be required. Thus, countries relying on agricultural production have either developed national guidelines and/or regulations or are implementing the criteria and standards set by the guidelines of the World Health Organization (WHO, 2006) and Food and Agriculture Organization of the United Nations (FAO, 1992) for the reuse of wastewater for agricultural irrigation (Shoushtarian and Negahban-Azar, 2020). The primary intention of all guidelines is human and public health while reusing the wastewater for restricted or unrestricted irrigation and raw consumed food crops, processed food crops and non-food crops. While the WHO guidelines focus on microbial health risks, including some maximum permissible soil concentrations of toxic inorganic and organic contaminants, the FAO guidelines focus on irrigation water-specific wastewater quality criteria and limit values for parameters to sustain soil and crop health. Thus, as the WHO guidelines contain microbiological details, the FAO guidelines define limit concentrations for ensuring good crop yield and minimal environmental risk. The primary parameter of concern of all guidelines and standards is always related to the pathogens. The WHO guidelines contain details on pathogenic microorganisms, whereas other guidelines and standards use *E. coli* or fecal coliform as the indicating parameter. To maintain both crop yield and soil health, major parameters for evaluating water quality for irrigation are salinity (electrical conductivity), total dissolved solids (TDS), sodium adsorption ratio (SAR) and specific ions like sodium, chloride, boron and bicarbonate. Some other wastewater quality parameters are biochemical oxygen demand (BOD), total suspended solids (TSS), pH and heavy metals. Water quality parameters are not restricted to these mentioned parameters. The number of parameters generally changes according to the strictness of competent authorities of countries. While Spain defined up to 90 parameters, France defined six parameters for reclaimed water with additional quality parameters on soil and sludge (European Commission, 2015).

Besides the quality of irrigation water, which may be treated wastewater, the irrigation system (e.g., drip or sprinkler) and regime with regard to meteorological and climatic variations becomes important in crop yield. An optimized irrigation system recognizing the exact crop-water requirement and adjusting the irrigation volume with respect to the precipitation and evaporation rate and soil type will avoid deficit irrigation and increase the crop yield. Regarding both human health and the environment, especially guidelines for agricultural reuse require regular monitoring of the quality of reclaimed wastewater, agricultural products and soil. WHO has defined maximum tolerable soil concentrations of toxic inorganic and organic chemicals. The influence of treated wastewater on soil and water resources is generally recognizable in the long term, indicating the importance of regular monitoring.

Treated wastewater reuse in the industrial sector can be as cooling water, boiler water and/or makeup water, which seems to have the highest share in industrial water consumption. Regarding cooling water, a certain percentage of blow-down water is generally replaced with some fresh water (makeup water) to maintain the water quality. Attempts are for the treatment of blow-down water for removing scaling minerals and chemicals used to avoid biofouling. As cooling waters are recycled in the cooling tower, the quality of cooling water and required makeup water differs with respect to the recycling number. Boiling water and boiling feed water qualities vary with the type of boilers (recirculation drum-type boiler; once-through boiler; low-, medium-, high-pressure boilers). Similarly, the water quality of drum-type boilers is maintained by the exchange of feed water with blow-down water. In order to prevent scaling and corrosion, high-quality water is used. Segregated wastewater streams that have been treated to meet cooling and boiling water quality and/or makeup water quality standards can be reused in industrial sectors like the chemical, iron and steel, refinery and power sectors. The water quality requirements of each industry, each manufacturing process and each operation are different, but by knowing the specific impurities that are unwanted, relevant treatments can be carried out to meet the exact water quality. Segregation in this sense will enable reduction in chemicals and treatment costs. Thus, segregation in both influent water qualities (excellent, high, medium, low quality) and wastewater qualities (high, medium, low strength quality) with respect to processes will ensure the most perfect treatment sequence. For example, the separation of post washing/rinsing/cleaning wastewater, which is accepted as low-strength wastewater in the textile industry, from the other wastewater streams, will enable higher recycling and/or reuse rates with appropriately determined easy treatments. Similarly, other industries with large water footprints can obtain economic and operational benefits. A well-structured strategy on the selection of the most appropriate streams and treatment sequences by combining conventional wastewater treatment processes with advanced processes will increase the share of reused wastewater.

13.7 Treatment Levels for Wastewater Reuse

Most guidelines cover suggestions for wastewater treatment level (e.g., secondary, tertiary treatments). Evaluation of wastewater characteristics and specific criteria or limit concentrations set for the reuse enable the determination of required removal efficiencies for each water quality parameter and the selection of the sequence of wastewater treatment unit operations. Primary treatment is generally not recommended but may be used for very low-strength wastewater for the removal of total suspended solids. Secondary treatment aims at organic carbon removal with treatment units relying on biological oxidation such as stabilization ponds, activated sludge, sequencing batch reactors and membrane bioreactors. The reclaimed wastewater can be used for the irrigation of non-food crops, stream augmentation and groundwater recharge for avoiding saltwater intrusion. Nutrient-containing wastewaters generally require tertiary treatment units with nitrification/denitrification and biological phosphorus removal. Tertiary treatment includes coagulation/flocculation and chemical precipitation for suspended solid removal. The treated wastewater can be used for the irrigation of food crops, unrestricted recreational areas, impoundment and so on. Recent publications show that both secondary and even tertiary treatment techniques are insufficient in removing dissolved salts and micropollutants (e.g., pharmaceuticals, household chemicals, industrial agents, etc.). Thus, wastewaters containing dissolved salts, heavy metals and other micropollutants generally require advanced treatment such as ion exchange, activated carbon adsorption, advanced oxidation processes and membrane processes. All treatment sequences should eventually end with disinfection to prevent risks from pathogenic microorganisms (Zhang et al., 2016; USEPA, 2012). Secondary wastewaters, for example, are known to contain pathogens that are assumed to survive for some hours or days when directly discharged to surface water resources. Risk evaluations suggest that resources, receiving upstream effluents at low rate, may be exceeding the irrigation water quality criteria for raw-eaten crops (Drewes et al., 2017). Surveys show that surface waters that are used for agricultural irrigation in the European Union are influenced by wastewater effluents. As treated wastewaters still contain organic constituents even at low levels, disinfection with chlorination is not preferred due to the formation of chlorinated organic compounds. Some advanced treatments enable disinfection and are also alternatives to chlorination. Advanced oxidation and membrane processes may reduce pathogen numbers and concentration of micropollutants.

13.8 Advanced Oxidation Processes

Advanced oxidation processes (AOPs) are the chemical treatment applications that can remove refractory organic pollutants in wastewaters resistant to conventional approaches by oxidation reactions with powerful and non-selective

radical species to enhance the biodegradability of wastewater (Rekhate and Srivastava, 2020). Advanced oxidation process can be applied under light (UV or visible light) or without light and in homogeneous or in heterogeneous environments, that is, in the absence or presence of catalysts (e.g., titanium dioxide [TiO_2], zinc dioxide [ZnO_2]). Although AOPs are based on different reaction mechanisms, they all rely on the formation of non-selective radicals such as hydroxyl radicals ($OH^•$) and sulfate radicals ($SO_4^•$). These are highly reactive oxidizing agents with redox potentials of 2.80 eV and 2.5–3.1 eV, respectively, higher than traditional oxidants such as hydrogen peroxide (1.78 eV under acidic conditions) or permanganate (1.68 eV under acidic conditions). These radicals react with most organic constituents of different complexity, including some persistent micropollutants in water and wastewater. AOPs may be classified based on the oxidizing agents and activating methods used for radical production as ozone-based processes (O_3/H_2O_2, O_3/metals, O_3/metal oxides, O_3/activated carbon), Fenton-based processes (Fe^{2+}/H_2O_2, $Fe^{2+}/H_2O_2/$UV, $Fe^{2+}/H_2O_2/$US, heterogeneous Fenton), electrochemical processes (anodic oxidation, electro-Fenton, photoelectro-Fenton), UV-based processes ($H_2O_2/$UV, $O_3/$UV, $O_3/H_2O_2/$UV, $TiO_2/$UV, $H_2O_2/TiO_2/$UV, $O_3/TiO_2/$UV), sulfate radical–based processes (PS/Fe^{2+}, PS/UV, PMS/Fe^{2+}, PMS/UV), sonochemical processes (sono/H_2O_2, sono/O_3, sono/$H_2O_2/$UV) and other processes (wet air oxidation, ionizing radiation, electron beam irradiation). Table 13.1 shows the main types of AOPs and reactive species generated in each process.

13.9 Ozone-Based AOPs

13.9.1 Ozonation

Ozone is one of the most powerful oxidants with standard oxidation potential of 2.07 V in acidic solution and 1.25 V in basic solution. In the ozonation process, oxidation occurs by means of direct molecular ozonation reaction and indirect ozonation reaction leading to the formation of hydroxyl radicals and ozone degradation. The pH of the solution is an important factor that determines the effectiveness of ozonation, as it changes

TABLE 13.1

Reactive Species Generated in AOPs

AOPs	Reactive Species
Fenton-Based AOPs Fenton, photo-Fenton, heterogeneous Fenton	$OH^•$, $HO_2^•$, $O_2^{•-}$
Ozone Based AOPs O_3, O_3/H_2O_2, O_3/catalyst	$OH^•$, $HO_2^•$, $HO_3^•$, $O_2^{•-}$, $O_3^{•-}$
UV-Based AOPs $O_3/$UV, $H_2O_2/$UV, $O_3/H_2O_2/$UV, $TiO_2/$UV	$OH^•$, $HO_2^•$, $O_2^{•-}$
Electrochemical AOPs Electro-Fenton, photoelectro-Fenton	$OH^•$, $HO_2^•$
Sonochemical AOPs Sonolysis, sono/H_2O_2, sono/O_3	$OH^•$, $HO_2^•$, $H^•$
Sulfate Radical–Based AOPs PMS, PMS/UV, PMS/catalyst	$OH^•$, $SO_4^{•-}$, $SO_5^{•-}$

the kinetics and routes of the reaction. At low pH values, the direct ozonation is dominant, while at high values, the indirect route prevails. In the ozonation process, OH• is generated by two different mechanisms at neutral pH or higher pH range. In neutral pH range, molecular ozone reacts with the hydroxide ions (OH^-) to produce superoxide radical anions ($O_2^{•-}$) and hydroperoxide radicals ($HO_2^•$) followed by hydroxyl radicals (OH•). In higher pH range, molecular ozone reacts with the hydroxide ions and then generates ozone radical anions ($O_3^{•-}$) and $HO_2^•$ to produce OH•. Furthermore, at very high pH values, OH• are also produced by the direct electron transfer reaction (Equations 13.1 and 13.2) (Wang and Xu, 2012).

$$O_3 + OH^- \rightarrow O_3^{•-} + OH^• \tag{13.1}$$

$$O_3^{•-} + H_2O \rightarrow OH^• + O_2 + OH^- \tag{13.2}$$

13.9.2 Catalytic Ozonation

In the catalytic ozonation process, a catalyst is used to produce hydroxyl radicals and cause ozone degradation. The catalytic ozonation process is used to oxidize components that show low reactivity with ozone. In the presence of catalyst, more radicals are produced compared to ozonation alone. In addition, it is thought that there is an increase in the nucleophilic areas of the adsorbed molecule during catalytic ozonation. Thus, a much better total organic carbon (TOC) and chemical oxygen demand (COD) removal occurs compared to only ozonation (Tong et al., 2003). Catalytic ozonation can be performed in homogeneous and heterogeneous environments. Homogeneous catalysis takes place in the presence of metal ions in aqueous solution, while heterogeneous catalysis takes place in a metal oxide or metal/metal oxide supported environment.

13.9.2.1 Homogeneous Catalytic Ozonation

In homogeneous catalytic reactions, the catalyst and reactants are in the same phase. Transition metals such as Fe^{2+}, Fe^{3+}, Mn^{2+}, Ni^{2+}, Cd^{2+}, Co^{2+}, Cu^{2+}, Zn^{2+}, Ag^+ and Cr^{3+} are homogeneous catalysts. The mechanism of homogeneous catalytic ozonation is based upon the decomposition of ozone initiated by metal species, followed by the formation of hydroxyl radicals. The metal ions initiate the decomposition of ozone to produce $O_2^{•-}$. The transfer of an electron from $O_2^{•-}$ molecule to O_3 results in the generation of $O_3^{•-}$, and subsequently HO•. The ozone concentration, catalyst dose and solution pH are the main factors affecting the efficiency and the mechanism of homogeneous catalytic ozonation process (Kasprzyk-Hordern et al., 2003).

Ozone decomposition in the presence of Fe^{2+} and Fe^{3+} generates OH• by the following reactions (Equations 13.3–13.7) (Wang and Xu, 2012):

$$Fe^{2+} + O_3 \rightarrow Fe^{3+} + O_3^{•-} \tag{13.3}$$

$$O_3^{•-} + H^+ \rightarrow O_2 + OH^• \tag{13.4}$$

$$Fe^{2+} + O_3 \rightarrow FeO^{2+} + O_2 \tag{13.5}$$

$$FeO^{2+} + H_2O \rightarrow Fe^{3+} + HO^• + OH^- \tag{13.6}$$

$$Fe^{3+} + O_3 + H_2O \rightarrow FeO^{2+} + H^+ + OH^• + O_2 \tag{13.7}$$

13.10 Heterogeneous Catalytic Ozonation

Heterogeneous catalytic ozonation is the mineralization of organic pollutants by combining the adsorptive and oxidizing properties of metal oxide catalysts in solid phase with ozone (Huang et al., 2005). Metal oxides (FeOOH, MnO_2, Al_2O_3, TiO_2 and CeO_2), metal oxide supported metal or metal oxides (Cu-Al_2O_3, Cu-TiO_2, Ru-CeO_2, FeO_3/Al_2O_3, TiO_2/Al_2O_3, V-O/TiO_2, V-O/silica gel and activated carbon), metal-modified zeolites and activated carbon were used as catalysts in the catalytic ozonation process. The effectiveness of the catalytic ozonation process is largely dependent on the catalyst and its surface properties as well as the pH of the solution, which affects the active sites on the catalyst surface and decomposition reactions of ozone in solution (Kasprzyk-Hordern et al., 2003).

Two mechanisms are proposed for heterogeneous catalytic ozonation – interfacial reaction mechanism and hydroxyl radical mechanism. In the interfacial reaction mechanism, organic pollutants are adsorbed on the catalyst surface and form active complexes with lower activation energy. Afterwards, ozone or OH• oxidizes the surface complex either in solution or at the catalyst surface. According to the OH• mechanism, the metal oxide catalysts initiate the decomposition of ozone and the reaction of ozone on reduced metal catalyst produces OH•. Also, the adsorption of ozone onto the catalyst surface initiates radical chain reactions to produce more OH• (Rekhate and Srivastava, 2020).

Activated carbon can also be used as a catalyst that promotes ozone oxidation. A low amount of activated carbon accelerates the degradation of ozone and leads to the formation of hydroxyl radicals. Two mechanisms are proposed for the decomposition of ozone: (i) activated carbon acts as an initiator of the decomposition of O_3 into OH• reacting with organics in aqueous solution and (ii) reduction of ozone on the activated carbon surface results in the formation of OH^- ions and H_2O_2 subsequently reacting with ozone to produce highly oxidative species, which are responsible for the degradation of organic compounds and the reduction of TOC (Kasprzyk-Hordern et al., 2003).

13.10.1 Peroxone (O_3/H_2O_2) Process

The peroxone process is based on the use of ozone combined with hydrogen peroxide and efficient catalytic process for the decomposition of refractory pollutants in wastewater. The peroxone process is more efficient than single ozonation, since H_2O_2 increases the decomposition rate of O_3 in water and the production of very reactive OH• (Oturan and Aaron, 2014). The peroxone process involves a radical chain mechanism initiated by the hydroperoxide anion (HO_2^-) for decomposition of ozone (Equations 13.8–13.12). A very fast reaction occurs between HO_2^- and ozone, which leads to the formation of OH•. However, excess H_2O_2 causes the scavenging of OH• and the formation of $HO_2^•$ (Rekhate and Srivastava, 2020).

$$H_2O_2 \rightarrow HO_2^- + H^+ \tag{13.8}$$

$$O_3 + HO_2^- \rightarrow HO_2^• + O_3^{•-} \tag{13.9}$$

$$O_3^{\bullet -} + H+ \rightarrow HO_3^{\bullet -} \tag{13.10}$$

$$HO_3^{\bullet -} \rightarrow O_2 + OH\bullet \tag{13.11}$$

$$OH\bullet + H_2O_2 \rightarrow HO_2\bullet + H_2O \tag{13.12}$$

13.11 UV-Based AOPs

Another way to decompose contaminants is by photochemical reactions using solar or UV light. As UV light is more powerful than solar light and has already found a market in water disinfection, UV-based AOP applications have been investigated for some time (Saraty and Mohseni, 2006; Shah, 2021). The degradation of contaminants in water depends on their ability (e.g., spectral and physical structure) to absorb UV light and its energy at such a level that initiates photochemical reactions. The formation of reactive species and radicals (e.g., hydroxyl radicals) after the photolysis of organic and inorganic constituents will enable the degradation of organic contaminants. The duration and spectrum of emitted light by UV lamp (low-pressure or medium-pressure UV lamp) was found to be important, as are other environmental conditions such as pH, temperature and water composition. In order to increase hydroxyl radical generation, UV radiation has been combined with hydrogen peroxide (UV/H_2O_2), ozone (UV/O_3) or both hydrogen peroxide and ozone ($UV/H_2O_2/O_3$).

13.11.1 UV/H_2O_2

Hydroxyl radical formation in the UV/H_2O_2 process occurs through the photolysis of peroxide (H_2O_2) as given in Equation 13.13. One mole of H_2O_2 forms two moles of hydroxyl radical, but as the molar absorption coefficient of H_2O_2 at 254 nm is low (ε_{254} = 19.6 M^{-1} cm^{-1}), the photolysis speed is slower than that of ozone photolysis (Saraty and Mohseni, 2006). Thus, higher concentrations and reaction times are required to achieve the desired level of hydroxyl radical formation and contaminant destruction. This in turn requires an optimization of H_2O_2 concentration and reaction time to overcome OH• scavenging. As can be seen from Equations 13.14 and 13.15, at elevated concentrations H_2O_2 reacts with OH• to produce less reactive radicals such as $HO_2\bullet$, which can react with H_2O_2 to generate OH•, water and oxygen or react with another $HO_2\bullet$ or OH• to produce H_2O_2 and superoxide ($O_2\bullet$).

$$H_2O_2 + h\nu \rightarrow 2\ OH\bullet \tag{13.13}$$

$$H_2O_2 + OH\bullet \rightarrow H_2O + HO_2\bullet \tag{13.14}$$

$$H_2O_2 + HO_2\bullet \rightarrow H_2O + OH\bullet + O_2 \tag{13.15}$$

13.11.2 UV/O_3

UV radiation in the presence of ozone similarly aims to increase the presence of hydroxyl radicals and thus the performance of complete oxidation. With ozone photolysis, H_2O_2 and O_2 are initially produced, further leading to the reaction of ozone with H_2O_2 and subsequent formation of hydroxyl radicals (Equations 13.16 and 13.17). The molar absorption coefficient of O_3 at 254 nm is high (ε_{254} = 3300 M^{-1} cm^{-1}), initiating

H_2O_2 formation and its catalytic effect in radical formation (Saraty and Mohseni, 2006).

$$O_3 + H_2O + h\nu \rightarrow H_2O_2 + O_2 \tag{13.16}$$

$$H_2O_2 + 2\ O_3 \rightarrow 2OH\bullet + 3O_2 \tag{13.17}$$

13.11.3 UV/H_2O_2/O_3

The combination of UV radiation with both peroxide and ozone increases the effect of peroxone (H_2O_2/O_3) and UV and thus enhances the hydroxyl radical formation (Equation 13.18). The degradation of refractory organic compounds increases within shorter timescales, reducing the cost of treatment.

$$O_3 + H_2O_2 + h\nu \rightarrow 2OH\bullet + 3O_2 \tag{13.18}$$

13.11.4 Photocatalytic Degradation

Heterogeneous photocatalytic processes rely on radical formation/activation on/of the surface of a semiconductor catalyst (e.g., titanium dioxide) exposed to UV irradiation. When photocatalysts absorb UV light, electrons excite to the conduction band (e^-), leaving positively charged holes in the valance band (h^+) (Equation 13.19). When electrons interact with molecular oxygen that have been adsorbed on the surface of the catalyst, $O_2^-\bullet$ formation prevails (Equation 13.20). Similarly, the interaction of holes with water molecules results in generation of OH• (Equation 13.21). One of the most often used semiconductor catalysts is titanium dioxide (TiO_2), which has a band gap energy of approximately 3.0–3.2 eV. UV irradiation with λ < 400 nm can easily excite the formation of electron-hole pairs.

$$TiO_2 + h\nu \rightarrow (e_{cb-} + h_{vb}^+) \tag{13.19}$$

$$e_{cb} + O_2 \rightarrow O_2^-\bullet \tag{13.20}$$

$$h_{vb}^+ + H_2O \rightarrow H^+ + HO\bullet \tag{13.21}$$

13.12 Fenton-Based AOPs

13.12.1 Homogeneous Fenton Process

The conventional Fenton process is a widely used catalytic method based on the production of hydroxyl radicals from hydrogen peroxide with ferrous iron (Fe^{2+}) ions acting as homogeneous catalysts under acidic conditions (Equation 13.22). The Fenton process is relatively inexpensive and easy to use compared to other AOPs and has been successfully applied for the treatment of different types of wastewaters. In the Fenton process, only a small amount of Fe^{2+} is required, since Fe^{2+} is regenerated by the reaction between Fe^{3+} and H_2O_2 (Equation 13.23). This reaction is called a Fenton-like reaction and is much slower than the Fenton reaction. Regeneration of Fe^{2+} may also occur by the reduction of Fe^{3+} with $HO_2\bullet$ as shown in Equation 13.24 and by the reactions of Fe^{3+} with organic radical (R•) and $O_2^-\bullet$ as shown in Equations 13.25 and 13.26 (Oturan and Aaron, 2014).

$$Fe^{2+} + H_2O_2 \rightarrow Fe^{3+} + OH\bullet + OH^- \tag{13.22}$$

$$Fe^{3+} + H_2O_2 \rightarrow Fe^{2+} + HO_2\bullet + H^+ \tag{13.23}$$

$$Fe^{3+} + HO_2\bullet \rightarrow Fe^{2+} + O_2 + H^+ \tag{13.24}$$

$$Fe^{3+} + R\bullet \rightarrow Fe^{2+} + R^+ \tag{13.25}$$

$$Fe^{3+} + O_2^{\bullet-} \rightarrow Fe^{2+} + O_2 \tag{13.26}$$

13.12.2 Photo-Fenton Process

In the photo-Fenton process, the degradation of pollutants using UV light increases. Under UV irradiation, Fe^{3+} complexes form more $OH\bullet$ and Fe^{2+} and provide the regeneration of ions (Equations 13.27 and 13.28):

$$Fe^{3+} + H_2O \rightarrow FeOH^{2+} + H^+ \tag{13.27}$$

$$FeOH^{2+} + h\upsilon \rightarrow Fe^{2+} + OH\bullet 410 \text{ nm} \tag{13.28}$$

In addition to the previous reactions in the photo-Fenton process, $OH\bullet$ is also generated with the following reactions (Equations 13.29 and 13.30) (Papic et al., 2009):

$$H_2O_2 + UV \rightarrow 2OH\bullet \tag{13.29}$$

$$Fe^{3+} + H_2O_2 + UV \rightarrow OH\bullet + Fe^{2+} + H^+ \tag{13.30}$$

13.12.3 Heterogeneous Fenton and Photo-Fenton Process

13.12.3.1 The Use of Activated Carbon (AC) as Catalyst

Except as an adsorbent, AC also acts as a catalyst owing to the presence of functional groups. AC can decompose H_2O_2 into oxygen without affecting the Fenton reaction. The general mechanism suggested for the interactions of AC with H_2O_2 is that surface hydroxyl groups are replaced by hydroperoxyl groups, which are stronger oxidizing agents. The reduction of these groups by H_2O_2 molecules in solution produces $OH\bullet$ and regenerates surface hydroxyl groups (Equations 13.31 and 13.32) (Navalon et al., 2011).

$$AC + H_2O_2 \rightarrow AC^+ + OH\bullet + OH^- \tag{13.31}$$

$$AC^+ + H_2O_2 \rightarrow AC + HO_2\bullet + H^+ \tag{13.32}$$

13.12.3.2 The Use of Iron-Containing Minerals as Catalysts

In the homogeneous Fenton process, there is no mass transfer limitation as iron species are in the same phase with the reactants. Despite their high oxidation efficiency, homogeneous Fenton and photo-Fenton process have a number of limitations. The major disadvantage is that large amounts of iron hydroxide sludge are formed for pH values above 4.0. By applying heterogeneous catalysts, these limitations can be overcome to some extent. As heterogeneous catalysts remain effective over a wide pH range, their use in the Fenton process is gaining increasing interest.

However, heterogeneous catalytic oxidation has a slower oxidation rate than the homogeneous oxidation, as a small iron fraction is present on the catalyst surface. Three possible mechanisms for the action of catalysts have been proposed in the heterogeneous Fenton process: (i) the transition of iron to the solution and the activation of hydrogen peroxide in a homogeneous way and/or (ii) the binding of H_2O_2 to the iron species on the catalyst surface to produce $OH\bullet$, or (iii) adsorption of the target molecule on the catalyst surface. In Fenton processes, a number of heterogeneous catalysts have been used. Among these, iron oxide minerals such as pyrite (FeS_2), hematite (Fe_2O_3), goethite (α-FeOOH) and lepidocrosite (γ-FeOOH) draw attention (Pouran et al., 2015).

13.13 Electrochemical AOPs

13.13.1 Electro-Fenton Process

The EF process is known as a highly efficient and economical process for removing toxic and/or persistent organic pollutants. The efficiency of the EF process largely depends on the H_2O_2 production rate, O_2 sparging rate, solution pH, anode material, catalyst type and applied current density. The type and concentration of catalyst significantly affect EF treatment efficiency. Various catalysts are used in electro-Fenton reaction, but Fe^{2+} or Fe^{3+} show good catalytic properties even at low concentrations (Meijide et al., 2018).

The EF process works as a mixture of H_2O_2 and Fe^{2+} to generate hydroxyl radicals through the electrogeneration of Fenton's reagent in solution. Hydrogen peroxide is produced continuously on the cathode surface by reducing oxygen in an acidic environment (Equation 13.33). Oxygen addition near the cathode is required during the process to achieve continuous hydrogen peroxide production. Adding a catalytic amount of iron salt to the solution leads to the generation of $OH\bullet$ according to the Fenton reaction (Equation 13.34).

$$O_2 + 2H^+ + 2e^- \rightarrow H_2O_2 \tag{13.33}$$

$$Fe^{2+} + H_2O_2 \rightarrow Fe^{3+} + OH^- + OH\bullet \tag{13.34}$$

Catalysis of the reaction with electrochemically regenerated ferrous ions is a significant advantage of the EF process over the traditional Fenton process. Optimum pH value for Fenton reaction is approximately 3. The dominant type of iron is $Fe(OH)^{2+}$ at this pH. Fe^{3+} ions produced by the Fenton reaction are in the form of $Fe(OH)^{2+}$ and subjected to cathodic reduction to produce Fe^{2+} ions according to Equations 13.35 and 13.36. In addition to this main source, iron ions are renewed by the reactions in Equations 13.37 and 13.38 in the EF process.

$$Fe^{3+} + H_2O \rightarrow Fe(OH)^{2+} + H^+ \tag{13.35}$$

$$Fe(OH)^{2+} + e^- \rightarrow Fe^{2+} + OH^- \tag{13.36}$$

$$Fe^{3+} + H_2O_2 \rightarrow Fe^{2+} + H^+ + HO_2\bullet \tag{13.37}$$

$$Fe^{3+} + HO_2\bullet \rightarrow Fe^{2+} + H^+ + O_2 \tag{13.38}$$

In this technique, Fenton reagents H_2O_2 and iron are continuously produced and regenerated in the electrochemical reactor (Meijide et al., 2018).

13.14 Sulfate Radical–Based AOPs

13.14.1 Peroxymonosulfate Process

The main radicals involved in the breakdown of organic substances in AOPs are hydroxyl and sulfate radicals. The sulfate radical is formed by the activation of persulfate ($S_2O_8^{2-}$) or peroxymonosulfate (HSO_5) anions. Persulfate and peroxymonosulfate anions are also strong oxidants, but they require activation since their direct reaction with most pollutants occurs at a very low speed (Ghanbari and Moradi, 2017). Although there are many activations to generate radicals, the most well-known of these are ferrous iron (Fe^{2+}), pH, UV and thermal activation (Devi et al., 2016). In neutral and alkaline conditions, the oxidation potential of the sulfate radical is higher than that of the hydroxyl radical. It has been reported that the half-life of the sulfate radical ($3–4 \times 10^{-5}$ s) is higher than the hydroxyl radical (2×10^{-8} s) (Ghanbari et al., 2016). It has been explained that the longer half-life of the sulfate radical is due to its selectivity for organic substances from which it will transfer electrons. On the other hand, the rapid reaction of the hydroxyl radical has been associated with its non-selectivity (Ghanbari and Moradi, 2017). The purpose of its application as peroxymonosulfate is to keep the electrical conductivity at a lower level by adding a lower level of sulfate ions to the environment. Additionally, peroxymonosulfate is a slightly stronger oxidant than hydrogen peroxide ($E°$ ($HSO_5^-/HSO_4^{•-}$) = +1.82 V; $E°$ (H_2O_2/H_2O) = + 1.76 V); Sharma et al., 2015).

In the peroxymonosulfate process, the activation of the oxidant is frequently performed using transition metals such as iron and cobalt. Iron is more environmentally friendly than other transition metals such as cobalt and nickel (Rodríguez-Chueca, 2019). It has been reported that the activation occurs through the following reactions (Equations 13.39–13.44) (Ghanbari and Moradi, 2017).

$$Fe^{2+} + HSO_5^- \rightarrow Fe^{3+} + SO_4^{•-} + OH^- \tag{13.39}$$

$$Fe^{3+} + HSO_5^- \rightarrow Fe^{2+} + SO_5^{•-} + H^+ \tag{13.40}$$

$$Fe^{2+} + SO_4^{•-} \rightarrow Fe^{3+} + SO_4^{2-} \tag{13.41}$$

$$HSO_5^- + SO_4^{•-} \rightarrow SO_4^{2-} + SO_5^{•-} + H^+ \tag{13.42}$$

$$HSO_5^- + H_2O \rightarrow SO_4^{2-} + 2OH^• + H^+ \tag{13.43}$$

$$SO_5^- + 2H_2O \rightarrow SO_4^{2-} + 3OH^• + H^+ \tag{13.44}$$

It is known that pH activation in chemical oxidation processes is another process that enables the oxidant to form radicals. While hydroxyl and sulfate radicals are effective in degrading organic substances in neutral pH conditions, it is reported that the sulfate radical is more effective in acidic pH conditions. In acidic conditions, while hydroxyl radicals decompose into water, sulfate radicals produce hydrogen sulfate radical anions (Equations 13.45 and 13.46) (Liang and Su, 2009).

$$HO^• + H^+ + e^- \rightarrow H_2O \tag{13.45}$$

$$SO_4^{•-} + H^+ + e^- \rightarrow HSO_4^{•-} \tag{13.46}$$

13.14.2 Photo-Peroxymonosulfate Process

Another activation used in the generation of radicals is the activation performed by UV. When UV application is performed with a low pressure mercury lamp (254 nm), a sulfate ($SO_4^{•-}$) and hydroxyl radical ($OH^•$) are formed as seen in the reaction (Equation 13.47) due to the degradation of the peroxide bond. Subsequently, as a result of the reaction with water, the transformation of SO_4^{2-}, H^+ and $OH^•$, shown in reaction (Equation 13.48), occurs (Sharma et al., 2015).

$$HSO_5^- + hv \rightarrow SO_4^{•-} + OH^• \tag{47}$$

$$SO_4^{•-} + H_2O \rightarrow H^+ + SO_4^{2-} + OH^• \tag{48}$$

13.14.3 Heterogeneous Peroxymonosulfate Process

Peroxymonosulfate interacting with the activated carbon surface forms $OH^•$ and $SO_4^{•-}$ radicals by the electron transfer mechanism. The main reactions suggested for peroxymonosulfate are given in Equations 13.49–13.51 (Devi et al., 2016).

$$HSO_5^- + AC \rightarrow OH^- + SO_4^{•-} + AC^+ \tag{13.49}$$

$$HSO_5^- + AC \rightarrow OH^• + SO_4^{2-} + AC^+ \tag{13.50}$$

$$OH^• + SO_4^{•-} + \text{organic compounds} \rightarrow \text{intermediate products} + CO_2 + H_2O + SO_4^{2-} \tag{13.51}$$

13.15 Combined Treatment Systems Including AOPs

The wastewaters produced in industrial processes often contain refractory organic compounds, which are very toxic and not susceptible to direct biological treatment. Textile, leather, pulp and paper, chemical, petrochemical and pharmaceutical industries generate wastewaters containing high amounts of refractory organic pollutants. Industrial wastewater characteristics vary not only with the industry that generates them, but also within the industry. Treatment of industrial wastewaters is a complex problem due to the wide variety of compounds and concentrations they may contain (Shah, 2020b).

For industrial wastewater reclamation, selection of the best process sequence requires comprehensive work and should consider the end use purpose, water quality targets and treatment objectives. The choice of treatment processes to be combined in certain conditions depends on the water quality standards to be met with the highest efficiency and lowest reasonable cost (Oller et al., 2011). A variety of conventional technologies, including biological, physical and chemical treatments, have been used for the treatment of industrial wastewaters to achieve discharge criteria. However, conventional wastewater treatment processes are not sufficient for the reclamation of wastewaters for agricultural and industrial applications requiring high water quality. Advanced treatment processes are needed to remove pollutants remaining in wastewater after primary, secondary and tertiary treatments and to provide wastewater meeting agricultural or industrial reuse criteria. Nowadays, numerous technologies available for the advanced treatment of wastewaters can be used in combined systems to achieve high water quality standards (Roccaro, 2018).

The efficiency of a combined process varies depending on the purpose of the treatment. Overall mineralization efficiency may be used to evaluate the process performance when high water quality is needed or specific organic carbon limit is targeted. The purpose of treatment may also be reduction of toxicity or elimination of a specific pollutant. In this case, biodegradability and toxicity of treated effluent may be the measure of the efficiency of a combined process. Determination of the treatment purpose is an essential step in selecting the process combination in order to define process efficiency and to optimize process (Oller et al., 2011).

Additionally, segregation of wastewaters into streams according to their biodegradability and treatment with the most appropriate techniques may be recommended as good practice. The only feasible option for recalcitrant wastewaters is the use of AOPs, widely recognized as a high efficiency treatment for refractory pollutants. Depending on the biodegradability of the streams, AOPs and biological processes can be used as a pre-treatment or post-treatment process. In the case of a biodegradable stream, biological processes are more appropriate as pre-treatment, followed by AOPs if necessary. In the case of a nonbiodegradable stream, AOPs may be applied for the pre-treatment of recalcitrant pollutants before biologic treatment (Pazdzior et al., 2019).

Another important issue in the process selection is cost of treatment. AOPs are efficient, versatile and environmentally compatible for the decomposition of biorefractory contaminants, but they are less applicable for the treatment of large volumes of low concentration pollutants (Ganiyu et al., 2015). Considering that AOPs pose high costs due to the use of chemicals and energy, AOPs are more feasible for the treatment of wastewater with low flow rates and containing nonbiodegradable pollutants. Therefore, segregation of wastewater streams and application of AOPs to wastewater streams with low biodegradability will reduce the treatment cost. Although AOPs are effective in the removal of recalcitrant compounds and in the reduction of toxicity, they are ineffective in removing salts. Membrane processes (nanofiltration [NF] or reverse osmosis [RO]) may be used before or after AOPs for the removal of salts and reuse of wastewater (Pazdzior et al., 2019).

Possible treatment sequences that can be applied for wastewater reuse are presented in Figure 13.2. AOPs may be applied as a pre-treatment step for the nonbiodegradable wastewaters to enhance their biodegradability or as a polishing step for biodegradable wastewaters to remove trace organic compounds. Reuse of wastewaters for agricultural applications generally requires a process sequence that includes conventional primary and secondary treatment, disinfection and salinity reduction, depending on the crop needs. The treatment sequence for industrial wastewater reuse may contain conventional primary, secondary and tertiary treatment, salinity reduction and trace organics removal, depending on the water quality requirements of the industry. As the treatment level increases, the achieved water quality and cost of treatment increases. End use–specific water quality requirements should be taken

into account for the selection of a suitable process sequence to achieve desired water quality with a reasonable cost.

13.16 Combinations of AOPs with Biological Processes

Among conventional treatment methods, biological treatment is the main technology capable of removing organic contaminants in wastewaters by transforming them into acceptable end products. Biological treatment of wastewaters is generally the most economical alternative compared to chemical treatment processes. However, industrial wastewaters most often contain a great amount of refractory organic compounds. Conventional biological methods are not capable of degrading and detoxifying refractory contaminants that are resistant to biological decomposition. In such cases, biological processes alone are not sufficient to achieve the water quality standards required for discharge or reuse of wastewaters; therefore, pre-treatment or post-treatment processes need to be applied (Tabrizi and Mehrvar, 2004).

The integration of AOPs with biological processes has been recognized as a promising approach for the treatment of recalcitrant wastewaters. AOPs can be applied as a pre-treatment step that destructs recalcitrant organic pollutants to improve subsequent biological treatment and inhibitory compounds that have toxic effects on biological treatment (Shah, 2020a). AOPs can also be used as the final polishing step to remove low concentrations of recalcitrant compounds that cannot be removed during biological treatment. AOP systems are implemented regarding the wastewater composition and pollutant concentrations and the required effluent quality (Kehrein et al., 2020). Although AOPs improve the biodegradability of refractory organic compounds and produce high-quality effluent in most cases, the main drawback of AOPs is their high capital and operating costs due to the consumption of chemicals and electricity. The general strategy to avoid high cost caused by AOPs is to apply these processes to the wastewaters containing refractory or toxic organics. In determining the process sequence, the overall goal of reducing costs is to minimize AOP treatment and maximize the biological stage, due to the large difference in the cost of the processes (Oller et al., 2011). The combination of AOP with biological treatments results in higher removal efficiencies and are therefore regarded as cost effective compared to AOP alone and less time consuming compared to biological processes (Tabrizi and Mehrvar, 2004). Thus, the determination and segregation of the wastewater stream with high refractory or toxic organics and its subsequent treatment with AOP before its combination with the other streams may be an excellent optimization scheme.

In the selection of AOPs to be employed in a combined treatment system, wastewater characteristics should also be taken into account. The combined systems, including ozone-based and the Fenton process, are cost effective, feasible and easy to operate. The main limitation of the Fenton process is the need to maintain the pH value at 3 for optimum operation. Therefore, the Fenton process is appropriate for wastewaters with an acidic pH value. The ozone-based processes may be appropriate for alkaline wastewaters. At high pH values, ozone

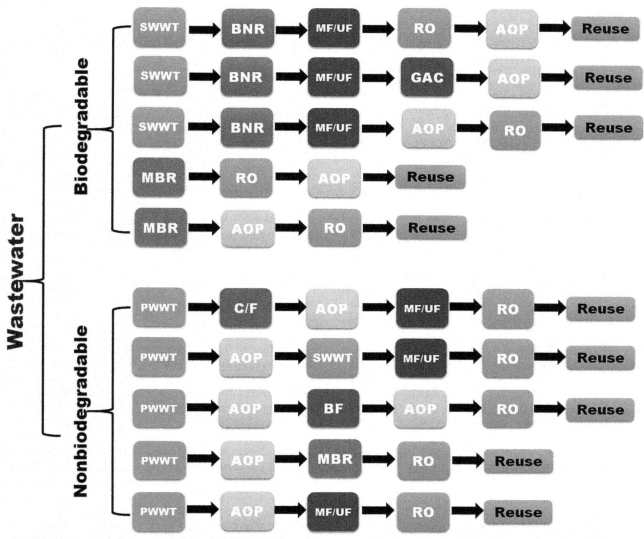

PWWT: Primary wastewater treatment, **SWWT:** Secondary wastewater treatment; **BNR:** Biological nutrient removal;
MBR: Membrane bioreactor; **BF:** Biofiltration; **C/F:** Coagulation/Flocculation; **MF:** Microfiltration; **UF:** Ultrafiltration;
RO: Reverse osmosis; **GAC:** Granular activated carbon; **AOP:** Advanced oxidation process

FIGURE 13.2 Combinations of AOPs with treatment processes for wastewater reuse.

generates reactive radical species that can oxidize recalcitrant organic pollutants. The photochemical AOPs are not effective for wastewaters that have intense color and turbidity due to reduced light penetration into bulk medium. Furthermore, the photochemical AOPs have higher operational costs compared to the other AOPs. The solar AOPs have generally lower operational costs but large surface area and require stable solar light for effective irradiation (Pazdzior et al., 2019).

13.16.1 AOP Pre-Treatment

For wastewaters that mostly contain nonbiodegradable compounds, chemical pre-treatment is usually recommended. AOPs are capable of converting the biorecalcitrant compounds into more biodecomposable intermediate products that can be successively treated by biological processes. The $BOD_5/$ COD ratio is one of the simplest and most used parameters for wastewater biodegradability estimation. AOPs are recommended for wastewaters with a BOD_5/COD ratio below 0.20. However, sometimes chemical oxidation may cause the formation of toxic products, so biodegradability and toxicity assessments are needed before and after AOP pre-treatment (Pazdzior et al., 2019). AOP pre-treatment improves the biodegradability of recalcitrant or toxic organic compounds due to the changes in their molecular structure. AOP pre-treatment probably decreases the aromaticity and enables the decomposition of large molecular structures, which causes the generation of functional groups like hydroxyl, carboxyl and aldehyde (Tabrizi and Mehrvar, 2004).

AOP pre-treatment may be used either for the complete mineralization of all the organic contaminants to carbon dioxide, water and inorganic ions or for the partial oxidation of refractory pollutants to more biodecomposable intermediates, enabling their treatment in a biological process in a shorter time and with a considerably lower cost. Total decomposition of contaminants is usually quite expensive, as oxidation intermediates generated during treatment tend to be more resistant to complete decomposition, and the longer treatment times required for complete mineralization increase consumption of energy and chemical reagents. During AOP pre-treatment, keeping mineralization to a minimum will reduce the chemical and energy consumption and thus reduce the operating cost (Oller et al., 2011).

13.16.2 AOP Post-Treatment

AOPs can also be used as a post-treatment process for polishing of wastewater after biological treatment. AOP post-treatment is recommended for wastewaters that contain both biodecomposable and biorefractory compounds. A biological pre-treatment step is used to remove the biodecomposable organic fraction followed by an AOP post-treatment as a polishing step. The main advantage of AOP post-treatment is that of avoiding the use of chemicals or energy for the biodegradable organic pollutants, which may be removed earlier by a biological process (Dewil et al., 2017). For wastewaters with high organic pollution, use of AOPs as both a pre-treatment and a post-treatment step seems to be a favorable option. AOP pre-treatment is applied for partial oxidation of the biologically persistent compounds to biodecomposable intermediates. AOP post-treatment is applied to convert biorecalcitrant compounds remaining after the initial biological step into smaller biodecomposable molecules, which can be mineralized by a second biological step (Pazdzior et al., 2019).

13.17 Combinations of AOPs with Membrane Processes

Membrane processes have been increasingly employed in municipal and industrial wastewater treatment and accepted as a reliable treatment technology for wastewater reuse. Membrane technologies are capable of producing a very high-quality effluent that allows reuse of any type of water, but their operational cost is relatively high. Also, membrane processes have some limitations such as membrane fouling and the necessity for disposal or treatment of membrane concentrate (Ganiyu et al., 2015). Membrane fouling is considered an important problem especially at high membrane fluxes. The application of lower fluxes decreases operational costs but raises capital costs, due to the increased need for membrane units. In order to reduce potential fouling and clogging, successful operations require efficient pre-treatment of effluents before membrane processes (Kehrein et al., 2020).

The combination of membrane filtration and AOPs provides both separation and degradation of pollutants. AOPs can be combined with membrane process either as a pre-treatment step for the degradation of organic compounds in the membrane feed stream; a post-treatment step for the mineralization of unretained pollutants in the permeate stream and highly concentrated organics in the concentrate stream; or a hybrid process in which separation and decomposition of contaminants by membrane filtration and AOPs are achieved simultaneously (Ganiyu et al., 2015).

13.17.1 AOP Pre-Treatment

AOPs are proper pre-treatment processes for the membrane applications. AOPs employed as pre-treatment prior to membrane filtration usually aims to decrease membrane fouling by the oxidation of natural and synthetic organic contaminants that cause blocking of membrane pores. AOP pre-treatment also reduces the organic load in the membrane concentrate. Ozonation and photolysis in the presence or absence of H_2O_2 are feasible AOPs that are accepted as a pre-treatment step for membrane filtration. AOPs such as Fenton oxidation, photocatalysis and EAOPs are not sustainable for high wastewater flow rates with low contaminant levels, usually observed for real situations (Ganiyu et al., 2015).

13.17.2 Hybrid Treatment

Integrated or hybridized membrane filtration and AOPs combine filtration and decomposition of contaminants in a single step. Hybrid membrane reactors and bioreactors seem to have various benefits with regard to higher treatment efficiency, reduced chemical and energy costs, enhanced membrane flux and reduced fouling. Different kinds of wastewaters can be treated and reclaimed by hybridization of a membrane reactor and a bioreactor with AOPs. The main advantages of membrane-integrated oxidation systems are their higher permeate flux, reduced space requirement, self-antifouling and self-cleaning features, in addition to concurrent oxidation and membrane filtration (Rostam and Taghizadeh, 2020). Hybrid membrane filtration AOP processes consist of a self-cleaning membrane, suspended catalyst and photocatalyst such as TiO_2 or Fenton's reagents that are able to produce hydroxyl radicals by photochemical or electrochemical processes for the oxidation of organic compounds (Ganiyu et al., 2015). In recent years, photocatalytic membranes, involving photocatalyst particles, have received increasing attraction in hybrid membrane applications.

13.17.3 AOP Post-Treatment

AOPs can be employed as a post-treatment step to support and improve membrane filtration, particularly for the mineralization of unretained pollutants in the permeate stream. In some cases, low molecular weight pollutants may not be retained by the membranes. AOP post-treatment of permeate completes the treatment to achieve high-quality water. An AOP post-treatment step can also be used for the reclamation of concentrate streams prior to their discharge to the environment or recycling back into the process. The elevated micropollutant concentrations within the concentrates create a suitable condition for a successful performance of the process, as AOPs show higher efficiencies at higher pollutant concentrations.

13.18 Combinations of AOPs with Physical/Chemical Processes

The integration of AOPs and separation processes into the treatment sequence is beneficial for effluents with high solids concentration. Filtration, sedimentation or coagulation processes may be initially used to remove solids in order to prevent their dissolution during AOP, the increase of organic load of wastewater and the oxidant consumption. Additionally, higher effluent turbidity may influence light penetration and thus process efficiency of photochemical AOPs. Therefore, removal of suspended and colloidal solids causing turbidity is an obligatory pre-treatment step before photochemical AOPs (Dewil et al., 2017).

The coagulation/flocculation (C/F) process removes pollutants from wastewater by entrapping them into flocs. Coagulation can be regarded as an efficient pre-treatment process for industrial wastewaters. However, it can only remove suspended and colloidal solids and is not effective in removing soluble organic compounds. The coupling of AOPs with the coagulation-flocculation process can be a suitable alternative to accomplish higher levels of wastewater reclamation. The combination of these technologies is expected to improve the efficiency of AOPs, to reduce the amounts of reagents and operation time required for the treatment and to decrease the operational costs considerably (GilPavas et al., 2017).

13.19 Combined Systems Including AOPs in Wastewater Treatment and Reuse

In various applications, AOPs have been integrated to wastewater treatment systems as a pre-treatment or post-treatment process for recycling and reuse of wastewaters, according to the wastewater composition and the water quality requirements. The applications of AOPs in combined systems for the reclamation of domestic, industrial and agricultural wastewaters are given in Table 13.2.

13.20 Fenton-Based Combined Systems

Aydıner et al. (2019) employed different combinations of ultrafiltration (UF) and Fenton process for the reclamation of textile washing wastewater. They compared a simultaneous and sequential hybrid membrane oxidation system with an externally integrated (AOP/UF) system. The simultaneous or sequential hybrid systems represented better performances (74.5% TOC, 85.9% COD and 99.4 L/m² h, and 70.5% TOC, 84.5% COD and 155.6 L/m² h, respectively) than the externally integrated system (47.9% TOC, 79.0% COD and 60.9 L/m² h). Hybrid treatments also enabled lower reagent consumption by the continuation of catalytic activity of Fe^{2+} and avoidance of oxidant escape from the reactor with consecutive operations. It was reported that the UF-hybridized Fenton operations are technologically favorable because of better treatment performance and lower reagent, time and land requirements.

Ribeiro et al. (2017) assessed textile wastewater reuse after treatment with conventional activated sludge followed by Fenton, photo-Fenton, and UV/H_2O_2 processes. Fenton process, with 68% COD and 88% turbidity removal, was found to be the best option for fulfilling the water quality guideline requirements for textile industrial use. The reclaimed wastewater represented partial conformity with the physicochemical quality guidelines for reuse. Removal and transformation of high and medium molecular weight compounds into lower molecular weight compounds was obtained, as a result of the destruction of various organic substances found in the wastewater. The reuse of treated effluent was additionally carried out at pilot scale to investigate its influence on fabric quality. Reuse tests confirmed the applicability of wastewater reuse after treatment.

Feng et al. (2010) investigated the efficiency of Fenton oxidation and membrane bioreactor (MBR) combined processes for the reclamation of dyeing wastewater treatment plant effluent. The average TOC and color removal efficiencies of Fenton oxidation under optimum conditions were determined to be 39.3% and 69.5%, respectively. The Fenton process was also found to enhance the biodegradability of the wastewater. TOC and color removal efficiencies of about 88.2% and 91.3% were observed by the Fenton oxidation and MBR combined process, respectively. The obtained MBR effluent was found to fulfill the reuse criteria of urban recycling water – water quality standard for miscellaneous water consumption.

Abdel-Shafy et al. (2016) evaluated the applicability of the combined Fenton process followed by microfiltration (MF) for the treatment of leather industry wastewater. They achieved 97.1% COD, 85.2% BOD_5, 99.3% TSS, 99.9% oil and grease, 99.9% sulfide, 84.5% TKN and 94.3% TP removals by Fenton oxidation. Fenton process was also effective in increasing the biodegradability of the leather industry wastewater. The BOD_5/COD ratio of raw wastewater increased from 0.15 to 0.80 after Fenton oxidation. They reported that leather industry wastewater was efficiently treated by Fenton oxidation followed by membrane filtration, and the final wastewater met the standards for unlimited water reuse.

Blanco et al. (2014) investigated, in a bench-scale study, the application of single and combined sequences of the photo-Fenton oxidation and aerobic sequencing batch reactor (SBR) processes for the treatment and reuse of real textile wastewater. The highest degradation efficiency was observed when the photo-Fenton process was applied as a polishing step after biological treatment. The TOC removal efficiency observed for aerobic biological treatment was 75% after 25 cycles. The combined photo-Fenton process, on the other hand, achieved COD and TOC reductions of 97% and 95%, respectively. When RO was additionally used as a final polishing step, the efficiency increased to 100% delivering directly reusable effluent. It was stated that the use of aerobic biological treatment together with photo-Fenton and RO can be regarded as an appropriate process combination capable of treating textile wastewaters, to obtain an effluent quality that meets 100%

Table 13.2

AOPs Including Combined Systems for the Recycling and Reuse of Wastewaters

Combined Process	Wastewater	Characteristic of Wastewater (mg/L)	Process Efficiency	Reuse Purpose	Reference
Fenton-Based Combined Systems					
Fenton/UF	Textile washing wastewater	COD: 2830 TOC: 1030	COD: 85.9% TOC: 74.5%	Reuse in textile industry	Aydıner et al., 2019
Fenton oxidation + MBR	Textile wastewater	TOC: 120–150 COD: 220–300	TOC: 88.2% COD: 91.3%	Reuse for miscellaneous water consumptions	Feng et al., 2010
Precipitation + electrodeposition + oxidation	Electroplating industry wastewater	COD: 1360 Ag^+: 5600 Cu^{2+}: 9069	COD: 72.2% Ag^+: 99.9% Cu^{2+}: 89.3%	Reuse as deplating solution	Gu et al., 2020
Fenton oxidation + MF	Tannery industry wastewater	COD: 23,400 BOD: 3650 TSS: 3680	COD, BOD, TSS, oil and grease > 99% TKN and TP > 96%	Reuse in leather industry	Abdel-Shafy et al., 2016
Conventional activated sludge + Fenton	Textile wastewater	COD (mean): 182 (61.2–299)	COD: 45%–66%	Reuse in textile industry	Ribeiro et al., 2017
Fenton-like oxidation + adsorption	Textile wastewater	TOC: 30–290 COD: 280–300	TOC: 86% COD: 58%	Reuse for irrigation	Bayrakdar et al., 2021
Solid-phase AOP + RO	Manufacturing industry	TOC: 1.691	TOC: 92%	Reuse in the industry	Choi and Chung, 2015
Conventional activated sludge + photo-Fenton	Textile wastewater	COD (mean): 182 (61.2–299)	COD: 37%–68%	Reuse in the textile industry	Ribeiro et al., 2017
Aerobic sequencing batch reactor (SBR) + photo-Fenton oxidation	Real textile wastewater	COD: 1560 TOC: 390	COD: 97% TOC: 95%	Reuse in textile industry for dyeing	Blanco et al., 2014
UV-Based Combined Systems					
UV/H_2O_2 + UF	Effluent of activated sludge system treating domestic wastewater	OMP (Organic micropollutants)	OMPs > 80%	Reuse for general purposes	Li et al., 2020
BAC (biological activated carbon) + H_2O_2/UV	Refinery wastewater MBR effluent	TOC: 15–50	TOC: 76%	Reuse in oil refinery	Souza et al., 2011
H_2O_2/UV + biofilter	Physicochemical and biologically treated oil refinery wastewater	TOC: 6.3–17.5	TOC: 64%–78%	Reuse in oil refinery as boiling and cooling water	Nogueira et al., 2016
MF or RO + UV/H_2O_2	Conventional activated sludge-based wastewater treatment effluent	Micropollutants (Pesticides: metaldehyde and others)	Metaldehyde: 45% Other pesticides: 85%–99%	Reuse for general purposes	James et al., 2014
UV/H_2O_2 + UF	Secondary effluent from urban wastewater treatment plant	DOC: 11.1 Turbidity: 3.9	DOC: 41% Turbidity: 96%	Reuse for general purposes	Benito et al., 2017
NFC-doped photocatalytic oxidation	Industrial and domestic WWTP's secondary effluent	COD: 67–160 TOC: 48–83 (domestic wastewater) COD: 229 TOC: 208 (industrial wastewater)	COD: 73%–79% TOC: 73%–86% (domestic wastewater) COD: 62% TOC: 62% (industrial wastewater) 99%–100% removal of antibiotic residues	Reuse for irrigation	Ata and Töre, 2019
H_2O_2/UVC	Biologically treated textile wastewater	COD: 65 Color: 134 Pt-Co	COD: 23% Color: 83.6%	Reuse in textile industry for scouring, bleaching or dyeing process	Silva et al., 2018
$UV/O_3/H_2O_2$	Textile wastewater	COD: 185 Color: 2150 (ptCo)	COD: 55% Color: 95%–100%	Reuse in textile industry	Güyer et al., 2016
$UV/H_2O_2/O_3$	Biologically treated textile wastewater	COD: 186.5 Color (DFZ): 100.8	COD: 60.1% Color: 100%	Reuse in textile industry for dyeing	Nadeem et al., 2017
Photocatalytic oxidation and membrane separation	Simulated dye wastewater	COD: 247 Congo red dye: 40	COD: 90.5% Congo red dye: 99.7%	Reuse in the textile industry	Ou et al., 2015
Natural sunlight and TiO_2/ $Na_2S_2O_8$	Agro-wastewater	100 µg/L of each pesticide	Pesticides: 99.5%.	Reuse for agricultural irrigation	Aliste et al., 2020

Combined Process	Wastewater	Characteristic of Wastewater (mg/L)	Process Efficiency	Reuse Purpose	Reference
Ozone Based Combined Systems					
Oxic/anoxic biodegradation + ozone oxidation	Mixed wastewater (industrial and domestic)	COD: 1432 BOD_5: 184 TN: 149 PO4-P: 32.3	COD: 92.1% BOD_5: 90.6% TN: 83.3% PO4-P: 83.8%	Reuse in agriculture	Egbuikwem et al., 2020
Ozonation + UF	Effluent of activated sludge system treating domestic wastewater	Organic micropollutants (OMP)	OMPs: 56%	Reuse for general purposes	Li et al., 2020
O_3/H_2O_2	Petroleum refinery effluent treated by the traditional techniques	TOC: 236	TOC: 91%	Reuse in the plant for cooling, firefighting or cleaning purposes	Demir-Düz et al., 2020
O_3/H_2O_2 + biofiltration + UVC/H_2O_2	Tertiary municipal wastewater	COD: 26.1 Total mean micropollutant concentration (ng/L): 1681.8	COD: 62% Total mean micropollutant: 99%	Reuse for general purposes requiring potable water quality	Piras et al., 2020
Electrocoagulation + catalytic ozonation	Textile wastewater	COD: 3440 TOC: 1790	COD: 62% TOC: 61%	Reuse in textile industry for dyeing	Bilinska et al., 2020
Electrochemical Based Combined Systems					
Electro-Fenton	Carwash wastewater	COD: 625 Anionic surfactants: 199	COD: 96% Anionic surfactants: 96%	Reuse as carwash water	Ganiyu et al., 2018
Electrochemical oxidation	Tannery saline wastewater	COD: 10175 TKN: 352	COD: 89.1% TKN: 96.3%	Reuse for pickling process	Sundarapandiyan et al., 2010
Electrochemical oxidation	Textile wastewater	COD: 4800	COD: 96%	Reuse in textile industry for dyeing	Mohan et al., 2007
Electrochemical oxidation	Industrial park wastewater	COD: 154 Color 125 Hazen	COD: 62% Color: 92%	Reuse for miscellaneous water consumption and scenic purposes	Zhu et al., 2015
Anodic oxidation with boron-doped diamond (BDD) anodes	Paper mill wastewater	COD: 300	COD: 85%	Reuse in the industry	Klidi et al., 2018
Sulfate Radical–Based Combined Systems					
Peroxymonosulfate (PMS/ Co^{2+})	Elderberry wastewater	TOC: 459 COD: 1320	TOC: 99% COD: 96%	Reuse for agricultural irrigation	Amor et al., 2021
Sunlight-activated persulfate	Agro-wastewater	Acrinathrin: 0.02 Fluopyram: 1.17	Nearly complete degradation (> 97%) of the parent molecules	Reuse for agricultural irrigation	Vela et al., 2019
UV/persulfate + UF	Effluent of activated sludge system treating domestic wastewater	Organic micropollutants (OMP)	OMPs: > 80%	Reuse for general purposes	Li et al., 2020

specific water reuse criteria and additional internal conditions for the optimum dyeing processes.

Bayrakdar et al. (2021) examined the consecutive treatment of textile wastewater with photo-Fenton-like oxidation and adsorption process for potential reuse for irrigation. In the photo-Fenton-like oxidation process, iron containing a walnut shell–based catalyst was used and TOC removal of about 20%–30% was observed under optimum conditions. When adsorption was used after photo-Fenton-like oxidation, the determined TOC and COD removal efficiencies were 86% and 58%, respectively. It was stated that textile wastewater treated with photo-Fenton-like oxidation and adsorption processes were meeting the irrigation water standards (e.g., TOC, color, turbidity and TSS) and can be used for the growing of plants such as barley, cotton, sugar beet, grass, spinach, long wheat grass, date palm tree, asparagus and Bermuda grass sorghum.

Choi and Chung (2015) evaluated the reuse of wastewater of display manufacturing industry by combining solid phase AOP (metal immobilized catalyst) with RO in a pilot system. The removal efficiency of TOC by the combined system was reported as 92%. In terms of water quality and operational cost, the solid phase AOP/RO combined processes was found to be superior to the other combined processes such as AC/RO and UV AOP/anion polisher/coal carbon. The TOC concentration (<200 μg L^{-1}) and conductivity (<5 μS cm^{-1}) in the solid phase AOP/RO combined processes were lower when compared to the other process combinations (AC/RO or UV-AOP/anion polisher/coal carbon). Hybrid processes such as solid phase AOP/RO can find practical applicability for the reuse of wastewater in terms of high performance and cost-effectiveness.

13.21 UV-Based Combined Systems

Nogueira et al. (2016) employed H_2O_2/UV oxidation and RO after biofiltration for tertiary treatment of oil refinery wastewater. Removal efficiencies observed for TOC and UV absorbance at 254 nm (UV_{254}) by biofiltration alone were 46% and 23%, while efficiencies for the combined biological/chemical oxidation treatments were 88% and 79%, respectively. The silt density index (SDI) was also efficiently decreased by H_2O_2/UV oxidation, which was beneficial for subsequent RO process. It was reported that the treatment sequence was suitable for long-term operation of RO by reducing membrane fouling and producing water for reuse in the industry. The characteristics of final effluent after the RO step ($NH_4 < 10$ mg N/L; COD < 50 mg/L; negligible conductivity and salinity) showed that treated water was of a suitable quality to be employed in high-pressure boilers and cooling towers.

Li et al. (2020) employed ozonation, UV/H_2O_2 and UV/PS as a pre-treatment step before ultrafiltration (UF) process for reuse of secondary wastewater. UV/PS represented the highest performance followed by UV/H_2O_2 and ozonation. UF alone was insufficient for the removal of organic micropollutants (OMPs). Ozonation was capable of OMP degradation but represented low efficiency in the decomposition of ozone-refractory OMPs (e.g., caffeine, atrazine and ibuprofen). UV/H_2O_2 and UV/PS effectively decreased 14 OMP concentrations ($> 80\%$), representing exceptional performances in OMPs decomposition. Additionally, oxidation pre-treatments enabled genotoxicity disappearance in the secondary effluent. The polysaccharides group in dissolved effluent organic matters (dEfOM) were effectively decreased with oxidation pre-treatments, consequently improving membrane flux maintenance. The combination of UV/PS and UF processes was recommended for wastewater reuse because of its effectively improving water quality and decreasing membrane fouling.

Benito et al. (2017) evaluated UV or UV/H_2O_2 pre-treatments for the UF process, which is applied as the tertiary treatment of an urban wastewater treatment plant (WWTP) for effluent reuse. UV or UV/H_2O_2 processes were applied as a pre-treatment step to control UF membrane fouling caused by effluent organic matter. Results presented that both pre-treatments reduced significantly membrane fouling and transmembrane pressure (TMP) by 30%–44%. The UV/H_2O_2 pre-treatment enabled an improvement in dissolved organic carbon (DOC) removal of about 31% by the membrane. Both UV-based pre-treatments increased the turbidity removal efficiencies more than 30%. Applying UV-based pre-treatments enable 64%–77% savings in operational costs in relation with membrane cleaning and maintenance. Thus, UV-based pre-treatments can be accepted as promising techniques capable of reducing membrane fouling and in improving the water quality for reuse.

James et al. (2014) investigated the removal of micropollutants (MPs) from secondary municipal wastewater by UV/H_2O_2 process integrated MF and RO through pilot-scale system. Under optimum conditions, AOP accomplished important removal efficiencies ($> 99\%$) for N-nitrosodimethylamine and endocrine-disrupting compounds. The comparison of different process sequences with MF, RO, AOP and AC presented that the MF/RO/AOP combination was the most cost efficient, on condition that the management of the RO concentrate stream does not cause important cost. The MF/RO/AOP combination was about 20% lower in operational cost than the MF/AOP and MF/AC/AOP combinations. MF/RO/AOP provided better water quality, but also a significantly higher volume of concentrate stream. It was reported that AOPs effectively decrease the levels of the more difficult recalcitrant MPs to attain stringent water quality standards for wastewater reuse. However, practical limitations and important costs exist.

Souza et al. (2011) investigated the treatment of a refinery wastewater with O_3/UV and H_2O_2/UV processes combined with biological activated carbon (BAC) aiming to produce water for reuse. It was reported that O_3/UV and H_2O_2/UV combined processes led to relatively low TOC removals. When the effluents of the mentioned processes were treated by granulated AC biofiltration, TOC removals increased. The lowest TOC levels (4–8.5 mg/L) were observed for the H_2O_2/UV+BAC treatment that attained the requirements for some water reuse options in the oil refinery industry. The treatment of MBR effluent with BAC filter, without pre-oxidation, revealed elevated TOC removal efficiencies. However, these removal efficiencies were still lower than the efficiencies with pre-oxidation, and TOC concentrations were higher than that of the H_2O_2/UV+BAC system.

Ata and Töre (2019) investigated the removal of antibiotic residues from effluents of industrial and domestic wastewater treatment plants by photocatalytic oxidation with nitrogen-fluoride-carbon (NFC)–doped titanium dioxide photocatalyst. By using NFC-doped photocatalytic oxidation under visible light illumination for 7 h, 99% to 100% removal of antibiotic residues (erythromycin, ciprofloxacin, sulfamethoxazole) was achieved together with approximately 73%–79% COD and 73%–86% TOC removals for domestic and 62% TOC and COD removals for industrial WWTP's secondary effluent. It was reported that it is possible to treat the secondary effluents with AOP by employing NFC-doped photocatalyst at the level to be reused for irrigation.

Silva et al. (2018) employed chemical and electrochemical AOPs (AOPs/EAOPs) as a polishing step of textile wastewater treatment after biological treatment, clarification and filtration. AOPs applied were hydrogen peroxide (H_2O_2), ultraviolet C (UVC) radiation, (H_2O_2/UVC), anodic oxidation (AO), AO with electrogenerated H_2O_2 (AO-H_2O_2) and AO-H_2O_2 with UVC radiation (AO-H_2O_2/UVC). Considering the similar efficiencies of the AO-H_2O_2/UVC and H_2O_2/UVC processes and the higher operating cost of AO-H_2O_2/UVC process, the H_2O_2/UVC process was preferred in the reuse tests. They achieved complete decolorization despite the maximum mineralization of 39% due to presence of recalcitrant compounds. The reuse of recycled textile wastewater was approved for scouring, bleaching and dyeing processes. It was stated that recycled textile wastewater can be reused successfully, even if not complying with all reuse requirements specified in the literature.

Aliste et al. (2020) investigated the degradation of pesticide residues in agro-wastewater by AOPs in a pilot plant using solar radiation. Treatment of agro-wastewater with TiO_2/$S_2O_8^{2-}$ process in pilot plant for about 1 day (400 kJ m^2 of accumulated UV-A) provided degradation efficiency >99.5% for 13 pesticides (azoxystrobin, boscalid, chlorpropham, flutolanil, flutriafol, isoxaben, methoxyfenozide, myclobutanil, napropamide, prochloraz, propamocarb, propyzamide and triadimenol). Finally, the reclaimed water was reused for lettuce irrigation without a reduction in product quality. It was reported that the use of solar advanced oxidation technology enables the reuse of agro-wastewater, consequently decreasing the risk for human health and the environment.

Nadeem et al. (2017) used various AOPs for the treatment of real textile dyeing wastewater as a final polishing step after biological treatment for its recirculation back to the textile wet processing for dyeing. When chemical usage, process efficiency and sludge production are considered to prioritize applicable water recycling processes, the most acceptable and efficient AOPs were selected as O_3, O_3/UV, UV/H_2O_2/O_3 and O_3/H_2O_2. Textile wastewater treated by O_3, O_3/UV, UV/H_2O_2/O_3 and O_3/H_2O_2 processes was securely recycled for dyeing intentions with 35% and 20% reduced NaCl and Na_2CO_3 consumptions, respectively. UV/H_2O_2/O_3 was evaluated as the most effective process with a water reuse potential of 663,000 m^3/year (85% water recovery), 100% color and 60.1% COD removal.

Güyer et al. (2016) investigated recycling and direct reuse of washing-bleaching wastewater from reactive dyeing of cotton fabric by employing AOPs like O_3, UV/O_3, O_3/H_2O_2 and O_3/H_2O_2/UV. All investigated AOPs were capable of decreasing the COD and color levels to the reuse limit values except conductivity. The O_3/H_2O_2/UV combination removed COD with a better efficiency (55%) than the other employed AOPs, with similar color removal efficiencies (> 95%). By recycling the total wastewater from the batch washing-rinsing process, which is equivalent to about 40% of the total water requirements at facility, freshwater utilization for washing purposes could be reduced.

13.22 Ozone-Based Combined Systems

Piras et al. (2020) performed a comprehensive pilot study to evaluate the performance of treatment sequences consisting of two AOP steps and biofiltration to bring the municipal wastewater to potable water quality. The initial AOP (O_3/H_2O_2) served as a pre-treatment to biofiltration (AC or limestone) and the second AOP (UVC/H_2O_2) served as a post-biofiltration polishing step to assure enhanced disinfection. Out of 219 suspected micropollutants, 13 organic pollutants were determined. The O_3/H_2O_2 + BAC + UVC/H_2O_2 combined treatment showed excellent performance by removing 99% of total micropollutants. It was reported that effluents obtained from the tested trains generally did not reveal any ecotoxicological observation, even when initial concentrations were increased 100–1000 times to raise the sensitivity of the bioassay methods.

Blinska et al. (2020) applied a catalytic ozonation process after industrial-scale electrocoagulation treatment to recycle highly polluted textile wastewater within the industrial plant. Although the industrial electrocoagulation treatment enabled 84% color removal within 8 min, the wastewater showed extremely high absorbance. However, electrocoagulation treatment caused the generation of troublesome by-products (e.g., aniline derivative). COD and TOC removals with electrocoagulation were 38% and 41%, respectively. Catalytic ozonation was applied as a polishing step after the electrocoagulation treatment. The catalytic ozonation was capable of effectively removing residual color/colorless by-products. After catalytic ozonation, 63% COD and 61% TOC removals were achieved. The average oxidation state, spectral analysis and toxicity assay represented the efficiency of catalytic ozonation in the oxidation of by-products. Thus, the catalytic influence of AC has been proven for the ozonation of textile wastewater. When the treated wastewater was recycled into the dyeing process, very good color quality was observed for textile samples.

Demir-Düz et al. (2020) employed an O_3/H_2O_2 system to improve the mineralization of refinery wastewater from the secondary treatment and to achieve water quality criteria for recycling and reuse purposes and thus decrease the refinery water requirements. It was reported that the peroxone treatment could be regarded as a promising post-treatment technique that enables water recycling in refineries. The finally observed water characteristics enabled its reuse as cooling, firefighting and cleaning water, which are the primary water-consuming operations of a refinery. The reduction of TOC below the reuse target value (4 mg/L) was achieved with 0.9 g ozone/h and 80 mg H_2O_2/L.

13.23 Electrochemical-Based Combined Systems

Ganiyu et al. (2018) applied electrooxidation (EO), electrooxidation with hydrogen peroxide production (EO-H_2O_2) and electro-Fenton process (EF) as alternative treatment techniques for total destruction of anionic surfactants and organic compounds in real carwash wastewater. The electrochemical processes were carried out with boron-doped anode and carbon felt cathode. Total destruction of anionic surfactants and enhanced removal of organic compounds were obtained. Elevated COD and surfactant removals were accompanied with increased current efficiency and reduced energy utilization in EF treatments when compared with EO or EO-H_2O_2 treatments. EO-H_2O_2 and EF processes provided the complete degradation of the anionic surfactant within 4 and 2 h, respectively, whereas the EO process degraded approximately 96% of the surfactant in 4 h, under similar conditions. It was reported that electrochemical treatment was an effective technology for total destruction of organic compounds in carwash wastewater for potential reuse.

Sundarapandiyan et al. (2010) investigated electrochemical treatment of synthetic and real tannery saline wastewater using graphite electrodes. Wastewater reuse was performed at commercial scale. Pickling wastewater, which is one of the saline effluents of leather manufacturing, was treated and reused three times. The treatment of saline effluents from tannery for reuse was found to be effective and sufficient. In reuse trials,

85%–88% TKN and 80%–81% COD removal was achieved. Repetitive recycling and reuse of treated wastewater did not cause contaminant accumulation. The effluent characteristic and the leather quality revealed that reuse of saline effluents by intermittent electrochemical application is possible.

Mohan et al. (2007) evaluated the removal of organic pollutants from textile effluent using three different anode materials during electrochemical treatment. Ruthenium oxide–coated titanium anode presented higher removal efficiencies than lead and tin oxide–coated titanium electrodes. It was reported that the effluent COD was reduced substantially by electrochemical treatment. The influence of numerous cycling of treated textile wastewater on dye uptake in dyeing operations and generated water quality was examined. Results indicated the feasibility of electrochemical treatment for textile wastewater and consequent reuse in dyeing applications. The water requirements of textile industries can be effectively minimized with the integration of this technique.

Zhu et al. (2015) investigated the use of a plug flow electrooxidation reactor with meshed plate electrodes containing titanium as cathodes and Ti/PbO_2 as anodes for the reclamation of industrial park wastewater. Under optimal operating conditions, 30 min of electrolysis provided 63.5% COD removal and reduced the COD and color values below 60.0 mg L^{-1} and 20 Hazen, respectively. The energy consumption was 4.12 kWh and operating cost was found to be \$0.57 for the reclamation of 1 ton of effluent. With advanced treatment, recycling water quality was attained by the effluent, enabling its reuse for various water utilization and scenic purposes, thus conserving water resources and preventing pollution.

13.24 Sulfate Radical–Based Combined Systems

Vela et al. (2019) investigated the use of sodiumpersulfate ($Na_2S_2O_8$) at pilot scale for the destruction of 17 pesticides (pymetrozine, flonicamid, imidacloprid, acetamiprid, cymoxanil, thiachloprid, spinosad, chlorantraniliprole, triadimenol, tebuconazole, fluopyram, difenoconazole, cyflufenamid, hexythiazox, spiromesifen, folpet and acrinathrin) present in agro-wastewater generated by washing of vessels and phytosanitary treatment equipment. Almost total destruction (> 97%) of parent molecules was accomplished, although 13% of initial DOC was found to remain. Results presented that PS is an efficient and low-cost process capable of degrading pesticides found in agro-wastewater in reasonable times. The reuse of purified water for broccoli irrigation did not reduce the quality of harvested product or expose risk to human health.

Amor et al. (2021) investigated the efficiency of an UV-A LED system for the decomposition of elderberry wastewater using two transition metal activators (Fe^{2+} and Co^{2+}) and various oxidants (hydrogen peroxide, peroxymonosulfate and persulfate). The PMS/Co^{2+} process revealed TOC removal efficiency of 99% within 45 min and an increase in the BOD_5/COD ratio from 0.30 to 0.53, making it the best alternative. It was concluded that oxidation processes involving hydroxyl and sulfate radicals have an application potential for agro-industrial wastewaters, enabling the reuse of purified effluent in irrigation.

REFERENCES

Abdel-Shafy, H.I., El-Khateeb M.A., Mansour, M.S. M. (2016) Treatment of leather industrial wastewater via combined advanced oxidatón and membrane filtration, *Water Science & Technology* 74, 586–594.

Adegoke, A.A., Amoah, I.D., Stenström, T.A., Verbyla, M.E., Mihelcic, J.R. (2018) Epidemiological evidence and health risks associated with agricultural reuse of partially treated and untreated wastewater: A review, *Frontiers in Public Health* 6, 337.

Aliste, M., Garrido, I., Flores, P., Hellín, P., Vela, N., Navarro, S., Fenoll, J. (2020) Reclamation of agro-wastewater polluted with thirteen pesticides by solar photocatalysis to reuse in irrigation of greenhouse lettuce grown, *Journal of Environmental Management* 266, 110565.

Amor, C., Fernandes, J.R., Lucas, M.S., Peres, J.A. (2021) Hydroxyl and sulfate radical advanced oxidation processes: Application to an agro-industrial wastewater, *Environmental Technology & Innovation* 21, 101183.

Ata, R., Yıldız Töre, G. (2019) Characterization and removal of antibiotic residues by NFC-doped photocatalytic oxidation from domestic and industrial secondary treated wastewaters in Meric-Ergene Basin and reuse assessment for irrigation, *Journal of Environmental Management* 233, 673–680.

Aydiner, C., Kiril Mert B., Can Dogan E., Yatmaz, H. C., Dagli, S., Aksu, S., Tilki, Y.M., Goren A.Y., Balci, E. (2019) Novel hybrid treatments of textile wastewater by membrane oxidation reactor: Performance investigations, optimizations and efficiency comparisons, *Science of the Total Environment* 683, 411–426.

Bayrakdar, M., Atalay, S., Ersoz, G. (2021) Efficient treatment for textile wastewater through sequential photo Fenton-like oxidation and adsorption processes for reuse in irrigation, *Ceramics International* 47, 9679–9690.

Benito A., Garcia, G., Gonzalez-Olmos, R. (2017) Fouling reduction by UV-based pretreatment in hollow fiber ultrafiltration membranes for urban wastewater reuse, *Journal of Membrane Science* 536, 141–147.

Bilinska, L., Blus, K., Foszpanczyk, M., Gmurek, M., Ledakowicz, S. (2020) Catalytic ozonation of textile wastewater as a polishing step after industrial scale electrocoagulation, *Journal of Environmental Management* 265, 110502.

Blanco, J., Torrades, F., Morón, M., Brouta-Agnésa, M., García-Montaño, J. (2014) Photo-Fenton and sequencing batch reactor coupled to photo-Fenton processes for textile wastewater reclamation: Feasibility of reuse in dyeing processes, *Chemical Engineering Journal* 240, 469–475.

Choi, J., Chung, J. (2015) Evaluation of potential for reuse of industrial wastewater using metal-immobilized catalysts and reverse osmosis, *Chemosphere* 125, 139–146.

Demir-Duz, H., Aktürk, A.S., Ayyildiz, O., Alvarez, M.G., Contreras, S. (2020) Reuse and recycle solutions in refineries by ozone-based advanced oxidation processes: A statistical approach, *Journal of Environmental Management* 263, 110346.

Devi, P., Das, U., Dalai, A.K. (2016) In-situ chemical oxidation: Principle and applications of peroxide and persulfate treatments in wastewate rsystems, *Science of the Total Environment* 571, 643–657.

Dewil, R., Mantzavinos, D., Poulios, I., Rodrigo, M.A. (2017) New perspectives for advanced oxidation processes, *Journal of Environmental Management* 195, 93–99.

Drewes, J.E., Hübner, U., Zhiteneva, V., Karakurt, S. (2017) *Characterization of unplanned water reuse in the EU final report*, the European Commission DG Environment, Contract No. 07001/2017/758172/SER/EMW.C.1.

Egbuikwem, P.N., Mierzwa, J.C., Saroj, D.P. (2020) Evaluation of aerobic biological process with post-ozonation for treatment of mixed industrial and domestic wastewater for potential reuse in agriculture, *Bioresource Technology* 318, 124200.

European Commission (2015) *Optimizing water reuse in the EU, final report-part I*, Publishing Office of the European Union, ISBN 978-92-79-46835-3.

European Commission (2018) *Proposal for a regulation of the european parliament and of the Council on minimum requirements for water reuse*, COM (2018) 337 final. https://ec.europa.eu/environment/water/pdf/water_reuse_regulation.pdf

Feng, F., Xu, Z., Li, X., You, W., Zhen, Y. (2010) Advanced treatment of dyeing wastewater towards reuse by the combined Fenton oxidation and membrane bioreactor process, *Journal of Environmental Sciences* 22, 1657–1665.

Food and Agriculture Organization of the United Nations (FAO) (1992) *Wastewater treatment and use in agriculture*, FAO, ISBN 92-5-103135-5.

Food and Agriculture Organization of the United Nations (FAO) (2011) *The state of the world's land and water resources for food and agriculture: Managing systems of risk*. Earthscan/FAO. www.fao.org/nr/solaw/solaw-home/en/.

Ganiyu, S.O., Dos Santos, E., De Araújo Costa, E.C.T., Martínez-Huitle, C.A. (2018) Electrochemical advanced oxidation processes (EAOPs) as alternative treatment techniques for carwash wastewater reclamation, *Chemosphere* 211, 998–1006.

Ganiyu, S.O., Van Hullebusch, E. D., Cretin, M., Esposito, G., Oturan, M.A. (2015) Coupling of membrane filtration and advanced oxidation processes for removal of pharmaceutical residues: A critical review, *Separation and Purification Technology* 156, 891–914.

Ghanbari, F., Moradi, M. (2017) Application of peroxymonosulfate and its activation methods for degradation of environmental organic pollutants: Review, *Chemical Engineering Journal* 310, 41–62.

Ghanbari, F., Moradi, M., Gohari, F. (2016) Degradation of 2,4,6-trichlorophenol in aqueous solutions using peroxymonosulfate/activated carbon/UV process via sulfate and hydroxyl radicals, *J. Water Process Eng.* 9, 22–28.

GilPavas, E., Dobrosz-Gomez, I., Gomez-García, M.A. (2017) Coagulation-flocculation sequential with Fenton or Photo-Fenton processes as an alternative for the industrial textile wastewater treatment, *Journal of Environmental Management* 191, 189–197.

Gu, J., Liang, J., Chen C., Li, K., Zhou, W., Jia, J., Sun T. (2020) Treatment of real deplating wastewater through an environmental friendly precipitation-electrodeposition-oxidation process: Recovery of silver and copper and reuse of wastewater, *Separation and Purification Technology* 248, 117082.

Güyer, Tezcanlı G., Nadeem, K., Dizge, N. (2016) Recycling of pad-batch washing textile wastewater through advanced oxidation processes and its reusability assessment for Turkish textile industry, *Journal of Cleaner Production* 139, 488–494.

Hamilton, A.J., Stagnitti, F., Xiong, X., Kreidl, S.L., Benke, K.K. and Maher, P. (2007) Wastewater irrigation: The state of play, *Vadose Zone Journal, Review and Analysis* 6(4), 823–840.

Huang, W.-J., Fang, G.-C., Wang, C.-C. (2005) A nanometer-zno catalyst to enhance the ozonation of 2.4.6-trichlorophenol in water, *Colloids and Surfaces A: Physicochemical and Engineering Aspects* 260, 45–51.

James, C. P., Germain, E., Judd, S. (2014) Micropollutant removal by advanced oxidation of microfiltered secondary effluent for water reuse, *Separation and Purification Technology* 127, 77–83.

Kasprzyk-Hordern, B., Ziółek, M., Nawrocki, J. (2003) Catalytic ozonation and methods of enhancing molecular ozone reactions in water treatment, *Appl Catal B-Environ* 46, 639–669.

Kehrein, P., van Loosdrecht, M., Osseweijer, P., Garfí, M., Dewulf, J., Posada, J. (2020) A critical review of resource recovery from municipal wastewater treatment plants – market supply potentials, technologies and bottlenecks, *Environmental Science Water Research & Technology* 6, 877–910.

Klidi, N., Clematis, D., Delucchi, M., Gadri, A. Ammar, S., Panizza M. (2018) Applicability of electrochemical methods to paper mill wastewater for reuse. Anodic oxidation with BDD and TiRuSnO$_2$ anodes, *Journal of Electroanalytical Chemistry* 815, 16–23.

Kocamemi A.B., Kurt, H., Sait, A., Sarac, F., Saatci, A.M., Pakdemirli, B. (2020) SARS-CoV-2 Detection in Istanbul wastewater treatment plant sludge, *medRxiv* doi.org/10.1101/2020.05.12.20099358.

Lahrich, S., Laghrib, F., Farahi, A., Bakasse, M., Saqrane, S., El Mhammedi, M.A. (2021) Review on the contamination of wastewater by COVİD-19 virus: Impact and treatment, *Science of the Total Environment* 751, 142325.

Lautze, J., Stander, E., Drechsel, P., da Silva, A.K. (2014) *Global experiences in water reuse*. International Water Management Institute (IWMI). CGIAR Research Program on Water, Land and Ecosystems (WLE). 31. Resource Recovery and Reuse Series.

Lazarova, V. (2017) *Water reuse in europe, status and recent trends in policy development, LIFE+ReQpro final conference*, 23 February, http://reqpro.crpa.it/media/documents/reqpro_www/eventi/20170223_FinalMeeting_RE/Lazarova_LIFE+ReQpro.pdf

Li, M., Wen, Q., Chen, Z., Tang, Y., Yang, B. (2020) Comparison of ozonation and UV based oxidation as pre-treatment process for ultrafiltration in wastewater reuse: Simultaneous water risks reduction and membrane fouling mitigation, *Chemosphere* 244, 125449.

Liang, C., Su, H.W. (2009) Identification of sulfate and hydroxyl radicals in thermally activated PS, *Industrial Engineering Chemistry Research* 48, 472–475.

Meijide, J., Rodríguez, S., Sanromán, M.A., Pazos, M. (2018) Comprehensive solution for acetamiprid degradation: Combined electro-Fenton and adsorption process, *Journal of Electroanalytical Chemistry* 808, 446–454.

Moazeni, M., Nikaeen, M., Hadi, M., Moghim, S., Mouhebat, L., Hatamzadeh, M., Hassanzadeh, A. (2017) Estimation of health risks caused by exposure to enteroviruses from agricultural application of wastewater effluents, *Water Research* 125, 104–113.

Mohan, N., Balasubramanian, N., Basha, C.A. (2007) Electrochemical oxidation of textile wastewater and its reuse, *Journal of Hazardous Materials* 147, 644–651.

Nadeem, K., Tezcanli Guyer, G., Dizge, N. (2017) Polishing of biologically treated textile wastewater through AOPs and recycling for wet processing, *Journal of Water Process Engineering* 20, 29–39.

Navalon, S., Dhakshinamoorthy, A., Alvaro, M., Garcia, H. (2011) Heterogeneous Fenton catalysts based on activated carbon and related materials, *Chem Sus Chem* 4, 1712–1730.

Nogueira A.A., Bassin J. P., Cerqueira A.C., Dezotti M. (2016) Integration of biofiltration and advanced oxidation processes for tertiary treatment of an oil refinery wastewater aiming at water reuse, *Environ Sci Pollut Res* 23, 9730–9741.

Oller, I., Malato, S., Sánchez-Pérez, J.A. (2011) Combination of advanced oxidation processes and biological treatments for wastewater decontamination – A review, *Science of the Total Environment* 409, 4141–4166.

Organisation for Economic Co-Operation and Development (OECD) (2012) *Environmental outlook to 2050: The consequences of inaction.* OECD Publishing, www.oecd-ilibrary. org/docserver/9789264122246-en.pdf?expires=1576513787 &id=id&accname=ocid177643&checksum=E5D1E6D4DB 78962941DAA08F2B58D805.

Oturan, M.A., Aaron, J.J. (2014) Advanced oxidation processes in water/wastewater treatment: Principles and applications. A review, *Critical Reviews in Environmental Science and Technology* 44, 2577–2641.

Ou, W., Zhang, G., Yuan, X., Su P. (2015) Experimental study on coupling photocatalytic oxidation process and membrane separation for the reuse of dye wastewater, *Journal of Water Process Engineering* 6, 120–128.

Papić, S., Vujević, D., Koprivanac, N., Šinko, D. (2009) Decolourization and mineralization of commercial reactive dyes by using homogeneous and heterogeneous Fenton and UV/Fenton processes. *J Hazard Mater* 164, 1137–1145.

Paździor, K., Bilińska, L., Ledakowicz, S. (2019) A review of the existing and emerging technologies in the combination of AOPs and biological processes in industrial textile wastewater treatment, *Chemical Engineering Journal* 376, 120597.

Piras, F., Santoro, O., Pastore, T., Pio, I., De E., Dominicis, E., Gritti, R., Caricato, M.G., Lionetto, G., Mele, D. (2020) Santoro, Controlling micropollutants in tertiary municipal wastewater by O₃/H₂O₂, granular biofiltration and UV254/H₂O₂ for potable reuse applications, *Chemosphere* 239, 124635.

Pouran, S.R., Aziz, A.R.A., Daud, W.M.A.W. (2015) Review on the main advances in photo-Fenton oxidation system for recalcitrant wastewaters, *Journal of Industrial and Engineering Chemistry* 21, 53–69.

Rekhate, C.V., Srivastava, J.K. (2020) Recent advances in ozone-based advanced oxidation processes for treatment of wastewater- A review, *Chemical Engineering Journal Advances* 3, 100031.

Ribeiro, M.C.M., Starling, M.C.V.M., Leão, M.M.D., de Amorim, C.C. (2017) Textile wastewater reuse after additional treatment by Fenton's reagent, *Environ Sci Pollut Res* 24, 6165–6175.

Roccaro, P. (2018) Treatment processes for municipal wastewater reclamation: The challenges of emerging contaminants

and direct potable reuse, *Current Opinion in Environmental Science & Health* 2, 46–54.

Rodríguez-Chueca, J., Alonso, E. (2019) D.N. Photocatalytic mechanisms for peroxymonosulfate activation through the removal of methylene blue: A Case Study, *Int. J. Environ. Res. Public Health* 16, 1–13.

Rostam, A.B., Taghizadeh, M. (2020) Advanced oxidation processes integrated by membrane reactors and bioreactors for various wastewater treatments: A critical review, *Journal of Environmental Chemical Engineering* 8, 104566.

Saraty, S.R., Mohseni, M. (2006) An overview of UV-based advanced oxidation processes for drinking water, *IUVA News*, 7(6), 1–12.

Schulte, P., Morrison, J. (2014) *Driving harmonization of water related terminology.* Discussion Paper. The CEO Water Mandate.

Shah, M.P. (2021) *Removal of emerging contaminants through microbial processes.* Springer.

Shah, M.P. (2020a) *Advanced oxidation processes for effluent treatment plants.* Elsevier.

Shah, M.P. (2020b) *Microbial bioremediation and biodegradation.* Springer.

Sharma, J., Mishra, I.M., Dionysiou, D.D., Kumar, V. (2015) Oxidative removal of Bisphenol A by UV-C/peroxymonosulfate (PMS): Kinetics, influence of co-existing chemicals and degradation pathway. *Chemical Engineering Journal* 276, 193–204.

Shoushtarian, F., Negahban-Azar, M. (2020) Worldwide regulations and guidelines for agricultural water reuse: A critical review, *Water* 12, 971.

Silva, L.G.M., Moreira, F.C., Souza, A.A.U., S. Souza, M.A.G.U., Boaventura, R.A.R., Vilar, V.J.P. (2018) Chemical and electrochemical advanced oxidation processes as a polishing step for textile wastewater treatment: A study regarding the discharge into the environment and the reuse in the textile industry, *Journal of Cleaner Production* 198, 430–442.

Souza, B.M., Cerqueira, A.C., Sant'Anna Jr., G.L., Dezotti, M. (2011) Oil-refinery wastewater treatment aiming reuse by advanced oxidation processes (AOPs) combined with biological activated carbon (BAC), *Ozone: Science & Engineering* 33, 403–409.

Sundarapandiyan, S., Chandrasekar, R., Ramanaiah, B., Krishnan, S., Saravanan, P. (2010) Electrochemical oxidation and reuse of tannery saline wastewater, *Journal of Hazardous Materials* 180, 197–203.

Tabrizi, G. B., Mehrvar, M. (2004) Integration of advanced oxidation technologies and biological processes: Recent developments, trends, and advances, *Journal of Environmental Science and Health, Part A, Toxic/Hazardous Substances and Environmental Engineering* A39, 3029–3081.

Tong, S., Liu, W., Leng, W., Zhang, Q. (2003) Characteristics of MnO₂ catalytic ozonation of sulfosalicylic acid and propionic acid in water, *Chemosphere* 50, 1359–1364.

Ungureanu, N., Vlădut, V., Voicu, G. (2020) Water scarcity and wastewater reuse in crop irrigation, *Sustainability* 12, 9055.

Unite Nations Development Programme (UNDP) (2016) *UNDP support to the implementation of sustainable development goal 6: Sustainable management of water and sanitation,* UNDP One United Nations Plaza.

US Environmental Protection Agency (2012) *Guidelines for water reuse,* Washington, DC.

Vela, N., Fenoll, J., Garrido, I., Pérez-Lucas, G., Flores, P., Hellín, P., Navarro, S. (2019) Reclamation of agro-wastewater polluted with pesticide residues using sunlight activated persulfate for agricultural reuse, *Science of the Total Environment* 660, 923–930.

Wang, J.L., Xu, L.J. (2012) Advanced oxidation processes for wastewater treatment: Formation of hydroxyl radical and application, *Critical Reviews in Environmental Science and Technology* 42, 251–325.

WHO (2006) *WHO guidelines for the safe use of wastewater, excreta and greywater*, 3rd ed. Vol. 2, Wastewater Use in Agriculture, World Health Organization.

World Water Assessment Programme (United Nation) (2018) *The United Nations world water development report 2018*. United Nations Educational, Scientific and Cultural Organization, www.unwater.org/publications/world-water-development-report-2018/.

Yıldız Töre, G., Ata, R. (2019) Assessment of recovery & reuse activities for industrial waste waters in miscellaneous sectors, *European Journal of Engineering and Applied Sciences* 2, 19–43.

Zhang, C.-M., Xu, L.-M., Xu., P.-C. (2016) Elimination of viruses from domestic wastewater: requirements and technologies, *World J. Microbiol Biotechnol* 32:69, 1–9.

Zhu, R., Yang, C. (2015) Mingming Zhou, Jiade Wang, Industrial park wastewater deeply treated and reused by a novel electrochemical oxidation reactor, *Chemical Engineering Journal* 260, 427–433.

2030 World Resources Group (2009) *Charting our water future*, www.2030wrg.org/wpcontent/uploads/2012/06/Charting_Our_Water_Future_Final.pdf

14

Advanced Oxidation Process for Leachate Treatment: A Critical Review

Shilpa Mishra, Baranidharan Sundaram and Muthukumar S

CONTENTS

DOI: 10.1201/9781003165958-14

List of Abbreviations

AOP	Advanced oxidation process
AS	Air stripping
BOD	Biological oxygen demand
CF	Coagulation-flocculation
CO	Chemical oxidation
COD	Chemical oxygen demand
CP	Chemical precipitation
F	Fair
FA	Fulvic acid
G	Good
GHMC	Greater Hyderabad Municipal Corporation
HA	Humic acid
HW	Hazardous waste
IE	Ion exchange
MBR	Membrane bioreactor
MF	Microfiltration
MSW	Municipal solid waste
NF	Nanofiltration
P	Poor
RO	Reverse osmosis
SBR	Sequencing batch reactor
SS	Suspended solid
TN	Total nitrogen
TOC	Total organic carbon
UASB	Upflow anaerobic sludge blanket
UF	Ultrafiltration
US	Ultrasound
UV	Ultraviolet
VFA	Volatile fatty acid

14.1 Introduction

Municipal solid waste (MSW) and hazardous waste (HW) have been generated in vast amounts in the last few decades due to rapid urban and economic development. Improper management of these wastes causes several environmental problems related to soil contamination, ground and surface water pollution, and air pollution, directly or indirectly affecting human beings and ecosystems. The safe disposal of these wastes is required to minimize the impact on the ecosystem. Commonly used methods for solid waste disposal are incineration, landfilling, composting, sea dumping and open burning. Among these methods, landfilling is one of the most popular MSW and HW disposal methods in developing countries as it involves low cost, easy operation and no area constraint for the disposal of this waste.

Solid waste dumped to various MSW and HW landfill sites consists of heavy metals, organic compounds, chlorinated organics, ammonia nitrogen and inorganic salts, and when in contact with water they undergo many physical, biological and chemical changes. A solid waste landfill site generates various gases like CO_2, CH_4, H_2S and CO. When solid wastes go through several decomposition stages, it will eventually result in liquid at the bottom of the landfill. This liquid is called leachate and has a yellow, brown or black color due

to the decomposition of some organic compounds such as HA (Naveen et al. 2016). The composition of leachate depends on many factors like waste composition (domestic or industrial waste), topography, site hydrology, rain pattern and seasonal weather variation, the availability of oxygen and moisture, and the design and operation of the solid waste landfill site and the age of the landfill (Abdulhussain et al. 2009; H. Luo et al. 2019). Leachate formation starts from the early stages of solid waste landfilling and continues even after the closure of the site. Some of the various factors that influence landfill leachate quality are different environmental conditions, moisture content present in waste, waste composition and available oxygen. Also, other climatic regions have various leachate quality for the same waste type (Kalčíková et al. 2011).

In general, leachate transfer, biological treatments, physical and chemical treatments, and membrane processes are commonly used for landfill leachate treatment. In leachate transfer, leachate is treated either in combination with domestic sewage or by recycling leachate. The main advantage of this method is low cost and easy maintenance. However, it may not be able to remove heavy metal and low-biodegradable organic compounds altogether. Also, if the amount of recycling is more than saturation, ponding and acid problems may occur. Biological treatment is used to remove the BOD present in most types of leachate under aerobic and anaerobic conditions. Physical and chemical processes reduce colloidal particles, floating material, suspended solids, toxic material and color present in landfill leachate. They are also used as pre-treatment and post-treatment techniques along with other leachate treatment methods for better results. Membrane processes such as microfiltration, ultrafiltration, membrane bioreactors, nanofiltration and reverse osmosis provide a high level of purification of leachate, but at high cost and with a membrane fouling problem (Sepehri and Sarrafzadeh 2018).

14.2 Composition of the Leachate

The amount of leachate generated from MSW and HW landfills depends on age of the landfill, the amount of water retained by the waste and the landfill's degree of compaction. The leachate quality generated depends on landfill age, waste composition and environmental conditions. The main components of leachate generated from MSW and HW landfill are as follows Peter Kjeldsen et al. (2002).

Organic matter: This may include dissolved organic matter, TOC, VFA and fulvic and humic acids. The organics concentration in leachate decreases with an increase in landfill age.

Inorganic matter: This may include calcium (Ca^{2+}), sodium (Na^+), magnesium (Mg^{2+}), hydrogen carbonate (HCO^{3-}), potassium (K^+), sulfate (SO_4^{2-}), iron (Fe^{2+}), chloride (Cl^-) and ammonium (NH^{4+}).

Heavy metal: This may include zinc (Zn^{2+}), copper (Cu^{2+}), chromium (Cr^{3+}), cadmium (Cd^{2+}), nickel (Ni^{2+}) and lead (Pb^{2+}).

Xenobiotic organic compounds: These are aromatic hydro-carbons, pesticides and benzene, usually present at deficient concentrations.

The characteristics of leachate generated from landfill are usually represented by the COD, BOD, BOD/COD ratio, pH, moisture content, heavy metals, humic acid and organic compounds present. The chemical composition of municipal solid waste generated in GHMC is shown in Table 14.1 (Vamsi Krishna et al. 2015). The composition and flow rate of leachate varies from one landfill to another, with the age of the landfill and seasonally at each landfill site. The comparison of various leachate parameters concerning landfill leachate age is given in Table 14.2. The landfill leachate is classified into three categories based on age, that is, young with age less than 12 months, medium with age between 12 to 60 months and old with age more than 60 months.

14.3 Landfill Leachate Treatment Methods

14.3.1 Leachate Transfer

14.3.1.1 Recycling

Leachate recycling is one of the most common methods and has been used for decades. The main advantage of this method is low cost and easy maintenance. However, it not able to remove heavy metal and low-biodegradable organic compounds altogether. Also, if the amount of recycling is more than saturation, ponding and acid problems may occur (Tu Anqi et al. 2020).

14.3.1.2 Treatment with Domestic Sewage

The most simple, conventional and inexpensive solution for leachate treatment is treating the landfill leachate in a municipal sewage treatment plant and domestic wastewater. This method's main disadvantage is the presence of low-biodegradable dissolved organic compounds and heavy metals in the leachate. This will decrease the throughput of the treatment plant, and

TABLE 14.1

Chemical Composition of Municipal Waste Generated in GHMC

Parameters	Unit	Range
pH		6.5–7.2
Moisture content in %		32–59
Carbon content in %		8–16
Nitrogen	mg/kg	High
Zn	mg/kg	High
Heavy metals	mg/kg	Low
Calorific value	kCal/kg	1250–2550

the effluent concentration will be higher (Tu Anqi et al. 2020). However, this method's advantage is that there is no need to add nitrogen, as it is one component of leachate, or phosphorous, as it is a component of sewage in the treatment plant.

14.3.2 Biological Treatment

This method is used for the removal of BOD present in most types of leachate. The leachate is degraded into organic compounds and sludge by microorganisms under aerobic and anaerobic conditions, respectively (Dario Bove et al. 2015). For example, the anaerobic biological treatment involves the following methods: suspended growth biomass process using SBR, digester and UASB reactor and attached-growth biomass processes that use a hybrid bed filter, anaerobic filter and fluidized bed reactor. Aerobic biological treatment involves the following methods: suspended growth biomass process that uses lagooning, activated sludge process, sequence batch reactor, and attached-growth biomass systems that use trickling filters and moving-bed biofilm reactor.

14.3.3 Physical and Chemical Treatments

These treatments reduce colloidal particles, floating material, suspended solids, toxic material and color present in landfill leachate. Commonly used treatment methods are listed next (Hongwei Luo et al. 2020).

Table 14.2

Landfill Leachate Classification vs. Age (Abdulhussain A. Abbas et al. 2009; Javier Tejera et al. 2019)

Parameters	Unit	Young	Medium	Old
Age	months	<12	12–60	> 60
pH		<6.5	6.5–7.5	> 7.5
COD	mg/L	> 15,000	3000–15,000	<3000
BOD_5	mgO_2/L	> 4000	1000–4000	<400
TOC	mg/L	> 2500	1000–2500	<1000
NH_3-N	mg/L	<450	450	> 450
Organic compounds		High VFA	Low–medium VFA, HA & FA	High HA & FA
Biodegradability		High	Medium	Low

14.3.3.1 Flotation

MSW and HW landfill leachate consists of nonbiodegradable HA, which this process can remove.

14.3.3.2 Coagulation-Flocculation (CF)

This method is used for the elimination of organic compounds and heavy metals present in leachate. The most commonly used coagulants are $Al_2(SO_4)_3$, $FeSO_4$, $FeCl_3$ and ferric chlorosulfate. CF is commonly used as a pre-treatment and/or post-treatment method to eliminate nonbiodegradable organic compounds present in MSW and HW leachate. For better performance, high pH has to be maintained. The main drawback of CF is the generation of sludge and high aluminum or iron concentration in the liquid phase. Alimoradi et al. (2018) stated that the combination of coagulation or adsorption with membrane process removed 100% SS, 97.3% TOC, 99.2% COD and 90% Al.

14.3.3.3 Chemical precipitation (CP)

This is a pre-treatment method used to eliminate ammonium nitrogen.

14.3.3.4 Adsorption

In terms of COD removal, this method has better performance than the chemical method by adsorption of pollutants using activated carbon or powder form.

14.3.3.5 Chemical Oxidation (CO)

CO includes ozonation, hydrogen peroxide, UV radiation, ultrasound or sonochemical and catalysts with or without UV radiation. This method oxidizes various organic compounds to CO_2 and H_2O so that complete mineralization is achieved. It also improves the biodegradability of an organic compound.

14.3.3.6 Air Stripping (AS)

This method is used for leachate with a high concentration of NH_4-N (ammonium nitrogen). Performance of this method depends on pH values and H_2SO_4 or HCl during the contaminated gas phase. This method's significant disadvantages are the release of NH_3 into the atmosphere and scaling of the stripping tower by $CaCO_3$.

14.3.3.7 Ion Exchange (IE)

This method is commonly used for removing nonbiodegradable compounds from landfill leachate and metal impurities from discharge.

14.3.4 Membrane Process

Various conventional treatment methods such as biological treatments, physical and chemical treatments for elimination and degradation of different toxic and organic compounds present in landfill leachate were already discussed. However, better purification is required to minimize the impact of leachate on the environment, and new treatment methods such as membrane process is being used for landfill leachate treatment (Abbas et al. 2009; Tu Anqi et al. 2020). Some of the most commonly used membrane processes are as follows.

14.3.4.1 Microfiltration (MF)

Microfiltration is commonly used for the removal of small particles, microbial cells and large colloidals. For this purpose, microfilters with pore sizes ranging from 0.05 μ to10 μ are used.

14.3.4.2 Ultrafiltration (UF)

This method is used to remove suspended solids and large molecular weight compounds present in wastewater and leachate, which will choke membranes of reverse osmosis. It uses pressures of up to 10 bar to work. It can be used as a direct filtration method or in combination with biological treatment without a sedimentation unit.

14.3.4.3 Membrane Bioreactors (MBR)

MBR is a combination of bioreactor and membrane separation methods, commonly used for industrial wastewater treatment. The main advantage of this method is a very compact system with low sludge production. Currently, it is also used for leachate treatment.

14.3.4.4 Nanofiltration (NF)

This method is used for the removal of heavy metal and refractory organic compounds present in landfill leachate. Using this method, COD removal of 60% to 70% and 50% reduction in NH_3-N present in landfill leachate can be achieved (Tu Anqi et al. 2020).

14.3.4.5 Reverse Osmosis (RO)

This is a relatively new method but the most efficient one for COD removal among the various physical-chemical methods. This method's main disadvantage is clogging of the membrane, which decreases the overall system performance by reducing the reject concentration. However, efficiency can be improved by continuous cleaning of such membranes, but it will reduce their life. The increase in the residual quantity requires further treatment.

Researchers such as Niazi (2018) and Güvenç and Güven (2019) report that coagulation-flocculation and biological treatment methods are promising pre-treatment and post-treatment methods. This will improve the biodegradability of leachate and reduce heavy metals, COD and organic compounds. A comparison of various landfill leachate treatment methods for different leachate ages is given in Table 14.3.

TABLE 14.3

Effect of Leachate Age on Various Treatment Methods (Abbas et al. 2009)

S. No	Treatment Method	Age of Leachate (years)		
		Recent (<5)	Intermediate (5–10)	Old (> 10)
1	Combined treatment with domestic sewage	G*	F*	P*
2	Recycling	G	F	P
3	Aerobic processes	G	F	P
4	Anaerobic processes	G	F	P
5	CF	P	F	F
6	CP	P	F	P
7	Adsorption	P	F	G
8	CO	P	F	F
9	AS	P	F	F
10	IE	G	G	G
11	MF	P	-	-
12	UF	P	-	-
13	NF	G	G	G
14	RO	G	G	G

*G: Good; F: Fair; P: Poor

In the past, various studies (Abbas et al. 2009; Hongwei Luo et al. 2020; Meina Hana et al. 2020; Pratibha Gautam et al. 2019; Reda Elkacmi and Mounir Bennajah 2019; Tu Anqi et al. 2020) have been done on MSW and HW landfill to gather information about leachate generation and its composition and treatment. The most common method for MSW and HW landfill leachate treatment is the conventional biological treatment process, but such an approach is not useful for leachate generated from hazardous waste landfills. These methods are adequate for the elimination of nitrogenous compounds from leachate. Other conventional methods used for landfill leachate treatment are expensive because of high energy requirements and chemical doses (Tu Anqi et al. 2020).

Leachate treatment methods such as active carbon adsorption, ion exchange and membrane only help transfer pollutants, do not degrade them completely and are expensive. In recent years, the main focus is on AOP for treating MSW and HW landfill leachate as it is an effective method for the degradation of recalcitrant organics present in leachate. If an advanced oxidation process is used in combination with biological treatment, then a significant reduction in overall landfill leachate treatment cost is observed with improved efficiency. In this chapter, we will review various AOPs used for the treatment of landfill leachate. The performance comparison of different AOPs for MSW and HW landfill leachate is compared for the following outcomes:

First is the degradation of organic substances present in MSW and HW landfill leachate into their highest oxidation state, and the second is the breakdown of refractory organic compounds into simpler biodegradable substances for subsequent biological treatment. The effects of various additives such as graphitic carbon nitride, polymers, ammonia nitrogen

(NH_3-N), pH and alkalinity on the AOP are also discussed in this chapter.

14.4 Advanced Oxidation Process

This process generates a powerful oxidizing agent in the form of $\cdot OH$ (hydroxyl radical). In various AOPs (Jelonek and Neczaj 2012; Bourechech et al. 2017), the toxic pollutants react with $\cdot OH$ radical and produce simple compounds. These radicals are short-lived, non-selective and strong oxidizing agents; therefore, they react with almost all organic materials that produce intermediate compounds. Those intermediate compounds themselves react with $\cdot OH$ and produce stable compounds. The principle of AOPs in simple form is represented as follows:

AOPs → Oxidative Agent

AOPs + Catalyst → Oxidative Agent

AOPs + Photon → Oxidative Agent

Oxidative Agent + Organic Compound → Intermediate Compounds

Intermediate Compounds + Oxidative Agent → CO_2, H_2O and Inorganic Salts

Hydroxyl radicals are the main driving forces behind many AOPs. Some of the main characteristics of hydroxyl radicals is that they are non-selective, react rapidly, have a short life cycle, are harmless, are strong oxidants and easily degrade pollutants to simpler compounds such as CO_2 and H_2O. Ozonation (O_3), a combination of O_3 with hydrogen peroxide (H_2O_2), a combination of ultraviolet light with H_2O_2, Fenton, photo-Fenton, electrochemical, wet air oxidation, ultrasound and photocatalysis and so on are often used to produce ·OH in sufficient quantities to degrade inorganic and organic compounds.

AOPs are powerful treatment processes that can treat nearly all organic compounds and remove some heavy metals as they can generate ·OH radicals at room temperature and ambient pressure compared to other conventional methods that require high temperature and high pressure (M'Arimi et al. 2020). They also do not require a large area to process the needed flow rate for the system. One of the major advantages of AOPs is that they convert organic materials into stable inorganic compounds, such as H_2O, CO_2 and salts, and do not introduce any new hazardous compound.

The main disadvantage of this process is high operation and maintenance costs. Variants of AOPs need to be carefully selected for efficient degradation of organic compounds. AOPs are also dosage-dependent processes, so an adequate amount of hydroxyl radicals must achieve the desired level of treatment. Various AOPs available for producing ·OH radicals are list next.

14.4.1 Non-Photochemical AOPs

In this method, ·OH is generated without using light energy. There are four well-known methods – O_3 based on AOP at high pH value (> 8.5), O_3 with H_2O_2, O_3 with catalyst and the Fenton method.

14.4.1.1 Ozonation at Elevated pH

Ozone-based AOPs use ozone to generate ·OH radicals. Because of the high oxidative power of ozone, it is one of the most commonly used leachate treatment methods. The decomposition of O_3 increases as the value of pH increases. The main limitation of this method is the presence of O_3-resistant compounds or hydroxyl radical scavengers. The overall chemical equation for ozonation is given in Equation 14.1:

$$3O_3 + OH^- + H^+ \rightarrow 2\cdot OH + 4O_2 \quad (14.1)$$

The most common scavengers of ·OH radicals are bicarbonate and carbonate. When they react with ·OH radicals, they produce passive carbonate or bicarbonate radicals. Now, these passive radicals do not react anymore with organic compounds or O_3. Ozonation at elevated pH is not considered a very effective process for leachate treatment, as high doses of O_3 are required to treat leachate. Because of the slow reaction time, it is not economical to use this method to treat complex leachate (Mujtaba Hussain et al. 2020). It can be made more effective when O_3 is used with UV or with H_2O_2 or catalyst. These AOPs are discussed in subsequent sections of this chapter.

14.4.1.2 Peroxonation

·OH is also generated by adding H_2O_2 to O_3, which can initiate the decomposition of O_3 (Mujtaba Hussain et al. 2020). The

generation of hydroxyl radical using H_2O_2 and O_3 is given in Equation 14.2.

$$2O_3 + H_2O_2 \rightarrow 2\cdot OH + 3O_2 \quad (14.2)$$

When we compare Equation 14.1 with Equation 14.2, we find that for the generation of 1 mole of ·OH, 1.5 moles and 1 mole of O_3 is required without and with H_2O_2, respectively. Hence, it can be concluded that O_3 with H_2O_2 generates a high concentration of ·OH radical.

14.4.1.3 Ozone with Catalyst

Another method to accelerate the decomposition of O_3 and generation of ·OH is to use homogeneous or heterogeneous catalysts. In the homogeneous catalytic ozonation process, the decomposition of O_3 is accelerated by use of transition metal ions. The organic compound and the catalyst react with each other and produced a complex compound, which O_3 oxidizes. Several metal oxides (Fe_2O_3, Al_2O_3, MnO_2, CeO_2, TiO_2, etc.) are used to accelerate the organic compounds in the heterogeneous catalyst ozonation process decomposition. This method will increase the efficiency of ozonation and accelerate the generation of ·OH at lower pH.

14.4.1.4 Fenton Process

Fenton first used the Fenton system (H_2O_2/Fe^{2+}) for the oxidation of maleic acid. In this process, hydrogen peroxide readily oxidizes Fe^{2+} into Fe^{3+}, which is used to oxidize waste (Pradeep Kumar Singa et al. 2018; Seda Koçak et al. 2013).

The use of the Fenton system in AOP for treatment of leachate is attractive because iron is a nontoxic element and is highly abundant (Agata Krzysztoszek et al. 2012; Mohammad Ali Zazouli et al. 2012; Umar et al. 2010).

14.4.2 Photochemical AOPs

The main drawback of AOP with ozonation is the incomplete oxidation of organic compounds into CO_2 and H_2O. In some reactions, this incomplete oxidation of organic compounds produces intermediate products, which may be more toxic than the parent organic compound. So, ultraviolet (UV) radiation is used with O_3 or H_2O_2 to destroy an organic compound. O_3 with UV, O_3 with H_2O_2, H_2O_2 with UV and O_3 with H_2O_2 and UV, Fenton and Fenton-like systems with UV radiation and catalytic oxidation with UV radiation are typical examples of this method.

14.4.2.1 Ozone with UV

This method uses O_3 with UV to produce hydroxyl radicals. Ozone absorbs UV radiation from 0.2 µm to 0.3 µm wavelength, and maximum absorbance is observed at 0.2537 µm.

$$O_3 + UV \rightarrow O_2 + O(^1D) \quad (14.3)$$

$$O(^1D) + H_2O \rightarrow 2\cdot OH \quad (14.4)$$

Generation of hydroxyl radicals using ozone with UV is conceptually the simplest method and generates more hydroxyl radicals than the simple ozonation process (Mujtaba Hussain et al. 2020).

14.4.2.2 Combination of Hydrogen Peroxide and UV

The combination of H_2O_2 with UV produces a greater degradation rate of contaminant in leachate. When UV light hits hydrogen peroxide, the bonds between oxygen and oxygen are broken and give 2 moles of hydroxyl radical for each hydrogen peroxide mole.

$$H_2O_2 + UV \rightarrow 2\cdot OH \tag{14.5}$$

This method's performance will depend on the quality of wastewater, type of contamination and contaminant concentration, and the removal efficiency required. Treatment of highly turbid water such as landfill leachate requires a significantly large number of doses of UV and H_2O_2. The use of high amounts of UV and H_2O_2 is expensive and energy and time consuming, making it less viable for landfill leachate treatment.

14.4.2.3 Ozone with Hydrogen Peroxide and UV Radiation

The combination of O_3 with H_2O_2 and UV not only speeds up the degradation of O_3 but also increases the generation rate of $\cdot OH$. Again, the amount of O_3-H_2O_2 and UV required for the generation of $\cdot OH$ will depend on the quality of wastewater, type and concentration of contaminants and the amount of removal efficiency required (Mujtaba Hussain et al. 2020).

14.4.2.4 Fenton and Fenton-Like Process with UV Radiation

In Fenton and Fenton-like processes with UV radiation, Fe^{3+} ions react with H_2O_2 in the presence of UV radiation to generate OH (Agata Krzysztoszek et al. 2012; Mohammad Ali Zazouli et al. 2012; Umar et al. 2010). Chemical equations for various AOPs are given in Figure 14.1.

14.4.2.5 Photocatalytic Oxidation

The main principle of the photocatalytic oxidation process is to generate electron-hole pairs that are used to generate hydroxyl radicals. Electromagnetic radiation generally in the near UV spectrum is used for electron-hole pair generation in semiconductor materials. The most commonly used photocatalyst is titanium dioxide (TiO_2) as it has high stability, is biologically and chemically inert in nature and is easy to produce.

$$TiO_2 + UV \rightarrow Electron\ (e^-) + Hole\ (h^+) \tag{14.6}$$

$$Electron\ (e^-) + O_2 \rightarrow O_2^-. \tag{14.7}$$

$$Hole\ (h^+) + H_2O \rightarrow \cdot OH + H^+ \tag{14.8}$$

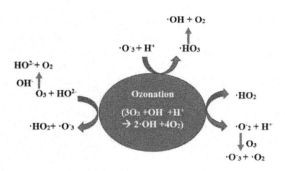

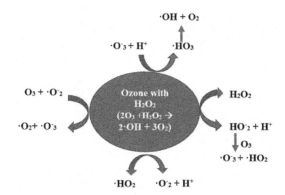

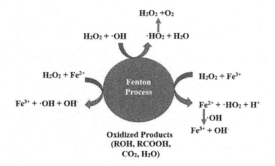

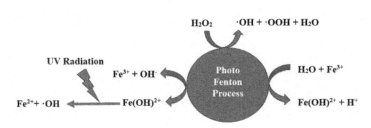

FIGURE 14.1 Chemical equations for various AOPs.

The efficiency of photocatalyst oxidation depends on several factors such as catalyst loading, dissolved oxygen concentration, pH, intensity of UV light, turbidity and ionic profile of H_2O. It is also used for the treatment of various toxic inorganic ions such as sulfates, bromate, nitrite and cyanide present in the leachate.

14.5 Electrochemical and Miscellaneous Process

14.5.1 Ozone with Ultrasound

In this process, ultrasounds are used along with the O_3 treatment process. This will increase the degradation of many non-biodegradable refractory organics present in the wastewater, which is not possible with O_3 treatment alone. Ozone with ultrasound is more effective and efficient than only ozonation and only ultrasound (US) processes.

$$O_3 + US \rightarrow O_2 + O(^3P) \tag{14.9}$$

$$O(^3P) + H_2O \rightarrow 2\,^{\bullet}OH \tag{14.10}$$

The ozone with ultrasound increases the discoloration of dyes, has higher COD removal and increases biological toxicity reduction. It is also useful for the degradation of the pesticide nitrobenzene, pharmaceutical compounds and chlorophenol present in the wastewater (Rossi et al. 2020).

14.5.2 Electrochemical Oxidation

The electrochemical oxidation process is a beneficial, cost-effective and efficient process that works on water's electrolysis process to oxidize the various contaminants present in wastewater. Anode oxidation reaction occurs, and the cathode reduction reaction occurs, which degrade the wastewater's pollutants. Electrochemical oxidation processes are classified into two main categories. These are direct and indirect electrochemical oxidation process. In the direct oxidation process, the organic compound's mineralization is achieved by the direct e^- transfer reactions on the anode surface. But in the indirect electrochemical oxidation process, the *in situ* electron generation is done by a reactive oxidant such as H_2O_2, O_3, hypochlorite or chlorine (Mayra S. de Oliveira et al. 2018).

14.5.3 Wet Air Oxidation

This process is used for the treatment of recalcitrant compounds and high molecular toxins present in leachate. It works on the principle of a chemical reaction between the organic compounds and the dissolved oxygen. In aqueous media, the dissolved oxygen at high pressure and high temperature causes the oxidation and degradation of organic compounds and converts them into H_2O, CO_2, inorganic salts and/or small biodegradable compounds (Antal Tungler et al. 2015). This process is commonly used for the oxidation of dissolved and suspended organic pollutants present in leachate. This process is generally used to treat wastewater that is considered to either be too diluted for incineration or have very large molecular recalcitrant compounds.

14.5.4 Electrocoagulation

The electrocoagulation process's main principle is the formation of coagulants from a sacrificial electrode using electrolytic oxidation after that contaminant is destabilized, followed by suspension of particulates and emulsion breakdown and then flocs formation from the aggregation of the destabilized phases. Both aluminum and iron are commonly used to form active coagulant species for the degradation of the phenolic and organic compounds present in leachate. Table 14.4 summarizes the various aspects of different AOPs (Reda Elkacmi et al. 2019; Meina Hana et al. 2020).

14.6 Discussion

In this section, the performance of various landfill leachate treatment methods is discussed, and results from past literature are consolidated. Kow Su-Huan et al. (2016) have reviewed various AOPs and found that AOPs improved the leachate's biodegradability. However, unassisted oxidants (ozone, H_2O_2 and S_2O_2) alone are not sufficient for the effective decomposition of various organic compounds into the simple compounds of CO_2 and H_2O. The use of UV with unassisted oxidants (ozone, H_2O_2 and S_2O_2) can enhance the efficiency of AOPs. Order of various AOP in terms of performance efficiency are as follows:

Fenton with UV > O_3 / S_2O_2 > Fentozone > Peroxone > Fenton > O_3/UV/Fe^{2+} >

O_3 with UV > Persulfate > O_3

The Fenton process has better removal efficiency as compared to all processes considered in this study. However, this process requires pH adjustment and treatment of sludge generated during the process. Four different advanced oxidation processes were used to treat the leachate sample obtained from the Perungudi dumping site, Chennai. A leachate sample of 20 L was collected for characterization in the form of pH, BOD, COD, turbidity, and total solids by Sharu et al. (2016). The various reagents used for different AOPs were as follows:

for solar photo-Fenton = 6.75 ml of H_2O_2 + 0.5 g of ferric chloride

for solar photocatalysis = 0.125 g of TiO_2

for solar with H_2O_2 = 13.5 ml of H_2O_2

for Fenton dark process = 0.5 g of ferric chloride and 6.75 ml of H_2O_2

For leachate characterization, 100 ml of leachate with 900 ml of distilled water and respective reagent was prepared. As per the study, the performance efficiency of various AOPs for COD, turbidity, total solids and BOD reduction was as follows:

H_2O_2 with solar < solar photocatalyst < Fenton < solar photo-Fenton

TABLE 14.4

Details of Various Advanced Oxidation Processes

Sl. No	AOPs	Advantage	Disadvantage
1	Ozonation	Reduce COD and BOD in leachate extensively Simple and cost effective	For better results, ozone AOP is used after the biological treatment
2	Ozone with hydrogen peroxide	Higher COD removal efficiency as compared to only ozonation Rapid decomposition of O_3 and an increase in hydroxyl radicals	Low solubility of ozone in aqueous solution Expensive reagents requirement High energy consumption Low performance efficiency at high pollutant concentrations
3	Hydrogen peroxide with UV	Oxidation of various organic compounds. No sludge formation Simple and less expensive	Difficulty in storage and transport of hydrogen peroxide Short lifetime of the UV source Incomplete degradation
4	Fenton process	The most popular method of AOP Easy implementation on an industrial scale	UV radiation is used to avoid the formation of ferric sludge waste Requires a narrow pH range High doses of reagents are required, which limits its applicability on a large scale
5	Wet air oxidation	No production of secondary pollutants Mild operating conditions Widely used for oxidation of dissolved and suspended organic pollutants	High energy consumption and high capital cost
6	Electrochemical oxidation	Higher removal efficiency of COD and ammonia nitrogen No chemical reagents or catalysts are required Low sludge production	Expensive and high energy consumption High operating cost because of the use of expensive metallic electrodes Effluents need to be diluted Formation of chlorinated organics
7	Sonochemical	Simple and environmentally friendly No toxic by-products	Expensive and low hydroxyl radical production
8	Photocatalyst	Simple and operates at a wide range of pH Oxidation of a wide range of organic compounds Better efficiency when used with UV	Formation of dark catalytic sludge Difficulty in the removal of the catalyst after treatment

Sharu et al. (2016) found that maximum COD removal was achieved by solar photo-Fenton at a reduced treatment time of 30 min. Graphical comparison of various AOPs considered in this study is shown in Figure 14.2.

Klauck et al. (2017) considered an approximately 50-year-old landfill for the study. The leachate sample of 50 L was analyzed using photoelectrooxidation and photoelectrooxidation with activated carbon with pre- and post-treatment techniques. Treatment with photoelectrooxidation uses a 6 L cylinder with two concentric electrodes. The anode surface area was 475.2 cm², while the cathode was 118 cm². The material used for anode and cathode was Ti/70TiO$_2$30RuO$_2$ and Ti/TiO$_2$, respectively, along with a UV lamp of 400 W and current density of 21.6 mA/cm². It was found that 70.3% removal of COD, 58.4% removal of TN and 58.4% of NH$_4$ was achieved by combination of the photoelectrooxidation process with activated carbon (Klauck et al. 2017). This process also eliminated toxicity and minimized pollution load and toxic compounds.

Meng Qiao et al. (2018) discussed AOPs with Fe^{2+} and NaClO. Ti/RuO$_2$ was used as an anode for the photoelectrooxidation process, and Ti was used as a cathode. The anode's and cathode's surface area were each 60 m², and the separation between the cathode and anode was 0.05 m. A 9 W UV lamp was placed between two electrodes. Leachate analysis was performed in 180 min at 200 A/m², 300 A/m² and 400 A/m². It was found that the post-treatment of intermediate compounds generated by the photoelectrooxidation using NaClO/

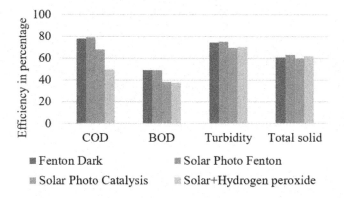

FIGURE 14.2 Removal efficiency of treated leachate at treatment time of 60 min.

Fe^{2+} coagulation gives better COD removal efficiency as compared to Fenton with enhanced coagulation. Optimal conditions, such as Fe^{2+} = 90 mmol/L, NaClO = 60 mmol/L and Fe^{2+}:NaClO ratio = 1.5:1 and photoelectrooxidation process for the duration of 180 min with current density = 400 A/m², give 95%, 86% and 95% removal efficiencies for UV$_{254}$, COD and color.

Fenton AOP performance was evaluated in Pradeep Kumar Singa et al. (2018) for HW landfill leachate. H$_2$O$_2$ was used as oxidizing reagent with Fe^{2+} as a catalyst. Under optimum operating conditions, such as pH = 3.0, reaction time = 150 min, molar ratio of H_2O_2:Fe^{2+} = 3:1, ferrous sulfates = 0.04 mol/L and H_2O_2 = 0.12 mol/L, a COD removal of 56.5% was achieved.

Sruthi et al. (2018) consider Fenton and electro-Fenton process for landfill leachate treatment. Fe-Mn binary oxide with zeolite was used as a catalyst for radical hydroxyl production at low pH. Under optimum conditions, such as catalyst concentration = 25 mg/L, pH = 3.0 at 4V for heterogeneous electro-Fenton, the COD removal of 87.5% was achieved, and biodegradability of landfill leachate was improved from 0.03 to 0.52. Wang et al. (2019) analyzed COD removal efficiency using Fenton with electrolysis and reported COD removal of 70% and HA from leachate. Fe^0 (zero-valent iron) was used for an iron source to generate Fe^{2+}. Fe^{2+}, along with H_2O_2, then generates the hydroxyl radical required for COD removal.

The optimum conditions used for leachate analysis were 0.187 mol/L of H_2O_2, current density = 20.6 A/cm^2 and spacing between the electrodes = 1.8 cm. AOPs based on O_3, UV, H_2O_2 and persulfate were studied for landfill leachate treatment (Poblete et al. 2019). The integrated UV, O_3 and H_2O_2 under optimum conditions required 140 min of irradiation and consumption of 0.67 g/L of H_2O_2 for removal of 17% COD and 56% color. The COD and color removal efficiency was improved to 77% and 29%, respectively, under optimum conditions when persulfate ($S_2O_8^{2-}$) of 0.2 g/L was also used for irradiation time of 250 min in combination with UV + O_3 + H_2O_2. Use of pre-and post-treatment methods such as adsorption with natural zeolite result in COD removal of 30% to 36%, NH_4 reduction of 90% and chloride reduction of 18% to 20%.

The combination of AOPs with membrane process was discussed in Pan et al. (2019) and Santos et al. (2019). According to Pan et al. (2019), the combination of AOPs and membrane process improves the treatment system's performance and reduces the membrane fouling issue. Integration of the Fenton process with NF and MF results in 94% to 96% reduction in COD and 96% to 99% reduction in leachate color (Santos et al. 2019). Compared with Fenton and Fenton with CF, it was found that COD removal was better in Fenton with MF and NF, as shown in Figure 14.3.

Chen et al. (2019) found that the combination of AOPs with coagulation decreases the concentration of various organic compounds. It also improves wastewater biodegradability. Fenton with adsorption, photocatalysis with adsorption and US with adsorption were studied (Bell and Raman 2019). According to this study, some AOPs cannot degrade complex organic compounds completely. Some intermediate compounds may form, which may be more toxic than their parent compound. Therefore, the integration of AOP with adsorption helps mineralize these compounds into CO_2, H_2O and inorganic compounds. Various AOPs for landfill leachate treatment were reviewed in Pratibha Gautam et al. (2019). Figure 14.3 shows the percentage reduction in COD in leachate from HW landfill using different AOPs reviewed in this chapter. The maximum percentage reduction in COD is achieved by using electrochemical oxidation when compared with other AOPs.

Integrated acidification/coagulation and Fe^0/H_2O_2 process was used for the treatment of leachate collected from Polish landfill. Jan Bogacki et al. (2019) found that acidification was less effective than coagulation. The main advantages of coagulation was a lower reagent dose requirement, which allowed for a larger removal of pollutant load. The pre-treatment performance at different pH values was as follows:

Coagulation (pH = 6.0) >> Coagulation (pH = 9.0) > Acidification (pH = 3.0)

When the amount of reagent increases, the leachate treatment efficiency improves. Under optimum conditions, the TOC removal of 75% was achieved after a 60 min process time. Some researchers (He et al. 2020; Xia et al. 2020) found that biodegradability of refractory contamination can be improved using ozonation, electrooxidation and photocatalyst as pre-treatment methods. According to (Xia et al. 2020),

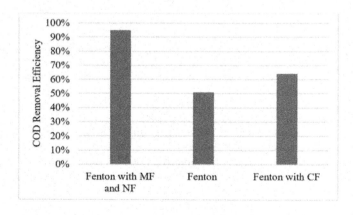

Santos *et al* (2019)

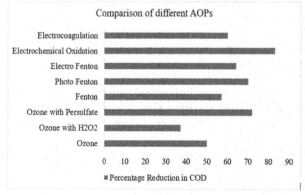

Pratibha Gautam *et al* (2019)

FIGURE 14.3 Percentage reduction in COD for different AOPs used for HW landfill leachate treatment.

70% or more aromatic organic compounds can be eliminated by using ozone as pre-treatment. Lastre-Acosta et al. (2020) listed that the combination of ozonation with membrane bioreactor removed 97% TOC, 97% COD, 94% BOD_5 and 100% sulfadiazine.

Mujtaba Hussain et al. (2020) reviewed applications of ozone-based AOP for wastewater treatment. Some of the study outcomes were that ozone/UV is cheaper than ozone alone and UV alone for treatment plant of capacity more than 38,000 m³/day. Ozone with UV generates more hydroxyl radicals than H_2O_2 with UV and ozone with H_2O_2 at 0.254 μm. Ozone with UV was a better option for the degradation of pesticides and phenolic compounds. Ozone with US had better performance than ozone with UV and ozone with H_2O_2.

Ardak Makhatova et al. (2020) used photo-Fenton with H_2O_2 and UV to treat leachate generated from MSW landfill located in Nur-Sultan city of Kazakhstan. The pre-treatment methods such as regulation of pH value and air stripping were used to remove inorganic carbon and inorganic nitrogen. The photo-Fenton-like process has better performance than does photo-Fenton. Under optimal conditions, 98.2% removal of color was achieved. The removal of total organic and inorganic carbon was 89% and 100%, respectively. Complete nitrogen removal of 96.5% was achieved with this method.

The effect of using pre-treatment and post-treatment methods on leachate treatment was discussed in Aziz et al. (2020). Studies show that adsorption followed by ozonation results in 92% removal of color, 82% COD removal and 75% NH_4-N removal. When ozonation followed by adsorption was used for landfill leachate, then 93% color, 80% COD and 71% NH_4-N removal were achieved. Also, the leachate's biodegradability was improved using the ozonation/adsorption and adsorption/ozonation process. This will increase BOD_5/COD from an initial value of 0.016 to 0.195 and 0.194, respectively. The graphical comparison of various methods is discussed in Aziz et al. (2020) is shown in Figure 14.4.

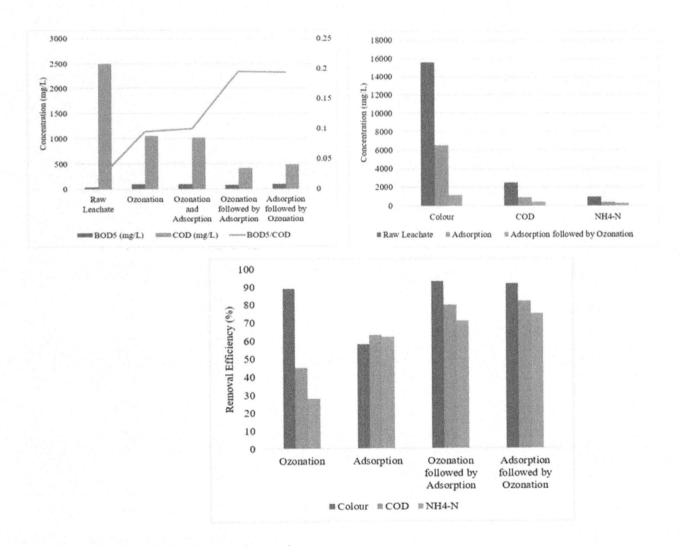

FIGURE 14.4 Comparison of various parameters for ozonation.

14.7 Conclusion

Nowadays, AOPs are gaining popularity in landfill leachate treatment and wastewater treatment. This chapter discussed briefly various landfill leachate treatment methods, composition and characterization of leachate, and details about various aspects of the advanced oxidation process used for landfill leachate treatment. These include $Fe^{2+}/Fe^{3+}/H_2O_2$, O_3/H_2O_2, hydrogen peroxide with UV radiation, ozonation with UV radiation, Fenton, Fenton-like processes with and without UV radiation and photocatalyst, electrochemical, wet air and sonochemical AOPs as well as combined systems. In all these AOPs, the main focus is on the generation of $\cdot OH$, which is used to suppress various organic compounds. Among all known AOPs, the suitability of an appropriate oxidation process depends on several factors such as ease of operation and maintenance, cost and safety, amount of reagent required, performance efficiency and intermediate products and by-products produced during leachate treatment.

The performance of AOPs can be improved when they are used in combination with various pre-treatment and post-treatment methods. Therefore, to achieve a further reduction in effluents and meet various quality and discharge standards, it will be better if various AOPs are combined with other conventional landfill leachate treatment methods. More future studies are required to optimize the reagent dosage and reaction time needed to treat landfill leachate. With an increase in focus on sustainability, there is a need to explore advanced oxidation processes that are eco-friendly, have a minimum environmental footprint, are economical, have a user-friendly operation, result in low sludge volume and use the least amount of chemical doses, such as electrocoagulation and microalgae-based techniques for MSW and HW landfill leachate treatment.

14.8 Competing Interests

All the authors declare that they have no competing interests.

REFERENCES

Abbas, Abdulhussain A., Guo Jingsong, Liu Zhi Ping, Pan Ying Ya, & Wisaam S. Al-Rekabi (2009) Review on landfill leachate treatments. *Journal of Applied Sciences Research*, 5(5), 534–545.

Alimoradi, S., R. Faraj, & A. Torabian (2018) Effects of residual aluminum on hybrid membrane bioreactor (coagulation-MBR) performance, treating dairy wastewater. *Chemical Engineering and Processing – Process Intensification*, 133, 320–324. https://doi.org/10.1016/j.cep.2018.09.023.

Aziz, H. A., H. R. AlGburi, M. Y. D. Alazaiza et al. (2020) Sequential treatment for stabilised landfill leachate by ozonation – adsorption and adsorption – ozonation methods. *International Journal of Environmental Science Technology*. https://doi.org/10.1007/s13762-020-02891-x.

Bello, M. M., & A. A. A. Raman (2019) Synergy of adsorption and advanced oxidation processes in recalcitrant wastewater treatment. Environment Chemistry Letter, 17, 1125–1142. https://doi.org/10.1007/s10311-018-00842-0.

Bogacki, Jan, Piotr Marcinowski, & Balkess El-Khozondar (2019) Treatment of landfill leachates with combined acidification/coagulation and the Fe^0/H_2O_2 process. *Water*, 1–17.

Bourechech, Z., F. Abdelmalek, M. R. Ghezzar, & A. Addou (2017) Treatment of leachate from municipal solid waste of Mostaganem district in Algeria: Decision support for advising a process treatment. *Waste Management & Research*, 1–11.

Bove, Dario, Sara Merello, Davide Frumento, Saleh Al Arni, Bahar Aliakbarian, & Attilio Converti (2015) A critical review of biological processes and technologies for landfill leachate treatment. *Chemical Engineering and Technology*, 2115–2126. https://doi.org/10.1002/ceat.201500257.

Chen, W., A. Zhang, G. Jiang, & Q. Li (2019) Transformation and degradation mechanism of landfill leachates in a combined process of SAARB and ozonation. *Waste Management*, 85, 283–294. https://doi.org/10.1016/j.wasman.2018.12.038.

Elkacmi, Reda, & Mounir Bennajah (2019) Advanced oxidation technologies for the treatment and detoxification of olive mill wastewater: A general review. *Journal of Water Reuse and Desalination*, 1–43.

Gautam, Pratibha, Sunil Kumar, & Snehal Lokhandwala (2019) Advanced oxidation processes for treatment of leachate from hazardous waste landfill: A critical review. *Journal of Cleaner Production*, 237, 1–14.

Güvenç, S. Y., & E. C. Güven (2019) Pre-treatment of food industry wastewater by coagulation: Process modeling and optimisation. *Celal Bayar University Journal of Science*, 15 (3), 307–316. https://doi.org/10.18466/cbayarfbe.581611.

Hana, Meina, Xiaoguang Duanb, Guoliang Caoa, Shishu Zhuc, & Shih-Hsin Hoa (2020) Graphitic nitride-catalysed advanced oxidation processes (AOPs) for landfill leachate treatment: A mini review. *Process Safety and Environmental Protection*, 139, 230–240.

He, H., H. Ma, & L. Liu (2020) Combined photocatalytic peroxidation reactor and sequencing batch bioreactor for advanced treatment of industrial wastewater. *Journal of Water Processing Engineering*, 36, 101259. https://doi.org/10.1016/j.jwpe.2020.101259.

Hongwei Luo, Yifeng Zeng, Ying Cheng, Dongqin He, & Xiangliang Pan (2020) Recent advances in municipal landfill leachate: A review focusing on its characteristics, treatment, and toxicity assessment. *Science of the Total Environment*, 703, 135468. https://doi.org/10.1016/j.scitotenv.2019.135468.

Hussain, Mujtaba, Mohd Salim Mahtab, & Izharul Haq Farooqi (2020) The applications of ozone-based advanced oxidation processes for wastewater treatment: A review. *Advances in Environmental Research*, 9 (3), 191–214.

Kalčíková, G., M. Vávrová, J. Zagorc-Končan, & A. Žgajnar Gotvajn (2011) Seasonal variations in municipal landfill leachate quality. *Management of Environmental Quality*, 22 (5), 612–619. https://doi.org/10.1108/14777831111159734.

Kjeldsen, Peter, Morton A. Barlaz, Alix P. Rooker, Anders Baun, Anna Ledin, and Thomas H. Christensen (2002) Present and long-term composition of MSW landfill leachate: A review. *Critical Reviews in Environmental Science and Technology*, 32 (4), 297–336.

Klauck, C. R., A. Giacobbo, C. G. Altenhofen, L. B. Silva, A. Meneguzzi, A. M. Bernardes, & M. A. S. Rodrigues

(2017) Toxicity elimination of landfill leachate by hybrid processing of advanced oxidation process and adsorption. *Environmental Technology & Innovation*, 8, 246–255. https://doi.org/10.1016/j.eti. 2017.07.006.

Koçak, Seda, Cansu Güney, M. Tuna Argun, Begüm Tarkin, E. Özlem Kirtman, Deniz Akgül, & Bülent Mertoğlu (2013) Treatment of landfill leachate by advanced oxidation processes. *Fen Bilimleri Dergisi*, 25 (2), Marmara University, 51–64.

Kow Su-Huan, Muhammad Ridwan Fahmi, Che Zulzikrami Azner Abidin, & Ong Soon-An (2016) Advanced oxidation processes: process mechanisms, affecting parameters and landfill leachate treatment. *Water Environment Research*, 88 (11), 2047–2058.

Krzysztoszek, A., & Jeremi Naumczyk (2012) Landfill Leachate treatment by Fenton, photo-Fenton processes and their modification. *Journal of Advanced Oxidation Technology*, 15 (1), 53–63.

Lastre-Acosta, A. M., P. H. Palharim, I. M. Barbosa, J. C. Mierzwa, & A. C. S. C. Teixeira (2020) Removal of sulfadiazine from simulated industrial wastewater by a membrane bioreactor and ozonation. *Journal of Environmental Management*, 271, 111040. https://doi. org/10.1016/j. jenvman.2020.111040.

Luo, H., Y. Zeng, Y. Cheng, Dongqin He, & Xiangliang Pan (2019) Recent advances in municipal landfill leachate: A review focusing on its characteristics, treatment, and toxicity assessment. *Science of the Total Environment*. https://doi. org/10.1016/j.scitotenv.2019.135468.

Makhatova, Ardak, Birzhan Mazhit, Yerbol Sarbassov, Kulyash Meiramkulova, Vassilis J. Inglezakis, & Stavros G. Poulopoulos. (2020) Effective photochemical treatment of a municipal solid waste landfill leachate. *PLoS One*, 15 (9), e0239433. https://doi.org/10.1371/journal pone.0239433:1–22.

M'Arimi, M. M., C. A. Mecha, A. K. Kiprop, & R. Ramkat (2020) Recent trends in applications of advanced oxidation processes (AOPs) in bioenergy production: Review. *Renewable and Sustainable Energy Reviews, Elsevier*, 121 (C).

Naveen, B. P., Sivapullaiah, P. V., & Sitharam, T. G. (2016) Effect of aging on the leachate characteristics from municipal solid waste landfill. *Japanese Geotechnical Society Special Publication*, 2 (56), 1940–1945. doi:10.3208/jgssp.ind-06

Niazi, S. (2018) *Coagulation effects of biological sludge reject water treatment*. Master Thesis, Energy and Environment Technology, University of South-Eastern Norway, Norway, 63.

Oliveira, Mayra S. de., Larissa F. da Silva, Andreia D. Barbosa, Lincoln L. Romualdo, Geraldo Sadoyama, & Leonardo Santos Andrade (2019) Landfill Leachate treatment by combining coagulation and advanced electrochemical oxidation techniques. *ChemElectroChem*, 6 (5), 1427–1433. https://doi.org/10.1002/celc.201801677.

Pan, Z., C. Song, L. Li, H. Wang, Y. Pan, C. Wang, J. Li, T. Wang, & X. Feng (2019) Membrane technology coupled with electrochemical advanced oxidation processes for organic wastewater treatment: Recent advances and future prospects. *Chemical Engineering Journal*, 376, 120909. https://doi.org/10.1016/j.cej.2019.01.188.

Paulina, Jelonek, & Ewa Neczaj (2012) The use of advanced oxidation processes (AOP) for the treatment of landfill leachate. *Inżynieria i Ochrona Środowiska*, 203–217.

Poblete, R., I. Oller, M. I. Maldonado, & E. Cortes (2019) Improved landfill leachate quality using ozone, UV solar radiation, hydrogen peroxide, persulfate and adsorption processes. *Journal of Environmental Management*, 232, 45–51. https://doi.org/10.1016/j. jenvman.2018.11.030.

Qiao, Meng, Xu Zhao, & Xiaoyun Wei (2018) Characterisation and treatment of landfill leachate membrane concentrate by Fe^{2+}/NaClO combined with advanced oxidation processes. *Scientific Reports*, 8, 12525. https://doi.org/10.1038/ s41598-018-30917-5: 1–9.

Rossi, G., M. Mainardis, E. Aneggi, L. K. Weavers, & D. Goi (2020) Combined ultrasound-ozone treatment for reutilization of primary effluent-a preliminary study. *Environmental Science Pollution Research*, 28 (1), 700–710.

Santos, A. V., L. H. de Andrade, M. C. S. Amaral, & L. C. Lange (2019) Integration of membrane separation and Fenton processes for sanitary landfill leachate treatment. *Environmental Technology*, 40 (22), 2897–2905. https://doi.org/10.1080/09 593330.2018.1458337.

Sepehri, A., & M. H. Sarrafzadeh (2018) Effect of nitrifiers community on fouling mitigation and nitrification efficiency in a membrane bioreactor. *Chemical Engineering Process: Process Intensification*, 128, 10e18. https://doi. org/10.1016/j.cep.2018.04.006.

Sharu, E., M. Sudha Kani, R. Sajitha, & C. Jenifa Latha (2016) Leachate treatment by advanced oxidation process. *International Journal of Advances in Mechanical and Civil Engineering*, 3 (3), 117–120, ISSN:2394-2827.

Singa, Pradeep Kumar, Mohamed Hasnain Isa, Yeek-Chia Ho, & Jun-Wei Lim (2018) Treatment of hazardous waste landfill leachate using Fenton oxidation process. *E3S Web of Conferences*, 34, 1–6.

Sruthi, T., R. Gandhimathi, S. T. Ramesh, & P. V. Nidheesh (2018) Stabilized landfill leachate treatment using heterogeneous Fenton and electro-Fenton processes. *Chemosphere*, 210, 38–43.

Tejera, Javier, Ruben Miranda, Daphne Hermosilla, Iñigo Urra, Carlos Negro, & Ángeles Blanco (2019) Treatment of a mature landfill Leachate: Comparison between homogeneous and heterogeneous photo-Fenton with different pretreatments. *Water*, 11 (9), 1849. https://doi.org/10.3390/ w11091849.

Tu Anqi, Zhang Zhiyong, Hao Suhua, & Li Xia (2020) Review on landfill leachate treatment methods. *6th International Conference on Energy Science and Chemical Engineering, Earth and Environmental Science*, 565 (2020), 012038. https://doi.org/10.1088/1755-1315/565/1/012038.

Tungler, Antal, Erika Szabados, & Arezoo M. Hosseini (2015) Wet air oxidation of aqueous wastes. *Wastewater Treatment Engineering*, Chapter 6, 153–178.

Umar, Muhammad, Hamidi Aziz, & Mohd Suffian Yusoff (2010) Trends in the use of fenton, electro-fenton and photo-fenton for the treatment of landfill leachate. *Waste Management*, 30, 2113–2121. https://doi.org/10.1016/j. wasman.2010.07.003.

Vamsi Krishna, K., Venkateshwar Reddy, & P. Rammohan Rao (2015) Muncipal solid waste management using landfills in Hyderabad city. *International Journal of Engineering Research & Technology (IJERT)*, 4 (2), 1047–1054, ISSN:2278-0181.

Wang, Z., J. Li, W. Tan, X. Wu, H. Lin, & H. Zhang (2019) Removal of COD from landfill leachate by advanced Fenton process combined with electrolysis. *Separation and Purification Technology*, 208, 3–11.

Xia, J., H. Sun, X. Ma, K. Huang, & L. Ye (2020) Ozone pretreatment of wastewater containing aromatics reduces antibiotic resistance genes in bioreactors: The example of p-aminophenol. *Environ Int*, 142, 105864. https://doi.org/10.1016/j.envint.2020.105864.

Zazouli, Mohammad Ali, Zabihollah Yousefi, Akbar Eslami, & Maryam Bagheri Ardebilian. (2012) Municipal solid waste landfill leachate treatment by Fenton, photo-Fenton and Fenton-like processes: Effect of some variables. *Iranian Journal of Environmental Health Sciences & Engineering*, 1–9.

15

Recent Trends in Nanomaterial-Based Advanced Oxidation Processes for Degradation of Dyes in Wastewater Treatment Plants

Samuel S. Mgiba, Vimbai Mhuka, Nomso C. Hintsho-Mbita, Nilesh S. Wagh, Jaya Lakkakula and Nomvano Mketo

CONTENTS

15.1 Introduction

Dyes are substances that provide color when applied on a certain substrate and the crystal structure of the colored substance temporally changes. These organic compounds are used in various industries, which include textile, cosmetics, painting, printing, food and plastics. Some of these industries use a substantial amount of water during the manufacturing process. For example, in textiles, a high amount of water is used during the dyeing and finishing operations in the plants, and 10%–50% of the colorants used during the dyeing processes is lost to the environment [1]. Furthermore, an amount of dye is lost in the final processing, which involves washing in a bath to remove the excess original dye not fixed on the fiber [2]. Therefore, wastewater produced in the textile industry is regarded as a major contributor to environmental pollution in the industrial sector. In addition, there is an increasing demand for textile products, thus an increase in production is directly proportionally to an increase in dye-contaminated wastewater [3].

Dyes have a complex aromatic structure; therefore, they are difficult to decolorize. These aromatic compounds are resistant to fading when exposed to oxidizing agents, sweat, water, light and so forth. They are also difficult to degrade because they are designed to meet the durability market in relation to

consumer demands [4]. The negative impacts of dyes include reduction of photosynthesis by hindering light penetration to plants. This in turn affects the production of oxygen by aquatic plants, thus having a negative impact on breathing in animals. These respiratory problems can sometimes affect the immune system of a human being. Other types of dyes such as azo dyes are capable of affecting the chemical properties of the soil, deteriorating water bodies and causing harm to flora and fauna. The toxicity of dyes can also lead to the death of microorganisms, which in turn can affect agricultural productivity. Furthermore, dyes are potentially carcinogenic and highly toxic, thus they are related to various disease in animals and human beings [5].

In water-scarce developing countries like South Africa, dye removal is very important for water recycling. Even though different technologies have been previously reported for dye treatment in wastewater samples, physical methods such as adsorption have been extensively reported for dye degradation in aqueous solutions [6]. Electrocoagulation, ozonation and biological, chemical and membrane filtration are some of dye removal methods that have been employed previously; however, these methods are not preferable for water scintillation. They are relatively costly, have limited applicability and produce waste products that are toxic and difficult to dispose of [7]. Recently, advanced oxidation processes (AOPs) have gained more attention from researchers because of their advantages over conventional methods. Photocatalytic oxidation is one of the advanced oxidation processes that needs to be enhanced to obtain a faster degradation rate and high efficiency [8]. Nanomaterials have been used to enhance photocatalytic AOPs due to their high surface area to volume ratio [6]. It has been reported that nanomaterials have significant positive impact on the adsorption mechanism for the target analyte [9]. Therefore, this chapter will investigate various nanomaterials used to enhance the AOPs for degradation of different dye compounds in wastewater. The current trends, efficiency, drawbacks and advantages of these processes will be discussed and summarized.

15.2 Dye Classifications

Dyes are classified in two ways: by their structure and/or by their applications or usage. These organic compounds have a chemical structure of chromophores. The latter is a group of unsaturated organic radicals that absorb and reflect incident electromagnetic radiation within a very narrow band of visible light, which provides the color, and they are usually functionalized with an electron-withdrawing group. Examples of chromophores include $-C=C-$, $-C=O$ and $-N=N-$. Dyes also contain auxochromes ($-NH_3$, $-COOH$, $-SO_3H$ and $-OH$), which are electron-donating substituents that increase the overall polarity of dye molecules. The intensity of the color depends on the auxochromes [10]. One of the commonly reported class of dyes are azo dyes. These dye molecules are characterized by the presence of one or more azo bonds ($-N=N-$) together with an aromatics structure. Azo dyes are mostly acidic and are the largest class of dyes found in nature. These compounds are mainly used for dyeing nitrogen-containing fabrics like wool, polyamide, silk and modified acyl by binding to the NH_4^+ ions of those fibers. It is well reported that azo dyes are responsible for 60%–70% of dye stuff produced in textile industries [11]. The second largest abundant class of dyes are anthraquinone dyes. This class of dye has a wide range of colors present in the visible region of the electromagnetic spectrum. Anthraquinones cannot be easily degraded due to their aromatic character, hence they are observed to last longer when applied in any substrate. These types of dyes cause liver and kidney cancer in animals [12].

Metal complex dyes are other types of dyes that are a combination of dye molecule and metal salt, generally chrome and other transition metals. These dye compounds are also known as pre-metallized dyes [13]. The structure of these dyes is usually a monoazo structure that consists of different groups such as hydroxyl, carbonyl and amino groups. When dyeing using metal complex dyes, a pH regulator, electrolytes and levelling agents in the form of ionic and nonionic surfactants are used as chemicals and auxiliaries. The metals can also be found in waste due to unfixed dye [13].

Direct dyes are used for dyeing cottons, rayon, linen, jute, silk and poly fibers. These dyes are regarded as substantive dyes. Direct dyes are a class of poly azo compounds along with oxazines and stilbenes. Direct dyes are water insoluble and contain different functional groups such as sulfur, carboxy and

hydroxyl groups. Treatment agents such as quaternary ammonium compounds with long carbon chains are used to improve the wet fastness during the dyeing process [13].

Vat dyes are types of dyes used for dyeing cellulose fibers. Most of these dyes are anthroquinones and are insoluble in water. The process of dyeing occurs by first reducing the dyes using sodium dithionite ($Na_2S_2O_4$) to form a soluble vast dye, which is termed leuco. The reduced form is then impregnated into the fabric after which it goes under another oxidation process to be brought back to its insoluble form. An example of a vat dye is indigo dye. Other types of dyes are also used, such as solvent dyes, sulfur dyes, fluorescent brighteners used for soaps and detergents and mordant dyes used for natural fibers, aluminum and wool [14–15].

15.3 Advanced Oxidative Processes

Advanced oxidation processes are chemical processes employed to remove or degrade different pollutants such as dyes in aqueous medium through the production of highly reactive chemical oxidizing agents. These processes rely on the use of different substances that include catalysts, ozone (O_3), hydrogen peroxide (H_2O_2) and ultraviolet light for maximum efficiency [16], as shown in Table 15.1. Highly reactive radicals such as hydroxyl ($\cdot OH$), sulfate ($SO_4^-\cdot$), and superoxide ($O_2^-\cdot$) are produced during these processes. Advanced oxidation processes result in mineralization, partial mineralization, transformation of organic pollutants to a biodegradable substance or inorganic species in a form of carbon dioxide (CO_2) and water (H_2O). Different types of advanced oxidation processes are summarized in Table 15.1. The general chemical reaction that represents the mineralization of dye organic molecules using AOPs can be represented as follows [17]:

$$R\text{-}H + \cdot OH \rightarrow H_2O + R\cdot$$

15.4 Nanomaterial-Based Advanced Oxidation Process

Nanomaterials contain at least one external dimension in the region between 1 and 100 nm. These materials have a wide range of good properties that make them extraordinary. These properties include larger surface area, more surface active sites and unique electron conduction properties. Nanomaterials can be used as catalysts, sensors or other applications because of these attractive properties. Therefore, nanomaterials bring a new dimension to the field of dye degradation in wastewater [8]. In the next section, different nanomaterials used to develop nanophotocatalytic materials and dye degradation mechanisms will be discussed.

15.4.1 Titanium Oxide (TiO_2)–Based Nanomaterials in AOPs

Titanium oxide (TiO_2) is a nanostructured metal oxide that has been widely used as a photocatalyst for different applications

TABLE 15.1

Commonly Reported AOPs Based on the Use of Chemicals, Equipment, Reactors, Catalysts and *In Situ* Radical Generation

AOP Type	Reagent	Commonly Produced Oxidative Species	General Comment
Ozonation	O_3	$\cdot OH$, $HO_2\cdot$, $HO_3\cdot$, $O_2^-\cdot$, $O_3^-\cdot$	The mechanism of this process takes place in two ways: the direct reaction with dissolved ozone and the indirect reaction through the formation of radicals. The path taken depends on factors such as the nature of contaminants and pH of the medium of ozone [18].
Ultraviolet	UV/O_3	$\cdot OH/O_2^-\cdot$	The mechanism of this process takes place in the presence of ultraviolet light. The presence of UV light generates a large concentration of highly reactive hydroxyl species at a faster rate. This is one of the most-used AOPs. This process inhibits the formation of bromate. When UV light is absorbed, energy is released from a molecule through a physical and chemical process that includes a compound breakdown [19].
Perozonation	O_3/H_2O_2	$\cdot OH$, $O_2^-\cdot$, $O_3^-\cdot$	This process takes place by direct oxidation of substances by liquid ozone or by oxidation of compounds by hydroxyl radicals. This process is faster than ozonation. The use of hydrogen peroxide helps to improve the efficiency of the process, since the reaction of H_2O_2 and O_3 results in the formation of hydroxyl radicals [18].
Cavitation	H_2O/O_2	$\cdot OH/HO_2\cdot$	This involves the formation, growth and collapse of cavities in a short interval with the release of energy. Solid surfaces release highly reactive free radicals and create hotspots resulting from the collapse of the cavities. The released radicals diffuse into the liquid bulk medium where they react with organic pollutants to produce an oxidized product. The most-used cavitations are acoustic and hydrodynamic cavitation [20].
Fenton	H_2O_2/Fe^{2+}	$\cdot OH/HO_2\cdot$	This is a simple process that has high oxidizing power. It is relatively expensive because it requires a large amount of salt. The Fenton process takes place through Fenton oxidation and coagulation [21].
Photo-Fenton	$H_2O_2/Fe^{2+}/UV$	$\cdot OH$	The presence of UV light accelerates the Fenton process with great effectiveness at an acidic pH. This process produces more hydroxyl radicals compared to the Fenton process. It allows sunlight to be applied as the source of UV light. The reduction of Fe^{3+} to Fe^{2+} by UV light also contributes to the formation of highly reactive radicals. Fe^{3+} reacts faster in the presence of UV light with hydrogen peroxide (H_2O_2) to form Fe^{2+} and $\cdot OH$ [18].
Photocatalysis	Semiconductor catalyst/UV	$\cdot OH$, h+, $O_2^-\cdot$, electrons, $HO_2\cdot$, $HOO\cdot$	This deals with the use of a solid catalyst that is activated by UV light. Photocatalysis is divided into homo- and heterogeneous catalysis. The latter refers to the organic pollutants being in liquid form and the catalyst in solid form. Different catalysts are used for this process. These include TiO_2 and ZnO, with the former being the mostly widely used due to its chemical stability and relatively low cost. Hollow-electron pair formation takes place when the light is absorbed by the catalyst. The photocatalytic process takes place in the presence of oxygen absorbed in the surface of the catalyst [22].
Persulfate	$S_2O_8^{2-}/UV$	$SO_4^-\cdot$	This alternative to AOP involves the use of peroxymonosulfate and peroxydisulfate instead of H_2O_2. Persulfate is a strong oxidant that can be activated by heat, transition metal and electricity to generate sulfate radicals. Persulfate has gained attention because of its long lifetimes in the surface. Persulfate-based AOP can enhance performance with collaboration of energy inputs such as ultrasound, UV and electrochemistry [23].
Electron/ electricity mediated	$H_2O/$electrons	$\cdot OH$	An electric current between 2 and 20 A is applied between two electrodes in water to produce highly reactive hydroxyl radicals. This process is also regarded as an anodic oxidation. The contaminants are oxidized though an electrolytic cell by direct electron transfer to the anode or indirect oxidation with oxidizing species from electrolysis of water at the anode [18].

under UV-light irradiation. It is regarded as a good photocatalyst for dye degradation. This is due to its strong resistance to corrosion, relative affordability and low operation temperature, among other advantages [24]. Researchers enhanced the photocatalytic activity of TiO_2 in different ways, including the use of polymers to sensitize the catalyst, doping of the catalyst with nonmetals, impregnating using transition metals and metal ions, among other ways. The possible pathway followed for the TiO_2 photocatalysis mechanism is represented in Figure 15.1 [25]. Yan and co-workers used reduced graphene oxide and meso-TiO_2/gold nanoparticles for the degradation of methyl blue dye. The photocatalytic mechanism of this process involved the introduction of etheylene diamine tetra acetate (EDTA), benzoquinone (BQ) and t-BuOH as scavengers for holes, $O_2^-\cdot$ and hydroxyl radical. The holes and $O_2^-\cdot$ were the oxidative species rather than the free hydroxyl radicals. The study concluded that ternary photocatalyst RGO/meso-TiO_2 AuNPs showed high photocatalytic activity (0.014 min^{-1}) [26]. Ternary chalcogenides possess a low band gap and have an ability to absorb solar photons. Therefore, the use of $CuInSe_2$ has to be explored by researchers. For ternary chalcogenides to be a good photocatalyst, the energy gap should be altered. Since $CuInSe_2$ is a small gap semiconductor, it can be coupled to TiO_2 with high band gap for photocatalytic enhancement. This process was applied for degradation of organic dyes where oxidative species where formed during

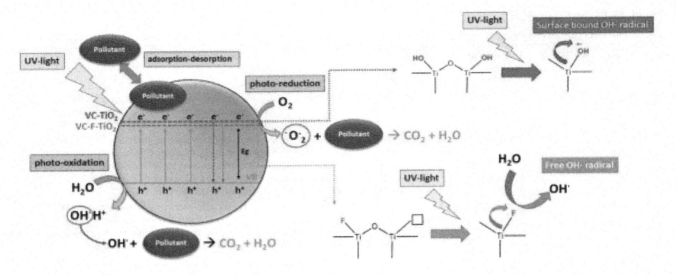

FIGURE 15.1   Mechanism for TiO_2 photocatalysis advanced oxidation processes of dye degradation [30]

the process. The formation of holes for oxidation was represented by:

$$H_2O + h^+ \rightarrow H^+ + \cdot OH$$

Furthermore, hydroxyl peroxide radicals were also formed during this process and their formation is represented by

$$O_2 + e+ \rightarrow O_2^-\cdot$$
$$O_2^-\cdot + H+ \rightarrow \cdot O_2H$$
$$2HO_2\cdot \rightarrow H_2O_2 + O_2$$
$$H_2O_2 + e^- \rightarrow OH^- + \cdot OH$$

It was then concluded that the product formed was due to oxidative cleavage using hydroxyl radicals [27].

However, the separation of TiO_2 from the heterogeneous solution has been a challenging task. This encouraged researchers to synthesize nanocatalysts with magnetic and photocatalytic properties for easy separation. Iron and cobalt exhibit very good magnetic and photocatalytic properties. Therefore, $CoFe_2O_4/TiO_2$ magnetic nanoparticles were synthesized using the coprecipitation method for photodegradation of reactive dyes 120. The degradation occurred by nanoparticles excitation under visible light due the band gap of $CoFe_2O_4/TiO_2$ and holes were generated. During this study, there was an enhanced production of a powerful oxidizing agent in a form of $SO_4^-\cdot$ and OH^-. The peroxomonosulfate (PMS), peroxodisulfate (PDS) and H_2O_2 receptors increased the photocatalytic degradation of reactive red dye [28]. Another way of sensitizing photocatalysts is through the use of conducting polymers. These polymers provide a high to moderate mobility of charge carriers through the π-conjugated electron system that can be coupled to TiO_2 electronically. The conducting polymers are regarded as sensitizers when coupled with semiconductors. Polycarbazole (PCz) is one conducting polymer with high electrical conductivity, thermal stability, low band gap

and low toxicity. This polymer was synthesized and decorated with TiO_2 nanohybrids for amido black dye degradation. The conducting polymer was illuminated with UV light to promote the transfer of electrons. The lowest unoccupied orbitals of the molecules of PCz to the TiO_2 conduction band reacts with oxygen and hydroxyl radicals. Thus, hydroxyl radicals were the active species that participated in the degradation process. The latter occurred by hydroxyl radicals attacking the azo and sulfonate groups. This results in the cleavage of the −N=N− group, and almost 100% degradation was achieved using these nanohybrids [29].

15.4.2 Zinc Oxide–Based Nanomaterials in AOPs

Zinc oxide (ZnO) is more efficient when compared to TiO_2 because it can absorb a large fraction of the solar spectrum than the latter and is a low-cost semiconductor [31]. Because of its photosensitivity and good catalytic properties, a lot of interest has been drawn towards this semiconductor. Velmurugan and Swaminathan used nanostructured ZnO for reactive dye degradation using solar light. Recyclable zinc oxide nanoparticles were synthesized and applied on red dyes. During this study, it was established that the degradation of red dye molecules without the use of the catalyst is negligible (0.2%). When the process was conducted in the presence of a catalyst and also solar light, 77% degradation occurred. For this to happen, dye molecules were absorbed on the surface of the semiconductor. It is worth noting that there was a photoelectron transfer from the dye molecule to the conduction band of ZnO. The superoxides that were produced from the sensitization degraded the dye molecule [32]. The same technique was also applied for reactive orange dye [33].

Zinc oxide has good chemical and thermal stability. It has also a high recombination ration of photo-induced electron-hole pairs. However, ZnO has a poor response to visible light and a possibility of photo corrosion. This has an impact on the photocatalytic activities of ZnO. Researchers used doping to minimize these drawbacks. Europium (Eu) was used in different concentrations with ZnO to degrade methylene blue and methylene orange under solar radiation. A hole was created by exciting an electron on the conductive band. Since it was coated with europium, the Eu^{3+} ion prevented the recombination of the hole by trapping the electron to create superoxide radicals. A high amount of superoxide radical was produced by introducing more of the Eu^{3+} ion to the ZnO structure. Dyes were attacked by the hydroxyl radical through a hole transfer. The formed intermediates during degradation also proceeded with self-degradation or were degraded with reactive oxidation species. This yielded 62% degradation of dye in 150 minutes [34].

The most suitable sensitizer of the UV-excited ZnO is with cadmium sulfate (CdS). CdS can form a type-II heterojunction with ZnO, and its lattice structure is also similar to that of ZnO. Oxidative species are produced when the ZnO conduction band is placed between the valence band and the conduction band of CdS. The bottom of the conduction band is occupied by charge carriers and electrons that are generated through photolysis using solar irradiation. The electron hole pairs are prevented from recombination. Hydroxyl radicals are generated by holes when CdS reacts with water. The degradation of dye under solar irradiation depends on the superoxides and hydroxyl radicals produced. The combination of CdS and ZnO produced an efficient 91% degradation of methylene blue in a period of 240 minutes [35]. Catalytic semiconductors are combined with nanomaterials of different component for higher potential and catalytic activity in the presence of light. Some of the nanomaterials combined easily because of the high surface energy from the surface area. However, this

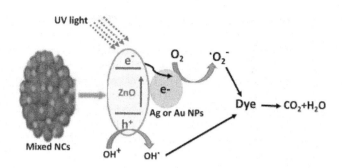

FIGURE 15.2 Mechanism for dye degradation using ZnO catalyst in the presence of ultraviolet light [37].

can also result in their activity being reduced [36]. Figure 15.2 represents the mechanism for ZnO photocatalysis of dyes.

15.4.3 Iron Oxide–Based Nanomaterials in AOPs

Bare iron oxide nanoparticles exhibit high chemical activity and can be oxidized in the air. These materials have super magnetic properties with high surface-to-volume ratio. Furthermore, the iron-based catalyst has high electrical conductivity and optical properties. The iron-based nanoparticles should be forged with templates to avoid loss of photocatalytic activity caused by clogging that can minimize the interactions between the dye molecules and nanoparticles. A possible mechanism for dye degradation using iron oxide–based nanomaterials is shown in Figure 15.3. A graphene-based photocatalyst was used in the presence of sunlight to degrade dye

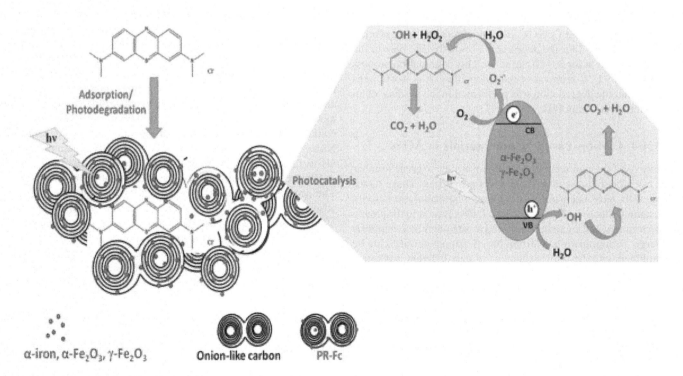

FIGURE 15.3 Iron oxide nanoparticles–based AOP mechanism for dye degradation [42].

molecules in both the presence and the absence of hydrogen peroxide. The processes involved the absorption of photons by Fe_3O_4/RGO (reduced graphene oxide) nanocomposites in the presence of sunlight. This resulted in the transfer of electrons from the conduction band of Fe_3O_4 and creating holes in the valence band. Therefore, the RGO acted as an electron carrier. Superoxide radicals are produced when electrons in the RGO attract dissolved oxygen and the holes and radicals effectively degrade the dye molecules. The mechanism for the degradation of dye molecules in the presence of hydrogen peroxide resulted in the addition of more free radicals in the degradation process. The latter can be a possible photo-Fenton process [38]. The Fenton process is another AOP used for rapid dye degradation in the presence of H_2O_2. The process can be described using the following equations:

$$Fe^{2+} + H_2O_2 \rightarrow Fe^{3+} + \cdot OH + OH^-$$

$$Fe^{3+} + H_2O + light \rightarrow Fe^{2+} + \cdot OH + H^+$$

The Fenton process was carried out by using Fe_3O_4 as both a catalyst and adsorbent for rhodamine B removal. Again, RGO was used to prevent possible agglomeration. In the study, the effectiveness of the magnetic nanocomposites resulted in 98% degradation of rhodamine B [39]. The H_2O_2 was also used as an oxidant under ultrasound to degrade xylenol orange. It was observed that slow degradation was taking place when the concentration of the oxidant was increased without the presence of the catalyst. This was highlighted by the little decrease on the absorption peak. On the contrary, when Fe_3O_4 nanosphere catalyst was added, the absorption peak disappeared completely after 45 minutes, thus making this nanosphere catalyst very efficient [40].

Magnetic recoverable nanocatalysts are also gaining attention from researchers in dye degradation. This was notable when recyclable Fe_3O_4@C@Ru nanocomposite was synthesized and applied for the degradation of methylene blue in the presence of light for 30–170 minutes. This degradation process resulted in 92.7% removal of methylene blue. It is worth indicating that the degradation was performed in the absence of an oxidant or reductant [41].

15.4.4 Carbon-Based Nanomaterials in AOPs

Carbon-based nanomaterials (CBNs) attract a great number of researchers to explore their novel properties. These nanomaterials have been used to enhance photocatalytic activity of semiconductors. This is because CBNs have a hydrophobic environment for localization of active sites and also suppress charge recombination. Carbon-based nanomaterials can be combined with the semiconductor through different ways such as sol-gel, solvothermal, hydrothermal and so on. Different CBNs, which include activated carbon, carbon nanotubes, carbon nanosheets and carbon nanodots, have been used [43]. These CBNs have low toxicity, unique optical properties and a large range of absorption in visible light. Their size is less than 10 nm, thus they are potential replacements for semiconductor catalysts. For example, Sun [44] used carbon nanodots to degrade organic dye and methylene blue, in the presence of light. The carbon nanodots excited the electrons to produce

an electron-hole pair when irradiated by light. The electrons generated through irradiation react with oxygen on the surface of the carbon nanodots to produce superoxide radicals, and the holes react with water to produce hydroxyl radicals. These radicals are responsible for degrading methylene blue to form carbon dioxide and water. The concentration of methylene blue decreased within 30 minutes of irradiation with visible light [44]. A 91.8% degradation efficiency of diazo dye was achieved using multiwalled carbon nanotubes (MWCNTs). These nanotubes have a capacity to conduct electrons with almost no resistance, that is, with a ballistic effect. MWCNTs have a good combination of mechanical, chemical, thermal and electronic properties [45]. A sol-gel method was applied to decorate an already doped titanium oxide with multiwalled carbon nanotubes. These tubes act as a source of carbon and introduce a new sub-band gap state in TiO_2 gap. Electrons are excited by visible light from the sub-band gap state to the conduction band of TiO_2, where they are absorbed by oxygen to produce reactive superoxide radicals [46]. Magnetic property is one aspect that is considered for separation processes. Therefore, Mahmoodi synthesized magnetic CNTs for photocatalytic dye degradation in the presence of hydrogen peroxide. During this process, the mineralization of dye yielded inorganic anions because heteroatoms are generally converted into anions, in which they are at their highest degradation degree. Nitrate and sulfate anions were detected as the mineralization products of dyes during the degradation process [47].

15.4.5 Other Nanomaterials in AOPs

Other nanomaterials have been used for degradation of dye using AOPs. Ash-based nanomaterials are important waste material–based nanomaterials that have been investigated because of low cost, abundant chemical composition and high adsorption capacity. Rice husk ash is a cheap adsorbent and was applied for the removal of dye. The silica from rice husk ash offers an environmentally friendly option compared to the commercially available one [48]. Adam and co-workers used rice husk silica as a support for the preparation of tin and titanium photocatalysts, prior to adsorptive removal of methylene blue and the obtained results showed 99% mineralization under UV light [49]. In another study, a heterogeneous Ni-based layered double hydroxide (Ni-LDH) nanomaterials were applied for methyl orange dye degradation and the AOP process was enhanced by the addition of ozone gas [50]. The degradation process started when ozone molecules were adsorbed on the surface of Ni-LDHs, to produce week bonds of hydroxyl groups. This interaction promoted the decomposition of ozone and the nickel ions were responsible for catalyzing the ozone degradation. Then, a peroxide complex, $Ni^{(iv)}O(OH)_2$ was formed from the oxidation of nickel and the methyl orange molecules were mineralized by the peroxide complex through hydroxyl reduction to form NO_3^- and SO_4^{2-} [50].

15.5 Summary

A wide range of literature has been recorded in relation to nanomaterials used for advanced oxidation processes over the past decades for dye degradation. Heterogeneous photocatalysis is the most-used AOP with nanomaterials. This can be due to its simplicity, high efficiency and use of solar light. Different semiconductors (WO_3, Fe_2O_3, ZnO, CdS, $SrTiO_3$) have been employed for photocatalysis of dye molecules. From Table 15.2, it can be seen that TiO_2 and ZnO has been a preferred semiconductor compared to the others. This semiconductor is relatively inexpensive, highly effective and has chemical stability. Zinc oxide exhibits a high absorption efficiency in different light spectra. However, there is still some research to be done on the zinc oxide modification to expand its absorption range and maximize its performance as a catalyst. Titanium oxide is preferred over zinc oxide in order to circumvent the drawbacks attributed with the latter.

Iron oxide (Fe_2O_3) nanomaterials are also the second most-used semiconductors. Table 15.2 indicates that hydroxyl radical (•OH) is one highly reactive species that has been generated in most studies for dye degradation using nanomaterials. High dye degradation efficiency has been achieved using this generated reactive species. The efficiency obtained for •OH is above 70% as seen from the table. The presence of UV light is one key factor that is present on most AOPs when nanomaterials are used. The UV enhances the processes and hence its preferences. During this enhanced processes, carbon dioxide (CO_2), water (H_2O) and other species with low molecular weight depending on the organic pollutant are always formed as intermediate or final products of the process.

TABLE 15.2

Nanomaterial-Based Advanced Oxidation Process for Dye Degradation

AOP type	Nanomaterials	Particle Size (nm)	Dye	Oxidative Species	Initial Concentration (ppm)	pH	Degradation Time (min)	Degradation Efficiency (%)	Degradation Products	Ref.
Photocatalysis	ZnO nanostructures	40–100	Reactive red 120	O_2^-•	267.6	5	60	95.8	H_2O, CO_2 and mineral acids	[32]
Photocatalysis	ZnO nanocrystals	40–100	Reactive orange 4	-	390.8	7	40	70.0	H_2O	[33]
Photocatalysis	ZnO/Eu nanoparticles	-	Methylene blue and orange	•OH and O_2^-•	10	10.1	150	62.0	H_2O	[34]
Photocatalysis	Fe_3O_4@C@Ru nanocomposite	> 5	Methylene blue	•OH and O_2^-•	31.98	-	30–170	92.7	H_2O, CO_2 and by-products	[41]
Photocatalysis	Nanostructure CdS/ZnO	50–100	Methylene blue	O_2^-• and •OH	-	-	240	91.0	CO_2 and H_2O	[35]
Photocatalysis	Magnetic carbon nanodots	5–10	Methylene blue	O_2^-• and •OH	1.6	-	30	83	CO_2 and H_2O	[44]
Photocatalysis	Reduced graphene oxide/meso-TiO_2/AuNPs	10.9	Methylene blue	h+, O_2^-• and •OH	30	-	50	~100	CO_2 and H_2O	[26]
Photocatalysis	Fe_3O_4 nanospheres	11.8	Xylenol orange	-	62.27	-	45	> 75	-	[40]
Photo-Fenton	Fe_3O_4/reduced graphene oxide nanocomposite	-	methyl green (MG), rhodamine B (RhB) and methyl blue (MB)	•OH and O_2^-•	60.9, 47.9 and 80.0	11 and 5	120	99.3, 98.8 and 87.1	Cl^-, NO_3^-, SO_4^{2-}, H_2O and CO_2	[38]
Photocatalysis	$CoFe_2O_4$/TiO_2 nanocatalysts	-	Reactive Red 120	•OH and SO_4^-•	13.4	-	60	-	CO_2 and H_2O	[28]
Photocatalysis	$CuInSe_2$/TiO_2 hybrid hetero-nanostructures	-	Methylene blue, methyl orange and Rhodamine B	HO_2• and •OH	100	7	60	86.9	$C_{16}H_{19}O_5N_3S$, $C_{15}H_{19}O_5N_3S$, $C_{14}H_{15}O_6N_3S$, $C_{12}H_{13}O_4N_3S$, $C_{16}H_{17}O_7N_3S$, $C_{13}H_{15}O_5N_3S$	[27]
Photocatalysis	TiO_2 nanohybrids	43.4	Amido black 10B	•OH, O_2^-	500	-	60–90	100	CO_2, H_2O	[29]
Photocatalysis	Multiwalled carbon nanotubes	15–22	Naphthol blue black	•OH	100	-	210	95.7	NO^{3-}, SO^{2-}_4, H_2O and CO_2	[46]

15.6 Conclusion and Future Recommendations

This review seeks to summarize advanced oxidation process (AOPs) based on nanomaterials. As is well known, water is a scarce commodity, hence significant research has been conducted to remove various pollutants in water. This chapter will help in developing different techniques that are efficient in tackling degradation of dye in wastewater. This includes the use of nanomaterials that are easy to synthesize, recyclable, economic and environmentally friendly. This chapter has demonstrated that nanomaterial-based advanced oxidation processes are effective technologies for dye treatment in wastewater. There is still need for development of low-cost nanomaterials that will be effective and easily regenerated for degradation of organic pollutants such as dyes. Natural materials such as mud and fly ash have not been intensively explored for degradation of organic pollutants. Furthermore, there is limited literature on the exploration of materials such as domestic wastes, fly ash, agro waste and coal. The materials can be used with iron-based nanoparticles for degradation of different organic pollutants.

15.7 Acknowledgement

The authors would like to acknowledge the University of South Africa and NRF THUTHUKA (113951) for their financial assistance.

REFERENCES

[1] Forgacs, E. and T. Cserha, "Removal of synthetic dyes from wastewaters: a review," *Environ Int.*, 2004, doi: 10.1016/j.envint.2004.02.001.

[2] Igor, P., M. Firmino, M. Erick, R. Silva, F. J. Cervantes and B. André, "Bioresource technology colour removal of dyes from synthetic and real textile wastewaters in one- and two-stage anaerobic systems," *Bioresour. Technol.*, vol. 101, no. 20, pp. 7773–7779, 2010, doi: 10.1016/j.biortech.2010.05.050.

[3] Hema, N. and S. Suresha, "Optimization of culture conditions for decolorization of acid red 10b by Shewanella putrefaciens," *Int. J. Curr. Microbial. Appl. Sci.*,vol. 4, no. 5, pp. 675–686, 2015.

[4] Chakrabarti, S. and B. K. Dutta, "Photocatalytic degradation of model textile dyes in wastewater using ZnO as semiconductor catalyst," *J. Hazard. Mater.*, vol. 112, no. 3, pp. 269–278, 2004, doi: 10.1016/j.jhazmat.2004.05.013.

[5] Santos, B., F. J. Cervantes and J. B. Van Lier, "Review paper on current technologies for decolourisation of textile wastewaters: Perspectives for anaerobic biotechnology," *Bioresour. Technol.*, vol. 98, pp. 2369–2385, 2007, doi: 10.1016/j.biortech.2006.11.013.

[6] Review, W. A., A. A. Yaqoob, T. Parveen and K. Umar, "Role of nanomaterials in the treatment of," *Water 2020*, vol. 12, p. 495, 2020.

[7] Daneshvar, N., M. Ayazloo, A. R. Khataee and M. Pourhassan, "Biological decolorization of dye solution containing Malachite Green by microalgae Cosmarium sp.," *Bioresour. Technol.*, vol. 98, pp. 1176–1182, 2007, doi: 10.1016/j.biortech.2006.05.025.

[8] Cai, Z., Y. Sun, W. Liu, F. Pan, P. Sun and J. Fu, "An overview of nanomaterials applied for removing dyes from wastewater," *Env. Sci. Pol. Res.*, pp. 15882–15904, 2017, doi: 10.1007/s11356-017-9003-8.

[9] Rajput, D., S. Paul and A. D. Gupta, "Green synthesis of silver nanoparticles using waste tea leaves," *Adv. Nano Res.*, vol. 3, no. 1, pp. 1–14, 2020.

[10] El Sikaily, A., A. Khaled and A. El Nemr, "Textile dyes xenobiotic and their harmful effect," *Non-Conventional Textile Waste Water Treatment*, pp. 31–64, 2012.

[11] Carliell, C. M., S. J. Barclay, N. Naidoo, C. A. Buckley, D. A. Mulholland and E. Senior, "Microbial decolourisation of a reactive azo dye under anaerobic conditions," *Water*, vol. 21, no. I, pp. 61–69, 1995.

[12] Islam, M. R. and M. G. Mostafa, "Textile dyeing effluents and environment concerns – a review," *J. Environ. Sci. & Natural Resources*, vol. 11, pp. 131–144, 2018.

[13] Chavan, R. B., *Environmentally friendly dyes*. Woodhead Publishing Limited, 2011.

[14] Burkinshaw, S. M., D. S. Jeong and T. I. Chun, "The coloration of poly(lactic acid) fibres with indigoid dyes: Part 2: Wash fastness," *Dye. Pigment.*, vol. 97, no. 2, pp. 374–387, 2013, doi: 10.1016/j.dyepig.2012.12.026.

[15] Benkhaya, S., S. El Harfi and A. El Harfi, "Classifications, properties and applications of textile dyes: A review," *Appl. J. Envir. Eng. Sci.*,vol. 3, pp. 311–320, 2017.

[16] Crittenden, J. C., R. R. Trussell, D. W. Hand, K. J. Howe and G. Tchobanoglous, "Advanced oxidation," *MWH's Water Treatment.* pp. 1415–1484, 2012, doi: 10.1002/9781118131473.ch18.

[17] Machulek, Amilcar Jr., Silvio C. Oliveira, Marly E. Osugi, Valdir S. Ferreira, Frank H. Quina, Renato F. Dantas, Samuel L. Oliveira, Gleison A. Casagrande, Fauze J. Anaissi, Volnir O. Silva, Rodrigo P. Cavalcante, Fabio Gozzi, Dayana D. Ramos, Ana P.P. da Rosa, Ana P.F. Santos, Douclasse C. de Castro and Jéssica A. Nogueira. Application of different advanced oxidation processes for the degradation of organic pollutants, organic pollutants - monitoring, risk and treatment, chapter 6, M. Nageeb Rashed, IntechOpen, 2013, DOI: 10.5772/53188.

[18] Cuerda-correa, E. M., M. F. Alexandre-franco and C. Fern, "Antibiotics from water: An overview," *Water*, vol. 12, pp. 1–50, 2020.

[19] Schneider, M. and L. Bláha, "Advanced oxidation processes for the removal of cyanobacterial toxins from drinking water," *Environ. Sci. Eur.*, vol. 32, no. 1, 2020, doi: 10.1186/s12302-020-00371-0.

[20] Rajoriya, S., J. Carpenter, V. K. Saharan and A. B. Pandit, "Hydrodynamic cavitation: An advanced oxidation process for the degradation of bio-refractory pollutants," *Rev. Chem. Eng.*, vol. 32, no. 4, pp. 379–411, 2016, doi: 10.1515/revce-2015-0075.

[21] Chen, Y., Y. Cheng, X. Guan, Y. Liu, J. Nie and C. Li, "A rapid Fenton treatment of bio-treated dyeing and finishing wastewater at second-scale intervals: Kinetics by stopped-flow technique and application in a full-scale plant," *Sci. Rep.*, vol. 9, no. 1, pp. 1–11, 2019, doi: 10.1038/s41598-019-45948-9.

[22] Gogate, P. R. and A. B. Pandit, "A review of imperative technologies for wastewater treatment I: Oxidation technologies at ambient conditions," *Adv. Environ. Res.*, vol. 8, no. 3–4, pp. 501–551, 2004, doi: 10.1016/S1093-0191(03)00032-7.

[23] Song, W., J. Li, Z. Wang and X. Zhang, "A mini review of activated methods to persulfate-based advanced oxidation process," *Water Sci. Technol.*, vol. 79, no. 3, pp. 573–579, 2019, doi: 10.2166/wcc.2018.168.

[24] Hashimoto, K., H. Irie and A. Fujishima, "TiO 2 photocatalysis: A historical overview and future prospects," *Japanese J. Appl. Physics, Part 1 Regul. Pap. Short Notes Rev. Pap.*, vol. 44, no. 12, pp. 8269–8285, 2005, doi: 10.1143/JJAP.44.8269.

[25] Etogo, A. et al., "Facile one-pot solvothermal preparation of Mo-doped Bi2WO6 biscuit-like microstructures for visible-light-driven photocatalytic water oxidation," *J. Mater. Chem. A*, vol. 4, no. 34, pp. 13242–13250, 2016, doi: 10.1039/c6ta04923k.

[26] Yang, Y., Z. Ma, L. Xu, H. Wang and N. Fu, "Preparation of reduced graphene oxide/meso-TiO 2 /AuNPs ternary composites and their visible-light-induced photocatalytic degradation n of methylene blue," *Appl. Surf. Sci.*, vol. 369, pp. 576–583, 2016, doi: 10.1016/j.apsusc.2016.02.078.

[27] Kshirsagar, A. S., A. Gautam and P. K. Khanna, "Efficient photo-catalytic oxidative degradation of organic dyes using CuInSe2/TiO2 hybrid hetero-nanostructures," *J. Photochem. Photobiol. A Chem.*, vol. 349, pp. 73–90, 2017, doi: 10.1016/j.jphotochem.2017.08.058.

[28] Sathishkumar, P., R. V. Mangalaraja, S. Anandan, and M. Ashokkumar, "CoFe2O4/TiO2 nanocatalysts for the photocatalytic degradation of reactive red 120 in aqueous solutions in the presence and absence of electron acceptors," *Chem. Eng. J.*, vol. 220, pp. 302–310, 2013, doi: 10.1016/j.cej.2013.01.036.

[29] Kashyap, J., S. M. Ashraf and U. Riaz, "Highly efficient photocatalytic degradation of amido black 10B dye using polycarbazole-decorated TiO2 nanohybrids," *ACS Omega*, vol. 2, no. 11, pp. 8354–8365, 2017, doi: 10.1021/acsomega.7b01154.

[30] Díaz-Sánchez, M. et al., "Ionic liquid-assisted synthesis of F-doped titanium dioxide nanomaterials with high surface area for multi-functional catalytic and photocatalytic applications," *Appl. Catal. A Gen.*, vol. 613, no. January, p. 118029, 2021, doi: 10.1016/j.apcata.2021.118029.

[31] Behnajady, M. A., N. Modirshahla and R. Hamzavi, "Kinetic study on photocatalytic degradation of C.I. Acid Yellow 23 by ZnO photocatalyst," *J. Hazard. Mater.*, vol. 133, no. 1–3, pp. 226–232, 2006, doi: 10.1016/j.jhazmat.2005.10.022.

[32] Velmurugan, R. and M. Swaminathan, "An efficient nanostructured ZnO for dye sensitized degradation of reactive red 120 dye under solar light," *Sol. Energy Mater. Sol. Cells*, vol. 95, no. 3, pp. 942–950, 2011, doi: 10.1016/j.solmat.2010.11.029.

[33] Velmurugan, R., K. Selvam, B. Krishnakumar and M. Swaminathan, "An efficient reusable and antiphotocorrosive nano ZnO for the mineralization of reactive orange 4 under UV-A light," *Sep. Purif. Technol.*, vol. 80, no. 1, pp. 119–124, 2011, doi: 10.1016/j.seppur.2011.04.018.

[34] Trandafilović, L. V., D. J. Jovanović, X. Zhang, S. Ptasińska and M. D. Dramićanin, "Enhanced photocatalytic degradation of methylene blue and methyl orange by ZnO: Eu nanoparticles," *Appl. Catal. B Environ.*, vol. 203, pp. 740–752, 2017, doi: 10.1016/j.apcatb.2016.10.063.

[35] Velanganni, S., S. Pravinraj, P. Immanuel, and R. Thiruneelakandan, "Nanostructure CdS/ZnO heterojunction configuration for photocatalytic degradation of Methylene blue," *Phys. B Condens. Matter*, vol. 534, no. November 2017, pp. 56–62, 2018, doi: 10.1016/j.physb.2018.01.027.

[36] Liang, H. et al., "Multifunctional Fe3O4@C@Ag hybrid nanoparticles: Aqueous solution preparation, characterization and photocatalytic activity," *Mater. Res. Bull.*, vol. 48, no. 7, pp. 2415–2419, 2013, doi: 10.1016/j.materresbull.2013.02.066.

[37] Pathak, T. K., R. E. Kroon and H. C. Swart, "Photocatalytic and biological applications of Ag and Au doped ZnO nanomaterial synthesized by combustion," *Vacuum*, vol. 157, no. July, pp. 508–513, 2018, doi: 10.1016/j.vacuum.2018.09.020.

[38] Boruah, P. K. et al., "Sunlight assisted degradation of dye molecules and reduction of toxic Cr(VI) in aqueous medium using magnetically recoverable Fe3O4/reduced graphene oxide nanocomposite," *RSC Adv.*, vol. 6, no. 13, pp. 11049–11063, 2016, doi: 10.1039/c5ra25035h.

[39] Qin, Y., M. Long, B. Tan, and B. Zhou, "RhB adsorption performance of magnetic adsorbent Fe 3 O 4 /RGO composite and its regeneration through A Fenton-like reaction," *Nano-Micro Lett.*, vol. 6, no. 2, pp. 125–135, 2014, doi: 10.5101/nml.v6i2.p125-135.

[40] Zhu, M. and G. Diao, "Synthesis of porous Fe3O4 nanospheres and its application for the catalytic degradation of xylenol orange," *J. Phys. Chem. C*, vol. 115, no. 39, pp. 18923–18934, 2011, doi: 10.1021/jp200418j.

[41] Zhang, Q. et al., "Preparation of highly efficient and magnetically recyclable Fe 3 O 4 @C@Ru nanocomposite for the photocatalytic degradation of methylene blue in visible light," *Appl. Surf. Sci.*, vol. 483, no. March, pp. 241–251, 2019, doi: 10.1016/j.apsusc.2019.03.225.

[42] Renda, C. G., L. A. Goulart, C. H. M. Fernandes, L. H. Mascaro, J. M. De Aquino and R. Bertholdo, "Novel onion-like carbon structures modified with iron oxide as photocatalysts for the degradation of persistent pollutants," *J. Environ. Chem. Eng.*, vol. 9, no. 1, p. 104934, 2021, doi: 10.1016/j.jece.2020.104934.

[43] Ng, Y. H., S. Ikeda, M. Matsumura and R. Amal, "A perspective on fabricating carbon-based nanomaterials by photocatalysis and their applications," *Energy Environ. Sci.*, vol. 5, no. 11, pp. 9307–9318, 2012, doi: 10.1039/c2ee22128d.

[44] Sun, A. C. "Synthesis of magnetic carbon nanodots for recyclable photocatalytic degradation of organic compounds in visible light," *Adv. Powder Technol.*, vol. 29, no. 3, pp. 719–725, 2018, doi: 10.1016/j.apt.2017.12.013.

[45] Trakakis, G., G. Tomara, V. Datsyuk, L. Sygellou, A. Bakolas, D. Tasis, J. Parthenios, C. Krontiras, S. Georga, C. Galiotis, K. Papagelis, "Mechanical, electrical, and thermal properties of carbon nanotube buckypapers/epoxy nanocomposites produced by oxidized," *Materials*, vol. 13, pp. 4380, 2020.

[46] Mamba, G., X. Y. Mbianda and A. K. Mishra, "Photocatalytic degradation of the diazo dye naphthol blue black in water using MWCNT/Gd,N,S-TiO2 nanocomposites under

simulated solar light," *J. Environ. Sci. (China)*, vol. 33, pp. 219–228, 2015, doi: 10.1016/j.jes.2014.06.052.

[47] Mahmoodi, N. M. "Synthesis of magnetic carbon nanotube and photocatalytic dye degradation ability," *Environ. Monit. Assess.*, vol. 186, no. 9, pp. 5595–5604, 2014, doi: 10.1007/s10661-014-3805-7.

[48] Han, R. et al., "Use of rice husk for the adsorption of congo red from aqueous solution in column mode," *Bioresour. Technol.*, vol. 99, no. 8, pp. 2938–2946, 2008, doi: 10.1016/j.biortech.2007.06.027.

[49] Adam, F., J. N. Appaturi, Z. Khanam, R. Thankappan and M. A. M. Nawi, "Utilization of tin and titanium incorporated rice husk silica nanocomposite as photocatalyst and adsorbent for the removal of methylene blue in aqueous medium," *Appl. Surf. Sci.*, vol. 264, pp. 718–726, 2013, doi: 10.1016/j.apsusc.2012.10.106.

[50] El Hassani, K., D. Kalnina, M. Turks, B. H. Beakou and A. Anouar, "Enhanced degradation of an azo dye by catalytic ozonation over Ni-containing layered double hydroxide nanocatalyst," *Sep. Purif. Technol.*, vol. 210, pp. 764–774, 2019, doi: 10.1016/j.seppur.2018.08.074.

16

Advanced Oxidation of Phenolic Pollutants in Wastewater

Reshmi Sasi and T.V. Suchithra

CONTENTS

16.1 Introduction

Water detoxification is a critical issue that necessitates rapid research and implementation of effective technologies capable of abating deadly contaminants in wastewater. Among various pollutants, phenolic pollutants need more attention due to their high toxicity, broad uses and resistance to complete degradation. Phenol is a raw material for synthesizing many chemicals such as resins, pharmaceuticals and dyes [1, 2]. The uncontrolled exposure of untreated phenolic compounds to nature can lead to their transformation into more hazardous products. Many conventional methods (coagulation, filtration, biodegradation, sedimentation and centrifugation) offer removal of phenolic compounds, but they do not represent a promising option for the complete mineralization of pollutants. In such a scenario, the advanced oxidation processes (AOPs) can be considered as a better alternative. Glaze and co-workers in 1987 established the concept of "advanced oxidation processes" [3].

16.2 Advanced Oxidation Processes – Water Treatment Processes of the 21st Century

AOPs are a set of chemical oxidation procedures used to treat wastewater, drinking water and groundwater contaminated with organic compounds by oxidation through reactions with hydroxyl radicals (OH^*) [4]. This method is based on the *in situ* production of highly reactive OH^*. They are strong oxidants and can oxidize any organic compound present in the water matrix. Once formed, the OH^* can react non-specifically with the contaminants, and then the pollutants will quickly be converted effectively into small inorganic molecules. Primary oxidants, energy sources and catalysts are the common sources of OH^*. Commonly employed primary oxidants are hydrogen peroxide, ozone and oxygen; energy sources include ultraviolet (UV) radiation, and the most widely used catalyst is titanium oxide. The perfect combinations of these reagents can be applied to get maximum OH^* yield. Generally, AOPs can reduce the pollutant concentrations from several thousand ppm to less than 5 ppb, causing a significant reduction in chemical oxygen demand and total organic carbon, which gained them the title of "water treatment processes of the 21st century" [5].

The major mechanisms of the advanced oxidation process are the formation of OH^*, the initial attack of OH^* on target molecules and their subsequent breakdown and, finally, successive attacks by OH^* until complete mineralization. AOPs generate various reactive oxygen species via different processes such as the Fenton, ozonation, sonolysis, UV, wet air oxidation and photo-Fenton reactions (Figure 16.1). These

DOI: 10.1201/9781003165958-16

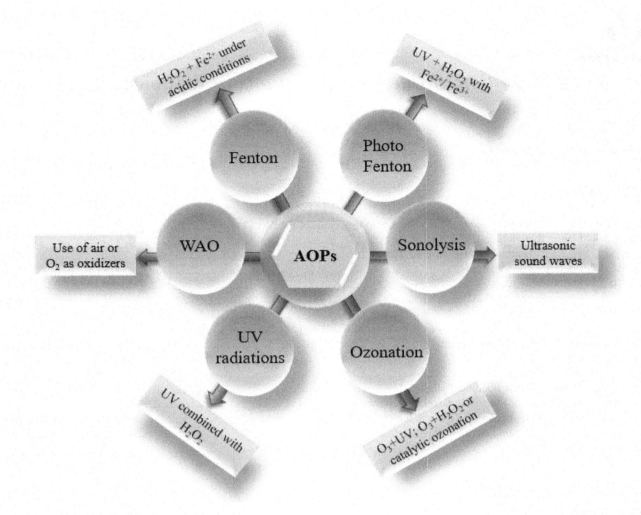

FIGURE 16.1   Types of advanced oxidation processes (AOPs).

reactive oxygen species are subsequently used for the degradation of pollutants [6].

16.2.1 Fenton Reaction

The Fenton reaction was first reported by H. J. Fenton in 1894 [7]. He described the reaction as enhanced oxidative potential of H_2O_2, when iron (Fe) is used as a catalyst under acidic conditions. Major events during the Fenton reaction are the following (Equations 16.1–16.4) [8].

$$Fe^{2+} + H_2O_2 \rightarrow Fe^{3+} + OH^* + OH^- \tag{16.1}$$

$$OH^* + H_2O_2 \rightarrow HO_2^* + H_2O \tag{16.2}$$

$$Fe^{2+} + HO_2^* \rightarrow Fe^{2+} + O_2 + H^+ \tag{16.3}$$

$$OH^* + OH^* \rightarrow H_2O_2 \tag{16.4}$$

16.2.1.1 Fenton-Like Reactions

The reaction in which hydrogen peroxide reacts with ferric ion is referred to as a Fenton-like reaction. Metals such as cobalt

and copper can also be used and are used at low oxidation states (Equation 16.5) [8].

$$Cu^+ + H_2O_2 \rightarrow Cu^{2+} + OH^* + OH^- \tag{16.5}$$

16.2.2 Photo-Fenton Reaction

This process uses a combination of UV radiation and hydrogen peroxide with ferrous (Fe^{2+}) or ferric (Fe^{3+}) ion and generates more hydroxyl radicals, increasing the rate of pollutant degradation. In the Fenton reaction system, the process completely stops after some time due to the depletion of ferrous ions (Fe^{2+}) and the accumulation of ferric ions (Fe^{3+}). In contrast, photochemical regeneration of Fe^{2+} ions happens in the photo-Fenton reaction by the photoreduction of ferric ions (Equations 16.6 and 16.7). Hence, the newly formed Fe^{2+} ions continue to react with hydrogen peroxide and produce hydroxyl radicals and Fe^{3+}, and the cycle continues [9].

$$Fe^{3+} + H_2O + h\nu \rightarrow Fe^{2+} + OH^* + H^+ \tag{16.6}$$

$$Fe^{3+} + H_2O_2 + h\nu \rightarrow Fe^{2+} + HO_2^* + H^+ \tag{16.7}$$

The photo-Fenton process is reported to be more efficient than the Fenton process. It has optimum performance at pH_3 (low acidic pH), when the $Fe(OH)^{3+}$ complexes are more soluble and hydroxy-Fe^{2+} complexes are more photo-active. Similarly, acidic conditions favor the conversion of carbonate and bicarbonate ions into carbonic acid, which has less reactivity with OH^*. Sunlight can be used instead of UV radiation, which helps to reduce the cost of treatment; however, this could be associated with a low degradation of pollutants [8].

16.2.3 Sonolysis

Sonolysis is an eco-friendly method that does not involve any hazardous chemicals and is based on the utilization of ultrasonic sound with 20 KHz–10 MHz frequency range. The ultrasonic sound waves create cavitation bubbles inside the water matrix, and organic pollutants are degraded as a result of the explosion of these cavitation bubbles. Bubbles are formed according to their ultrasound frequencies. The size of the bubbles fluctuates until they are collapsed at their resonance size, which leads to the release of stored energy and causes explosion. However, high frequencies lead to the production of a smaller number of reactive species due to the collision of active bubbles, and lower frequencies produce fewer bubbles due to high vapor content inside the collapsing bubbles. Degradation of pollutants happens due to the explosion of bubbles at extreme pressure (500–10,000 atm pressure) and temperature (3000–5000 K). The explosion causes the generation of OH^* in the system by the dissociation of water molecules. These highly reactive OH^* react with pollutants and degrade them to their simplest forms (Figure 16.2). These OH^* migrate from the bubble-liquid interface into the bulk liquid, causing secondary reactions and complete mineralization of the pollutants [10].

16.2.4 Ozone-Based AOPs

Ozone is a strong oxidant and is highly selective towards pollutants. Usually, ozonation is used in wastewater treatment to kill microbes, to remove micropollutants, taste and odor and to decolorize. In water, the ozone decomposition mechanism involves complex sequences of atoms, a single electron transport with intermediate hydroxyl radical formation and subsequent production of H_2O, OH and so on. This process can be accelerated by increasing the pH or by adding H_2O_2. Ozone is a good disinfectant, while OH and O_3 both are good oxidants in the oxidation process. Hence, ozonation can be simultaneously used for oxidation and disinfection. Even though ozonation has several advantages, it has many disadvantages such as high cost of ozone reagent, low efficiency in ammonia removal and formation of toxic compounds during reaction with OH^*. Later, ozonation combined with AOPs emerged as a new dimension in wastewater treatment, which could effectively oxidize both inorganic and organic compounds better than standalone processes could [11].

16.2.4.1 Ozonation with UV Radiation (O₃/UV)

A combination of ultraviolet radiation and ozone is an efficient oxidation system for the degradation of refractory compounds in wastewater. Photolysis of ozone leads to the production of OH^* due to the reaction of O^* with water. The combined effect of ozone and ultraviolet radiation causes the decomposition of ozone by the direct or indirect production of OH^*. The resulting hydroxyl radicals attack organic molecules and convert them into the simplest forms of inorganic molecules [12]. Hydroxyl radical formation is shown in Equations 16.8–16.10).

$$O_3 + UV \rightarrow O_2 + O^* \tag{16.8}$$

$$O^* + H_2O \rightarrow 2\,OH^* \tag{16.9}$$

$$2O^* + H_2 \rightarrow OH^* + OH^* \rightarrow H_2O_2 \tag{16.10}$$

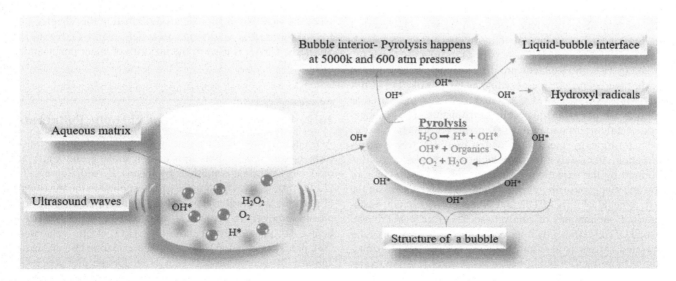

FIGURE 16.2 Sonolytic degradation of organic pollutants in water.

The following reaction of hydroxyl radicals' formation is also possible (Equations 16.11 and 16.12).

$$O_3 + H_2O \rightarrow O_2 + H_2O_2 \tag{16.11}$$

$$H_2O_2 \rightarrow 2\ OH^* \tag{16.12}$$

16.2.4.2 Ozonation with Hydrogen Peroxide (O_3/H_2O_2)

This method is also known as the "peroxone" method (adding hydrogen peroxide to ozonated water for treatment). It involves a radical chain mechanism based on the decomposition of O_3 by the hydroperoxide anion HO_2^-. The combined effect of O_3 and hydrogen peroxide causes the production of OH^*, and these OH^* will act on organic pollutants in the wastewater (Equations 16.13 and 16.14). Ozone reacts with excess HO_2^- and forms OH^* (Equation 16.15–16.17).

$$H_2O_2 \rightarrow HO_2^- + H^+ \tag{16.13}$$

$$HO_2^- + O_3 \rightarrow HO_2^* + O_3^{*-} \tag{16.14}$$

$$O_3^{*-} + H^+ \rightarrow HO_3^{*-} \tag{16.15}$$

$$HO_3^{*-} \rightarrow O_2 + OH^* \tag{16.16}$$

$$OH^* + H_2O_2 \rightarrow HO_2^* + H_2O \tag{16.17}$$

Neither low nor high hydrogen peroxide concentrations are desirable as they can cause the failure of the pollutant degradation system. High or excess H_2O_2 in the system can lead to OH^* scavenging and the formation of hydroperoxide ions (HO_2^-). Similarly, in low hydrogen peroxide concentration, H_2O_2 competes for OH^* and decomposes without oxidizing the pollutants [13].

16.2.4.3 Catalytic Ozonation

Catalytic ozonation uses a catalyst to accelerate the decomposition of O_3, which in turn causes the production of hydroxyl radicals. The addition of heterogeneous or homogeneous catalysts during ozonation can enhance the oxidation process. The heterogeneous catalytic ozonation uses a solid catalyst for improving the process to remove the refractory pollutants. Several metal oxides such as Fe_2O_3, $MgFe_2O_4$, $CuFe_2O_4$, $ZnFe_2O_4$, TiO_2 and Ru/CeO_2 are the commonly used catalysts for the heterogeneous process, while metal ions like Fe^{2+}, Cu^{2+}, Cr^{2+}, Mn^{2+}, Ni^{2+}, Co^{2+}, Cd^{2+}, Ag^+ and Zn^{2+} are widely used as catalysts for homogeneous catalytic ozonation. The homogeneous catalytic ozonation has two major mechanisms: ozone decomposition by metal ions and subsequent free radicals formation, and catalyst–organic molecule complex formation followed by the oxidation of the complex. A combination of iron and manganese along with ozonation is very useful for the easy removal of TOC, COD and organochloride from wastewaters. Combinations of catalysts such as $Ru/CeO/O^{33}$, Alp/O_3 and O/TiO_2 efficiently remove organic pollutants. Granular activated carbon is also an excellent catalyst for the ozonation of biorefractory pollutants in wastewater [11]. An example of iron-mediated catalytic ozonation is given as follows (Equations 16.18–16.20):

$$Fe^{2+} + O_3 \rightarrow FeO^{2+} + O^* \tag{16.18}$$

$$FeO^{2+} + H_2O \rightarrow Fe^{3+}\ OH^* + OH^- \tag{16.19}$$

$$FeO^{2+} + Fe^{2+} + 2H^+ \rightarrow 2Fe^{3+} + H_2O \tag{16.20}$$

16.2.5 UV-Based AOPs

Ultraviolet radiation is widely used in wastewater treatment to remove organic compounds that can absorb UV and kill microbes. When the pollutants absorb UV, the electrons will be transformed to an excited state. The excited electrons are then transferred to an oxygen molecule to convert both pollutant molecules and O_2 into radicals. These radicals will oxidize other molecules to attain a stable form. Photolysis of water with UV light (wavelength less than 190 nm) can produce OH^*, which will subsequently oxidize the organic pollutants [3].

UV is used in combination with many substances for AOPs. UV/H_2O_2 treatment methods are commonly used for the treatment of drinking water. This method is based on the production of OH^* by the cleavage of H_2O_2 with UV (Equation 16.21).

$$H_2O_2 + h\nu \rightarrow 2\ OH^* \tag{16.21}$$

The resulting OH^* will react with the organic pollutants present in the water. This reaction will cause the production of organic radicals. These organic radicals react with oxygen to form peroxyl radicals, which initiate thermal oxidation, leading to the production of less harmful pollutants like carbonyl compounds and superoxide anions [14].

16.2.6 Wet Air Oxidation

WAO involves the utilization of air or molecular oxygen as oxidizers under high temperature and pressure (Figure 16.3). The extreme environmental conditions cause the production of free radicals, which will further degrade the pollutants. Temperature and pressure are the major controlling factors of WAO processes. Acids, except propionic acid and acetic acid, are converted to CO_2 at higher temperature and pressure [15].

For the efficient removal of refractory organic contaminants, all of the AOP-based approaches described in this chapter are frequently used in wastewater treatment plants (Table 16.1). Similarly, UV/O_3 is used in the majority of water purification systems.

16.3 Advanced Oxidation of Phenolic Pollutants

AOPs have received huge attention due to their ability to transform organic compounds into carbon dioxide and water at moderate operation conditions by means of oxidizing agents [29, 30]. AOPs have been successfully applied in the treatment of phenolic wastewater. Some of the common AOPs work for phenol removal are Fenton [31], photon-Fenton [32], electro-Fenton [1], photoelectro-Fenton [33], H_2O_2 electro-generated [34], peroxicoagulation [35–38] anodic oxidation [39] and ozonation [40]. Fenton process uses H_2O_2 for the production of OH^*, which is considered a nontoxic reagent as its decomposition produces only inert compounds such as oxygen and water.

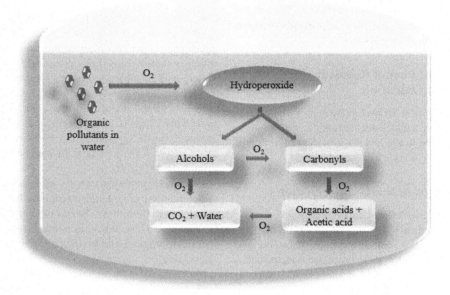

FIGURE 16.3 Mechanism of wet air oxidation process for the removal of organic pollutants in water.

TABLE 16.1

Pollutants Degraded Using Various AOPs

No.	AOPs	Pollutants	Ref.
1.	Fenton/Photo-Fenton reaction	Bisphenol A	[16]
		Pharmaceutical wastewater	
		Micropollutants from municipal wastewater	[17]
		Metronidazole	[18]
		Paracetamol	
		Rhodamine B	[19]
2.	Ozonation	Ethylene diamine tetra acetate	[20]
		Carbamazepine	
		Sulfamethoxazole	[21]
		Roxithromycin	
3.	UV	Sulfamethoxazole	[22]
		Oxytetracycline	
		Ciprofloxacin	
		Oiuron	[23]
		Alachlor	
		Pentachlorophenol	
		Atrazine	
		Boldenone	
		Chlorphenvinphos	
4.	Sonolysis	Bisphenol A	[24]
		1,4-dichlorobenzene	[25]
		Phenol	[26]
		Ciprofloxacin	[27]
5.	Wet air oxidation	Quinolone	[28]
		Glucose-glycine melanoidins	
		Ammonia	
		Phenol	
		Tert-amyl-methyl ether	
		Resin effluent	

Electrochemical oxidation involves the production of hydroxyl radicals through anodic/cathodic reactions taking place at the anode and cathode electrodes. All these techniques work on the basis of hydroxyl radical formation.

16.3.1 Mechanism of Phenol Removal by Advanced Oxidation

Phenol oxidation by Fenton reaction has two major stages: formation of aromatic intermediates and organic acid formation. The organic acids formed in the last stage of degradation will be converted into carbon dioxide and water.

16.3.1.1 Formation of Aromatic Intermediates

Oxidation of phenol starts with the hydroxylation of an aromatic ring by free hydroxyl radicals. Hydroxylation causes the formation of dihydroxybenzenes such as catechol, resorcinol and hydroquinone, depending on the position of the hydroxylation in the aromatic ring. Resorcinol and hydroquinone are the products of meta- and parahydroxylation respectively, while catechol is the primary orthohydroxylation product. This is followed by the aromatic ring opening of the hydroxylation products with the subsequent formation of organic acids. These organic acids, except acetic acid, are finally oxidized to CO_2 and H_2O [41].

16.3.1.2 Organic Acid Formation

Once the aromatic intermediates are formed, they will be quickly converted to organic acids. Initially, the aromatic intermediates are converted to muconic acid by C-C cleavage, which is then oxidized to maleic acid by the action of OH^*. Maleic acid is subsequently oxidized to oxalic acid followed by malonic acid, glyoxylic acid and formic acid. Malonic acid is then oxidized to form formic acid, oxalic acid and acetic acid. Thus formed formic acid and oxalic acid are finally oxidized into CO_2 and H_2O, whereas acetic acid remains in its native form as the final product of phenol oxidation (Figure 16.4) [41, 42].

Advanced oxidation carries such advantages as complete mineralization of organic pollutants, less or no harmful by-product formation, low cost of operation and easy availability of reactants. These advantages make them preferable to other chemical, physical and biological methods of phenol removal (Table 16.2).

The majority of the industrial phenolic wastewater treatment systems follow AOPs for water processing. In recent years, they have been widely accepted and practiced, and rightfully so. The scientific community is still working on it to rectify the minor disadvantages of these techniques and make them the most efficient wastewater treatment technology.

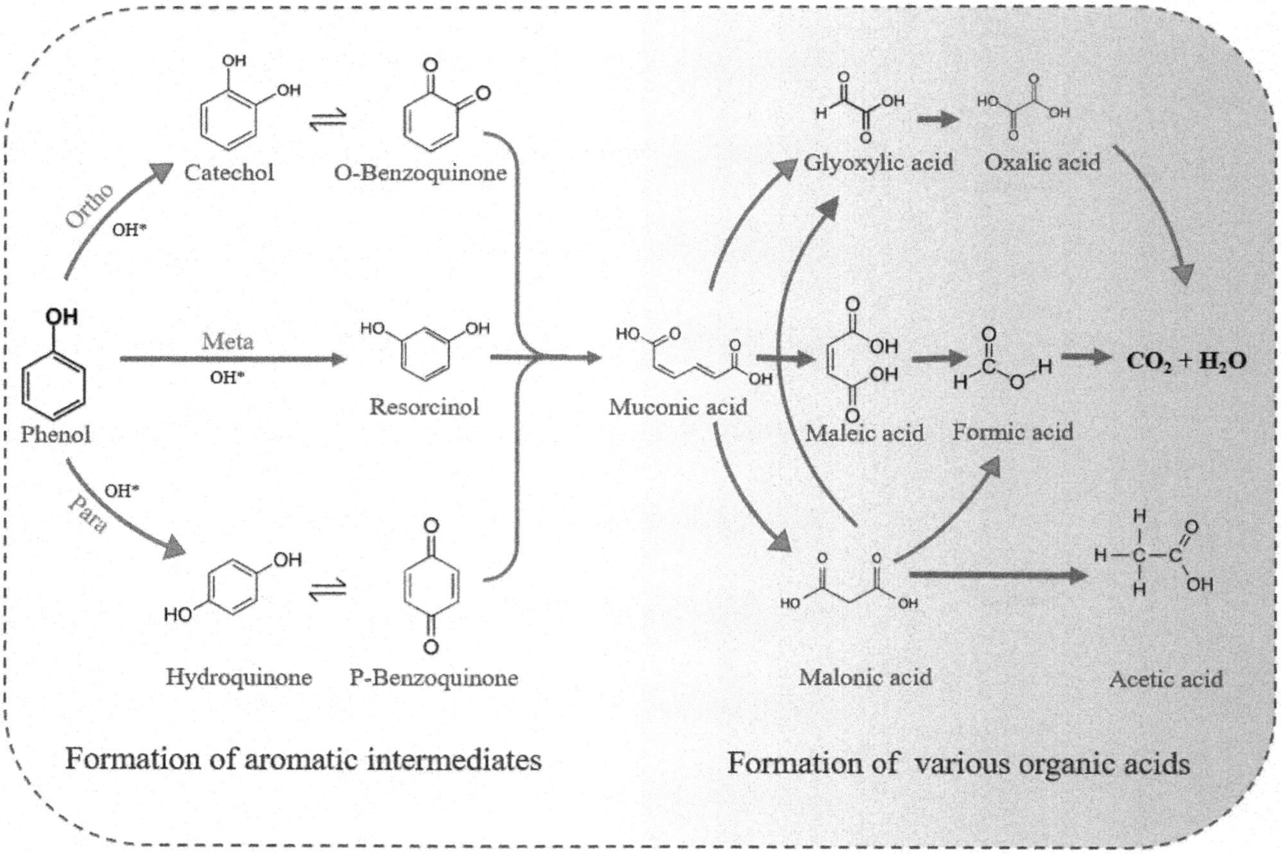

FIGURE 16.4 Phenol degradation via Fenton reaction.

TABLE 16.2

AOPs Used for the Removal of Various Phenolic Pollutants

No.	Phenolic Compounds	AOPs Used for Treatment	Ref.
1.	Phenol 2,4-Dichlorophenol Nonylphenol decaethoxylate Phenolic xenoestrogens Bisphenol A	UV/H_2O_2 treatment	[43]
2.	Polyphenol Phenol 4-chlorophenol	Fenton and Fenton-like treatment	[43]
3.	Volatile phenols Phenol 2-chlorophenol	Wet air oxidation and catalytic wet air oxidation	[43]
4.	p-nitrophenol Nonylphenols Phenol	Ozone	[43]
5.	Phenol	Sonolysis	[26]
	Bisphenol A		[24]

16.4 Conclusion

The advanced oxidation processes consist of an array of techniques sharing a common feature of OH* production. The significant advantage of AOPs over other water treatment techniques is that they offer complete mineralization of organic pollutants, including phenolic compounds, without the production of any lethal by-products and end products. Many water purification systems use AOPs for water processing, and the processes are found effective in removing the pollutants and harmful microbes in drinking water. Their low cost of operation, eco-friendly procedures and hazardless nature make them more acceptable than any other water treatment method.

REFERENCES

[1] Babuponnusami, A. and K. Muthukumar, "Advanced oxidation of phenol: A comparison between Fenton, electro-Fenton, sono-electro-Fenton and photo-electro-Fenton processes," *Chem. Eng. J.*, vol. 183, pp. 1–9, 2012. https://doi.org/10.1016/j.cej.2011.12.010

[2] Hurwitz, G., P. Pornwongthong, S. Mahendra and E. M. V Hoek, "Degradation of phenol by synergistic chlorine-enhanced photo-assisted electrochemical oxidation," *Chem. Eng. J.*, vol. 240, pp. 235–243, 2014. https://doi.org/10.1016/j.cej.2013.11.087

[3] Glaze, W. H., J. W. Kang and D. H. Chapin, "The chemistry of water treatment processes involving ozone, hydrogen peroxide and ultraviolet radiation," *Ozone Sci. Eng.*, vol. 9, no. 4, pp. 335–352, Sept. 1987. https://doi.org/10.1080/01919518708552148

[4] Alnaizy, R. and A. Akgerman, "Advanced oxidation of phenolic compounds," *Adv. Environ. Res.*, vol. 4, no. 3, pp. 233–244, 2000. https://doi.org/10.1016/S1093-0191(00)00024-1

[5] Munter, R., "ChemInform abstract: Advanced oxidation processes: Current status and prospects," *ChemInform*, vol. 32, no. 41, Oct. 2001.

[6] Fernández-Castro, P., M. Vallejo, M. F. San Román and I. Ortiz, "Insight on the fundamentals of advanced oxidation processes: Role and review of the determination methods of reactive oxygen species," *J. Chem. Technol. Biotechnol.*, vol. 90, no. 5, pp. 796–820, May 2015. https://doi.org/10.1002/jctb.4634

[7] Fenton, H. J. H., "LXXIII. – Oxidation of tartaric acid in presence of iron," *J. Chem. Soc. Trans.*, vol. 65, pp. 899–910, 1894.

[8] Ameta, R., A. K. Chohadia, A. Jain and P. B. Punjabi, *Chapter 3 – Fenton and photo-Fenton processes*, Academic Press, 2018, pp. 49–87. https://doi.org/10.1016/B978-0-12-810499-6.00003-6

[9] Kim, S. M. and A. Vogelpohl, "Degradation of organic pollutants by the photo-Fenton-Process," *Chem. Eng. Technol.*, vol. 21, no. 2, pp. 187–191, 1998. https://doi.org/10.1002/(SICI)1521-4125(199802)21:2 < 187::AID CEAT187>3.0.CO;2-H

[10] Joseph, C. G., G. Li Puma, A. Bono and D. Krishnaiah, "Sonophotocatalysis in advanced oxidation process: A short review," *Ultrason. Sonochem.*, vol. 16, no. 5, pp. 583–589, 2009. https://doi.org/10.1016/j.ultsonch.2009.02.002

[11] Rekhate, C. V. and J. K. Srivastava, "Recent advances in ozone-based advanced oxidation processes for treatment of wastewater- A review," *Chem. Eng. J. Adv.*, vol. 3, p. 100031, 2020. https://doi.org/10.1016/j.ceja.2020.100031

[12] Emam, E. A., "Effect of ozonation combined with heterogeneous catalysts and ultraviolet radiation on recycling of gas-station wastewater," *Egypt. J. Pet.*, vol. 21, no. 1, pp. 55–60, 2012. https://doi.org/10.1016/j.ejpe.2012.02.008

[13] Prieto-Rodríguez, L., I. Oller, N. Klamerth, A. Agüera, E. M. Rodríguez and S. Malato, "Application of solar AOPs and ozonation for elimination of micropollutants in municipal wastewater treatment plant effluents," *Water Res.*, vol. 47, no. 4, pp. 1521–1528, 2013. https://doi.org/10.1016/j.watres.2012.11.002

[14] Legrini, O., E. Oliveros and A. M. Braun, "ChemInform abstract: Photochemical processes for water treatment," *ChemInform*, vol. 24, no. 28, 2010.

[15] Debellefontaine, H., M. Chakchouk, J. N. Foussard, D. Tissot and P. Striolo, "Treatment of organic aqueous wastes: Wet air oxidation and wet peroxide oxidation®," *Environ. Pollut.*, vol. 92, no. 2, pp. 155–164, 1996. https://doi.org/10.1016/0269-7491(95)00100-X

[16] Huang, W., M. Luo, C. Wei, Y. Wang, K. Hanna, and G. Mailhot, "Enhanced heterogeneous photo-Fenton process modified by magnetite and EDDS: BPA dedgraation," *Environ. Sci. Pollut. Res.*, vol. 24, no. 11, pp. 10421–10429, 2017. http://doi.org/ 10.1007/s11356-017-8728-8

[17] Villegas- Guzman, P., O. Giannakis, S. Rtimi, S. Grandjean, D. Bensimon, M. Alencastro, L. Felippe, Torres-Palma, R. Pulgarin and Cesar, "A green solar photo-Fenton process for the elimination of bacteria and micropollutants in municipal wastewater treatment using mineral iron and natural organic acids," *Appl. Catal. B Environ.*, vol. 219, pp. 538–549, 2017. https://doi.org/10.1016/j.apcatb.2017.07.066

[18] Velichkova, F., H. Delmas, C. Julcour and B. Koumanova, "Heterogeneous Fenton and photo-Fenton oxidation for paracetamol removal using iron containing ZSM-5 zeolite as catalyst," *AIChE J.*, vol. 63, no. 2, pp. 669–679, Feb. 2017. https://doi.org/10.1002/aic.15369

[19] Guo, S., N. Yuan, G. Zhang and J. C. Yu, "Graphene modified iron sludge derived from homogeneous Fenton process as an efficient heterogeneous Fenton catalyst for degradation of organic pollutants," *Microporous Mesoporous Mater.*, vol. 238, pp. 62–68, 2017. https://doi.org/10.1016/j.micromeso.2016.02.033

[20] Pryor, W. A., D. H. Giamalva and D. F. Church, "Kinetics of ozonation. 2. Amino acids and model compounds in water and comparisons to rates in nonpolar solvents," *J. Am. Chem. Soc.*, vol. 106, no. 23, pp. 7094–7100, Nov. 1984.

[21] Muñoz, F., E. Mvula, S. E. Braslavsky and C. von Sonntag, "Singlet dioxygen formation in ozone reactions in aqueous solution," *J. Chem. Soc. Perkin Trans.*, vol. 2, no. 7, pp. 1109–1116, 2001.

[22] Avisar, D., Y. Lester and H. Mamane, "pH induced polychromatic UV treatment for the removal of a mixture of SMX, OTC and CIP from water," *J. Hazard. Mater.*, vol. 175, no. 1, pp. 1068–1074, 2010. https://doi.org/10.1016/j.jhazmat.2009.10.122

[23] Sanches, S., M. T. Barreto Crespo and V. J. Pereira, "Drinking water treatment of priority pesticides using low pressure UV photolysis and advanced oxidation processes," *Water Res.*, vol. 44, no. 6, pp. 1809–1818, 2010. https://doi.org/10.1016/j.watres.2009.12.001

[24] Kitajima, M., S. Hatanaka and S. Hayashi, "Mechanism of O_2-accelerated sonolysis of bisphenol A," *Ultrasonics*, vol. 44, pp. e371–e373, 2006. https://doi.org/10.1016/j.ultras.2006.05.062

[25] Selli, E., C. L. Bianchi, C. Pirola, G. Cappelletti and V. Ragaini, "Efficiency of 1,4-dichlorobenzene degradation in water under photolysis, photocatalysis on TiO_2 and sonolysis," *J. Hazard. Mater.*, vol. 153, no. 3, pp. 1136–1141, 2008. https://doi.org/10.1016/j.jhazmat.2007.09.071

[26] Lim, M., Y. Son and J. Khim, "The effects of hydrogen peroxide on the sonochemical degradation of phenol and bisphenol A," *Ultrason. Sonochem.*, vol. 21, no. 6, pp. 1976–1981, 2014. https://doi.org/10.1016/j.ultsonch.2014.03.021

[27] De Bel, E., J. Dewulf, B. De Witte, H. Van Langenhove and C. Janssen, "Influence of pH on the sonolysis of ciprofloxacin: Biodegradability, ecotoxicity and antibiotic activity of its degradation products," *Chemosphere*, vol. 77, no. 2, pp. 291–295, 2009. https://doi.org/10.1016/j.chemosphere.2009.07.033

[28] Sushma, M. Kumari and A. K. Saroha, "Performance of various catalysts on treatment of refractory pollutants in industrial wastewater by catalytic wet air oxidation: A review," *J. Environ. Manage.*, vol. 228, pp. 169–188, 2018. https://doi.org/10.1016/j.jenvman.2018.09.003

[29] Oturan, M. A. and J. J. Aaron, "Advanced oxidation processes in water/wastewater treatment: principles and applications: A review," *Crit. Rev. Environ. Sci. Technol.*, vol. 44, no. 23, pp. 2577–2641, Dec. 2014. https://doi.org/10.108 0/10643389.2013.829765

[30] Martínez-Huitle, C. A. and M. Panizza, "Electrochemical oxidation of organic pollutants for wastewater treatment," *Curr. Opin. Electrochem.*, vol. 11, pp. 62–71, 2018. https://doi.org/10.1016/j.coelec.2018.07.010

[31] Babuponnusami, A. and K. Muthukumar, "A review on Fenton and improvements to the Fenton process for wastewater treatment," *J. Environ. Chem. Eng.*, vol. 2, no. 1, pp. 557–572, 2014. https://doi.org/10.1016/j.jece.2013.10.011

[32] Catrinescu, C., D. Arsene, P. Apopei and C. Teodosiu, "Degradation of 4-chlorophenol from wastewater through heterogeneous Fenton and photo-Fenton process, catalyzed by Al – Fe PILC," *Appl. Clay Sci.*, vol. 58, pp. 96–101, 2012. https://doi.org/10.1016/j.clay.2012.01.019

[33] Nidheesh, P. V. and R. Gandhimathi, "Trends in electro-Fenton process for water and wastewater treatment: An overview," *Desalination*, vol. 299, pp. 1–15, 2012. https://doi.org/10.1016/j.desal.2012.05.011

[34] Cho, S. H., A. Jang, P. L. Bishop and S. H. Moon, "Kinetics determination of electrogenerated hydrogen peroxide (H_2O_2) using carbon fiber microelectrode in electroenzymatic degradation of phenolic compounds," *J. Hazard. Mater.*, vol. 175, no. 1, pp. 253–257, 2010. https://doi.org/10.1016/j.jhazmat.2009.09.157

[35] Vasudevan, S., "An efficient removal of phenol from water by peroxi-electrocoagulation processes," *J. Water Process Eng.*, vol. 2, pp. 53–57, 2014. https://doi.org/10.1016/j.jwpe.2014.05.002

[36] Shah, M. P. *Removal of emerging contaminants through microbial processes.* Springer. 2021.

[37] Shah, M. P. *Advanced oxidation processes for effluent treatment plants.* Elsevier. 2020.

[38] Shah, M. P. *Microbial bioremediation and biodegradation.* Springer. 2020.

[39] Brillas, E., R. Sauleda and J. Casado, "Degradation of 4-chlorophenol by anodic oxidation, electro-Fenton, photoelectro-Fenton, and peroxi-coagulation processes," *J. Electrochem. Soc.*, vol. 145, no. 3, pp. 759–765, 1998. http://dx.doi.org/10.1149/1.1838342

[40] Elghniji, K., O. Hentati, N. Mlaik, A. Mahfoudh and M. Ksibi, "Photocatalytic degradation of 4-chlorophenol under P-modified TiO_2/UV system: Kinetics, intermediates, phytotoxicity and acute toxicity," *J. Environ. Sci.*, vol. 24, no. 3, pp. 479–487, 2012. https://doi.org/10.1016/ S1001-0742(10)60659-6

[41] Zazo, J. A., J. A. Casas, A. F. Mohedano, M. A. Gilarranz and J. J. Rodríguez, "Chemical pathway and kinetics of phenol oxidation by Fenton's reagent," *Environ. Sci.*

Technol., vol. 39, no. 23, pp. 9295–9302, Dec. 2005. http://doi.org/10.1021/es050452h

[42] Jones, C. W., J. H. Clark and M. J. Braithwaite, *Applications of hydrogen peroxide and derivatives*. The Royal Society of Chemistry. 1999.

[43] Villegas, L. G. C., N. Mashhadi, M. Chen, D. Mukherjee, K. E. Taylor and N. Biswas, "A short review of techniques for phenol removal from wastewater," *Curr. Pollut. Reports*, vol. 2, no. 3, pp. 157–167, 2016. https://doi.org/10.1007/s40726-016-0035-3

17

Advanced Oxidation Processes for Remediation of Persistent Organic Pollutants

Sadia Noor, Ambreen Ashar, Muhammad Babar Taj and Zeeshan Ahmad Bhutta

CONTENTS

17.1 Introduction

The last decades have shown a reevaluation of the issue of environmental pollution, under all aspects, both at regional and at international levels. The progressive accumulation of more and more organic compounds in natural waters is mostly due to the development and extension of chemical technologies for organic synthesis and processing. Population explosion and expansion of urban areas has increased the adverse impacts on water resources, particularly in regions in which natural resources are still limited. Currently, water use and reuse has become a major concern. Population growth leads to significant increases in default volumes of wastewater, which makes it an urgent imperative to develop effective and affordable technologies for wastewater treatment (Baig, Hansmann, and Paolini 2008; Alaoui et al. 2013; Chunhong Nie and Wang 2017; Parsons 2004; Shah 2020b, 2021).

The common treatment of wastewater (coagulation and flocculation) consists of physicochemical processes using various chemical reagents (aluminum chloride or ferric chloride, polyelectrolytes, etc.) and generates large amounts of sludge. Increasing demands for water quality indicators and drastic change regulations on wastewater disposal require the emergence and development of processes more efficient and more effective (ion exchange, ultrafiltration, reverse osmosis and chemical precipitation, electrochemical technologies).

Each of these treatment methods has advantages and disadvantages. Water resources management exercises ever more pressing demands on wastewater treatment technologies to reduce industrial negative impact on natural water sources (Fujishima, Zhang, and Tryk 2008; Bazrafshan, Alipour, and Mahvi 2016; Bousselmi, Geissen, and Schroeder 2004; Hasanbeigi and Price 2015). Thus, new regulations and emission limits are imposed, and industrial activities are required to seek new methods and technologies capable of effectively removing heavy metal pollution loads and reducing wastewater volume, closing the water cycle, or reusing and recycling water waste (Fujishima, Zhang, and Tryk 2008; Chunhong Nie and Wang 2017; Shah 2021).

Advanced technologies for wastewater treatment are required to eliminate pollution and may also increase pollutant destruction or separation processes, such as advanced oxidation various media. These technologies can be applied successfully to remove pollutants that are partially removed by conventional methods, for example biodegradable organic compounds, suspended solids, colloidal substances, phosphorus and nitrogen compounds, heavy metals, dissolved compounds and microorganisms that thus enable recycling of residual water. Special attention is paid to electrochemical technologies, because of the following advantages: versatility, safety, selectivity, possibility of automation, environmentally friendly and low investment costs (Sanches, Crespo, and Pereira 2010; Parsons 2004; Pignatello, Oliveros, and MacKay 2006).

The technologies for treating wastewater containing organic compounds fall within one of the following categories:

1. *Nondestructive procedures* – based on physical processes of adsorption, removal, stripping and other related processes.
2. *Biological destructive procedures* – based on biological processes using active materials such as mud and clay.
3. *Oxidative destructive processes* – based on oxidative chemical processes. The most significant treatment is known as advanced oxidation process – AOP. The operating conditions may vary in terms of temperature and pressure and various oxidative agents, namely ozone, peroxides (H_2O_2) and even O_2, catalysts and/or ultraviolet radiation.

The increasing levels of recalcitrant organic pollutants in air and wastewater streams have made environmental laws and regulations more vigilant and stringent. The main causes of surface water and groundwater contamination are industrial effluents (even in small amounts), excessive use of pesticides, fertilizers (agrochemicals) and domestic waste landfills. Wastewater treatment is usually based on physical and biological processes. After elimination of particles in suspension, the usual process is biological treatment (natural decontamination), but unfortunately, some organic pollutants, classified as biorecalcitrant, are not biodegradable (Deshmukh et al. 2009; Poulopoulos 2019; Mottaleb 2015; Bourgin et al. 2017; Chandra et al. 2018; Shah 2020a). In this way, AOPs may become the most widely used water treatment technologies for organic pollutants not treatable by conventional techniques due to their high chemical stability and/or low biodegradability (Bourgin et al. 2017). AOPs are indicated for removal of organic contaminants such as halogenated hydrocarbons (trichloroethane, trichlorethylene), aromatics (benzene, toluene, and xylene), pentachlorophenol (PCP), nitrophenol, detergents and pesticides. These processes can also be applied to oxidation of inorganic contaminants such as cyanides, sulfides and nitrites. A general classification of AOPs is based on the source which provides the radicals (Teoh, Scott, and Amal 2012).

Electrochemical and photochemical technologies may offer an efficient means of controlling pollution. Their effectiveness is based on the generation of highly reactive and non-selective hydroxyl radicals, which are able to degrade many organic pollutants. Electrolysis, heterogeneous photocatalysis, and photo-assisted electrolysis may be regarded as AOPs and are used in the supplementary treatment of wastewaters. The efficiency of the electrochemical oxidation depends on the anode material and the operating conditions, for example current density or potential. In general, in most applications of photoelectrocatalysis in the degradation of organics, the applied anodic bias potential is lower than the oxidation potential of organics on the electrode, which, due to direct electrooxidation, does not complicate the photocatalytic mechanism (Wang et al. 2014).

The efficiency of photoelectrochemical degradation for organic pollutants depends not only on the selection of a suitable supporting electrolyte and pH values, but also on the electrode potential and preparation conditions of the semiconductors involved. In a photoelectrochemical system, photoelectrons and photoholes can be separated under the influence of an applied electric field. The problem of the separation of semiconductor particles from the treated solution, so persistent in heterogeneous photolysis, is not an issue in photoelectrochemical systems. Numerous semiconductors can be used as photoelectrocatalytic materials, such as TiO_2, WO_3, SnO_2, ZnO, CdS, diamond and others.

17.2 Advanced Oxidation Technologies (AOTs)

Chemical oxidation processes comprise reaction mechanisms that generally change the chemical properties and structures of the toxic organic substances. During the oxidation process, molecules decompose into smaller fragments with a higher content of oxygen. The oxidized forms of molecules may be alcohols, aldehydes or carboxylic acids. Oxidation of organic compounds with oxidants like ozone or OH• radicals produce oxidized compounds, which are more easily biodegradable than are other compounds. This is the main idea that led to the combination of a chemical oxidation processes. Advanced oxidation technologies (AOTs), including AOPs and many other methods in use, are shown in Figure 17.1.

17.2.1 Advanced Oxidation Processes

AOPs are widely used for the removal of recalcitrant organic constituents from industrial and municipal wastewater (Wang et al. 2014). In this sense, AOP-type procedures are very promising technologies for treating wastewater containing nonbiodegradable or hardly biodegradable organic compounds with high toxicity. These procedures are based on generating highly oxidative HO radicals in the reaction medium.

The preferential use of H_2O_2 as oxidative agent and HO radical generator is justified by the fact that the hydrogen peroxide is easy to store, transport and use, and the procedure is safe and efficient. The technologies developed so far indicate the use of zeolites, active coal, structured clay, silica textures, Nafion membranes or Fe under the form of goethite (α-FeOOH) as support materials for the catalytic component (Glaze, Kang, and Chapin 1987; Venkatadri and Peters 1993).

The AOPs can be successfully used in wastewater treatment to degrade the persistent organic pollutants, the oxidation process being determined by the very high oxidative potential of HO• radicals generated in the reaction medium by different mechanisms. AOPs can be applied to fully or partially oxidize pollutants, usually using a combination of oxidants. Photochemical and photocatalytic AOPs, including UV/H_2O_2, UV/O_3, $UV/H_2O_2/O_3$, $UV/H_2O_2/Fe^{2+}(Fe^{3+})$, UV/TiO_2 and $UV/H_2O_2/TiO_2$, can be used for oxidative degradation of organic contaminants. A complete mineralization of the organic pollutants is not necessary, as it is more worthwhile to transform them into biodegradable aliphatic carboxylic acids followed by a biological process (Wang et al. 2017; Fujishima, Zhang, and Tryk 2008; Poulopoulos 2019; Shinde et al. 2017; Hassan et al. 2017).

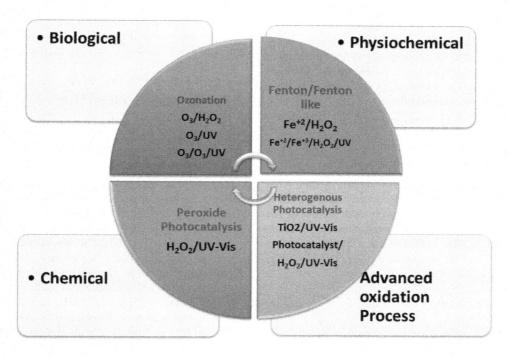

FIGURE 17.1 Classification of oxidation processes.

The idea of advanced oxidation processes was presented by Glaze in 1987 (Glaze, Kang, and Chapin 1987). AOPs can be defined as a set of chemical and photochemical reactions using reactive oxygen species (ROS), especially hydroxyl radicals (HO$^\bullet$), produced *in situ* for the oxidation of organic contaminants in water at ambient temperature that cannot be oxidized by conventional oxidants like chlorine (Poulopoulos 2019; Glaze, Kang, and Chapin 1987). These highly reactive species (OH$^\bullet$) react with the majority of organic compounds with rate constants of 10^{-6} to 10^{-9} M^{-1}s^{-1}. Different reactive species that are used in different process of AOPs are given in Table 17.1.

The overall reactions involved in photocatalytic mineralization of organic pollutants are concisely represented by the following equation (Chong et al. 2010; Wang et al. 2017).

Persistent organic pollutants + AOPs → CO_2 + H_2O + mineral acids

TABLE 17.1

Several Reactive Oxygen Species used in AOPs

AOPs	Reactive Species
Fenton process	HO$_2$$^\bullet$ and OH$^\bullet$
O$_3$/H$_2$O$_2$	O$_3$$^{\cdot\bullet}$, OH$^\bullet$ and O$_2$$^{\cdot\bullet}$
Ozone (O$_3$) treatment	HO$_2$$^\bullet$, OH$^\bullet$, O$_3$$^{\cdot\bullet}$, O$_2$$^{\cdot\bullet}$
Photo-Fenton process	OH$^\bullet$
UV/O$_3$ and UV/H$_2$O$_2$	OH$^\bullet$, HO$_2$$^{\bullet-}$, O$_2$$^{\cdot\bullet}$ and O$_3$$^{\cdot\bullet}$
Photocatalytic treatment: UV/Vis light by TiO$_2$, ZnO (catalysts)	OH$^\bullet$, O$_2$, HO$_2$, HOO$^\bullet$
Ultrasonic treatment	OH$^\bullet$ and H$^\bullet$
Gamma radiolysis	OH$^\bullet$ and H$^\bullet$

can be carried out in both homogeneous and heterogeneous phases (Abu Hatab, Oran, and Sepaniak 2008; Yue et al. 2012).

17.2.1.1 Photolysis (Vacuum UV)

The catalyst capable of catalyzing the reaction at ambient temperature and pressure upon irradiating it with photons of suitable energy is called a photocatalyst (Venkatadri and Peters 1993; Malato et al. 2003). Hence, photocatalysis is a process where light and catalysts are concurrently used to trigger and speed up a chemical reaction. So, photocatalysis can be defined as catalysis which is driven by the acceleration of a light to carry out light-induced reactions (Fujishima, Zhang, and Tryk 2008). Photocatalysis is a comparatively new science that has attracted the attention of environmentalists and that involves "green" chemistry. These photocatalytic reactions

17.2.1.2 Photocatalysis

A wide variety of AOPs are available for wastewater treatment based on photocatalytic systems such as TiO$_2$/UV and Fe^{+2}/UV-Vis or reactions that produce hydroxyl radicals by the action of radiation on metals or metal oxides (Védrine 2017; Getsoian, Zhai, and Bell 2014). Photocatalytic reactions can be divided into heterogeneous and homogeneous processes that use catalyst particles either in immobilized form or in suspension form (soluble catalyst). The light energy necessary can be provided by either lamps or solar radiation. Photocatalysis processes have been found to be effective for the degradation of organic pollutants (Sherrard et al. 1996).

17.2.1.2.1 Homogeneous Photocatalysis

Homogeneous catalysis involves a series of reactions being catalyzed by a catalyst in the same phase as that of the reactants. Homogeneous catalysts are supposed to be dissolved along with the reactants in a solvent, mostly water. The most common homogeneous photocatalytic reactions are Fenton and photo-Fenton (Zhou et al. 2012; Fujishima, Zhang, and Tryk 2008).

17.2.1.2.2 Fenton and Photo-Fenton AOP

The degradation of organic pollutants increases by improving the efficiency of oxidation of hydrogen peroxide to produce hydroxyl radicals. A catalyst can also be used to enhance the decomposition of H_2O_2. A suitable catalyst, which was discovered by Fenton in 1894, is ferrous ion. Fenton's reagent is a mixture of H_2O_2 and ferrous ion and it generates hydroxyl radicals according to the following reaction.

$$Fe^{2+} + H_2O_2 \rightarrow Fe^{3+} + OH^- + OH^{\cdot} \qquad (17.1)$$

The ferrous ion (Fe^{+2}) initiates and catalyzes the decomposition of H_2O_2, causing the production of hydroxyl radicals, and H_2O_2 can act as an $OH^{\cdot}$ scavenger and also as an initiator (Venkatadri and Peters 1993). The activity of the Fenton's reagent depends on the sample characteristics, iron concentration, pH, H_2O_2 dosage and reaction time. The Fenton reaction is mostly followed by the precipitation of $Fe(OH)_3$ due to the formation of Fe^{+3} during the reaction. Ferrous iron and hydrogen peroxide are used for the oxidation of organic substances. Fenton's oxidation process consists of five main stages, including neutralization, pH adjustment, oxidation reaction, precipitation and coagulation (Munawar et al. 2014; Bourgin et al. 2017; Sanches, Crespo, and Pereira 2010; Glaze, Kang, and Chapin 1987; Parsons 2004; Pignatello, Oliveros, and MacKay 2006).

The photo-Fenton process is the most recent process that makes the combined use of UV-Vis radiation and Fenton's reagent, generally increases the generation of hydroxyl radical by the photo reduction of Fe^{3+} to Fe^{2+} ions. In the photo-Fenton process, the production of the sludge waste generated in the original Fenton process is decreased to a great extent (Mendez-Arriaga, Esplugas, and Gimenez 2010).

The optimum pH range suggested for this process is 2.6–3 to achieve best performance. The homogeneous photocatalytic photo-Fenton ($Fe^{2+}/H_2O_2/UV$) method is among the most applied AOPs in the recent application of a wastewater treatment process. Photo-Fenton is an efficient and lower-cost process under solar irradiation as compared to UV light sources and its application at pilot scale have been already mentioned in the chapter for the removal of large range of organic pollutants in wastewaters (Malato et al. 2003).

The oxidation process is determined by the very high oxidative potential of the $HO^{\cdot}$ radicals generated in the reaction medium by different mechanisms. In the case of the AOP Fenton-type procedure (hydrogen peroxide and Fe^{2+} as catalyst) (Mendez-Arriaga, Esplugas, and Gimenez 2010; Venkatadri and Peters 1993; Pignatello, Oliveros, and MacKay 2006), the generation of hydroxyl radicals takes place through

a catalytic mechanism in which the iron ions play a very important role the main reactions involved, as presented in Equations 17.2–17.5.

$$Fe^{2+} + H_2O_2 \rightarrow Fe^{3+} + HO^- + HO \qquad (17.2)$$

$$Fe^{3+} + H_2O_2 \leftarrow\rightarrow H^+ + [Fe(OOH)]^{+2} \qquad (17.3)$$

$$[Fe(OOH)]^{+2} \rightarrow Fe^{+2} + HO_2 \qquad (17.4)$$

$$HO_2 + Fe^{+3} \rightarrow Fe^{+2} + H^+ + O_2 \qquad (17.5)$$

17.2.1.2.3 Heterogeneous Photocatalysis

A photo-induced reaction accelerated in the presence of a catalyst is termed photocatalysis. These reactions are based on absorption of a photon by a photocatalyst, usually a semiconductor having energy equal to or higher than its band-gap energy (E_{bg}) (Parsons 2004). This absorption causes a charge separation generating holes (h^+) in the valence band due to excitation of electrons (e^-) from the valence band of the semiconductor catalyst to the conduction band.

The photocatalyzed reaction can be carried out by preventing the recombination of the electron and the hole pair. The photogenerated electron-hole pairs get separated from each other and migrate to active sites at the surface of the semiconductor photocatalyst and then react with adsorbed species.

Heterogeneous photocatalysis has proven to be of real interest as an efficient tool for degrading both aquatic and atmospheric organic contaminants because this technique involves the acceleration of photoreaction in the presence of a semiconductor photocatalyst. Thus, these processes can be classified on the basis of the source of oxidation such as ozone-based AOP, H_2O_2, photocatalytic process, AOP "hot" technology based on ultrasound, electrochemical oxidation process and oxidation process with electron beam. These processes involve generation and subsequent reaction of hydroxyl radicals ($\cdot OH$), which is one of the most powerful oxidizing species. Photocatalytic reaction is initiated when a photoexcited electron is promoted from the filled valence band of a semiconductor photocatalyst to the empty conduction band as the absorbed photon energy, hole (h^+), equals or exceeds the band gap of the semiconductor photocatalyst leaving behind a hole in the valence band. Thus, in concert, an electron and a hole pair ($e^- - h+$) is generated. An ideal photocatalyst for photocatalytic oxidation is characterized by the following attributes: photostability, chemically and biologically inert nature, availability and low cost (Punzi, Mattiasson, and Jonstrup 2012; Baig, Hansmann, and Paolini 2008; Munawar et al. 2014; Bourgin et al. 2017; Sanches, Crespo, and Pereira 2010; Venkatadri and Peters 1993; Getsoian, Zhai, and Bell 2014; Parsons 2004; Pignatello, Oliveros, and MacKay 2006).

Many metal oxide and sulfide-based semiconductors, such as TiO_2, ZnO, ZrO_2, CdS, MoS_2, Fe_2O_3 and WO_3, have been explored as photocatalysts for the degradation of organic contaminants (Fujishima, Zhang, and Tryk 2008; Chunhong Nie and Wang 2017; Venkatadri and Peters 1993; Malato et al. 2003). TiO_2 is the most preferred one due to its chemical and biological inertness, high photocatalytic activity, photodurability, mechanical robustness and cheapness. Thus, these materials were used in the degradation of phenol, 1,4-dichlorobenzene, methanol, azo dye, trichloromethane, hexachloro cyclohex-ane, trichloroethylene and dichloropropionic acid. To avoid the problem of filtration, many methods were proposed to immo-bilize the photocatalysts, but in these conditions the photocata-lyst is expected to be used for a relatively long time, especially for industrial applications. Various substrates have been used as a catalyst support for the photocatalytic degradation of pol-luted water. The glass materials – glass mesh, glass fabric, glass wool, glass beads and glass reactors – were very commonly used as a support for titania. Other uncommon materials such as microporous cellulosic membranes, alumina clays, ceramic membranes, monoliths, zeolites and even stainless steel were also experimented with as a support for TiO_2.

17.2.1.3 Basic Principles of Heterogeneous Photocatalysis

Heterogeneous photocatalysis reaction is preferred to homoge-neous photocatalysis reaction because the catalyst is in a dif-ferent phase than the reactant and hence more easily separated. Heterogeneous photocatalysis generally involves the process of photosensitization, in which a photochemical reaction occurs in a chemical species due to the absorption of photonic energy by another species called a photo sensitizer. The *in situ* production of OH˙ radicals can cause the oxidation of organic reactant only if they become adsorbed on the surface of the photocatalyst. The reaction that involves the photocatalytic mineralization of organic pollutants is given as follows (Teoh, Scott, and Amal 2012; Punzi, Mattiasson, and Jonstrup 2012):

Organic Contaminant → Intermediates → $CO_2 + H_2O$

In general, five steps are required for a heterogeneous reaction to occur.

1. Transfer the organic pollutants (A) from aqueous phase to photocatalyst surface.
2. These pollutants are adsorbed on the active site of photocatalyst surface.
3. Photocatalytic reaction (adsorbed molecule, i.e., pol-lutants breakdown) occurs on the surface of the pho-tocatalyst (A→B).
4. After breakdown, molecule (B) will be desorbed from the surface of the photocatalyst.
5. These molecules (B) are transferred from the inner surface to the outer (bulk) surface.

Different semiconductor nanomaterials such as ferric oxide (Fe_2O_3), magnesium oxide (MgO), titanium dioxide (TiO_2), copper oxide (CuO), calcium oxide (CaO), cadmium sulfide (CdS) and zinc oxide (ZnO) are used as the photocatalyst in heterogeneous photocatalysis under sunlight. From various semiconductor photocatalysts, ZnO nanoparticles are preferred to others because of their high photostability, absorption of a wide range of the radiation, chemical inertness, cost effective-ness, and unique physical and chemical properties such as opti-cal properties, electrical properties and high chemical stability.

17.2.2 Types of Advanced Oxidation Processes

17.2.2.1 Hydrogen Peroxide–Based Homogeneous Fenton Method

The photo-Fenton process ($UV/Fe^{3+}/H_2O_2$) was applied for treating a mixture of three cationic dyes. Effect of initial concentration of oxidants, that is, Fe^{3+} and H_2O_2 on the dyes' degradation, was investigated (Bouafia-Chergui et al. 2012). Synthetic textile wastewater was degraded with a homoge-neous and a heterogeneous photo-Fenton oxidation method and the results were compared. Synthetic wastewater was prepared by mixing remazol red RR and remazol blue RB. Complete decolorization was obtained with homogeneous oxidation, which showed a higher efficiency than the hetero-geneous photo-Fenton method did (Punzi, Mattiasson, and Jonstrup 2012).

17.2.2.2 AOPs Based on Ozone

17.2.2.2.1 O_3 (Direct Ozone Feeding)

Ozone is a strong oxidant, with an oxidation potential of 2.08 V, and can be used for water and wastewater treat-ment. Dissolved ozone reacts with many organic com-pounds and gives two different mechanisms, including direct oxidation as molecular ozone and an indirect mech-anism through generation of hydroxyl radicals (secondary oxidants) during O_3 decomposition (Gottschalk, Libra, and Saupe 2009). The rate of the attack by ˙OH radical is faster than molecular ozone. Oxidation with ozone or hydrogen peroxide is considered a safe alternate to chlorination, as the oxidation does not produces toxic chlorinated organic compounds.

Ozone can be used as a strong disinfectant and is an efficient oxidant to remove odor, color and toxic synthetic organic pol-lutants. The combination of O_3 with clays and activated car-bon can further enhance the production of OH• (Pignatello, Oliveros, and MacKay 2006).The disadvantage of the process is that it requires a complex setup. Ozone is an unstable oxi-dant with a short lifetime that can be used after its synthesis. Due to the high expenses of production, it is generally used on effluent that has been pre-treated by a biological or a physico-chemical process.

17.2.2.2.2 O_3 + UV (Photo-Ozone Feeding)

In many cases, the complete mineralization of organic con-taminants cannot be done by conventional ozonation. Thus, metabolites like alkylphenol ethoxylates (APEOs) produced can have a more toxic effect than their parent compounds do. UV radiation can complete the oxidation by enhancing the generation of OH• radicals to carry out the secondary oxida-tion reaction. UV photons are used to activate ozone mole-cules. The use of the O_3/UV process favors the formation of hydroxyl radicals (Baig, Hansmann, and Paolini 2008). Ozone frequently absorbs UV radiation at 254 nm wavelength (the extinction coefficient ε of O_3 at 254 nm = 3300 M^{-1} cm^{-1}) pro-ducing H_2O_2 as an intermediate, which then decomposes to OH• radical.

17.2.2.2.3 O_3 + catalysts (Catalytic Ozone Feeding)

Ozonation carrying out a reaction by a radical system makes the oxidation of refractory molecules possible, but they can be further oxidized faster through heterogeneous or homogeneous catalysts, for example metal oxides and metal ions like Al_2O_3, Fe_2O_3, MnO_2, Ru/CeO_2, TiO_2, Mn^{2+}, Fe^{2+} and Fe^{3+}. Using an ozone/catalyst system removal of total organic carbon (TOC) and chemical oxygen demand (COD) is faster than using simple ozone at high pH value in wastewater.

The modification of oxidative properties of ozone in the presence of various types of catalysts is considered to be one of the most effective AOPs for the removal of a wide variety of organic compounds. Catalytic ozonation can increase the efficiency and decrease the reaction time with respect to separate ozonation and catalytic treatments by affecting the reaction mechanisms. The application of catalysts combined with ozone results in a noticeable decrease in ozone consumption by decreasing the reaction time and increasing the efficiency of ozonation.

17.2.2.3 AOPs Based on Combined Oxidation of H_2O_2 + O_3 (Peroxone)

The combined effect of both ozone and hydrogen peroxide speed up the decomposition of ozone and increase the generation of the hydroxyl radical in wastewater. It was examined that at acidic pH, H_2O_2 reacts very slowly with O_3, while pH values above 5 increase the rate of O_3 decomposition with H_2O_2 (Zhang et al. 2008).

17.2.2.3.1 TiO2 + UV (Photocatalysis)

TiO_2 is a semiconductor most commonly used as a photocatalyst for environmental purification, showing considerable potential in this process. This photocatalyst is relatively inexpensive and is commercially available under more than one trademark. Degussa P-25, BDH, Sachtleben Hombikat UV100 and Millennium TiONA PC50 are examples of commercial TiO_2 materials commonly used for scientific investigation. TiO_2 is chemically inert and stable and has relatively low toxicity. This semiconductor mainly occurs in nature in three forms called anatase, rutile and brookite. Anatase seems to be the most photoactive form, while rutile has the most stable structure, such that other forms convert to rutile when heated or processed at temperatures higher than 600 °C.

The fundamental mechanisms of TiO_2 photocatalysis is based on the photoexcitation of the TiO_2 surface by photons (hv) with an energy level providing the band gap energy of TiO_2, which causes the generation of holes and electrons on the catalyst surface. After this photogeneration, hole-electron recombination is assumed to take place immediately, releasing a certain amount of energy, even though the existence of adsorbed electrophilic and nucleophilic molecules on the surface promotes oxidative-reductive reactions between electrons and electrophilic substances on one side and holes and nucleophilic elements on the other side.

AOPs based on the photoactivity of semiconductor-type materials can be successfully used in wastewater treatment for destroying the persistent organic pollutants that are resistant to biological degradation processes. TiO_2 is the most attractive semiconductor because of its higher photocatalytic activity and can be used suspended into the reaction medium (slurry reactors) or immobilized as a film on solid material. A very promising method for solving problems concerning the photocatalyst separation from the reaction medium is to use the photocatalytic reactors in which TiO_2 is immobilized on support. The immobilization of TiO_2 onto various supporting materials has largely been carried out via physical or chemical route.

The application of photocatalysis in water and wastewater treatment has been well established, particularly in the degradation of organic compounds into simple mineral acids, carbon dioxide and water. Titanium dioxide (TiO_2), particularly in its anatase form, is a photocatalyst under UV light. A reactor refers to TiO_2 powder that is suspended in the water to be treated, while the immobilized catalyst reactor has TiO_2 powder attached to a substrate that is immersed in the effluents to be treated. Immobilized TiO_2 has become more popular due to the complications in the TiO_2 suspension systems.

The mechanism of photocatalysis is based on the photoexcitation of electron (e^-) present in the valence band (VB) of semiconductor (SC) photocatalyst to the conduction band (CB) upon irradiating with light (UV, visible or solar radiation) leaving hole (h^+) VB in Equation 17.6. Presence of such catalysts in an aqueous media directs h^+ to be trapped by water molecules to generate hydroxyl radicals (˙OH), a powerful ($E^° = +3.02$ eV) and indiscriminate oxidizing agent (Equations 17.7 and 17.8). There is a probability that photogenerated e^- either degrade organic molecules directly (Equation 17.13) or react with oxidants such as O_2, adsorbed on the surface of the catalyst or dissolved in water, reducing it to $O_2^{-·}$ superoxide radical anion (Equation 17.9) (Deng et al. 2013). On the other hand, the photogenerated h^+ are also capable of oxidizing the organic molecule R to $R^{·+}$ (Equation 17.12). Along with other highly reactive oxidizing species, hydroperoxyl free radicals (˙HO_2) are also produced and are converted to (OH˙) on reacting with H^+ (Equation 17.10). The (OH˙) can reduce or oxidize the organic molecules to mineralize them ultimately into water and CO_2 (Equation 17.11) and is shown in Figure 17.2 (Anjum et al. 2016).

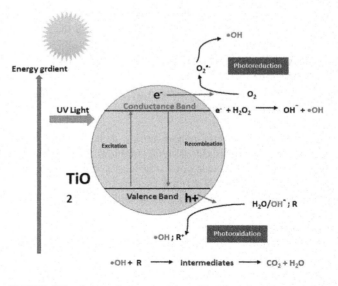

FIGURE 17.2 Schematic representation of the photocatalytic process.

$$TiO_2 + hv \rightarrow SC \ (e^-_{CB} + h^+_{VB}) \tag{17.6}$$

$$TiO_2 \ (h+_{VB}) + H_2O \rightarrow SC + H^+ + OH^{\bullet} \tag{17.7}$$

$$TiO_2 \ (h+_{VB}) + OH^- \rightarrow SC + OH^{\bullet} \tag{17.8}$$

$$TiO_2 \ (e^-_{CB}) + O_2 \rightarrow SC + O_2^{-}\bullet \tag{17.9}$$

$$O_2^{-\bullet} + H^+ \rightarrow HO_2^{\bullet} \tag{17.10}$$

$$R + OH^{\bullet} \rightarrow \text{degradation products} \tag{17.11}$$

$$R + h^+_{VB} \rightarrow \text{oxidation products} \tag{17.12}$$

$$R + e^-_{CB} \rightarrow \text{reduction products} \tag{17.13}$$

17.2.3 AOPs in Wastewater Treatment

AOPs are based on the action of highly reactive and unselective species hydroxyl radicals, which are able to promote oxidation and mineralization of organic matter at high reaction rates. Although the treatment of wastewater is the most common area for research and development, AOPs have also found several other applications such as groundwater treatment, soil remediation, ultrapure water production and treatment of organic volatile compounds. These are among the top most used industrial methods to remove pollutants from various effluents. AOPs operate in the presence of different strong oxidizing compounds through a physicochemical approach for treating industrial effluents (Erkurt 2010).

There are four main types of oxidation processes, including the following:

i. Chemical oxidation, which is based on peroxide photocatalysis
ii. Biological oxidation based on ozonation
iii. Physicochemical oxidation during Fenton or Fenton-like processes
iv. AOP in the presence of solar irradiated nanophotocatalysts, especially metal oxides. This is the most important and widely implemented process.

Combination of ozone with hydrogen peroxide has opened new vistas in the remediation, purification and removal of persistent pollutants from effluents due to effective oxidization of both organic and inorganic materials. Many compounds are oxidized with great difficulty by ozonation due to the presence of certain saturated ring systems. This issue was resolved by combining the oxidation potential of ozone with homogeneous/heterogeneous catalyst systems (metal oxides: MOs) to further enhance the oxidation reaction for complete mineralization of pollutants and contaminants (Bourgin et al. 2017).

Photochemical AOPs make use of solar radiation sources, including UV (B, C) and visible, either independently or in synergy with nanomaterials and chemicals to degrade pollutants (Bousselmi, Geissen, and Schroeder 2004; Teoh, Scott, and Amal 2012; Chunhong Nie and Wang 2017; Malato et al. 2003). The solar irradiation for remediation is generally termed photolysis and comes into action at the tertiary step of removing microbes and degrading organic compounds (Sanches, Crespo, and Pereira 2010). This process was advanced further when solar light was combined with the revolutionized oxidation technology. This technique has provoked researchers to

extend its applications. The ultraviolet region of solar light on encountering O_3 generates free radicals in aqueous medium (Hassan et al. 2017).

These advanced AOP-based technologies are major role players that ensure complete degradation of persistent pollutants from water in a mineralization process. AOPs have some very handy advantages to offer in such treatments, including high degradation efficiency and rates, ambient conditions for operation, reduction of the hazards and toxicity of recalcitrant organic compounds and complete mineralization of recalcitrant into "green" end products (Chunhong Nie and Wang 2017).

The appropriate and economic treatment of drinking water and wastewater can surmount the prevailing issue, but the traditional techniques of water treatment have not proved to be efficient enough to remove these contaminants completely, according to strict water quality standards (Qu, Alvarez, and Li 2013). Conventional methods of water treatment such as sedimentation, flocculation, filtration, chemical methods and membrane technologies involve high operating costs. In most of these treatment methods, toxic secondary pollutants are generated in the ecosystem (Martínez-Quiroz et al. 2017). The textile industry generates large quantities of colored wastewater, regarded as highly toxic, and its treatment by conventional physical, chemical and microbiological methods is difficult. In fact, existing wastewater treatment technologies suffer from several drawbacks, such as incomplete removal of dyes and other chemical residues in addition to the generation of toxic sludge (Anjum et al. 2016).

Dyes are unsaturated aromatic complex compounds with distinct characteristics like color, intensity, solubility, fastness and substantial applications. Synthetic dyes are one of such pollutants being used to impart color to diverse materials such as paper, textile, leather and food in one way or the other (Muhd Julkapli, Bagheri, and Bee Abd Hamid 2014). During the process of dyeing, up to 50% of the dyes applied are not fixed to substrates and are subsequently washed out as excessive volumes of colored effluents. In fact, the discharge of synthetic dyes in wastewater is undesirable for multiple reasons. Azo dyes are considered recalcitrant pollutants owing to their xenobiotic nature, as they accumulate in ecosystems, transfer along food chains and cause harmful effects on human health (Pandey, Singh, and Iyengar 2007; Engel et al. 2008). Effluents containing azo dyes are intensely colored, which not only affects the receiving water bodies aesthetically but also reduces the solubility of oxygen in water and hinders the penetration of sunlight.

When these effluents are mixed with clean water of natural resources, they cause retardation in the rate of photosynthesis and disruption in ecosystems. Hence, aesthetic perspectives and biodiversity issues are highly disturbed by colored water streams. Moreover, most of the synthetic dyes are rigidly recalcitrant in the natural environment. Since they are not naturally degraded by microbes in the environment, they are considered nonbiodegradable. Furthermore, a majority of synthetic dyes are not only thermostable but are also resistant to photolysis under sunlight; consequently, they usually tend to accumulate in the environment, imparting harmful effects on the whole biosphere. Although adsorption and coagulation methods have been applied to treat acid dyes, this has always resulted in the generation of secondary pollutants (Wang et al. 2013). Furthermore, owing to the presence of $-COOH$, $-SO_3^-$, $-OH$ and hydrophilic groups on acid dyes, they exhibit excellent solubility in the water stream (Han et al. 2009).

Due to the environmental hazards of synthetic dye wastewater discharge, it is crucial for industries to treat wastewater to acceptable levels fixed by environmental agencies before being released to the environment. Previous studies have shown that if discharge is untreated, recalcitrant azo dyes gradually undergo some partial biotransformations. In addition to dyes themselves, their biotransforms may also prove to be toxic and carcinogenic in nature (Mansour et al. 2009). Benzidine moiety of azo dyes and their other biotransformed products possess a genotoxic effect, which can often be attributed to planar structure enhancing the ability to intercalate in between DNA double helices (Mansour et al. 2009).

Effluent treatment cannot be considered an easy task owing to its huge volume. For example in Pakistan, it is estimated that approximately 150 million tons of dye-contaminated wastewater is produced annually (Ozkan-Yucel and Gokcay 2014). Therefore, the proper treatment of effluent containing synthetic dyes in addition to other pollutants from industries such as tanneries, textile, food, paper, pharmaceuticals and cosmetics is important environmentally and legally. Recently, an increase in environmental awareness has led to stringent legislation pertaining to untreated industrial effluent release into the natural environment (Shinde et al. 2017).

17.2.4 Future Perspective

Advanced oxidation technologies are going be the dominating technologies to treat a variety of hazardous waste materials and industrial effluents. The use of freely available solar energy can be harnesses for solar photocatalysis to get maximum benefit from this amazing oxidation technology. The advanced oxidation process leads to the decomposition and mineralization of many groups of organic materials in both liquid and gas phases. AOPs have attracted significant attention in recent years. Depending on the chemical structure of the pollutant molecules, AOPs mineralize numerous pollutants into ultimately harmless substances like CO_2 and H_2O and therefore avoid the issue of pollution shifting. AOPs of Fenton and photo-Fenton type can be used for antibiotic degradation from wastewater or for increasing their biodegradability in biological wastewater treatment. The particular importance of these technologies appears to be in destroying biologically nondegradable chemical structures as well as ozone-resistant substances such as organic pesticides and herbicides, aromatic structures, organo-halogens and petroleum constituents in wastewaters and other sources.

17.3 CONCLUSION

This chapter covers the advanced oxidation processes for the treatment of recalcitrant organic pollutants and methods to remove them from wastewaters. During the past few decades, the amount of organic pollutants has increased manifold, so water reclamation and recycling must be implemented rapidly worldwide to offset anthropogenic effects and avoid water scarcity. Climate changes and poor management of water resources have aggravated this problem and caused an imbalance in the availability and consumption of water resources. One of the best and most attractive solutions is to avoid water pollution in the reclamation of wastewater to ensure sustainable development and management of water resources. The major concerns of water pollution arise from the involvement of persistent or recalcitrant organic pollutants, even in water treated by conventional technologies. Environmentalists are striving in search of nontoxic, cost-effective, environmentally friendly and multifaceted new materials to remediate pollutants from all ecosystems. In terms of irradiation sources, there are some other issues that need to be dealt with aptly, especially the generation of renewable energy. The sustainability of the environment needs to be ensured, and the production of innovatively effective nanomaterials and modifications already existing in materials should infuse more attributes in the forthcoming era.

"Advanced oxidation processes" are defined in general as physicochemical procedures that promote *in situ* generation of free hydroxyl radicals as highly oxidative reagents for the decomposition of pollutants in water or air. These oxidation processes basically use three different reagents – ozone, hydrogen peroxide and oxygen – in many combinations, either combined with each other or applied with UV irradiation and/or various kinds of catalysts homogeneously and heterogeneously. Due to the generation of increased amounts of OH radicals, the combination of two or more AOPs usually leads to higher oxidation rates. With promising results observed on the laboratory scale, compared with conventional water and wastewater treatment methods, these technologies will likely be more essential for real applications in the near future. However, the reactivity of hydroxyl radicals with radical scavengers (carbonate, phosphate, etc.), which exist in real wastewaters, is the main disadvantage of all oxidative degradation processes based on hydroxyl radical reactions.

REFERENCES

Abu Hatab, Nahla A, Jenny M Oran, and Michael J Sepaniak. 2008. Surface-enhanced Raman spectroscopy substrates created via electron beam lithography and nanotransfer printing. *ACS Nano* 2 (2):377–385.

Alaoui, Mdaghri S, J Ghanam, M Merzouki, MJ Penninckx, and M Benlemlih. 2013. Immobilisation of Pycnoporus coccineus laccase in ca alginate beads for use in the degradation of aromatic compounds present in olive oil mill wastewaters. *Journal of Biotechnology Letters* 4 (2):91.

Anjum, Muzammil, R Miandad, Muhammad Waqas, F Gehany, and MA Barakat. 2019. Remediation of wastewater using various nano-materials. *Arabian Journal of Chemistry* 12 (8): 4897–4919.

Baig, S, G Hansmann, and B Paolini. 2008. Ozone oxidation of oestrogenic active substances in wastewater and drinking water. *Water Science and Technology* 58 (2):451–458.

Bazrafshan, Edris, Mohammad Reza Alipour, and Amir Hossein Mahvi. 2016. Textile wastewater treatment by application of combined chemical coagulation, electrocoagulation, and adsorption processes. *Desalination and Water Treatment* 57 (20):9203–9215.

Bouafia-Chergui, Souâd, Nihal Oturan, Hussein Khalaf, and Mehmet A. Oturan. 2012. A photo-Fenton treatment of a mixture of three cationic dyes. *Procedia Engineering* 33:181–187.

Bourgin, Marc, Ewa Borowska, Jakob Helbing, et al. 2017. Effect of operational and water quality parameters on conventional ozonation and the advanced oxidation process O_3/H_2O_2: Kinetics of micropollutant abatement, transformation product and bromate formation in a surface water. *Water Research* 122:234–245.

Bousselmi, L, S-U Geissen, and H Schroeder. 2004. Textile wastewater treatment and reuse by solar catalysis: Results from a pilot plant in Tunisia. *Water Science and Technology* 49 (4):331–337.

Chandra, Ram, Pooja Sharma, Sangeeta Yadav, and Sonam Tripathi. 2018. Biodegradation of endocrine-disrupting chemicals and residual organic pollutants of pulp and paper mill effluent by biostimulation. *Frontiers in Microbiology* 9:960.

Chong, Meng Nan, Bo Jin, Christopher WK Chow, and Chris Saint. 2010. Recent developments in photocatalytic water treatment technology: A review. *Water Research* 44 (10):2997–3027.

Chunhong Nie, Jing Dong, Pingping Sun, Chao Yan, Hongjun Wu, and Baohui Wang. 2017. An efficient strategy for full mineralization of an azo dye in wastewater: A synergistic combination of solar thermo- and electrochemistry plus photocatalysis. *RSC Advances* 7:36246.

Deng, Shujun, Hongjie Xu, Xuesong Jiang, and Jie Yin. 2013. Poly (vinyl alcohol)(PVA)-enhanced hybrid hydrogels of hyperbranched poly (ether amine)(hPEA) for selective adsorption and separation of dyes. *Macromolecules* 46 (6):2399–2406.

Deshmukh, NS, KL Lapsiya, DV Savant, et al. 2009. Upflow anaerobic filter for the degradation of adsorbable organic halides (AOX) from bleach composite wastewater of pulp and paper industry. *Chemosphere* 75 (9):1179–1185.

Engel, Eva, Heidi Ulrich, Rudolf Vasold, et al. 2008. Azo pigments and a basal cell carcinoma at the thumb. *Dermatology* 216 (1):76–80.

Erkurt, Hatice Atacag. 2010. *Biodegradation of azo dyes*. Springer.

Fujishima, Akira, Xintong Zhang, and Donald A. Tryk. 2008. TiO_2 photocatalysis and related surface phenomena. *Surface Science Reports* 63 (12):515–582.

Getsoian, Andrew "Bean", Zheng Zhai, and Alexis T Bell. 2014. Band-gap energy as a descriptor of catalytic activity for propene oxidation over mixed metal oxide catalysts. *Journal of the American Chemical Society* 136 (39):13684–13697.

Glaze, William H, Joon-Wun Kang, and Douglas H Chapin. 1987. The chemistry of water treatment processes involving ozone, hydrogen peroxide and ultraviolet radiation. *Ozone: Science and Engineering* 9(4).

Gottschalk, Christiane, Judy Ann Libra, and Adrian Saupe. 2009. *Ozonation of water and waste water: A practical guide to understanding ozone and its applications*. John Wiley & Sons.

Han, Fang, Venkata Subba Rao Kambala, Madapusi Srinivasan, Dharmarajan Rajarathnam, and Ravi Naidu. 2009. Tailored titanium dioxide photocatalysts for the degradation of organic dyes in wastewater treatment: A review. *Applied Catalysis A: General* 359 (1–2):25–40.

Hasanbeigi, Ali, and Lynn Price. 2015. A technical review of emerging technologies for energy and water efficiency and pollution reduction in the textile industry. *Journal of Cleaner Production* 95:30–44.

Hassan, Muhammad, Xiaoyuan Wang, Fei Wang, Dong Wu, Asif Hussain, and Bing Xie. 2017. Coupling ARB-based biological and photochemical (UV/TiO_2 and UV/$S_2O_8^{2-}$) techniques to deal with sanitary landfill leachate. *Waste Management* 63:292–298.

Malato, Sixto, Julián Blanco, Alfonso Vidal, et al. 2003. Applied studies in solar photocatalytic detoxification: An overview. *Solar Energy* 75 (4):329–336.

Mansour, Hedi Ben, Ridha Mosrati, David Corroler, Kamel Ghedira, Daniel Barillier, and Leila Chekir. 2009. In vitro mutagenicity of acid violet 7 and its degradation products by Pseudomonas putida mt-2: Correlation with chemical structures. *Environmental Toxicology and Pharmacology* 27 (2):231–236.

Martínez-Quiroz, Marisela, Eduardo A López-Maldonado, Adrián Ochoa-Terán, Mercedes T Oropeza-Guzman, Georgina E Pina-Luis, and José Zeferino-Ramírez. 2017. Innovative uses of carbamoyl benzoic acids in coagulation-flocculation's processes of wastewater. *Chemical Engineering Journal* 307:981–988.

Mendez-Arriaga, Fabiola, Santiago Esplugas, and Jaime Gimenez. 2010. Degradation of the emerging contaminant ibuprofen in water by photo-Fenton. *Water Research* 44 (2):589–595.

Mottaleb, MA, MJ Meziani, MA Matin, MM Arafat, and MA Wahab. 2015. Emerging micro-pollutants pharmaceuticals and personal care products (PPCPs) contamination concerns in aquatic organisms-LC/MS and GC/MS analysis. In *Emerging micro-pollutants in the environment: Occurrence, fate, and distribution*. American Chemical Society:43–74.

Muhd, Julkapli, Nurhidayatullaili, Samira Bagheri, and Sharifah Bee Abd Hamid. 2014. Recent advances in heterogeneous photocatalytic decolorization of synthetic dyes. *The Scientific World Journal* 90.

Munawar, Iqbal, Ijaz Ahmad Bhatti, Zia-ur-Rehman Muhammad, Haq Nawaz Bhatti, and Shahid Muhammad. 2014. Efficiency of advanced oxidation processes for detoxification of industrial effluents. *Asian Journal of Chemistry* 26 (14):4291–4296.

Ozkan-Yucel, Umay Gokce, and Celal Ferdi Gokcay. 2014. Effect of anaerobic azo dye reduction on continuous sludge digestion. *CLEAN – Soil, Air, Water* 42 (10):1457–1463.

Pandey, Anjali, Poonam Singh, and Leela Iyengar. 2007. Bacterial decolorization and degradation of azo dyes. *International Biodeterioration & Biodegradation* 59 (2):73–84.

Parsons, Simon. 2004. *Advanced oxidation processes for water and wastewater treatment*. IWA Publishing.

Pignatello, Joseph J, Esther Oliveros, and Allison MacKay. 2006. Advanced oxidation processes for organic contaminant destruction based on the Fenton reaction and related chemistry. *Critical Reviews in Environmental Science and Technology* 36 (1):1–84.

Poulopoulos, Stavros G, Azat Yerkinova, Gaukhar Ulykbanova, and Vassilis J Inglezakis. 2019. Photocatalytic treatment of organic pollutants in a synthetic wastewater using UV light and combinations of TiO_2, H_2O_2 and Fe(III). *PLoS One* 14 (5):e0216745.

Punzi, Marisa, Bo Mattiasson, and Maria Jonstrup. 2012. Treatment of synthetic textile wastewater by homogeneous and heterogeneous photo-Fenton oxidation. *Journal of Photochemistry and Photobiology A: Chemistry* 248:30–35.

Qu, Xiaolei, Pedro JJ Alvarez, and Qilin Li. 2013. Applications of nanotechnology in water and wastewater treatment. *Water Research* 47 (12):3931–3946.

Sanches, Sandra, Maria T Barreto Crespo, and Vanessa J Pereira. 2010. Drinking water treatment of priority pesticides using low pressure UV photolysis and advanced oxidation processes. *Water Research* 44 (6):1809–1818.

Shah, MP. 2020a. *Advanced oxidation processes for effluent treatment plants.* Elsevier.

Shah, MP. 2020b. *Microbial bioremediation and biodegradation.* Springer.

Shah, MP. 2021. *Removal of emerging contaminants through microbial processes.* Springer.

Sherrard, Kim B, Philip J Marriott, R Gary Amiet, Malcolm J McCormick, Ray Colton, and Keith Millington. 1996. Spectroscopic analysis of heterogeneous photocatalysis products of nonylphenol-and primary alcohol ethoxylate nonionic surfactants. *Chemosphere* 33 (10):1921–1940.

Shinde, Dnyaneshwar R, Popat S Tambade, Manohar G Chaskar, and Kisan M Gadave. 2017. Photocatalytic degradation of dyes in water by analytical reagent grades ZnO, TiO_2 and SnO_2: A comparative study. *Drinking Water Engineering and Science* 10 (2):109–117.

Teoh, Wey Yang, Jason A Scott, and Rose Amal. 2012. Progress in heterogeneous photocatalysis: From classical radical chemistry to engineering nanomaterials and solar reactors. *The Journal of Physical Chemistry Letters* 3 (5):629–639.

Védrine, Jacques C. 2017. Heterogeneous catalysis on metal oxides. *Catalysts* 7 (11):341.

Venkatadri, Rajagopalan, and Robert W Peters. 1993. Chemical oxidation technologies: Ultraviolet light/hydrogen peroxide, Fenton's reagent, and titanium dioxide-assisted photocatalysis. *Hazardous Waste and Hazardous Materials* 10 (2):107–149.

Wang, Hui, Lei Hu, Wei Du, et al. 2017. Two-photon active organotin (IV) carboxylate complexes for visualization of anticancer action. *ACS Biomaterials Science & Engineering* 3 (5):836–842.

Wang, Jian, Ling Kong, Wei Shen, Xiaoli Hu, Yizhong Shen, and Shaopu Liu. 2014. Synergistic fluorescence quenching of quinolone antibiotics by palladium (II) and sodium dodecyl benzene sulfonate and the analytical application. *Analytical Methods* 6 (12):4343–4352.

Wang, Xianbiao, Weiping Cai, Shengwen Liu, Guozhong Wang, Zhikun Wu, and Huijun Zhao. 2013. ZnO hollow microspheres with exposed porous nanosheets surface: Structurally enhanced adsorption towards heavy metal ions. *Colloids and Surfaces A: Physicochemical and Engineering Aspects* 422:199–205.

Yue, Weisheng, Zhihong Wang, Yang, et al. 2012. Electron-beam lithography of gold nanostructures for surface-enhanced Raman scattering. *Journal of Micromechanics and Microengineering* 22 (12):125007.

Zhang, Yaping, Xuan Zhou, Yixin Lin, and Xia Zhang. 2008. Ozonation of nonylphenol and octylphenol in water. *Fresenius Environmental Bulletin* 17 (6):760–766.

Zhou, Hailong, Yongquan Qu, Tahani Zeid, and Xiangfeng Duan. 2012. Towards highly efficient photocatalysts using semiconductor nanoarchitectures. *Energy & Environmental Science* 5 (5):6732–6743.

18 Treatment of Wastewater and Its Reuse

Ambreen Ashar, Noshin Afshan, Tanzila Aslam, Sadia Noor and Zeeshan Ahmad Bhutta

CONTENTS

18.1 Introduction

Water is a natural resource on this globe, and its accessibility in its pure form is critical to living beings, as life would be unimaginable without it. Due to its unique property of solubility strength, it is known as the universal solvent. It is a vital component for the existence of life, and its accessibility is critical for all organisms. Water contamination is caused by the addition of various microorganisms, gases and contaminants (chemicals or heavy metals) into the water through streaming water, rain and industrial effluents. Water with adversely affected quality through anthropogenic activities is referred to as wastewater. There are different sources of wastewater, ranging from municipal sewage to industrial effluents in liquid, gas and solid forms. Domestic wastewater is one of the most common waste resources and has a significantly negative impact on the environment.

Owing to the diversifying factors such as hazardous waste content, insufficient sewage treatment, marine disposal problems, industrial wastes and some agricultural perspectives, water pollution has become a major global concern, as it has harmful consequences to human health. Chemical, physical and biological contaminants pose a variety of health and environmental risks to humans and the ecosystem.[1–3]

Pure drinking water will become one of the scarcest resources late in this century because of the massive rise in the global population. Urbanization has made the simultaneous provision of freshwater resources and health services quite impossible. In this regard, the ties between infrastructure of municipal and water services need to be revived following strict regulations. It has been concluded that the rate of water

DOI: 10.1201/9781003165958-18

usage is three times higher than the growth rate of the world population. Though housing, medical care and social services provide basic infrastructure for human needs, the availability of clean water will be included in the list in the near future. Since the shortage of pure water and wastewater use has been posing the greatest threat, restricting economic and agricultural development, it is of great significance to reserve the water resources, recharge the groundwater resources, stop the discharge of industrial wastes into the water bodies and treat the wastewater just before its removal into the water bodies. Hence, wastewater reuse has undoubtedly been practiced, which compensated for several needs but also has a hazardous effects on human life and ecosystems. Hence, cleaning water through wastewater treatment facilities is a major challenge for engineers, planners and politicians. For this purpose, a series of remediation techniques such as advanced oxidation process and artificial intelligence have been applied, which will be discussed in detail in this chapter. However, the right treatment and proper sanitation of wastewater is essential to protect public health and the environment. Legislation regarding industrial wastewater is becoming more stringent, especially in developed countries, and requires wastewater to be treated before it is discharged into water reservoirs.

18.1.1 Standards of Wastewater

Effluent standards demonstrate pollutant concentrations in wastewater released by outfall pipes from publicly owned sewage treatment facilities or industrial plants and are expressed in parts per million. Each pollutant has its own set of standards, such as biochemical oxygen demand from organic matter or suspended solids, based on the technologies available at the time to minimize the particular pollutant in each sector, as determined by the Environmental Protection Agency (EPA). Three types of technologies have been proposed: the cost-based best feasible technology (BPT) supposed to be in place by 1977; highly restrictive best available technology (BAT) supposed to be in place by 1983; and the best traditional technology (BCT) for nontoxic water contaminants. The US National Pollutant Discharge Elimination System (NPDES) permit has been given to each point source of water contamination based on these criteria.[4]

18.1.2 Categories of Wastewater

Wastewater can be divided into five categories:

a. Domestic wastewater: Discharged from commercial and residential areas

TABLE 18.1

Characteristics of Wastewater[5]

Parameter	Raw Sewage	Treated Sewage
pH	6.50 to 8.00	7.0–8.0
TSS (mg/L)	200 to 250	<30.0
COD (mg/L)	30 to 450	<150
BOD at 27 °C in 3 days (mg/L)	150 to 300	<30.0
Grease and oil (mg/L)	20 to 50	<10.0

b. Industrial wastewater: Released from industries as effluent

c. Infiltration/inflow: Water enters the sewage system from cracks, underground drains, leaking joints, inlet walls, rooftops, foundation, storm drain connections or pitfalls

d. Rainwater/storm H_2O: Flow from floods or rain

e. Agricultural water: Wastewater from agriculture sector[6]

A variety of approaches have been utilized to purify wastewater, including ozonation, coagulation, precipitation, adsorption, extraction, filtration, ion exchange, reverse osmosis and advanced-oxidation processes.[7]

18.1.3 Sources of Pollutants in the Wastewater

A variety of hazardous and unsuitably treated wastewater products have been known to be toxic to plants, animals, humans and the ecosystem. The fundamental contamination in wastewater has been caused by nitrogen- and phosphorus-based supplements, microorganisms, heavy metals, hydrocarbons and endocrine disruptors derived from different sources such as domestic sewage, sanitary wastes, discharge from mines and industrial effluents. The natural organic matter and different kinds of toxins present in the wastewater make it a favorable place for natural life-forms like microbes, bacteria, protozoa, fungi and viruses. The presence of such species in contaminated water frequently causes several waterborne illnesses. For example, a high concentration of NH_3 in polluted water has put the life of marine species at high risk, and liquid nitrate has caused methemoglobinemia, or blue baby syndrome, in humans. Various endocrine disruptors, including estrone, 17-estradiol and testosterone, have caused reproduction failure in animals and humans. Heavy metals like Zn and Hg have caused protein degradation and, in turn, cancer.[8] Pollutants have been categorized as physical, chemical and biological and prove to be lethal for living beings and also the ecosystem.

18.1.3.1 Physical Pollutants

Physical pollutants refer to the contamination that affects the physical parameters of water, such as odor, color, taste and pH. It is injurious to health if the water used for drinking and domestic purposes is colored or smells unpleasant. The growth of living organisms is affected negatively when pH of water consumed is away from neutral value. Moreover, high acidity of soil or agricultural water inhibits the growth of plants. As a result, pH is a significant indicator of contamination in the atmosphere.

Metals, dye runoff, soil particles, radionuclides, electricity, plant debris and the phenomenon of water bloom all contribute to the physical pollutants of water bodies. Metals and microorganisms that create musty odors cause the offensive odor and taste of water. Metals such as Al, Cu, Fe and Mn contribute to pollutants that impart color. Radionuclide-based contamination of water is caused by elements such as radon, plutonium and radium. Mechanical pollution of water is caused by the ballast water and electricity coming out of ships and depleted of seawater. Bacteria and dead algae along with leached out heavy metals impart bad taste. Humic acid and humin are pollutants produced by microorganisms in soil transforming plant and animal remains into dark or blackish brown humus. Since humin is a color contamination agent that interacts with chlorine to form the carcinogen trihalomethane, it must be fully removed when a high concentration of humin is present in a water supply.

18.1.3.2 Chemical Pollutants

A range of chemicals such as fertilizers, pesticides, petroleum and related products, plastics and cosmetics used for agricultural, industrial and personal care purposes tend to pollute water if treated carelessly and disposed of in water bodies. For example, when a ship ruptures and spills oil, it seriously affects wildlife, even extending for miles. It kills countless fishes and binds to seabird feathers, wiping out ability to fly.

18.1.3.3 Biological Pollutants

Pharmaceutical products and pesticides have been used for humans and animals respectively to control and prevent diseases. Owing to their persistent effects on the marine ecosystem, such chemicals have been reported as emerging contaminants. Pharmaceutical products such as antiepileptics, antibiotics, anti-inflammatories, endocrine disruptors and analgesics, even at low concentration, penetrate the aquatic environment and spoil water quality, posing a major threat to the health of human beings and the ecosystem. For example, antibiotics not only prevent the growth of beneficial microbes obstructing their function in wastewater treatment plant operations but also render the microbes resistant under their persistent exposure. Therapeutic hormones such as synthetic estrogens, which are recommended for birth control pills and estrogen replacement therapy, and estrone, which influences the endocrine system of living beings, have been the most effective endocrine-disrupting chemicals in aquatic species. Analgesic drugs, including acetaminophen, diclofenac, ibuprofen, meprobamate and naproxen, are considered environmental contaminants. After administration, approximately 15% of ibuprofen and 26% of its metabolite have been excreted. It has been reported that ibuprofen's metabolites have been more toxic to marine species than the parent drug. Moreover, active metabolites like 10,11-dihydro-10,11-epoxy-carbamazepine, 4-hydoxydiclofenac and N4-acetylsulfamethoxazole accumulate in aquatic organisms and bind to their cellular protein through covalent bonding, thereby eliciting immune response or causing toxicity.[9]

18.1.4 Wastewater Treatment

The treatment of wastewater is the main issue for industry and society. The use and maintenance of wastewater treatment plants have become troublesome.[10] Many industries have been using hazardous and toxic chemicals without taking preventive measures. The industrial effluents negatively influence freshwater quality, making the water harmful for living beings to consume. Some treatment plants are unable to handle a bulk amount of industrial effluents, which deteriorate water quality and adversely affect the ecosystem.[11]

In the past few decades, several biological, chemical and physical treatment techniques have been practiced to purify wastewater such as oxidation, precipitation, flotation, evaporation, solvent extraction, ion exchange, adsorption, membrane filtration, phytoremediation and electrochemical biodegradation. In addition to advantages, every treatment has some limitations in terms of feasibility, cost, efficiency and environmental impact. Owing to the complex composition of industrial wastewater, there is currently no way to treat it properly. In practice, a suitable combination of compatible methods has been applied to obtain the required quality of water in the most economical way.[10]

Taking environmentally friendly treatment standards under consideration, several ways to treat wastewater have been formulated that include multiple technologies to get rid of toxicity. Once the appropriate treatment technology has been selected, a series of pilot-scale experiments must be conducted on real industrial wastewater. However, new directives and regulations are required to find the best treatment plan to reduce harmful compounds. The composition of industrial wastewater (IWW) is highly variable. The treatment techniques and disposal strategies for IWW widely differ from those utilized for municipal waste. According to the properties of IWW, the processing technology for IWW has been divided into the following categories.

18.1.4.1 Physical Techniques

Physical treatment technique involves isolating waste from the main stream either with little or no degradation of the waste such as coagulation and filtration, or the ingestion of organic waste by microorganisms. In this category, natural forces, such as van der Waals forces, gravity and electric gravity, and physical barriers, such as screens, fences, membranes, deep-bed filters, ion exchange and electrodialysis to complete the removal of matter, have been recommended for use. The physical techniques regarding wastewater treatment include adsorption, flotation, sedimentation screening, aeration and filtration.[12]

18.1.4.2 Chemical Techniques

These techniques have been practiced to remove chemicals by means of generating insoluble solids and gases, coagulating colloidal suspensions and removing biodegradable chemicals from nonbiodegradable, damaged or inactivated chelating agents. Along with the advanced oxidation process or biological technology to treat difficult-to-remove industrial wastewater, the removal rate has been greatly improved. In recent years, photocatalysis, ozonolysis and ultrasonic oxidation have been replaced by advanced oxidation processes with improved biodegradability and the enhanced ability to detoxify wastewater containing polar and hydrophilic chemical substances.[12]

18.1.4.3 Biological Techniques

In this category, carbon-containing organic materials have been applied for the removal of phosphorus, biochemical oxygen demand, nitrification, denitrification and stabilization. Aerobic biological processes are highly efficient, while anaerobic bacteria use the idea of recovery and utilization of the resources. Currently, highly aerobic and anaerobic bioreactors have been preferably used to degrade highly resistant IWW with the advantages of requiring minimal space, incredible COD removal efficiency and lower cost.[13]

Biological treatment techniques employed for the removal of organic debris from wastewater are generally considered affordable, eco-friendly and energy-saving procedures as compared with physicochemical methods. In this type of treatment technique, various species of microorganisms such as fungi and enzymes have been utilized to degrade dyes through reduction or oxidation schemes, for example oxidation of the dyes through metabolic reactions of bacteria and reduction of azo dyes by various microbes to form aminobenzenes. However, the biological effluent treatments being applied widely are slow and limited based on incomplete degradation of nonbiodegradable pollutants, with the appearance of some intermediates causing toxicity to microorganisms. Moreover, oxidation reaction of dyes carried out by bacteria has been found difficult due to the large size of dye molecules while passing through the cell membranes. The oxidation of azo compounds by fungal oxidases has been found more toxic than their parent dyes, and the fungal cultures require longer growth phases to produce high amounts of oxidase enzymes.[13] Hence, a highly efficient method that can decompose dye molecules is required.

18.1.4.4 Bioremediation Treatment Techniques

Phytoremediation is a treatment technique that employs the use of plants to eliminate or convert the contaminants into harmless by-products. To degrade, extract or immobilize pollutants from soil and water, the process employs a variety of plants. It has gained keen attention over the years, and it's generally thought of as a safe and inexpensive process. Since the method of remediation requires contact between the plant's root and the contaminant, either the plant must be able to expand its roots to the contaminant or the contaminated media must be moved inside the plant's vicinity. Although the procedure may be used in areas where contaminant concentrations at the root zone are low or medium, the plant's growth would be inhibited at high levels of contaminants. While several phytoremediation processes exist, the ones that are applicable in wastewater treatment are phytoextraction, phytodegradation and phytoimmobilization.[14–16]

18.1.4.5 Wastewater Treatment by Nanomaterials

Water is an essential component of life for all living beings, and its availability is crucial. Unfortunately, there has been found a severe shortage of pure drinking water. The addition of different gases, microorganisms and other contaminants to water during rain, flow from rivers, sanitary waste removal and industrial discharge causes water pollution. Nanomaterials have a considerable capacity to treat wastewater containing toxic metals and organic and inorganic pollutants due to their specific properties, such as high functional efficiency at low concentration and greater surface-to-volume ratio.

Although the employability of nanophotocatalysts, nano/micromotors, nanostructured membranes and nanosorbents to eliminate contaminants from wastewater is environmentally friendly and efficient, it requires more energy and high cost. A range of nanomaterials has been used, such as nanophotocatalyst, nano/micromotors, nanomembranes and nanosorbents applied for wastewater treatment.[17]

Nanophotocatalysts

A photocatalyst is a material that alters the rate of a chemical reaction after being stimulated by light photons without participating in the transformation process. Since nanophotocatalysts such as ZnO, SiO_2, TiO_2 and Al_2O_3 retain improved catalytic activity, greater surface-to-volume ratio and high sensitivity to light, their oxidation potential has significantly been increased, catalyzing efficient degradation of contaminants from wastewater. For example, Congo red, azo colors, phenolic toxins, dichlorophenol, toluene, trichlorobenzene and chlorinated ethene can be taken out from wastewater utilizing TiO_2-based nanotubes.

Heterogeneous Photocatalysis

Heterogeneous photocatalysis utilizing a light source and a photocatalyst is the most commonly employed AOP that differs from other treatment techniques such as redox reaction. There are many light-specific compounds that tend to catalyze redox reactions through the photolysis of water. Photogenerated non-specific oxidizing agents involved in photocatalysis may degrade several pollutants such as synthetic dyes. Such peculiar characteristics render the photocatalytic technique universally applicable, exhibiting remarkable efficiency in a wide range of ambient conditions. Heterogeneous nanophotocatalysts have several appealing features, including low chemical cost, ability to operate at low concentration, absence of additives and chemical constancy, for example TiO_2 is found stable in the aqueous medium.

In context to the problems associated with prevailing techniques, the need for some efficient, cost-effective and environmentally friendly technique for the treatment of industrial and municipal wastewater is increasing. This challenge can be overcome either by the development of new techniques or by modernizing the existing ones through interventions. Among the various novel and emerging technologies, nanotechnology has put forward a plausible potential for wastewater remediation.[18]

Nano/Micromotors

Advances in nanotechnology have resulted in a plethora of water treatment options. Nano/micromotors have been recognized in the modern era as devices that can transform energy from a variety of sources into a machine-driven force, allowing them to accomplish specific aims following different mechanisms. The cutting-edge motors that can be driven either with fuel or with other sources such as magnetic field, electric field or acoustics have a wide range of applications in addition to eliminating contaminants from wastewater. They exhibit high strength, increased speed, self-mixing capacity and precise control of movement.

Nanomembranes

Nanomaterials made up of nanofibers and readily used to filter nanoparticles out of aqueous solutions have been considered as nanomembranes. Nanomembranes have been utilized to treat wastewater because of having several properties, including high consistency, homogeneity potential, advancement, brief timeframe, simple operation and high response rate. Microorganisms have been effectively deactivated and natural impurities have been removed utilizing TiO_2-improved nanomembranes. Nanomaterials with various practical properties have been orchestrated utilizing an assortment of strategies.

Doping nanomaterials such as alumina, zeolite and TiO_2 into polymer ultrafiltration layer results in an enhanced film with expanded hydrophilicity and the fouling opposition on a superficial level. Antimicrobial materials like silver nanoparticles are doped with polymers to make a polymeric film that forestalls microscopic organisms' connection and biofilm development on the surface of the layer. They have been utilized to inactivate infections that were equipped for forestalling film biofouling. Nanoparticles affect a film's selectivity and porousness, and this is influenced by the structure, quantity, measurement and different attributes of nanoparticles. Nanocomposite films are produced using palladium acetic acid derivation and polyetherimide, and hydrogen-based Pd nanoparticles are consolidated in a novel way to enhance the effectiveness of water treatment. Nanotechnology has likewise empowered the advancement of various significant synergist films with high porosity, selectivity and fouling opposition.

Nanosorbents

Nanosorbents have peculiar properties that impart advantages and great effectiveness for water treatment, including high sorption capability. Their use on a commercial scale is extremely rare. In addition to commonly found metal oxide and polymeric nanosorbents, the most well-known nanosorbents are carbon nanosorbents such as graphite, carbon black, graphene oxide and reduced graphene oxide. Composites of different materials, such as Ag/polyaniline, Ag/biomass and C/TiO_2 are extremely critical for minimizing the influence of poisons during the process of wastewater treatment.

Carbonaceous substances such as carbon nanotubes (CNTs) have observable adsorption sites and large surface area without being aggregated. They are an effective adsorbent material for contaminants. Dendrimers, a form of polymeric nanoadsorbent, are effective in removing organic contaminants and heavy metals from wastewater. Copper ions have been minimized using a dendrimer ultrafiltration device that was regenerated with pH change, demonstrating biocompatibility, biodegradability and a toxin-free climate. Almost 99% of dyes and other organic contaminants are removed. Furthermore, magnetic nanosorbents play an important role in water treatment by removing various organic contaminants. Magnetic filtration has been used to eliminate certain organic contaminants. The bioadsorbents are cost-effective, biodegradable, biocompatible and nontoxic, making them a viable alternative to chemically synthesized nanosorbents.

18.1.4.6 AOPs for Wastewater Treatment

Advanced oxidation processes (AOPs) are the most commonly used physicochemical procedures, utilizing strong oxidants to remove industrial pollutants. Their basic mechanism includes the production of hydroxyl free radical ($^{\cdot}OH$), a non-selective chemical oxidant, as a potent catalyst for the destruction of compounds that can be mixed with ordinary oxidants such as ozone, oxygen and chlorine. Hydroxyl radicals are effective in the destruction of living organisms because they are active electrophiles with a high reaction rate and greater selectively in almost all electron-rich organic compounds. When a hydroxyl radical is produced, it may attack the chemical species via electron transfer, hydrogen extraction or redox pathways.[19] The reactions that may occur are summarized as follows:

$$R + {}^{\cdot}OH \rightarrow ROH \tag{18.1}$$

$$R + {}^{\cdot}OH \rightarrow R + H_2O \tag{18.2}$$

$$Rn + {}^{\cdot}OH \rightarrow Rn^{-1} + OH^- \tag{18.3}$$

18.1.4.7 Role of Biochar for Wastewater Treatment

Biochar is a cheaper remediation technology for the adsorption of organic and inorganic pollutants, dyes, heavy metals and toxins due to the availability of a range of cheap feedstocks for biochar production. When cheap biomasses like agricultural by-products are used as feedstocks, the cost of biochar production is primarily determined by necessary machinery and heating. Biochar matrix, along with its large surface area, high degree of porosity and strong affinity for organic compounds like furan and dioxin, plays an important role as a surface sorbent for pollutant control in the atmosphere. The surface of biochar has chemically active groups such as COOH, OH and ketones that can be physically triggered following pyrolysis and steam or CO_2 treatment to boost their ability to adsorb heavy metals and toxic pollutants, for example aluminum and manganese, arsenic, cadmium, copper, nickel and lead. The concentration and form of surface functional groups have a major impact on the adsorption of biochar potential and removal mechanism. It has net negative charge on its surface because of dissociation of oxygen-based functional groups. The contaminants of organic and inorganic nature are extracted from soil and water using biochar as a super-sorbent. Regardless of similarity in its preparation strategy and function with the activated carbon, its feedstock and physicochemical properties are quite different.[21]

18.1.4.8 Artificial Intelligence for Wastewater Treatment

In order to reduce water pollution, wastewater treatment is of vital importance. The presence of pollutants renders the composition of wastewater highly complex depending on its nature, concentration and treated effluent.[22] Strict policy has been formulated regarding emission, resource recycling and

Table 18.2

Advantages and Shortcomings of the Current Methods of Dye Removal from Industrial Wastewater[7, 18–20]

Process	Advantages	Disadvantages
Advanced oxidation-processes (AOPs)	Highly reactive free radicals formation; identifiable oxidants in decolorization	Production of contaminants as by-products; incomplete process of mineralization
UV/O₃	Applicable to almost all dyes; lack of sludge; short reaction time	pH-dependent deletion (neutral to slightly alkaline); dispersed dyes are difficult to remove; minimal or non-existent COD elimination; expensive operation; limited penetration of UV
UV/H₂O₂	Lack of sludge; short reaction time; COD reduction	Not suitable for all dyes; limited penetration of UV; radical scavenging at lower pH
Fenton's reagent	Decolorization of both soluble and insoluble dyes; simple equipment and easy operation; COD reduction	Sludge generation; pH-dependent within a narrow range of 3–5; longer reaction time
Ozonation	Sludge generation; applied in gaseous state	Longer reaction time; pH-dependent within a narrow range of 3–5; short half-life (20 min)
Photocatalysis	Lack of sludge; COD reduction; use of solar light	Limited light penetration; catalyst fouling; difficult isolation of fine catalyst from the treated liquid
Electrochemical	Efficient decolorization of soluble or insoluble dyes; COD reduction; nonhazardous by-products	Sludge generation; direct anodic oxidation required; high operational cost
Photochemical	Lack of sludge	Formation of by-products
Sonolysis	No chemical additives; lack of sludge	Excess O₂ (as dissolved gas) required; expensive operation
Ionization radiation	Lack of sludge; efficient oxidation at laboratory scale	Partial decolorization and mineralization; excess amount of dissolved O₂ required
Wet air oxidation (WAO)	Well-suited to effluents that are dilute for ignition but volatile or intense for biological treatment	Lack of complete mineralization; not applicable to low molecular weight complexes; high pressure and temperature required; high operational cost

efficiency of wastewater treatment plants (WWTPs).[23] Hence, it is imperative to promote and maintain water quality in order to overcome the shortage of pure water by following some efficient, cost-effective and commercially applicable treatment technique. For this reason, artificial intelligence (AI) has been the best approach to practically solve the problems, for example wastewater purification,[24] water quality enhancement and modeling,[25] recycling water resources,[26] aerospace integrated vehicle control[27] and machine fault diagnosis.[28] Therefore, it has been employed to simulate, predict and optimize removal of COD; BOD such as the effluent of the Tabriz WWTP;[29] nutrients; typical pollutants, for example heavy metals such as Cd^{2+}, As^{3+} and Pb^{2+};[30] and organic pollutants[31] in the physicochemical and biochemical treatment methods of WWTP. Moreover, AI technology has also been applied to evaluate, predict, diagnose, control and automate such that specially designed AI technology manages the control system of an entire WWTP and eradicates the main issues of traditional control pathways hovering around individual sets of information. AI prediction results in a higher level of accuracy in both physical and biological processes with enhanced membrane pollution predictability.[32]

18.1.5 Approaches for Wastewater Reuse

18.1.5.1 Primary Process Based on Screening Operation

Material that can float and settle out by gravity is removed during the primary treatment procedure. Different types of physical processes include screening, grit removal, comminution

and sedimentation. Long but narrow metal bars are used to build screens that prevent floating debris like rags, wood and bulky items from clogging pumps and pipes. The screens in modern plants are washed manually, and material is disposed of quickly on plant grounds. Debris that passes through screens is ground and shred using a comminutor. Sedimentation or flotation techniques are used to extract the shredded content.

Grit chambers are long and narrow tanks that slow down the flow of water, allowing waste solids like coffee grounds, sand and eggshells to settle out. Pumps and other plant machinery suffer from unnecessary grit wear and tear. Its removal is crucial in cities with combined sewage systems that move a lot of sand, silt and gravel washed off streets and land during rain or a storm.

In sedimentation tanks, suspended solids that pass through screens and grit chambers are separated from sewage. These tanks have about two hours of detention time for gravity settling. The solids eventually fall to the bottom as sewage flows steadily into them. Mechanical scrapers transfer settled solids around the tank floor. The sludge is stored in a hopper and pumped out for disposal. Grease and some other floating poisons are removed using mechanical surface skimming systems.

18.1.5.2 Secondary Process Based on Removal of Suspended Solids and Organic Matter

Secondary treatments are used to extract soluble organic pollutants that are not removed during primary treatment. Biological processes eliminate suspended solid organic impurities by heating them followed by their conversion into carbon dioxide and energy for reuse. This type of procedure takes place in a sewage treatment plant made up of steel and concrete. The trickling filter activated sludge process and oxidation ponds are the most common biological treatment methods in addition to the rotating biological contact process.

Trickling Filter

A trickling filter is a tank with a deep stone bed in it. Sewage settles on the top of stones and trickles down to the bottom where it is collected for treatment. Bacteria accumulate and multiply on stones as wastewater trickles down. The consistent flow of sewage over these growths allows microbes to consume dissolved organic compounds, lowering sewage's biochemical oxygen demand (BOD). For metabolic processes, air flowing upward through spaces between stones provides oxygen, and settling reservoirs, also known as secondary clarifiers, are used after trickling filters. Microbes that are washed off rocks by wastewater flow are removed by these clarifiers.

Activated Sludge

An aeration tank precedes a secondary clarifier in the activated sludge treatment system. The aeration tank is filled with settled sewage mixed with fresh sludge recirculated from the secondary clarifier. The mixture is then pumped with compressed air through porous diffusers at the tank's rim. The diffused air provides oxygen and quick mixing as it floats to the surface. The churning motion of mechanical propeller-like mixers positioned at the tank surface can be used to add air.

Microorganisms flourish in such oxygenated habitats, creating an active and stable suspension of biological solids like bacteria in activated sludge. In the aeration tank, there are about six hours of detention, which provides ample time for bacteria to consume dissolved organic matter from sewage lowering BOD. The mixture then flows from the aeration tank to the secondary clarifier where gravity separates activated sludge. The clear water from the clarifier's surface is disinfected, skimmed and discharged as secondary waste effluent. The sludge is pumped out of the tank through a hopper at the bottom. About 30% of the sludge is recirculated and mixed with primary effluent in the aeration tank. The activated sludge process relies heavily on recirculation. The recycled bacteria are well adapted to the sewage system and could easily absorb organic materials in primary effluent. The remaining 70% of secondary sludge must be treated and disposed of in a healthy and eco-friendly, sustainable manner.[34]

18.1.5.3 Advanced Engineering Process Based on TDS Removal

For total dissolved solids (TDS) reduction, in which coagulants may modify or destabilize dissolved, negatively charged particulate matter and colloidal contaminants, the dosage is an important criterion to be considered in order to optimize the conditions for the highest coagulant efficiency in pollutant removal. Inadequate dose or overdosing will result in poor flocculation efficiency. At an initial stage when pH of wastewater is 7.9, the elimination of TDS increases as the coagulant dose increases from 0.1 to 0.4 g/L. With an increase in dose up to 0.5 g/L, TDS removal decreases because particles are destabilized because of surface saturation. For example, the *Sesamum indicum* seed extract used as a coagulant effectively eliminated TDS from aquaculture wastewater. At a pH of 2 using 0.4 g/L dose at 303 K solution temperature with 60 minutes settling time, maximum 82% TDS removal was achieved. The kinetic results for coagulation/flocculation followed a second-order coagulation/flocculation model ($R^2 = 0.9837$), implying that the coagulation reaction order is 2. The reaction rate (Km) and coagulation/flocculation half-life were calculated to be 0.0002 and 20 minutes respectively.[35]

18.1.5.4 Artificial Intelligence for Wastewater Reuse

Artificial intelligence has made it possible to recover energy and clean water and several components from wastewater. The complex natural conditions bring about variation and uncertainty in wastewater treatment technology resulting in fluctuations in operational cost and effluent water quality. In order to minimize the complications and complexities in wastewater treatment technique, AI has emerged in wastewater treatment plants as a powerful technology tool facilitating cost reduction, removing pollutants and managing challenges in complex practical applications and water reuse. Wastewater reuse has been considered a life-saving factor as it restores water, brings about fruitful outcomes economically and upgrades the quality of the environment.[36] A well-organized neural network model has been designed to access the wastewater reuse potential. In order to improve the quality and lower the cost of wastewater reuse, the rainfall index was set as a compatible input in the model, and decisions varied in parallel with the weather conditions.[37–39]

Several applications have paved the way for wastewater reuse, including artificial recharge of groundwater, agricultural irrigation and industrial use. Agricultural reuse significantly influences soil texture, altering the microbiota and biomass. Such measures must be practiced for wastewater recycling in the near future worldwide as appropriate legislation, proper sanitation, investment in infrastructure and the use of efficient wastewater treatment technologies to make the disposal of pollutant an affordable as well as environmentally feasible, and to monitor the effect and after-effects and to spread awareness about the contaminants and their stages. Likely comprehensive and coordinated strategies lead to assess the sustainable development goals of wastewater reuse and water conservation. Research endeavors in the future must focus on the following points, including (1) designing some new models with improved capability to optimize the operation, boost the pollutant removal efficiency and decrease operational expense; (2) bringing about increased predictability, which is made possible considering variation in specific parameters of the wastewater treatment technique to ensure that the treated wastewater meets the quality standards set by the World Health Organization (WHO); (3) eliminating the limitations in its practicality due to its reduced size and narrow data sets; (4) offering a wider range of user-friendly applications in wastewater treatment with greater accuracy and faster speed of performance; and (5) giving rise to a model that would gather various aspects of wastewater treatment, for example technology tools, management, economics as well as wastewater reuse.[37-39]

18.2 Wastewater Reuse

Wastewater treatment helps to remove pollutants from wastewater, improving the water quality for its reuse in a variety of applications such as municipal, agricultural and industrial use. Highly efficient and safe management of water bodies is quite impossible unless a finely devised framework policy is put forward and in turn implemented successfully. Since various treatment techniques have been practiced on a commercial scale, AI technology renders very high prediction accuracy for the pollutants removal that is up to 1.00 along with a series of advantages such as low operational cost and less chance of error.

18.2.1 Wastewater Reuse in Agriculture

Wastewater reuse practice has great significance for Third World countries faced with water shortage because of sources, climate conditions and variability. According to the guidelines formulated by the Food and Agriculture Organization of the United Nations (FAO) regarding wastewater reuse in agriculture, almost 10% of the total land area has been irrigated with untreated or partially treated wastewater globally, accounting for 20 million hectares in 50 countries. Wastewater reuse, after being treated, results in a series of benefits for humans, economy and the ecosystem. This practice has been adopted as an alternative source of water for several purposes, of which agriculture reuse is of pivotal importance for life in various regions faced with water scarcity and the ever-growing population with increased water demands.[40-41]

Since almost 70% of the available water has been consumed for agriculture globally, freshwater resources are preserved as wastewater reuse serves as an alternative source of irrigation. Furthermore, it tends to increase the agriculture productivity in regions faced with water scarcity accounting for food safety.[42] In addition to waiving the high cost for extraction of groundwater resources, it saves the energy required to pump groundwater that accounts for ~65% of the total cost of irrigation.[43] Some nutrients have been found in wastewater that compensate the fertilizer need[40,42] ensuring an environmentally friendly nutrient cycle that stops the micro- and macronutrients from returning to water bodies. It facilitates water pollution prevention as it causes a profound decline in wastewater discharge, enhancing the quality of water resources.[44] In this way, the groundwater reservoirs are recharged with high-quality water and in turn are preserved. It is also advantageous economically such that it saves labor expenses in searching for new water resources using sophisticated and expensive means. Another profound economic benefit is the evaluation of quality standards of the water discharged from WWTPs for human consumption, as this has been given the highest priority.

18.2.2 Hazards of Wastewater Reuse in Agriculture

Reusing wastewater either untreated or after being treated in the field of agriculture has not been exempted from adverse environmental effects. In this regard, soil is the most vulnerable component subjected to the change in its physicochemical composition due to the varying nature of microbial biomass and the increasing activity of microbes.[41] The socioeconomic and sanitary conditions of a community dictate the categories and concentration of microbes and chemical entities in wastewater,[45] for example parasites, protozoa, helminths and viruses have been found up to 1000 times higher in concentration in underdeveloped countries and cause either acute or chronic wastewater-borne diseases.[46]

18.2.3 Hazards of Wastewater Reuse in the Medical Field

As enteric pathogens, for example *E. coli*, *Salmonella spp.*, intestinal giardia, hepatitis A and E viruses, adenovirus and blood flukes and substances of sanitary interest have most probably been found in the untreated wastewater used for agricultural irrigation. Aforementioned components and chemical substances such as mercury, arsenic, furans, dioxins, DDT and aldrin found in the wastewater appear as an alarming threat to human life. Another hazardous class of pollutants found in wastewater, air, soil and vegetation that affects the endocrine and immunological systems of organisms is emerging contaminants (ECs), including antihypertensive drugs, antibiotics, analgesics, pesticides and endocrine disruptors (EDs) such as estriol, 17β-estradiol and estrone.[47] In addition to the major source of superficial water pollution – the discharge of almost 500 tons of analgesics, for example salicylic acid and diclofenac, into water bodies, other sources of contamination by EDs include runoff from soils consisting of sludge, animal excreta and fertilizers.[48] Since the EDs induce hormone production such as thyroid stimulating hormone, luteinizing hormone and the follicle stimulating hormone, endocrine activity is severely affected leading to abnormalities in the normal functions of organisms.[49] It has been reported that EDs may not be removed easily through WWTPs. The use of EDs in food preservation has undoubtedly increased their production and consumption, but it has resulted in microbial resistance as their discharge into water resources has induced a high degree of adaptability in microbes such as *Staphylococcus aeromonas*, *Pseudomonas*, *Salmonella* and *Escherichia*. Hence, there arises the need for strong legislation and strict action to control the pollutants' concentration into water bodies and microbial resistance as far as public health and wastewater reuse scenarios are concerned.[50]

18.3 Conclusion

Wastewater from different sources such as household sanitation, industries and rainwater pile up threat to humans, aquatic life and ecosystems. Hence, it is essential either to take preventive measures before discharging wastewater into rivers or to follow treatment techniques to compensate for the impact of wastewater on humans and ecosystems. If pollutants in wastewater are left untreated before being discharged, they will be taken up by aquatic organisms, thereby entering our food chain and causing adverse effects on human health.

Although there are several remediation techniques for pollutants removal such as adsorption, precipitation, ion exchange, electrocoagulation, membrane filtration, solidification and aerobic, anaerobic and thermal treatment, they cannot be applied practically because of high cost and the generation of secondary pollutants as by-products that need further treatment. The advanced oxidation process based on solar photocatalysis technology has been reported as an efficient method for pollutant removal due to such advantages as high efficiency, cost-effectiveness, complete degradation of pollutants, low energy consumption and decreased production of secondary pollutants. Bioindicators, biomarkers, biosensors and analytical techniques also play their role in the monitoring and analysis of pollutants. Artificial intelligence has been prioritized over all the classical remediation techniques because of its potential to predict the nature of pollutants such as heavy metals, ECs and organic pollutants with high accuracy and optimize their removal. In addition to improving pump efficiency and aeration efficiency and addressing sludge expansion issues, AI also reduces operational costs, lowers consumption of chemicals and saves energy and labor with its use of models and controlled aeration. It also supports the sustainable development of wastewater treatment by means of water saving, water reuse, better environment quality and economic supremacy. Several preventive measures taking regulatory standards under consideration have been adopted to control the pollution caused by harmful metal contaminants present in the environment.[37]

Because of higher demand for purified clean water supplies and water supremacy principles, the trend of water purification and recycling has spread rapidly worldwide. Owing to the risk factors concerned with the concentrations of pollutants in the wastewater, exposed organisms and the ways of exposure, wastewater reuse is accompanied with various limitations. Despite its limitations, it has found various applications in different fields such as use in agriculture for irrigation, washing vehicles, floor cleaning, flushout and raw material in industries. It has vital importance worldwide as it ensures pollution reduction, food safety and preservation of natural water resources minimizing cost, energy and exhaustion of groundwater reserves. Its applicability as an alternative gateway of irrigation must ensure fruitful outcomes along with the downsizing of risks such as unaltered soil pH, balanced soil salinization, no pollutant transfer from wastewater to the soil, retention of increased organic matter in soil and lack of soil leaching and toxicity. Nevertheless, there are several risk factors of wastewater reuse, for example in agriculture ranging from peculiar characteristics of the soil to impact on human health. Hence, according to the guidelines formulated by WHO in 2006, a wise spectrum of guidance has been put forward for legislators to approve the applications of wastewater reuse in line with the set standards for the safe existence of humans and ecosystem.

REFERENCES

1. Gude, V.G., *Microbial Electrochemical and Fuel Cells*, 2016, 247–285.
2. Gude, V.G., *Journal of Cleaner Production*, 2016, **122**, 287–307.
3. Gude, V.G., N. Nirmalakhandan and S. Deng, *Renewable and Sustainable Energy Reviews*, 2010, **14**, 2641–2654.
4. Schellenberg, T., V. Subramanian, G. Ganeshan, D. Tompkins and R. Pradeep, *Frontiers in Environmental Science*, 2020, **8**, 30.
5. Mareddy, A. R., Technology in EIA, In *Environmental Impact Assessment 1st Edition*. Elsevier BV, 2017.
6. Crini, G. and E. Lichtfouse, *Environmental Chemistry Letters*, 2019, **17**, 145–155.
7. Asha, R. C. and M. Kumar, *Journal of Environmental Science and Health, Part A*, 2015, **50**, 1011–1019.
8. Akpor, O. B., D. A. Otohinoyi, D. T. Olaolu and B. I. Aderiye, *International Journal of Environmental Research and Earth Science*, 2014, **3**, 50–59.
9. Tiwari, B., B. Sellamuthu, Y. Ouarda, P. Drogui, R. D. Tyagi and G. Buelna, *Bioresource Technology*, 2017, **224**, 1–12.
10. Crini, G. and E. Lichtfouse, *Green Adsorbents for Pollutant Removal*, 2018, **18**, 1–21.
11. Nafees, M., A. Nawab and W. Shah, *Journal of Engineering and Applied Sciences*, 2015, **34**, 1–9.
12. Koli, S. K. and A. Hussain, *Advanced Treatment Techniques for Industrial Wastewater*, IGI Global, 2019, 238–250.
13. Chan, Y. J., M. F. Chong, C. L. Law and D. G. Hassell, *Chemical Engineering Journal*, 2009, **155**, 1–18.
14. Akpor, O. B., *Biological and Environmental Engineering*, 2011, **20**, 85–90.
15. Akpor, O. B. and M. Muchie, *African Journal of Biotechechnology*, 2011, **10**, 2380–2387.
16. Akpor, O. B., T. A. Adelani-Akande and B. I. Aderiye, *Microbiology and Applied Sciences*, 2013, **9**, 328–340.
17. Alvarez, P. J., C. K. Chan, M. Elimelech, N. J. Halas and D. Villagrán, *Nature Nanotechnology*, 2018, **13**, 634–641.
18. Yang, G., J. J. Zhu, K. Okitsu, Y. Mizukoshi, B. M. Teo, N. Enomoto, S. G. Babu, B. Neppolian, M. Ashokkumar, S. Shaik and S. H. Sonawane, Lipid-coated nanodrops and microbubbles. In *Handbook of Ultrasonics and Sonochemistry*, Springer, 2016.
19. Al-Mayyahi, A. and H. A. A. Al-Asadi, *Asian Journal of Applied Science and Technology*, 2018, **2**, 18–30.
20. Hai, F. I., K. Yamamoto and K. Fukushi, *Critical Reviews in Environmental Science and Technology*, 2007, **37**, 315–377.
21. Crini, G. and E. Lichtfouse, Springer Science and Business Media LLC, 2018.
22. Long, S., L. Zhao, H. Liu, J. Li, X. Zhou, Y. Liu and Z. Qiao, *Science of the Total Environment*, 2019, **647**, 1–10.
23. Huang, M., Y. Ma, J. Wan and X. Chen, *Applied Soft Computing Journal*, 2015, **27**, 1–10.
24. Al-Aani, S., T. Bonny, S. W. Hasan and N. Hilal, *Desalination*, 2019, **458**, 84–96.
25. Ahmed, A. N., F. B.Othman, H. A. Afan and A. Elsha, *Journal of Hydrology*, 2019, 578.
26. Xu, Y., Z. Wang, Y. Jiang, Y. Yang and F. Wang, *Resources, Conservation & Recycling*, 2019, **149**, 343–351.
27. Ezhilarasu, C. M., Z. Skaf and I. K. Jennions, *Progress in Aerospace Sciences*, 2019, **105**, 60–73.
28. St-Onge, X. F., J. Cameron, S. Saleh and E. J. Scheme, *IEEE Transactions on Industrial Electronics*, 2019, **66**, 7281–7289.
29. Nadiri, A. A., S. Shokri, F. T. C. Tsai and A. Asghari-Moghaddam, *Journal of Cleaner Production*, 2018, **180**, 539–549.

30. Peiman, S., G. Zaferani, M. Reza, S. Emami, M. Kiannejad and E. Binaeian, *International Journal of Biological Macromolecules*, 2019, **139,** 307–319.

31. Ranjbar-Mohammadi, M., M. Rahimdokht and E. Pajootan, *International Journal of Biological Macromolecules*, 2019, **134,** 967–975.

32. Bagheri, M., A. Akbari and S. A. Mirbagheri, *Process Safety and Environmental Protection*, 2019, 229–252.

33. McCarthy, S. and R. Meehan, *Code of Practice, Domestic Waste Water Treatment Systems*, Environmental Protection Agency, 2021.

34. Li, Y., B. Zhang, G. Li and W. Luo, *Current Pollution Reports*, 2018, **4,** 23–34.

35. Igwegbe, C. A. and O. D. Onukwuli, *The Pharmaceutical and Chemical Journal*, 2019, **6,** 32–45.

36. Bozkurt, H., M. C. M. van Loosdrecht, K. V. Gernaey and G. Sin, *Chemical Engineering Journal*, 2016, **286,** 447–458.

37. Zhao, L., T. Dai, Z. Qiao, P. Sun, J. Hao and Y. Yang, *Process Safety and Environmental Protection*, 2020, **133,** 169–182.

38. Shah, M.P. *Removal of Emerging Contaminants Through Microbial Processes*, Springer, 2021.

39. Shah, M.P. *Advanced Oxidation Processes for Effluent Treatment Plants*, Elsevier, 2020.

40. Winpenny, J., I. Heinz, S. Koo-Oshima, M. Salgot, J. Collado, F. Hernandez and R. Torricelli, *Reutilizacion del Agua en Agricultura: Beneficios para Todos*, **124,** 2013.

41. Becerra, C., A. Lopes, I. Vaz, E. Silva, C. Manaia and O. Nunes, *Environment International*, 2015, **75,** 117–135.

42. Corcoran, E., C. Nellemann, E. Baker, R. Bos, D. Osborn and H. Savelli, *Sick Water? The Central Role of Wastewater Management in Sustainable Development: A Rapid Response Assessment*, Earthprint, 2010.

43. Cruz, R., *Tecnicaña*, 2009, **34,** 27–33.

44. Toze, S., *Water Management*, 2006, **80,** 147–159.

45. Gerba, C. and J. Rose, *Water Science and Technology*, 2003, **3,** 311–316.

46. Jimenez, B., D. Mara, R. Carr and F. Brissaud, Wastewater Treament for Pathogen Removal and Nutrient Conservation: Suitable Systems for Use in Developing Countries. In *Wastewater Irrigation and Health: Assessing and Mitigating Risk in Low-Income Countries*, Earthscan, 2010.

47. Fent, K., A. Weston and D. Caminada, *Aquatic Toxicology*, 2006, **76,** 122–159.

48. Heberer, T., *Journal of Hydrology*, 2002, **266,** 175–189.

49. Hutchinson, T., G. Ankley, H. Segner and C. Tyler, *Environmental Health Perspectives*, 2005, **114,** 106–110.

50. Jimenez, C., *Revista Lasallista de Investigación*, 2011, **8,** 143–153.

19

Remediation of Metal Pollutants in the Environment

Ambreen Ashar, Nida Naeem, Noshin Afshan, Mohammad Mohsin and Zeeshan Ahmad Bhutta

CONTENTS

19.1 Introduction

Water is the most abundant valuable natural resource on earth and the basic unit for all attributes of human life such as health, food, energy and economy. Although two-thirds of the earth's surface is covered with water, the lack of pure fresh water has been a global issue for many years. The environment has its own methodology to recycle and supply a sufficient quantity of fresh water, but its contamination as a result of increasing levels of industrialization and an ever-growing human population is an immense problem worldwide. Since sanitation of contaminated water has social, ecological and economic impacts, the availability of freshwater sources is essential for the protection of humanity. Owing to fatal infection and waterborne diseases, approximately 10–20 million and 200 million people

DOI: 10.1201/9781003165958-19

are dying per annum respectively. Among offspring, nearly 5000–6000 deaths have been reported per day just because of water-related diseases. It has been found that 4 billion people have very little or no access to fresh water, and millions of people die annually from drinking water contaminated with toxic pollutants.

Contaminated water contains undesirable substances, which harms its value. Owing to rapid growth of industrialization, increasing population and long-term droughts, the availability and regular supply of clean water has become a major issue worldwide. The augmented utilization of pure water sources has been based on the increase in consumption of food, energy and so forth. Along with the increase in population, demand for fresh water has also been increasing. Water pollution contributed to the depletion of clean water or freshwater resources. Various practices such as saltwater intrusion, soil erosion, pesticides, heavy metals, inadequate sanitation, fertilizers, detergents and natural organic matter (NOM) have been found responsible for polluting freshwater resources, causing changes in water's physical, chemical and biological characteristics (Ghorani-Azam et al., 2016; Halder and Islam, 2015; Sweileh et al., 2018; Sun et al., 2020; Tyagi et al., 2013; Walakira and Okot-Okumu, 2011; Shah, 2021). Different sources such as inhabited areas, inorganic substances, gases, toxic metal pollutants, organic compounds, pathogenic organisms and nonpathogenic microorganisms also act as causative agents for water contamination. If the untreated effluent is disposed of, it causes uncountable threat to living organisms, severely affecting the environment.

Most water contaminants are microbial, organic and inorganic in nature. Microbial pollutants consist of various communities of microorganisms like protozoa, bacteria and viruses. Organic pollutants such as greases, plasticizers, medicinal residues and high concentrations of proteins, carbohydrates, fats and nucleic acids have been discharged into fresh water as domestic and industrial waste that pollutes water bodies. Inorganic pollutants mainly consist of heavy metals such as chromium, nickel, manganese, cadmium, arsenic, lead and zinc, which are introduced into fresh water either by disposing of industrial effluent on the land or by draining it into rivers or lakes. Such harmful, toxic and carcinogenic substances as persistent organic pollutants (POPs) and heavy metals present in water make it undrinkable, causing severe health issues. To meet the demand of a growing population, industries have increased their production and in turn residual wastes may accumulate in the tissues of living organisms, causing health disorders. Therefore, it becomes requisite to remove metal pollutants from wastewater before it is discharged into freshwater resources.

19.2 Environment

The environment has been considered a region with different types of interaction between living and non-living species. The term "environment" originated from the term "Environia," and means the interaction of organisms with one another. The two intricate elements of nature have been taken as environment and ecosystem, which collectively give rise to biosphere, that is, the region of interaction of living organisms with each other

and their abiotic environment. The environment tremendously affects the diversity of organisms living therein and their extinction. It plays an essential role for the survival of biotic species on earth and modulates atmospheric conditions. The environment on earth basically contains atmosphere (air), hydrosphere (water), biosphere (living organisms) and lithosphere (soil), with various gases surrounding them. The environment can be categorized into several types such as micro (like noise, vibration, air or water composition that surround living organism instantaneously), macro (influence human beings externally), physical (abiotic parameters like light, air, water, soil, temperature, etc.), biotic (flora, fauna and living organisms), cultural (faith, behavior and norms) and psychological (recognition and participation). There are three main categories of environment: natural (light, air and water), industrial (cosmopolis, hamlet and factories sector) and social (companies, institutions, organizations and administration) (Appannagari, 2017).

19.2.1 Environmental Pollution and Its Classification

Environmental pollution causing lethal effects to the health of humankind has become a serious concern of the modern world. The introduction of pollutants into the environment that brings harm to living organisms and natural sources is thought of as environmental pollution. The pollutants inducing toxic effects to biological systems have been classified as primary pollutants, for example dichlorodiphenyltrichloroethane; secondary pollutants, for example peroxyacylnitrate; quantitative pollutants, for example those available in the atmosphere like carbon dioxide; and qualitative pollutants, for example human-made chemical products such as pesticides. The pollutants can be either naturally occurring or derived from foreign sources. Three main types of pollutants based on their nature showing harmful impact on ecosystems are organic, inorganic and biological pollutants.

19.2.1.1 Organic Pollutants

Various biological compounds degrade into organic pollutants and act as persistent contaminants, for example chlordane, dieldrin, hexadrin, campheclor and divinyl oxide, due to their greater stability, enhanced toxicity and high solubility in lipids. They are water-insoluble compounds and are widely distributed in nature (Obinna and Ebere, 2019).

19.2.1.2 Inorganic Pollutants

The inorganic pollutants, for example Ca, Mg, Na, K, Cl$^-$, SO$^{2-}_4$ and NO^{-3}, influence the physical characteristics of water such as taste, smell and color. They are highly toxic, nonbiodegradable and injurious to health, aquatic and plant life depending on the interaction between biotic and abiotic components (Bernhoft, 2012; Sharma et al., 2015).

19.2.1.3 Biological Pollutants

Biological pollutants mostly come from living organisms and include bacteria, fungi, pollen, protozoa, microbes and viruses.

They enter through the food chain system and drinking water, severely affecting human life in the developing countries.

19.3 Distribution of Metal Pollutants in the Environment

Heavy metals have been widely distributed in the earth's crust due to their persistent nature. The distribution process chiefly depends on various factors such as bioavailability and toxicity of heavy metals. At the industrial scale, human activities have introduced pollutants into the atmosphere, soil, water and air through several ways.

19.3.1 Sources of Metal Pollutants

Agricultural Activities

Various agricultural processes such as the application of fertilizers and pesticides represent the major sources of metal pollutants in the environment. Although plants require both macro- and micronutrients for the proper nourishment of plants, some soils have an inadequate supply of heavy metals such as cobalt, nickel, manganese and zinc. Hence, these nutrients have been provided to the soil either by spraying or foliar feeding. Fertilizers in a large quantities have been supplied with nitrogen, potassium, phosphorous and so on for the better development of crops.

Coal and Petroleum Combustion

Metallic elements have been generated from coal-burning operations and petroleum process. Eighty heavy metals have been found distributed in various fractions of crude oil, coal, fuel and gasoline. As a result of the burning process, metallic pollutants have largely been released into surrounding areas, along with different types of organic and inorganic materials.

Effluent Stream

Water is one of the most important natural resources in the world and an essential component for the survival of living organisms. Untreated wastewater contains a large number of pathogens such as fungi, viruses, bacteria and so on. Owing to wastewater treatment operations, the effluent thus obtained has been considered one of the major causes of environmental contamination due to the presence of a range of toxic substances such as surfactants, pesticides, toxic heavy metals, sulfides, hydrocarbons, phosphorous and nitrogenous compounds.

Automobile Exhaust

A range of heavy metals are generally released from the emissions of vehicles. Dust particles are emitted from the several operations derived from the sources such as tires plates and abrasion of dynamic coating onto the controlled device (catalytic convertor). The heavy metals associated with the dust particles include chromium, copper, zinc, mercury, nickel, lead, cobalt and cadmium. Moreover, the particulate matters released from motor vehicles include PM-10 and PM-25, which severely pollute the environment (Lough et al., 2005; Shah, 2021).

Mining and Smelting Operations

A high risk of environmental contamination results from the mining process. A large number of metal pollutants are released in the form of minerals adsorbed onto the rock surface; for example, a famous park located in United States, "Great Smoky Mountains," contains toxic heavy metals and sulfide emitted during the process of weathering.

Atmospheric Deposition

A large number of pollutants accumulate in the atmosphere followed by their deposition through either wet process or dry process depending upon climate conditions and tailoring different properties such as altering the air, soil and water quality. Mostly rainwater and storms represent the wet form of atmospheric deposition.

Industrial Processes

A number of manufacturing activities are greatly involved in the generation of metal contaminants. Tanneries, which produce leather, are a main contributor of heavy metals such as Cr(VI) that have been found in large concentrations due to their use of chromium sulfate. The textile industry also generates several toxic substances such as dyes, toxins, organic compounds, sulfides, chlorides and VOCs.

Animal Manures

Livestock manure contains a large number of harmful metal pollutants. High amounts of heavy metals such as cadmium, lead, chromium, arsenic, zinc and copper have been found in poultry farms. Manures also contain trace amounts of various pathogens and antibiotics.

Airborne Activities

The major pollutants present in the atmosphere include particulate matters, smoke, dust particles, sulfur dioxide, nitrogen oxides, radioactive substances and several other types of air contaminants.

Pharmaceuticals and Surgical Instruments

The manufacturing industry has been used for the clinical devices developed in the 19th century in Pakistan. This sector produces a range of instruments such as surgical equipment, pharmaceuticals and therapeutics utilized for foreign trade. The devices are made up of stainless steel materials assembled with heavy metals such as Cr, Ni and many others.

Domestic Wastes

Heavy metals have largely been released from many household activities such as washing, utilization of medication, savage sludge and sanitation. The heavy metals mainly generated from the sludge include cadmium, mercury, manganese, nickel, lead, copper and chromium.

Weathering of Rocks

Primary rocks contain a huge number of chemical elements with metal contaminants, for example high concentration of chromium and nickel metals present therein as minerals, and are assembled into the earth. Olivines contain small amounts of pollutants such as cobalt, nickel, manganese, lithium and copper.

Leakage of Storage Tanks

Harmful substances such as benzene, polycyclic aromatic hydrocarbons, phenolics, organics, sulfides and toxic metal pollutants may be leaked from the storage tanks due to technical issues.

Dredged Sediments

Various types of contaminants have been found in sediments such as organic substances, toxins and lethal heavy metals like Al, Fe and Mn and their oxide-hydroxide complexes (Kumar and Katoria, 2013; Shah, 2020).

Urban Wastes

Composts produced from urban areas contain heavy metals in solid form; excessive amounts of copper, lead and zinc, for example, are found in these wastes.

Secondary Generation of Metal Pollutants and Their Recycling

Heavy metals such as cadmium, lead and mercury are abundantly found as toxic pollutants in many processes and released as a secondary generation. Automobile manufacturing units result in the disposal of chromium. The steel industries introduce elements like Pb and Cd as secondary metal pollutants (Wuana and Okieimen, 2011).

19.3.2 Environmental Metal Pollutants

Heavy metals such as chromium, nickel, zinc, copper, cadmium, lead, mercury and arsenic are known to be highly toxic contaminants in the environment. The extent of toxicity depends upon the amount of contaminant and its exposure time.

19.3.3 Types of Metal Pollutants

There is a direct relationship between human population in the world and contamination caused by heavy metals. Heavy metals have been classified into two main categories that show hazardous impact to the environment.

Essential Metal Pollutants

Owing to the human need for essential heavy metals such as iron, copper, zinc and manganese, the importance of dietary

supplements is increasing daily worldwide. Though they play an important role in the biological system for its proper function and well-being, their high concentration develops toxicity. Various elements such as magnesium, potassium, calcium, sodium and phosphorous are considered to be an essential component of the human diet as they carry out development of skeletal muscles, bones, metabolic functions and regulation of equilibria in terms of acid-base, but their higher concentration in the human body causes lethal diseases such as cardiovascular diseases and hepatic and kidney failure.

Non-Essential Metal Pollutants

This category includes heavy metals that are unimportant to living systems and are considered toxic contaminants, for example lead, mercury, cadmium and tin. They accumulate in living tissues rapidly in various organs of the human body, such as the kidney and liver, and cause their failure.

19.4 Advanced Oxidation Processes

Advanced oxidation processes (AOPs) utilizing various strong oxidants is the most common physicochemical approach used for the remediation of metal pollutants in the environment. The first-ever AOP-based wastewater treatment was proposed in the early 1980s and followed by significant achievements in the field. The process has been based upon the removal of metal ions and subsequent attack by free radicals such as hydroxyl radical (OH·), created by free radical generating chemical species such as O_3 and H_2O_2. Moreover, UV light has been used in combination with strong oxidizing agents in order to increase the efficiency of the remediation process via photolysis; combinations such as O_3/H_2O_2, UV/H_2O_2, and UV/O_3 are currently in use for treatment of heavy metal pollutants. AOPs have been categorized as environmental friendly processes because of their zero sludge production.

19.4.1 Homogeneous Photocatalysis

The Fenton process is one of the important AOPs and is termed as homogeneous photocatalysis, and reactions related to this include those of peroxides (H_2O_2) with Fe^{2+} to produce reactive oxygen species (ROS) that can mortify organic and inorganic pollutants in an aqueous medium. Fenton's reagent (Fe^{2+}/H_2O_2) has also been used to remediate the heavy metals present in contaminated effluent waters. The increase in efficiency is correlated to the photochemistry of Fe^{3+} complex

TABLE 19.1

Classification of Metal Pollutants along with Their Function

Essential Metal Pollutants	Non-Essential Metal Pollutants
Zn (Metabolic activity)	Pb (High toxicity)
Ca (Seed germination)	Hg (Neurotoxicity)
Fe (Cellular functions)	Cd (Carcinogenic, nephrotoxicity)
Cu (Development of plants)	Sn (Hepatoxic, neurotoxic)

dissociated into Fe^{2+}. The photochemistry of Fe^{3+} gives an advantage to Fenton processes because the reduced Fe^{2+} reacts with H_2O_2 forming $OH^•$ as per reaction.

19.4.2 Heterogeneous Photocatalysis

Heterogeneous photocatalysis is used as a mediator in redox reactions because of its electronic structure. During this process, first charge carriers form with the help of photon absorption. In the absence of a hole scavenger and suitable electrons, the excited electrons and holes combine together and energy is released in the form of heat. The photocatalytic activity is powerfully reliant on the rival between the electron–hole recombination and transfers of charge carriers. If the photogenerated electrons or holes are stuck with an effective scavenger or surface defect, recombination is restricted and additional photocatalytic reactions take place, which result in the formation of mineralization products during the reaction. The positive holes have strong oxidizing power and tend to oxidize water and single electrons of the system in order to produce $HO^•$ (Mohammed et al., 2020).

19.4.3 Basic Principle of Heterogeneous Photocatalysis

During heterogeneous photocatalysis, percentage removal of heavy metal is greatly dependent on the redox reaction. For example, the phenomenon of photocatalysis takes place on titanium dioxide MO (molecular orbital). It starts with photon absorption and leads to the production of electron and hole pairs, which tend to combine together releasing energy in the form of heat. Charges thus produced play their role in the oxidation-reduction reaction. On the outer surface, es^-/h^+ (electron and hole pairs) react with H_2O, hydroxide ion or oxygen. However, water reacts with pollutants fairly easily (Stanbury, 2020).

19.4.4 Mechanism of Photocatalysis

The mechanism of photocatalysis is based on the photoexcitation of electron (e^-) present in the valence band (VB) of semiconductor (SC) photocatalyst to the conduction band (CB) upon irradiating with light (UV, visible or solar radiation) leaving a hole (h^+) in the VB. The presence of such catalysts in an aqueous media directs h^+ to get trapped by water molecules to generate hydroxyl radicals ($•OH$), a powerful ($E° = +3.02$ eV) and indiscriminate oxidizing agent. There is a probability that photogenerated e^- either degrades the organic molecule directly or reacts with oxidants such as O_2, adsorbed on the surface of the catalyst or dissolved in water, reducing it to $O_2^{-•}$ (superoxide radical anion). On the other hand, the photogenerated h^+ is also able to oxidize the organic molecule R to $R^{•+}$. Along with other highly reactive oxygen species (ROS), hydroperoxyl free radicals ($HO_2^•$) have also been produced that are converted to $•OH$ on reaction with H^+. The $•OH$ may reduce or oxidize the organic molecules to mineralize them ultimately into water and CO_2 (Kow et al., 2016).

$$SC + h\nu \rightarrow SC\ (e^-_{CB} + h^+_{VB}) \tag{19.1}$$
$$SC\ (h+_{VB}) + H_2O \rightarrow SC + H^+ + OH^• \tag{19.2}$$
$$SC\ (h+_{VB}) + OH^- \rightarrow SC + OH^• \tag{19.3}$$
$$SC\ (e^-_{CB}) + O_2 \rightarrow SC + O_2^{-•} \tag{19.4}$$
$$O_2^{-•} + H^+ \rightarrow HO_2^• \tag{19.5}$$
$$R + OH^• \rightarrow degradation\ products \tag{19.6}$$
$$R + h^+_{VB} \rightarrow oxidation\ products \tag{19.7}$$
$$R + e^-_{CB} \rightarrow reduction\ products \tag{19.8}$$

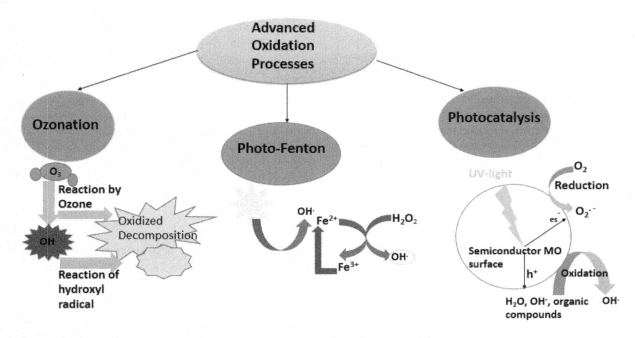

FIGURE 19.1 Classification of advanced oxidation processes (AOPs) and their possible mechanisms.

19.5 Effects of Contamination

Metals with the value of density greater than 5 g/cm^3 have been regarded as heavy metals and include chromium, copper, lead, arsenic, zinc, mercury, nickel and cadmium. A high concentration of heavy metals dissolved in water makes it unable to use. Water contamination thus caused through the release of industrial effluent is a major concern of the modern world. Different components of the environment undergoing heavy metal contamination are as follows.

19.5.1 Soil

Soil is polluted due to the accumulation of heavy metals from human activities. The most commonly found heavy metals in soil include chromium, cadmium, lead, arsenic and copper. These pollutants have been considered essential but they inhibit the growth of crops mainly due to the potential threat of bioaccumulation through the food chain process (Singh and Kalamdhad, 2011).

19.5.2 Aquatic Life

The toxicity of harmful metal pollutants poses a serious threat to the life of aquatic animals. The pollutants enter through the food chain process and sometimes even cause the death of marine life. Although copper is harmful to fish, it is required in trace quantity to perform metabolic functions of the body. The toxicity of this metal pollutant is altered greatly depending on many physicochemical properties of water. Compounds of zinc enter into water bodies such as rivers, lakes and streams as a result of industrial processes. However, it accumulates in fish and causes harmful effects (Shah, 2017).

19.5.3 Flora and Fauna

Plant and animal life are disturbed as a result of contamination of toxic heavy metals. If some of the pollutants such as lead, selenium, arsenic and cadmium found non-essential for the development of plants accumulate in the body, they cause lethal effects to the life of plants. Various metal pollutants such as zinc, copper, cobalt and manganese are known as essential elements, because they have been found necessary for the proper metabolism of many plants. But if they are present in high concentration, they may cause severe poisoning. Composts are widely used in agricultural fields to improve the quality of soil, but they contain harmful components that may cause negative effects (Verma et al., 2018).

19.5.4 Air

Toxic pollutants present in the air such as SO_2, NO_2, CO, Pb and O_3 have negative effects on the growth of plants (Shahid et al., 2017).

19.5.5 Human Health

Heavy metals accumulate in the food chain and may cause several diseases such as cancer; diabetes; asthma; cardiovascular, hematopoietic and immunologic disorders; and several neurological diseases (Jaishankar et al., 2014; Mahurpawar, 2015; Zahra and Kalim, 2017).

19.6 Mechanism of Toxicity of Metal Pollutants Based on Generation of Free Radicals

Chromium

Two basic forms of hexavalent chromium are chromate and dichromate ions. The hexavalent state is movable in water and soil and is its most toxic form and also considered as carcinogenic substance for human health while the trivalent state is believed as a micro-nutrient for human, compulsory element for lipid and sugar metabolism (H. Oliveira, 2012). The hexavalent state is hazardous and mutagen while Cr(III) is an important nutrient in small quantities. Cr(VI) can be agglomerated biologically in the food chain and thus poses a serious hazard to human health (Cervantes et al., 2001; Costa, 1997; Igiri et al., 2018).

Cadmium

Cadmium disrupts DNA, produces ROS and affects chromosomes. Osteoporosis is one of the most popular diseases caused by cadmium, which interacts with minerals like calcium that are essential for the development of bones and skeletal muscles (Oberdörster, 1989).

Iron

The toxicity of iron greatly affects the kingdom Plantae. Redox reactions play an important role to determine the toxic effects of iron in soil. In the first step, ferric ions are reduced to ferrous and the uptake process is enhanced. Then the ferrous ions are oxidized to Fe^{3+}, and thus its concentration greatly decreases in the plants. This process of oxidation-reduction is measured in units such as millivolt.

Arsenic

As(III) has the capability to release Fe from the protein called ferritin. Iron accelerates Fenton's process acting as a catalyst to break down H_2O_2 and thereby generates free radicals that deteriorate the proper functioning of enzyme and DNA strands. Hydrogen peroxide, being a highly reactive species produced inside the body, reacts with bases and causes mutagenesis (Roy and Saha, 2002).

Lead

Lead is a highly toxic heavy metal and causes severe problems in the biological system such as liver damage, asthma, plumbism, weakness, anxiety, anemia, high blood pressure, gastrointestinal tract disturbance, red blood cell damage, mental retardation, nervous problems, and sometimes even death, especially in children. Lead poisoning also causes serious side effects among individuals due to the use of its salts (Assi et al., 2016).

Nickel

The toxicity of nickel greatly depends on the rate of exposure, time, dosage of heavy metal and age of individuals. The basic mechanism of toxicity in skin involves the release of Ni^{2+} ions from the metal compounds. Nickel(II) interacts with the epithelial layers of the skin and thus inactivates the immune system completely (Buxton et al., 2019).

Mercury

The toxicity of mercury causes serious health problems such as tremors, joint weakness, nerve damage, difficulty in walking, kidney, spleen and liver disorder, and insomnia and in some cases it may be lethal. Minamata disease, which is widespread in Japan, arises because of eating food contaminated with mercury (Rice et al., 2014).

Cobalt

Respiratory problems such as asthma largely develop because of exposure to cobalt present in the air through inhalation. It also increases ROS production (Leyssens et al., 2017).

Vanadium

Vanadium is greatly harmful to humankind and causes toxic effects such as respiratory disorders, lung problems like

TABLE 19.2

Hazardous Metal Pollutants Influence the Major Organ Systems of Humans and Permissible Limits Provided by the World Health Organization

Heavy Metals	Major organ Targeted	Permissible Limit (mg/L)
Chromium	Lungs (respiratory system), skin (eczema)	0.05
Cadmium	Kidney, liver, lungs, respiratory system, skeleton	0.005
Iron	Heart, pancreas, cerebrospinal nervous system	0.1
Arsenic	Stomach (gastrointestinal tract), dermis, peripheral nervous system, immune, urinary system, etc.	0.05
Lead	Vascular endothelium, bones, kidney, reproductive, nervous system, blood, etc.	0.05
Nickel	Kidneys, spleen, testes, liver, lungs, cardiovascular system	0.2
Mercury	Brain, digestive tract, immune, vascular, circulatory system, eyes (cornea)	0.001
Cobalt	Respiratory system (lungs), blood cells, peripheral nervous system, heart	0.05
Vanadium	Dermis, eyes, lungs, digestive system, kidneys, immune and respiratory system	0.1
Zinc	Red blood cells, skeleton, brain, stomach, thymus, reproductive, immune system, etc.	5

asthma, cardiovascular diseases, damaged DNA and neurological issues. Its toxicity mostly arises from ROS production, which promotes cytotoxicity through the Fenton reaction (Treviño et al., 2019).

Zinc

The toxicity of this metal pollutant can be either acute or chronic depending on the symptoms such as nausea, weakness and vomiting. Such symptoms occur mostly while taking zinc in lesser amounts. But if the concentration is increased, then chronic conditions such as immune function disruption and cholesterol problems significantly appear among young people (Chasapis et al., 2020).

19.7 Techniques for the Remediation of Metal Pollutants

There are several technologies employed for the remediation of toxic heavy metals present in the environment. Some of them are given next.

19.7.1 Physicochemical Approach

Metal pollutants can be removed from industrial wastewater using adsorption, ion exchange and membrane filtration techniques, which tend to use fewer chemicals and are easy to operate. The main advantages include the use of fewer chemical reagents and low-cost equipment. Adsorption is a very simple technique for the remediation of heavy metals due to greater efficiency obtained in this process. The ion exchange method is employed to remove the small amount of contaminants from aqueous solution and is used in various fields. This method includes cation and anion groups that can be strongly/weakly acidic/basic in nature, depending on the type of functional group they carry. The membrane filtration method is widely used for the remediation of dead metal pollutants present in the effluent.

19.7.2 Bioremediation Approach

This method can be used for the remediation of harmful metal pollutants. Various bacterial strains are used in this method. Cadmium, nickel, mercury and chromium ions can be removed from aqueous solution by taking the following steps. First, metal ions must be able to bind to the surface. Second, the heavy metal moves into the cell and, third, eliminates the heavy metal. This method seems to be effective and inexpensive as compared to the physicochemical process. Biological treatment is further categorized into aerobic and anaerobic processes. In the first approach, the rate of decomposition achieved is comparatively higher than that of the second approach. It is regarded as the best method to treat industrial effluent.

19.7.3 Electrokinetic Approach

This is the most widely used method for the remediation of metal pollutants. The electrochemical oxidation method is highly suitable for the effective treatment of wastewater from industries containing deadly pollutants. Electrochemical oxidation does not require auxiliary chemicals, high pressure or high temperature and is found to be very effective. It does require huge electricity, however, which limits the process. Furthermore, in some cases, low selectivity and slow reaction rates in addition to the formation of by-products are also drawbacks of this technique, which must be overcome through some other technique.

19.7.4 Phytobial Approach

This is a green technology that has been extensively used for the remediation of organic and inorganic contaminants present in the environment. Depending on the mechanism, this technique has been further divided into various types such as phytofiltration (utilization of plants for the complete elimination of metal ions from effluent), phytoextraction (removal of heavy metals from the soil), phytovolatization (volatize contaminants using plants) and phytodegradation (degradation of pollutants using microorganisms). Heavy metals have been completely eliminated using various plant species. Metal ions such as Cr, Ni, Pb, Cd, Hg, Se, Ag and Zn may be extracted with the help of genetic breeding process in the case of increased plant growth. This is a clean, inexpensive and eco-friendly approach to remediate the high concentration of heavy metals (Ali et al., 2013; Burlakovs and Vircavs, 2012).

19.7.5 Thermal Approach

This method has been employed to remove toxic heavy metals by means of heat energy. Besides having great volatility, mercury may not be removed through chemical methods; rather, it has been completely removed from the soil by employing the thermal treatment method. Other heavy metals such as copper, lead and chromium ions have also been eliminated from the soil by applying this technique (Huang et al., 2011).

19.7.6 Solidification Approach

The solidification approach, also called stabilization, involves the addition of reagents to the polluted soil to remove harmful pollutants. It has been used worldwide for the complete remediation of metal contaminants. The basic mechanism of this technique involves the addition of additive agents such as kaolin, $CaCO_3$, oxides, zeolites and cement as inorganic binders to the polluted soil. The heavy metals are then immobilized onto the surface of soil by precipitation with $\cdot OH$. This technique is affected by the chemical composition of soil, temperature and amount of H_2O present, as these factors control the remediation efficiency of metal pollutants (Wuana and Okieimen, 2011).

19.7.7 Containment Approach

This technique has mostly been used to control the transport of harmful metal ions into the environment. It has been utilized for the remediation of heavy metals with low mobility and high toxicity that cause great risk of polluting the environment. The main advantages of this technique include simplicity, ability to treat large areas contaminated with hazardous pollutants and fast implementation. Metal pollutants such as cadmium, lead, chromium and arsenic are greatly remediated by employing this technology (Evanko and Dzombak, 1997).

19.8 Solar Photocatalysis for Remediation of Metal Pollutants

Solar energy, a combination of light and heat coming from sun, is the most profuse clean energy resource. The magnificent amount of solar energy striking the earth even in an hour has been estimated higher than the energy consumed per annum by human beings. Because the study regarding fabrication of advanced materials capable of harvesting solar radiation has been progressing, it is essential for green technology to abate environmental pollution. Solar photocatalysis, being the most promising technique, has attracted great interest in order to eliminate toxic metal pollutants from wastewater streams.

The nanophotocatalysts carrying out chemical oxidation have been greatly advantageous because they target recalcitrant compounds, shorten reaction time, produce no secondary sludge and transform wastes into valuable by-products. They are capable of degrading in wastewater non-specifically such POPs as detergents, dyes, pesticides and surfactants in addition to halogenated, non-halogenated and volatile compounds. Moreover, they may also remediate inorganic anions and heavy metal cations under specific conditions. One of the major advantages of the photocatalytic process over conventional technologies is that it eliminates the constraint of secondary sludge disposal. The nanophotocatalysts have mostly been regenerated, reused and recycled multiple times for the degradation of chemical and biological pollutants in wastewater, non-specifically. The solar photocatalytic technique may also be applied that excludes the use of UV lamps and specific kinds of reactors to destroy natural organic matter (NOM) to POPs. Usually the nanophotocatalysts act efficiently under mild conditions and are effective even in very low concentration (Ahmad et al., 2019; Al-Saad et al., 2012; Ashraf et al., 2019; Baragaño et al., 2020; Dinesh et al., 2016; Eskizeybek et al., 2012; Hemalatha et al., 2018; Khan et al., 2016; Mavinakere Ramesh and Shivanna, 2018; Mishra et al., 2018; Mustapha et al., 2020).

19.9 Monitoring and Analysis of Metal Pollutants in the Environment

After the removal of harmful pollutants, it is necessary to monitor and assess the water quality utilizing different techniques. Moss bag technology is a simple and low-cost method used for the biomonitoring analysis of various heavy metals (Maxhuni et al., 2016). The passive technique has been based on the use of various tree and plant species. In the active monitoring system, the mosses have been employed to determine the trace concentration of metal pollutants such as cadmium, lead, nickel and arsenic. The bioindicators such as phytoplankton, fish, aquatic plants and protozoa have been highly

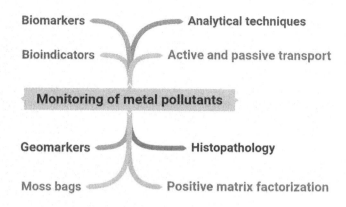

FIGURE 19.2 Different processes for the monitoring and analysis of metal pollutants.

effective to monitor metal pollutants. Some parasites have been employed to determine the concentration of pollutants as indicators because of their tendency to respond to air pollution. Histopathology has been applied to monitor heavy metals and detect their effects in many aquatic species such as fish (Belhadj et al., 2017). The biomarkers respond distinctively to various sources of metal pollutants. The main function of this technique is to determine the bioavailability of pollutants obtained from different sources. Similarly, positive matrix factorization and geomarkers have been used extensively for the monitoring of metal pollutants. The concentration of heavy metals, especially in the sea and in marine sediments, has been monitored by employing a large number of analytical techniques (Ekrami et al., 2021; Farzin et al., 2017; Giné et al., 2008; Kuerban et al., 2020; Mehana et al., 2020; Mussali-Galante et al., 2013).

19.10 Preventive Measures to Control Pollution Caused by Metal Pollutants

A significant number of policies and technologies such as solar light, hydroelectricity, wind energy and heating pumps have been utilized to control environmental pollution. Some control instruments are also used for reducing harmful air contaminants from manufacturing industry and effluents. They can demolish all the pollutants before they are released into the environment. Particulate matter can be removed by the cyclonic method, in which atmospheric particles from fluid, gas streams and air are eliminated (Kumar and Katoria, 2013).

The major challenges regarding the rising demand of water are based on growing population and anthropogenic effects influencing the continuous contamination of water reservoirs. A large number of problems result from polluted water and affect the population in multiple ways. In highly dense areas, government agencies must provide techniques to clean water bodies. Various strategies have been adopted by means of awareness campaigns to reduce water contamination, such as preserving water by turning off the tap, removing industrial wastewater into water bodies only after its treatment, converting toxic heavy metals present in wastewater into eco-friendly

by-products and applying biocompatible wastewater treatment techniques. Since the burning of fossil fuels produces a large number of hazardous pollutants, efforts have been made to deplete these fuels to control the pollution. An efficient innovative strategy is renewable energy that tends to prevent combustion and assists in refining the quality of environment. The reduction in the generation of plastic has been practiced mainly by a recycling strategy that provides an efficient management system. Planting trees tends to reduce soil erosion, lessen the greenhouse effect and protect water sources from contamination, thereby improving water quality. In the past, the rate of contamination increased rapidly in China due to the rise in human activities. Then the population was controlled and the Chinese government formulated a law entitled "Environmental Protection" to control the harmful pollutants and protect the land surface, plants and the whole biotic life. Two main proposals have been reported there, mostly organized on the national level. The first one has been called the 12th five-year scheme and the second one the strategy to control harmful contamination in river basins. The major goals that have been set in the first plan to control the pollution are to develop a complete system that would address the main problems of human health caused by metal pollutants. The second major goal is to reduce the release of harmful metals such as cadmium, mercury, chromium, nickel and lead. Hence, the management system has been developed to monitor and assess the heavy metal pollutants and take preventive measures to control pollution.

19.11 Future Recommendations

Metal pollutants are carcinogenic and lethal to living beings. Their discharge in water increases rapidly with time because of rapid industrial development such as extraction of different minerals from their ores in mining, coating of a metal especially in electroplating, and the production of leather, paper, fertilizers, pesticides and pharmaceuticals. AOPs have widely been applied to degrade organic pollutants found in effluents, but nowadays photocatalysis has been applied to remove the toxic metal pollutants employing sidewise oxidation at ambient temperature and pressure. Another technique to remediate metal pollutants is sonolysis, in which toxic heavy metals are eliminated by means of explosions of cavitation bubbles formed through ultrasonic hydrodynamics. A treatment technique for wastewater, ozonation, is applicable for degrading microbes, decolorization, micropollutant remediation and removal of non-protonated amines, taste and odor. Ozone in combination with hydrogen peroxide emerged as a new dimension in wastewater treatment, and it oxidizes both inorganic and organic substances effectively. Addition of homogeneous or/and heterogeneous catalyst to ozone also enhances oxidation efficiency. To a significant extent, the toxic metal pollutants may be transformed into steady inorganic compounds, for example salts, water and carbon dioxide, that is, enduring mineralization. The main objective of wastewater distillation through AOPs is to completely remove the chemical impurities, resulting in pure water. An efficient, low-cost and ideal wastewater technique still needs to be discovered that could be applied at the commercial level to purify water for the next generation.

19.12 Conclusions

Water is a major constituent for life on earth, as it plays a vital role in the ecosystem. It is popularly known as the universal solvent based on its ability to dissolve different types of materials. During rain rinses, flowing and surface water along with different substances (chemicals), toxic contaminants such as heavy metals, gases and microbes have caused water pollution. Water adulteration has become a foremost problem due to industrialization worldwide. Hence, the main objective is to control the emission of detrimental contaminants into the environment. In this chapter, water pollution, its cause through incorporation of heavy metal pollutants, sources, effects on biological systems and different techniques to control the level of metal pollutants have been discussed. Many organic and inorganic substances such as metal pollutants and biological contaminants have been reported in water. These compounds have been considered to be harmful, carcinogenic and dangerous for all living organisms on earth. In developed countries, it is thus compulsory to control the discharge of these hazardous and toxic substances in every concentration. The remediation of various metal pollutants from the environment has widely been studied to find cost-effective approaches. Several methods are used to remediate metal pollutants in the environment, including ion exchange, electrocoagulation, membrane filtration, adsorption, precipitation, aerobic, anaerobic, thermal, solidification, phytobial and containment. All of these methods, however, either have been costly or have produced secondary pollutants as by-products that require secondary treatment for their remediation. The advanced oxidation process based on solar photocatalysis technology has been reported as an efficient method for the remediation of various metal pollutants due to its potential of almost complete degradation of pollutants, low energy consumption, high efficiency, cost effectiveness and decreased production of secondary pollution. After remediation, monitoring and analysis of metal pollutants are performed by employing bioindicators, biomarkers, biosensors and analytical techniques. Several preventive measures such as regulatory standards have been adopted to control the pollution caused by harmful metal contaminants present in the environment.

REFERENCES

Ahmad, I., W.A. Siddiqui, T. Ahmad and V.U. Siddiqui. 2019. Synthesis and characterization of molecularly imprinted ferrite (SiO_2@ Fe_2O_3) nanomaterials for the removal of nickel (Ni^{2+} ions) from aqueous solution. *Journal of Materials Research and Technology*. 8: 1400–1411.

Ali, H., E. Khan and M.A. Sajad. 2013. Phytoremediation of heavy metals – concepts and applications. *Chemosphere*. 91: 869–881.

Al-Saad, K., M. Amr, D. Hadi, R. Arar, M. Al-Sulaiti, T. Abdulmalik, N. Alsahamary and J. Kwak. 2012. Iron oxide nanoparticles: Applicability for heavy metal removal from

contaminated water. *Arab Journal of Nuclear Sciences and Applications*. 45: 335–346.

Appannagari, R.R. 2017. Environmental pollution causes and consequences: A study. *North Asian International Research Journal of Social Science & Humanities*. 3.

Ashraf, S., A. Siddiqa, S. Shahida and S. Qaisar. 2019. Titanium-based nanocomposite materials for arsenic removal from water: A review. *Heliyon*. 5: e01577.

Assi, M.A., M.N.M. Hezmee, M.Y.M. Abd Wahid Haron and M.A.R. Sabri. 2016. The detrimental effects of lead on human and animal health. *Veterinary World*. 9: 660.

Baragaño, D., R. Forján, L. Welte and J.L.R. Gallego. 2020. Nanoremediation of as and metals polluted soils by means of graphene oxide nanoparticles. *Scientific Reports*. 10: 1–10.

Belhadj, H., D. Aubert and N.D. Youcef. 2017. Geochemistry of major and trace elements in sediments of Ghazaouet Bay (western Algeria): an assessment of metal pollution. *Comptes Rendus Geoscience*. 349: 412–421.

Bernhoft, R.A. 2012. Mercury toxicity and treatment: A review of the literature. *Journal of Environmental and Public Health*. 4096.

Burlakovs, J. and M. Vircavs. 2012. Heavy metal remediation technologies in Latvia: Possible applications and preliminary case study results/Technologie remediacji obszarów zanieczyszczonych metalami ciężkimi na Łotwie: możliwe zastosowania i wstępne wyniki badań. *Ecological Chemistry and Engineering S*. 19: 533–547.

Buxton, S., E. Garman, K.E. Heim, T. Lyons-Darden, C.E. Schlekat, M.D. Taylor and A.R. Oller. 2019. Concise review of nickel human health toxicology and ecotoxicology. *Inorganics*. 7: 89.

Cervantes, C., J. Campos-García, S. Devars, F. Gutiérrez-Corona, H. Loza-Tavera, J.C. Torres-Guzmán and R. Moreno-Sánchez. 2001. Interactions of chromium with microorganisms and plants. *FEMS Microbiology Reviews*. 25: 335–347.

Chasapis, C.T., P.-S.A. Ntoupa, C.A. Spiliopoulou and M.E. Stefanidou. 2020. Recent aspects of the effects of zinc on human health. *Archives of Toxicology*. 94: 1443–1460.

Costa, M. 1997. Toxicity and carcinogenicity of Cr(VI) in animal models and humans. *Critical Reviews in Toxicology*. 27: 431–442.

Dinesh, G.K., T. Sivasankar and S. Anandan. 2016. Metals oxides and doped metal oxides for ultrasound and ultrasound assisted advanced oxidation processes for the degradation of textile organic pollutants. *Handbook of Ultrasonics and Sonochemistry*. 1–27.

Ekrami, E., M. Pouresmaieli, P. Shariati and M. Mahmoudifard. 2021. A review on designing biosensors for the detection of trace metals. *Applied Geochemistry*. 104902.

Eskizeybek, V., H. Gülce, A. Gülce, A. Avcı and E. Akgül. 2012. Preparation of polyaniline/ZnO nanocomposites by using arc-discharge synthesized ZnO nanoparticles and photocatalytic applications. *J. Fac. Eng. Arch. Selcuk Univ.* 27: 111–120.

Evanko, C.R. and D.A. Dzombak. 1997. *Remediation of metals-contaminated soils and groundwater*. Ground-Water Remediation Technologies Analysis Center.

Farzin, L., M. Shamsipur and S. Sheibani. 2017. A review: Aptamer-based analytical strategies using the nanomaterials for environmental and human monitoring of toxic heavy metals. *Talanta*. 174: 619–627.

Ghorani-Azam, A., B. Riahi-Zanjani and M. Balali-Mood. 2016. Effects of air pollution on human health and practical measures for prevention in Iran. *Journal of Research in Medical Sciences: The Official Journal of Isfahan University of Medical Sciences.* 21.

Giné, M.F., A.F. Patreze, E.L. Silva, J.E. Sarkis and M.H. Kakazu. 2008. Sequential cloud point extraction of trace elements from biological samples and determination by inductively coupled plasma mass spectrometry. *Journal of the Brazilian Chemical Society.* 19: 471–477.

Halder, J.N. and M.N. Islam. 2015. Water pollution and its impact on the human health. *Journal of Environment and Human.* 2: 36–46.

Hemalatha, K., A. Manivel, M. Kumar and S. Mohan. 2018. Synthesis and characterization of Sn/ZnO nanoparticles for removal of organic dye and heavy metal. *Biological Chemistry.* 12: 1–7.

Huang, Y.-T., Z.-Y. Hseu and H.-C. Hsi. 2011. Influences of thermal decontamination on mercury removal, soil properties, and repartitioning of coexisting heavy metals. *Chemosphere.* 84: 1244–1249.

Igiri, B.E., S.I. Okoduwa, G.O. Idoko, E.P. Akabuogu, A.O. Adeyi and I.K. Ejiogu. 2018. Toxicity and bioremediation of heavy metals contaminated ecosystem from tannery wastewater: A review. *Journal of Toxicology,* 2018, 1–16.

Jaishankar, M., T. Tseten, N. Anbalagan, B.B. Mathew and K.N. Beeregowda. 2014. Toxicity, mechanism and health effects of some heavy metals. *Interdisciplinary Toxicology.* 7: 60–72.

Khan, M.A. and A.M. Ghouri. 2011. Environmental pollution: its effects on life and its remedies. *Researcher World: Journal of Arts, Science & Commerce.* 2: 276–285.

Khan, S.B., H.M. Marwani, A.M. Asiri and E.M. Bakhsh. 2016. Exploration of calcium doped zinc oxide nanoparticles as selective adsorbent for extraction of lead ion. *Desalination and Water Treatment.* 57: 19311–19320.

Kow, S.H., M.R. Fahmi, C.Z.A. Abidin and O. Soon-An. 2016. Advanced oxidation processes: process mechanisms, affecting parameters and landfill leachate treatment. *Water Environment Research.* 88: 2047–2058.

Kuerban, M., B. Maihemuti, Y. Waili and T. Tuerhong. 2020. Ecological risk assessment and source identification of heavy metal pollution in vegetable bases of Urumqi, China, using the positive matrix factorization (PMF) method. *PloS One.* 15: e0230191.

Kumar, S. and D. Katoria. 2013. Air pollution and its control measures. *International Journal of Environmental Engineering and Management.* 4: 445–450.

Leyssens, L., B. Vinck, C. Van Der Straeten, F. Wuyts and L. Maes. 2017. Cobalt toxicity in humans – A review of the potential sources and systemic health effects. *Toxicology.* 387: 43–56.

Lough, G.C., J. J. Schauer, J.-S. Park, M.M. Shafer, J.T. Deminter and J.P. Weinstein. 2005. Emissions of metals associated with motor vehicle roadways. *Environmental Science & Technology.* 39: 826–836.

Mahurpawar, M. 2015. Effects of heavy metals on human health. *International Journal of Reseacrh-Granthaalayah.* ISSN-23500530.2394–3629.

Mavinakere Ramesh, A. and S. Shivanna. 2018. Visible light assisted photocatalytic degradation of chromium(VI) by using nanoporous Fe_2O_3. *Journal of Materials,* 2018, 1–13.

Maxhuni, A., P. Lazo, S. Kane, F. Qarri, E. Marku and H. Harmens. 2016. First survey of atmospheric heavy metal deposition in Kosovo using moss biomonitoring. *Environmental Science and Pollution Research.* 23: 744–755.

Mehana, E.-S.E., A.F. Khafaga, S.S. Elblehi, A. El-Hack, E. Mohamed, M.A. Naiel, M. Bin-Jumah, S.I. Othman and A.A. Allam. 2020. Biomonitoring of heavy metal pollution using acanthocephalans parasite in ecosystem: An updated overview. *Animals.* 10: 811.

Mishra, P.K., P. Gahlyan, R. Kumar and P.K. Rai. 2018. Aerogel based cerium doped iron oxide solid solution for ultrafast removal of arsenic. *ACS Sustainable Chemistry & Engineering.* 6: 10668–10678.

Mohammed, H., S.K. Ali and M. Basheer. *Heavy metal ions removal using advanced oxidation (UV/H_2O_2) technique.* IOP Conference Series: Materials Science and Engineering, 2020. IOP Publishing, 012026.

Mussali-Galante, P., E. Tovar-Sánchez, M. Valverde and E. Rojas Del Castillo. 2013. Biomarkers of exposure for assessing environmental metal pollution: From molecules to ecosystems. *Rev. Int. Contam. Ambie.* 29: 117–140.

Mustapha, S., M. Ndamitso, A. Abdulkareem, J. Tijani, D. Shuaib, A. Ajala and A. Mohammed. 2020. Application of TiO_2 and ZnO nanoparticles immobilized on clay in wastewater treatment: A review. *Applied Water Science.* 10: 1–36.

Obinna, I.B. and E.C. Ebere. 2019. A review: Water pollution by heavy metal and organic pollutants: Brief review of sources, effects and progress on remediation with aquatic plants. *Analytical Methods in Environmental Chemistry Journal.* 2: 5–38.

Oliveira, H. 2012. Chromium as an environmental pollutant: insights on induced plant toxicity. *Journal of Botany,* 2012: 375843.

Rice, K.M., E.M. Walker Jr, M. Wu, C. Gillette and E.R. Blough. 2014. Environmental mercury and its toxic effects. *Journal of Preventive Medicine and Public Health.* 47: 74.

Roy, P. and A. Saha. 2002. Metabolism and toxicity of arsenic: A human carcinogen. *Current Science.* 38–45.

Shah, A.I. 2017. *Heavy metal impact on aquatic life and human health – an over view.* IAIA17 Conference Proceedings| IA's Contribution in Addressing Climate Change 37th Annual Conference of the International Association for Impact Assessment, 2017.4–7.

Shah, M.P. 2020. *Advanced oxidation processes for effluent treatment plants.* Elsevier.

Shah, M.P. 2021. *Removal of emerging contaminants through microbial processes.* Springer.

Shahid, M., C. Dumat, S. Khalid, E. Schreck, T. Xiong and N.K. Niazi. 2017. Foliar heavy metal uptake, toxicity and detoxification in plants: A comparison of foliar and root metal uptake. *Journal of Hazardous Materials.* 325: 36–58.

Sharma, H., N. Rawal and B.B. Mathew. 2015. The characteristics, toxicity and effects of cadmium. *International Journal of Nanotechnology and Nanoscience.* 3: 1–9.

Singh, J. and A.S. Kalamdhad. 2011. Effects of heavy metals on soil, plants, human health and aquatic life. *International Journal of Research in Chemistry and Environment.* 1: 15–21.

Stanbury, D.M. 2020. Mechanisms of advanced oxidation processes, the principle of detailed balancing, and specifics of the UV/chloramine process. *Environmental Science & Technology.* 54: 4658–4663.

Sun, J., Z. Zhou, J. Huang and G. Li. 2020. A bibliometric analysis of the impacts of air pollution on children. *International Journal of Environmental Research and Public Health*. 17: 1277.

Sweileh, W.M., S.W. Al-Jabi, H.Z. Sa'ed and A.F. Sawalha. 2018. Outdoor air pollution and respiratory health: A bibliometric analysis of publications in peer-reviewed journals (1900–2017). *Multidisciplinary Respiratory Medicine*. 13: 15.

Treviño, S., A. Díaz, E. Sánchez-Lara, B.L. Sanchez-Gaytan, J.M. Perez-Aguilar and E. González-Vergara. 2019. Vanadium in biological action: chemical, pharmacological aspects, and metabolic implications in diabetes mellitus. *Biological Trace Element Research*. 188: 68–98.

Tyagi, S., B. Sharma, P. Singh and R. Dobhal. 2013. Water quality assessment in terms of water quality index. *American Journal of Water Resources*. 1: 34–38.

Verma, R., K. Vijayalakshmy and V. Chaudhiry. 2018. Detrimental impacts of heavy metals on animal reproduction: A review. *Journal of Entomology and Zoology Studies*. 6: 27–30.

Walakira, P. and J. Okot-Okumu. 2011. Impact of industrial effluents on water quality of streams in Nakawa-Ntinda, Uganda. *Journal of Applied Sciences and Environmental Management*. 15.

Wuana, R.A. and F.E. Okieimen. 2011. *Heavy metals in contaminated soils: A review of sources, chemistry, risks and best available strategies for remediation*. International Scholarly Research Notices.

Zahra, N. and I. Kalim. 2017. Perilous effects of heavy metals contamination on human health. *Pakistan Journal of Analytical & Environmental Chemistry*. 18: 1–17.

20

Treatment Techniques of Industrial Effluents and Wastewater Treatment Plants

Noshin Afshan, Alina Bari, Ambreen Ashar and Nazia Saleem

CONTENTS

20.1 Introduction

Water is a necessity for life, as it accounts for 65% of our bodies and constitutes 71% of the earth. Everyone wants pure water for drinking, cooking and cleaning purposes. If the water is polluted, it will lose its worth to us aesthetically and economically, threatening not only our health but also the viability of aquatic organisms and wildlife depending on it.

Wastewater is a water-based solid or liquid discharged into water bodies that presents the waste of community life.[1] Any substance affecting water quality negatively is a possible source of water pollution.[2] The chemical and textile industries are the main sources of wastewater that contain pollutants, for example acids, alkalis, substances with high biological oxygen demand, dyes and suspended solids that are toxic, mutagenic, carcinogenic or difficult to biodegrade.[3] Moreover, sewage

and wastewater, pesticides, fertilizers, radioactive waste and urban development also cause water pollution. In the past few decades, water contamination by pollutants, such as pharmaceutical active compounds (PhAC), personal care products (PCP), artificial sweeteners (AS) and endocrine-disrupting chemicals (EDC), has gained great attention because these pollutants are ubiquitous in the environment and may cause severe problems for all living beings and their habitats.

Since the supply of clean water is a basic requirement for the maintenance of human activities, it has become a global challenge, especially in developing countries where industrial and municipal wastewater is discharged into the water reservoirs. Water reservoirs such as rivers polluted by chemical pollutants is one of the most critical environmental problems because they provide valuable food through aquatic organisms and irrigation.[4]

DOI: 10.1201/9781003165958-20

Due to the massive increase in the global population, water will become one of the scarcest resources in the 21st century. The core of the urbanization phenomenon is the main issue related to provision of infrastructure for municipal and water services, including simultaneous provision of freshwater resources and health services. Wastewater causes widespread public health problems and restricts economic and agricultural development, adversely affecting ecosystems. As the population has been increasing, the available resources have started to decline, threatening environmental resources. The report of Secretary-General of the United Nations Commission for Sustainable Development (UNCSD, 1997) that was conducted on fresh water is currently being followed in the developing or developed countries. It is concluded that the growth rate of water usage is three times more than the growth rate of the world population. Though housing, medical care and social services provide basic infrastructure for human needs, the availability of clean water will be included in the list in the near future. Wastewater treatment facilities to clean water is a major challenge for engineers, planners and politicians. The right treatment and proper sanitation of wastewater is essential to protect public health and the environment. Legislation on industrial wastewater is becoming more stringent, especially in developed countries, and requires wastewater to be treated before it is discharged into water reservoirs.[2]

20.2 Technologies for Wastewater Treatment

The production and treatment of wastewater is the main problem of industry and society. The use and maintenance of wastewater treatment plants is highly troublesome.[2] It is estimated that annual production of industrial wastewater in Pakistan is 4432.35 million cubic meters. Approximately 1% of it is treated before its disposal, and 3% of industries use hazardous and toxic chemicals without proper treatment and preventive measures. The industrial effluents have a negative impact on freshwater quality, making it harmful for humans and aquatic life. Some treatment plants cannot handle huge amounts of industrial effluents, which deteriorate water quality adversely and affect living beings and the environment.[5]

In the past 30 years, various biological, chemical and physical technologies have been employed to eliminate the pollutants from wastewater such as oxidation, precipitation, flotation, evaporation, solvent extraction, ion exchange, adsorption, membrane filtration, phytoremediation and electrochemical biodegradation. In addition to advantages, every treatment has some limitations in terms of feasibility, cost, efficiency and environmental impact. At present, due to the complexity of industrial wastewater, there is currently no method that can adequately treat it. In practice, a suitable combination of compatible methods is usually employed to obtain the required quality of water in the most economical way.[2]

Owing to environmentally friendly treatment standards, various pathways of industrial wastewater treatment consist of multiple technologies to reduce toxicity. After selecting the appropriate treatment technology, a series of pilot-scale experiments must be carried out on real industrial wastewater on a larger scale. However, new directives and regulations are required to find the best treatment plan to reduce harmful compounds. The composition of industrial wastewater (IWW) is highly variable. The methods for the treatment and disposal of IWW differ significantly from those utilized for municipal waste. According to the characteristics of IWW, the technology for processing IWW is usually divided into the following three categories.

1. Physical methods: These processes use natural forces such as van der Waals forces, gravity and electric gravity, and physical barriers, for example screens, fences, membranes, deep-bed filters, ion exchange and electrodialysis to complete the removal of matter. The physical methods regarding wastewater treatment include adsorption, flotation and sedimentation.[6]
2. Chemical methods: These processes can be used to remove chemicals by generating insoluble solids and gases, coagulate colloidal suspensions and remove biodegradable chemicals from nonbiodegradable, damaged or inactivated chelating agents. Combined with the advanced oxidation process or biological technology to treat difficult-to-remove industrial wastewater, chemical methods greatly improve the removal rate. In recent years, photocatalysis, ozonolysis and ultrasonic oxidation have been replaced by advanced oxidation processes with improved biodegradability and the ability to detoxify wastewater containing polar and hydrophilic chemical substances.[6]
3. Biological methods: In this type of treatment, carbon-containing organic materials are employed for the removal of phosphorus, biochemical oxygen demand (BOD), nitrification, denitrification and stabilization. Aerobic biological processes become highly efficient, while anaerobic bacteria use the idea of recovery and utilization of the resources. In recent years, highly aerobic and anaerobic bioreactors have been greatly used to degrade highly resistant IWW with the advantages of requiring minimal space, incredible COD removal efficiency (over 83%) and lower investment cost.[7]

20.2.1 Nondestructive Procedures Based on Physical Processes

Physical treatment involves isolating waste from the main stream either with little or no degradation of the waste such as coagulation and filtration or the ingestion of organic waste by microorganisms.[8]

Among the physical processes, filtration (such as membrane and depth filtration), sedimentation and adsorption have long been well-known treatments for wastewater discharged from the textile industry. Besides precipitation, which is an ineffective treatment technique, membrane filtration is an efficient technique, although it is accompanied by the recurring clogging of membranes by some odd matter, thereby increasing the overall cost of the process. Adsorption is also an efficient treatment method, depending on the nature of the adsorbent and preparation technology, and its price is slightly higher than other processes.[9] On the industrial scale, adsorption can be used to remove dyes and other organic and inorganic pollutants by employing activated carbon-based materials, magnetic nanomaterials with high surface-to-volume ratio and food residue such as seeds, peel and pulp, which can be used as effective adsorbents.[10]

20.2.2 Biological Destructive Procedures Based on Biological Processes

Biological treatments used for removing organic matter from wastewater are generally considered the most cost-effective, energy saving and environmentally friendly methods. Several microorganisms in addition to fungi and enzymes degrade dyes through reduction or oxidation, for example oxidation of the dyes through metabolic reactions of bacteria and reduction of azo dyes by various microbes to form aminobenzenes. However, the biological effluent treatments being applied widely are slow and limited based on incomplete degradation of nonbiodegradable pollutants, with the appearance of some intermediates causing toxicity to microorganisms. Moreover, the oxidation reaction of dyes carried out by bacteria has been found difficult due to the large size of dye molecules while passing through the cell membranes. The oxidation of azo compounds by fungal oxidases has been found more toxic than their parent dyes, and the fungal cultures require longer growth phases to produce high amounts of oxidase enzymes.[11] Hence, a highly efficient method that can decompose dye molecules is required.

20.2.3 Oxidative Destructive Processes Based on Oxidative Chemical Processes

In highly industrialized areas, water pollution caused by dyes is a serious issue. These distinctively colored components will stop reoxygenation capacity when dispensed in water bodies, and the wastewater deteriorates the ability of water bodies to suspend the sunlight, destroying aquatic activities.[12] It is found that almost 10,000 dyes and pigments are annually produced in bulk amount (> 700,000 tons) in the world, and approximately 20% of it is discharged in the form of industrial wastewater during the dyeing process in the textile industry. Based on the chemical structures of the dyes, they can be characterized as heterocyclic dyes, anthraquinone dyes and azo dyes. Based on their application methods, the dyes can be characterized as direct dyes, basic dyes, acid dyes, reactive dyes, disperse dyes and vat dyes. Decolorization of wastewater containing dyes is essential for either its reuse or its safe discharge into different water bodies. Hence, a combination of biological, chemical and physical processes is applied to textile wastewater by using advanced oxidation, microfiltration, ultrafiltration, nanofiltration and so on.

Oxidation with Hydrogen Peroxide for Textile Wastewater Treatment

Owing to H_2O_2's wide range of applicability for removing pollutants from wastewater, for example sulfide, hypochlorite, nitrite, cyanide and chlorine, it has also been used in the surface treatment industry, which involves metal cleaning, decoration, etching and protection. The advanced oxidation process with hydrogen peroxide (oxidant) to decompose non-biodegradable organic pollutants in textile wastewater is used as wastewater treatment in two systems:

1. Heterogeneous system based on using semiconductors with/without UV light, such as TiO_2 and stable iron-modified zeolite such as FeY_5 and $FeY_{11.5}$.
2. Homogeneous system based on visible light or ultraviolet light, soluble catalyst and other chemical activators such as ozone and peroxidase.

But high concentrations of contaminants, for example aromatic compounds, which are highly chlorinated, and inorganic compounds, for example cyanide, require a high concentration of H_2O_2 because of the low reaction rate. Therefore, the decolorization, decomposition or mineralization of dyes in textile wastewater requires activated hydrogen peroxide. Ozone UV light, transition metal salts and peroxidase can help in activating hydrogen peroxide by converting it into hydroxyl radicals, which act as strong oxidants retaining higher reaction rates than H_2O_2 does. The efficiency of the treatment depends on the peroxidase usage, its concentration, pH and the temperature of the aqueous medium.[12] This approach is just outstanding regarding water treatment, especially in countries facing water deficiency. Solar photocatalytic reactors of various types have also been implanted for this purpose, but still there is a huge potential in employing such reactors as standalone, time efficient and cost effective wastewater treatment technology.[13]

20.2.4 Incineration

Municipal wastewater is in fact a mixture of water in small factories, service facilities, trades and households, and it often depends on the intensity of rainfall causing mineral and organic pollutants to flush into rainwater pipes. Sewage sludge usually gathers metals such as Ni, Cu, Zn, Hg, Pb, Cd, Cr and Sb, which are found in suspended solid pollutants in wastewater.

Therefore, sewage sludge produced during wastewater treatment is supposed to be a sink of pollutants and minerals present in the water that is treated through incineration. The aim of incineration is to decline the volume, quality and negative effects of the sludge on the environment via high temperature detoxification and disinfection. The incineration process burning of organic matter results in volatile components and heat-resistant solids that are produced in the emission gas. The exhaust gas is monitored continuously to determine the extent of suspended particles and the composition of gases (CO, PM_{10}, NH_3, NO_x, hydrogen halide and oxides of S, and transition metals such as, Sb, Co, Cr and Pb, etc.) to control the air index in accordance with directive 2000/76/EC.[14]

20.2.5 Wet Oxidation (WO)

Wet Air Oxidation (WAO)

Most of the research about WAO concerns wastewater treatment. Emulsified wastewater is a typical high-concentration organic wastewater that usually comes from the machinery industry and is difficult to biodegrade. It contains various organic substances such as surfactants, additives and mineral oil. The traditional treatment methods employing the ordinary physical or chemical methods such as chemical demulsification, electrolytic flotation and membrane separation have obvious shortcomings in bringing secondary pollution and too high operating cost. Therefore, the wet air oxidation process is practiced for the treatment of emulsified wastewater. It is found that WAO depends mainly on the temperature in the wastewater, the partial pressure of oxygen and the concentration of organic matter, while other operating conditions have different effects on the oxidation efficiency. In this treatment technique, temperature is the key factor such that the increase in temperature directly promotes the oxidation reaction, and in turn, the removal rate of COD_{Cr} and TOC is significantly increased.[15]

Catalytic Wet Air Oxidation (CWAO)

The industrial methods produce wastewaters having organic pollutants, influencing the ecosystem negatively. The development of economically feasible and advanced oxidation methods is necessary to destroy organic pollutants. Catalytic wet air oxidation, a green process, is an effective and promising process for removing oxidative pollutants. It is used to treat wastewater that cannot be incinerated due to dilution and whose concentration is too high to consider any biological treatment. Its powerful function is credited to the high temperature ranging from 130 °C to 300 °C in combination with air at about 5–60 bar pressure and highly active catalysts. In this method, the catalyst acts as a key component for a wide range of industrial applications, and its selection must be favorable economically and ecologically, that is, stable, inexpensive and functionally compatible with the CWAO operating conditions.[16] This task is full with challenges leading to the selection of clay as the right choice. Therefore, modified clays are greatly emphasized in catalytic reactions involving selective oxidation, refineries/biorefineries and environmental protection. The clay is modified by a columnarization process, which means that a large amount of polyoxygenation is inserted, and a stable oxide column can be generated by calcination. The resulting interlayer gallery is molecularly accessible. The transition metal ions acting as intercalant species produce effective columnar clay. Hydrogen peroxide can be used as an oxidant to oxidize organic substrates in water convening Fenton-like and photo-Fenton-like processes.[16]

The great potential of aluminum-iron pillared clay has been reported for CWAO of phenolic wastewater. However, the column process must reduce time so that these materials can be mass produced, thereby increasing their industrial benefits. In fact, traditional pillared clay synthesis involves three main steps: preparation of column solution, insertion of the clay and formation of a column. The first two steps need long-term aging, but microwave radiation may shorten the aging time. Moreover, the resulting solid has a more uniform columnar distribution and higher surface area compared to the traditionally prepared clay. At the same time, microwave technology allows significantly reduced water consumption during the column process, forwarding it towards a green and environmentally

friendly procedure with low cost. Hence, microwave radiation has been suggested for the formation of column so that the heating and calcination procedures may slow down.[16]

Supercritical Water Oxidation (SCWO)

SCWO is an auspicious technology that can absolutely eliminate waste and recover energy. This process is strongly applicable to the actual wastewater on a laboratory scale. However, SCWO treatment in actual wastewater pilot scale has been highly restricted and its applicability on industrial wastewater has some disadvantages, including two-phase waste treatment, salt deposition, the existence of suspended solids, corrosion and high cost. In order to decrease its cost, recovery of the energy from the reactor effluent has been proposed. SCWO is a promising alternative to the incineration of wastewater streams and has attracted great interest in the past decade. However, it has some disadvantages that hinder the industrial application of this method. In addition to the technical problems caused by reactor corrosion and salt precipitation, the problem also comes from the fact that SCWO is generally considered to be a "universal" technology for all possible waste streams, which cannot be achieved.[17]

20.3 Liquid Oxidation: Advanced Oxidation Processes (AOPs)

Advanced oxidation is considered the most commonly used physicochemical procedure and utilizes strong oxidants to remove industrial dyes. The first water purification treatment based on AOPs was reported in the 1980s, with subsequent developments in later years. The azo bond or the aryl rings subjected to the free radicals such as hydroxyl radical (˙OH) undergo the oxidation of the azo bond. The free radicals are generated from various chemical species, for example O_3 and H_2O_2. Along with strong oxidizing agents, UV light is used to accelerate the efficiency of degradation, subsequently followed by the photolysis and combination strategies such as O_3-H_2O_2, UV-O_3 and UV-H_2O_2, which are still employed to carry out chemical treatment of azo dyes. The ultraviolet light coupled with ozone (O_3) catalyzes it into ˙OH in aqueous medium.[18]

Sonolysis is a technique in which the organic molecules are tainted because of explosions of cavitation bubbles formed through ultrasonic hydrodynamics. The treatment method, ozonation, tends to decolorize wastewater, degrade microbes and remove micropollutants, non-protonated amines, taste and odor.[19] Ozone coupled with hydrogen peroxide has become an effective dimension for water treatment carrying out oxidation of both inorganic and organic substances. But the compounds with a saturated ring system are difficult to oxidize through this treatment.[20] Addition of a homogeneous and/or a heterogeneous catalyst such as TiO_2-Me, Al_2O_3-Me and Fe_2O_3 to ozone also enhances the oxidation reaction.

Photochemical AOPs employ radiation sources like UV (A, B and C) and visible radiation independently or synergized with chemicals to degrade persistent organic pollutants (POPs). The UV treatment termed "photolysis" has been undergone as a tertiary step for the degradation of organic compounds and removal of the microbes. The treatment technique based on coupling of UV with O_3 has been found to be an advanced oxidation technique. Since this treatment technique combines the advantages of both, it is advantageous as compared with the individual O_3 and UV treatment technique.[21]

The Fenton process is one of the important AOPs termed homogeneous photocatalysis and likewise reactions includes those of peroxides with iron ions to produce reactive oxygen species (ROS), which can mortify inorganic and organic pollutants. The Fe^{2+}/H_2O_2, Fenton's reagent, oxidizes azo dyes and their derivatives present in the contaminated effluent water. Owing to its tendency to catalyze oxidation of H_2O_2 into ·OH, the efficiency of this process increases side by side with the photochemistry of iron complex.[22]

Many chemical AOPs have not been practical to apply on an industrial scale due to lack of cost efficiency, installation of closed reactors and generation of oxidants like ozone. Hence, some innovative treatment technique is urgently needed to treat wastewaters containing industrial dyes, so they can be discharged in a controlled manner, rendering our ecosystem safe from synthetic dyes.

Heterogeneous photocatalysis for treatment of textile effluent:

Heterogeneous photocatalysis is the most commonly employed AOP, which differs from other treatment techniques including redox reaction using light source and a photocatalyst. A range of light-specific compounds can catalyze redox reactions through the photolysis of water. Photogenerated nonspecific oxidizing agents involved in photocatalysis have a tendency to degrade various pollutants such as synthetic dyes. The fact renders the photocatalytic technique universally applicable, exhibiting remarkable efficiency in a wide range of ambient conditions.

In the context of the problems associated with prevailing techniques, requirement for some efficient, cost-effective and powerful technologies for the sake of treatment of industrial and municipal wastewater is increasing. This challenge can be overcome either by developing new techniques or by modernizing existing ones through interventions. Among various novel and emerging technologies, nanotechnology has been shown to have potential for remediating wastewater in addition to providing solutions for other environmental problems.[23]

20.4 Water Resources and Management

The production of wastewater is due to the processing and utilization of water. Depending on the water usage of the industry, wastewater contains a variety of suspended matter, for example, nonbiodegradable and degradable organic matter, oils, inorganic acids, alkalis, colorants and greases dissolved by heavy metal ions.[24]

20.4.1 Sources of Industrial Effluents

The characteristics of industrial wastewater are chemical oxygen demand (COD), conductivity, abnormal turbidity, hardness and total suspended solids (TSS), which give basic details of the aquatic habitat set by water streams and rivers. Most of the wastewater causes immeasurable harm to microbial entities. Waste generated by the textile industry usually has high concentration of chemical substances that have a high concentration of corrosive chemicals causing its pH value to vary between 10.0 and 11.0. The net effect is a change in the acidity or alkalinity of the water. Emissions from textile industries also carry strong coloration derived from dyes.[24] The sources of industrial effluents include mines and quarries, power-generating plants, and food processing, metallurgical, nuclear, chemical, textile, wood, paper and petroleum industries.

20.4.2 Phases of Wastewater Treatment

Once the water is contaminated, it must be decontaminated; an excellent purification method must be selected to achieve the goal (as regulated by law). The purification technique usually involves five consecutive procedures, which are given as follows.[2]

1. Pre-treatment, which involves mechanical and physical techniques
2. Primary treatment, which involves chemical and physical techniques
3. Secondary treatment, also called purification, which involves chemical and biological techniques
4. Tertiary treatment, which involves chemical and physical techniques
5. Treatment of the formed sludge, which manipulates supervised dumping, recycling or incineration

Depending on the situation, the first two treatments have usually been collected under the concept of pre-treatment,[2] during which all the suspended solid particles are eliminated from wastewater. Before considering secondary treatment, the pre-treatment stage is essential, because particulate contamination (such as SS, colloid, fat, etc.) will hinder subsequent treatment, reducing efficiency or destroying the decontamination ability of equipment.

20.4.2.1 Primary Treatment (Solid Removal)

In this phase, the solids – either organic and inorganic in nature – are removed from wastewater by means of physical processes like sedimentation and flotation. A clarifier or sedimentation tank removes settleable organic and inorganic solids from wastewater. Almost 50% of BOD, 70% of TSS and 65% of the grease are taken out during this phase. Some organic-based nitrogen and phosphorus and heavy metals are also removed from the wastewater, but the colloids and the dissolved components remain unaffected. Therefore, the wastewater is further subjected to secondary treatment. The effluents from primary sedimentation process are regarded as primary effluents.

20.4.2.2 Secondary Treatment (Bacterial Decomposition)

This phase is aimed to remove residual suspended organic solids. With respect to the size of solids, almost 30% of suspended solids, 6% colloids and 65% dissolved solids are found therein. During this phase of treatment, a variety of microorganisms are employed to treat wastewater in a controlled fashion following different types of aerobic procedures. The main difference lies in the way that oxygen is supplied to microorganisms and in turn the rate at which organic material is metabolized by the microorganisms.

20.4.2.3 Tertiary Treatment (Extra Filtration)

During primary and secondary phases of treatment, most of the BOD and suspended solids are removed from wastewater. Although in several cases, this phase of treatment is not necessarily required either to provide reusable water for industrial and household purposes or to keep the receiving water reservoir safe from contaminants. This phase of treatment has been added to remove the organic material, suspended solid, nutrients and toxic substances. Hence, the advanced wastewater treatment aims to produce higher quality water than is usually obtained through secondary treatment processes.

The tertiary treatment process aims to add unit operation to the flow scheme after regular secondary processing. Its addition to secondary treatment is either as simple as adding a filter for the removal of the suspended solids or as complicated as adding many units for the removal of organic matter, suspended solids, nitrogen and phosphorus.

20.5 Photocatalysis and Wastewater Treatment

Meeting sustainable development goals, including water treatment techniques, has been the area of continuous endeavors in order to improve water quality and to make the availability of pure drinking water accessible. These are the crucial issues, especially in regions where the water treatment systems lose their function, which leads to the accumulation of toxic metal ions and industrial waste in the water reservoirs. Hence, the advanced oxidation process, specifically heterogeneous photocatalysis, deals with the conversion of photon energy into chemical energy. Nanotechnology has enabled the fabrication of nanomaterials, which boosts their ability to exhibit the properties of photocatalytic materials.[25]

Pure water is crucial for human health. It must not contain any impurity such as carcinogenic substances, harmful bacteria and toxic materials. According to the report formulated by United Nations' World Water Development 2018, the need for pure water is expected to increase up to one-third in next 30 years.[25] Pure water is also a basic need of industries, for example pharmaceutical, food and electronic industries. Hence, in order to fulfill the water supply requirements, much more effort should be made to discover novel treatment techniques for wastewater. Innovative discoveries in the field of nanotechnology lead to improved water quality employing bioactive nanoparticles, nanocatalysts and nanosorbents.[26, 27] The nanomaterials show significant physical and chemical characteristics at nanoscale as compared to their counterparts.

Various nanomaterials show photocatalytic activity under visible light, for example CdS,[28] BiVO4,[29] ZnO,[30] and AgCl.[31]

The most widely studied TiO_2 nanomaterials have been spanning over three decades exhibiting excellent anti-reflection property, self-cleaning ability, photocatalytic activity, long-term stability, lower toxicity and inexpensive and good hydrophilicity.[32] The Fe_3O_4 nanoparticles of less than 20 nm particle size and exhibiting superparamagnetic behavior act as an excellent photocatalyst for wastewater treatment.[33, 34] Moreover, ZnO, g-C_3N_4 and graphene act as promising photocatalysts in this regard.[35–37] The distinguishing characteristics of graphene such as two-dimensional structure, greater surface area, enhanced conductivity and high electron mobility tend to promote recombination of electron-hole pair, thereby improving its photocatalytic efficiency.[37, 38]

The semiconducting materials tend to degrade the microorganisms and contaminants present in the water resources by means of heterogeneous photocatalysis. During this process, photon energy is converted into chemical energy, which carries out a range of applications including water treatment. In this way, various organic contaminants undergo degradation and in turn mineralization into carbon dioxide and water.[39] This treatment technique is more advantageous than the other techniques as it completely removes the contaminants rather than transforms them from one phase to another.

The solar photocatalytic system, for example a desktop compound parabolic collector (CPC)-based reactor known as P25/PET plates, has practically been performing excellently and, hence, can be further utilized in real life. Such a treatment technique would also be applied to treat industrial wastewater samples obtained from the dyeing and printing industries.[40]

20.5.1 Heterogeneous Photocatalysis

The problem of water pollution has been a critical issue due to the ever-increasing population and growing level of industrialization, leading to the discovery of treatment techniques including AOPs. The technique of photocatalysis utilizing a semiconductor as a photocatalyst is one of the promising techniques and retains potential for total mineralization of the toxic metal ions and organic pollutants.[41, 42]

The mechanism of heterogeneous photocatalysis follows a series of redox reactions that proceed on the surface of the photocatalyst. Generally, every semiconductor consists of two energy bands: the highest unoccupied energy band and the lowest occupied energy band and a band gap in between these two energy bands. When a beam of photons with an energy level greater than or equal to that of the semiconductor band gap is exerted on its surface, the electrons get photoexcited from its valence bands to the conduction band in femtoseconds. Hence, the valence band becomes empty and is treated as a hole (h^+), and it further forms an electron-hole pair. The electron-hole pair, after being trapped on the surface of the semiconductor, may initiate a chain reaction only if they undergo prevention of recombination.[43] The semiconductor's tendency to generate an electron-hole pair under light irradiation leads to the formation of free radicals, such as hydroxyl radicals, which catalyze further reactions, eventually forming CO_2 and H_2O.[44] The distinguishing characteristics of heterogeneous photocatalysis are as follows:[45]

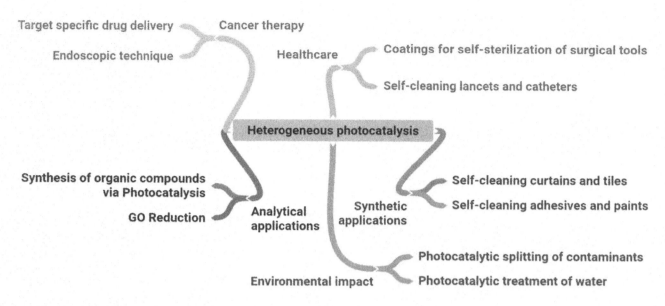

FIGURE 20.1 The potential applications of heterogeneous photocatalysis.

- The degradation of pollutants into inorganic components and CO_2
- The procession of the reaction under optimum conditions
- The availability of oxygen and ultra-band gap energy for initiation of the reaction
- Presence of inert matrices to adhere to the catalyst, for example glass, polymer, carbon nanotubes and graphene oxide[46]
- The catalyst is nontoxic, inexpensive and reusable

The photocatalytic character along with self-sterilization of TiO_2 can be employed in the manufacturing of tiles, paints, roads and surgical equipment and to preserve food. This type of photocatalysis is eco-friendly and the cheapest treatment technique for air and water purification. The distinct applications of heterogeneous photocatalysis in the synthetic, environmental and medical fields are shown in Figure 20.1.[47, 48]

20.6 Monitoring and Analysis of Industrial Effluents in the Environment

The tannery industry is one of the most polluting industrial sectors. Tannery industries use a significant amount of chemicals for the transformation of animal hides into leather. In Bangladesh, about 90% of tannery industries are engaged in the chrome tanning process because it is simple in operation and renders excellent properties to the leather. Tanning is a completely wet process that consumes significant amounts of water and generates about 90% of the used water as effluent.[49] Tannery effluents carry heavy pollution loads due to the massive presence of highly colored compounds, sodium chloride and sulfate, various organic and inorganic substances, toxic

metallic compounds and putrefying suspended matter that are biologically oxidizable, carcinogenic and toxic. Tannery effluents damage the normal life of receiving water bodies and land surfaces.

Generally, water consumption is the highest in the pretanning areas, but significant amounts of water are also consumed in post-tanning processes. The soaking stage, the most polluting stage of the tanning process, contributes around 50%–55% of the total pollution load of the tannery industry. In the liming stage, protein, hair, skin and emulsified fats are removed from the hides and are released in the effluent, which increase its total solids contents. The effluents from the tan-yard processes, de-liming and bating, contain sulfides, ammonium salts and calcium salts, leaving the effluent slightly alkaline. The pickling and chrome tanning effluents contain sulfuric acid, chrome, chlorides, sodium bicarbonate and sulfates. Only about 20% of the large number of chemicals used in the tanning process is absorbed by the leather, and the rest is released as wastes. The major pollutants of the post-tanning process are chrome salts, dyestuff residues, fat liquoring agents, syntans and other organic matter. Worldwide, it is estimated that discharged tannery effluents contain 300–400 million tons of heavy metals, solvents, toxic sludge and other wastes, which are dumped into water bodies each year.[50] Hence, human health is severely affected by toxic hazards generated through the unskilled and unprotected tanning chemicals, treated hides and skins and handling of pesticides.

Industrial effluents are considered extremely complex mixtures containing a variety of organic and inorganic pollutants. Standard targeted chemical analysis of the effluents stipulated in environmental protection and quality regulations may not be adequate for evaluating the toxic/genotoxic potential of complex mixtures present in the effluents. Analyses of samples collected from areas known to receive industrial wastes and effluents have shown that genotoxins can accumulate in the receiving aquatic ecosystems and can cause adverse effects on indigenous biota. Monitoring of toxic/genotoxic effects of chemicals and their mixtures by bioassays is a useful approach for hazard identification and risk assessment.[51]

In order to understand the impact of pollutants on health, it is important to assess and monitor the concentration of pollutants at emission sources and exposure to humans. Human health risk assessment is used to analyze the potential impact of pollutants (organic/inorganic) of specific populations under specific parameter sets. It is expected that risk assessment will provide comprehensive information to risk managers, especially policy and decision makers and regulators. Risk assessment tools can be used to examine individual or population risks. The risks faced by individuals may be related to individuals who are mildly or highly exposed or who are particularly susceptible. Health risks may be assessed for various exposure durations (for example, one year or a lifetime) or locations. Only hypothetical persons with expected characteristics can be assessed for personal health risks. Population health risks may be related to multiple harmful health effects (such as cancer, diabetes and death) of a certain population in a specific period or the rate of harmful effects in a specific area or on a specific population.[52]

Though tanneries are a revenue- and job-generating sector, the pollution from their effluent is of major concern. It is unfortunate that there are no reliable estimates of the quantity and types of hazardous wastes generated in most developing countries, like Bangladesh, and in some cases proper documentation is not available. A few reports have been found on composite and unit-wise tannery effluents. For most of the chemicals used, there are very few toxicological data, mainly about long-term and low-level exposure. Therefore, identifying potentially toxic organic and inorganic groundwater pollutants is a major issue. The lack of effective implementation of legislative control, poor processing practices and use of unrefined conventional leather processing methods have further aggravated the pollution problems. There is also a worldwide need to develop an action plan for emerging infectious biological agents.[53]

20.7 Preventive Measures to Control Industrial Effluents

In a range of industries, wastewater is generated by cooling water (due to the release of excessive heat into the surroundings), sewage effluents (generated by cleaning, drinking, etc.), processing drained water (including water used in the production and washing of products, to remove wastewater) and cleaning wastewater (including flushing and preservation of wastewater generated in industrial localities). Urban wastewater is distributed equally between commercial and metropolitan sources, with the exception of huge quantities of cooling water eliminated from the power industry. The qualitative and quantitative analysis of wastewaters differ from one industry to the other, and such changes have been caused by the bath liquid discharge, operation startup and shutdown, working time allocation and more.[6]

One of the main pollutants in fresh water is industrial wastewater. It can be said that this is the main problem in the developed countries. Industrial wastewater often contains several harmful elements, like radioactive substances, heavy metals, hydrocarbons and toxic organic compounds. Wastewater discharged from industry also contains various diffused and suspended contaminants, and their compositions depend on the type of industry and processes being used. For example, due to the industry in northern India, several places have reported high amounts of Cr, Ni, Cd, Pb and Zn in the groundwater. The pollution of groundwater by organic compounds caused by the unintentional release of industrial wastewater has caused major concern and posed safety hazards to drinkable water. The most common organic pollutants are petroleum hydrocarbons; however, because of their low water solubility, they do not pose a major threat to groundwater.[6]

In the past few years, rapid industrialization and economic growth have greatly promoted human well-being, aggravated industrial pollution and eliminated the world's natural resources. Particularly, the generation of a huge amount of industrial wastewater greatly limits the available water resources, so it has attracted great attention not only in developing countries but also all around the world. There is a lot of evidence that if there is no effective treatment method to discharge industrial wastewater into the environment, it will have subsequent dangerous and adverse impacts on aquatic life.[54]

Therefore, industrial wastewater treatment is found to be a reasonable approach to deal with this challenge, because it too provides the chance of water reuse, which is particularly important in countries with water shortages such as the Middle East.[55] In this regard, the production of a successful industrial wastewater treatment technology to solve this issue is the main focus of scientific research. Various sources of industrial wastewater contain difficult-to-treat pollutants, such as adsorbable organic halides (AOX) and phytotoxic compounds, which make it resistant to biodegradation, hence hindering the effectiveness of the biological treatment process. Choosing the most appropriate and feasible technology to treat organic and inorganic pollutants to meet strict environmental rules is a difficult task. The literature review allows an opinion on the criteria to be taken into account for assessing the durability of the membrane technologies used in engineering fluid solution (EFs).[56]

After adopting one or more treatment methods, the potential for reuse of industrial wastewater is very important, mainly with regard to the environment. This variable is the focus of recent research on the performance of mechanical biological treatment (MBT) method, and they are a key element of advanced wastewater treatment processes.[57]

Solid wastes caused by textile industries are found to be another environmental hurdle that must be handled. The four main methods for processing polymer solid wastes are normally source reduction, reusing techniques, incineration and landfill. Several studies in the scientific literature address the environmental impact of textile reuse and recycling, including various types of fibers, polymers, monomers and fabrics. Sandin and Peters reviewed 41 studies, of which 85% were recycled and 41% were reused.[58] They found that fiber recycling was the most studied type of recycling (57%), followed by polymer/oligomer recycling (37%), monomer recycling (29%) and textile recycling (14%). Cotton (76%) and polyester (63%)

were the most reported materials. Previously reported data strongly supports the claim that the reuse and recycling of textile water have reduced environmental effect compared with the incineration process and landfills. Minimizing sources of waste must be the first option to be considered in an integrated waste management system. Internal reuse of textile disposal materials in wastewater treatment may be a promising model for reducing emission sources.

It is generally believed that effluent regulations have been implemented, the quality of receiving water bodies is improving, and the competent authorities are taking necessary steps to solve these problems. As compared with major landfill, energy recovery and recycling waste management practices, waste elimination is widely reviewed to have greater environmental potential. In the United States (US EPA, 2016) and the European Union (European Commission, 2008), waste prevention is a policy priority, and more broadly, it is a suggestion of the OECD (2000) and the World Bank (2013).[59]

20.7.1 Regulation of Industrial Effluents

At the moment we need to improve the status of water quality and industrial wastewater management for effective treatment and control, it should be known that any effort or step taken for the improvement must be equipped with all the probable pollutant discharge layouts. The probable discharge layouts and scenarios are shown in Figure 20.2.[60]

Therefore, for a country's successful environmental laws and regulations, nationwide effective laws and regulations, market competitiveness through research and development, and industry innovation capabilities using lesser-waste techniques play an important part. In light of the previous discussion, it is concluded that in order to preserve the ecosystem from poisonous industrial wastewater, the following measures with two principal components are found very important:

20.7.2 Ecocentric Component

- Having capacity based emission levels based on local ecosystems
- Efficient monitoring procedures can prevent the industry from compromising wastewater treatment

20.7.3 Econocentric Component

- Motivating industries to reduce the harmful emissions through encouragement and subsidies

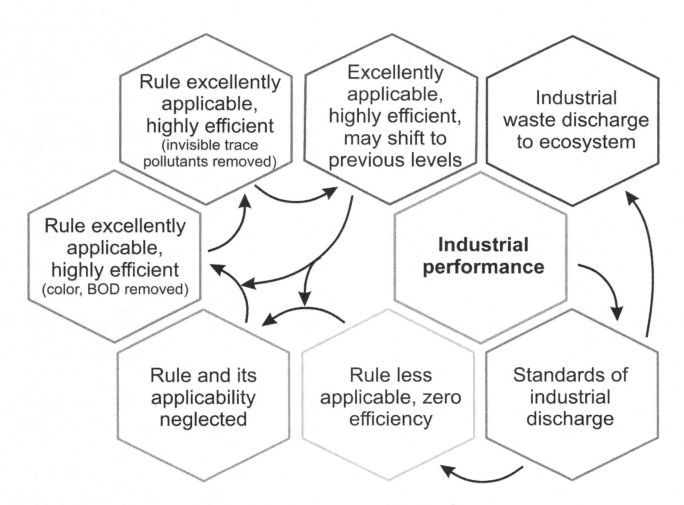

FIGURE 20.2 Flowchart showing standards of industrial waste discharge under law enforcement.[61]

TABLE 20.1

Various Effluents, Corresponding Parameters and Their Risk to Public Health[60]

S. Number	Effluent	Level of Contaminant Identification	Harmfulness to Ecosystem – Health Hazards	Worth of Applicability
1	Apparently clean and colorless but contains toxic wastes, e.g., pesticides	Expensive and complex to perform	Prolonged and high	Practically applied to some extent
2	Bright colored containing a variety of contaminants	Easy and rational to perform	May or may not be prolonged but high	Practically applied to some extent
3	Purely colorless containing toxic wastes, e.g., pesticides	Same as effluent 1	Prolonged and high	Practically applied on a large scale
4	Purely colorless containing very small amount of toxic wastes	Same as effluent 1	Low	Practical broad-range applicability
5	It lies in between effluent 2 and 4	Varying between effluent 2 and 4	May be prolonged and high	Practically applied on a large scale

- Through revenue- and development-based investment, technology transfer from developed countries encourages industry synergism and helps industries move towards low-waste technology[60]

20.8 Future Perspective

The supply and disposal areas of the settlement structure are currently experiencing great changes. At one point, the center of attention is the user's energy, water and food/commodity supply; while on the other hand, the focus is also on the treatment of effluents. Therefore, discharge comprises reusing and removal procedures and corresponding treatment techniques. In previous times, through the establishment of a sewerage system leading to a traditional wastewater treatment plant (WWTP) as a solution for wastewater treatment, the basic interests of health and water preservation have been observed. At least in industrial countries, the outbreaks of disease because of lack of sewage have been greatly prevented. With increased knowledge it is known that effluent components (such as carbon, nitrogen and phosphorus) can cause siltation, oxygen absorption and waterlogging of water bodies. Hence, wastewater treatment plants are in implementation, which greatly improves the peculiarity of water bodies. Anyhow, the point is that the goal of durable effluent treatment has been carried out in this way.

By merging the protection of health and water resources, the sewage treatment plant can alter its scope from (only) wastewater/effluent treatment to (system) service providers:

- Sewage treatment and water protection service: Maximizes protection of sewage from the discharge and wastewater treatment of settlement structure to safeguard the receiving water body. Wastewater treatment is an important component to ensure water ecosystem services (basic, supply, control and cultural services, etc.).

- Energy service provider: Engaging with the energy industry in the energy sector for energy consumption, production and storage.

- Manufacturers: Providing water and fertilizers.

Sewage treatment plants play an important role in the circulation of the economy. New ideas in urban water handling can also make important contributions. With the new challenge of "resource conservation," wastewater treatment companies will face a new self-image. Services and products have to be deliberately adjusted in accordance with the requirement such as water bodies, energy industries and agriculture. As a result, detailed solutions have led to interdisciplinary collaborations among toxicologists,[61–63] power grid and virtual power plant operators,[64] and experts in the fertilizer industry. In relation to inter-city cooperation, new models of organizational structure and business models may necessarily become the need of the hour.

In the long-term plan regarding sewer treatment plants, the establishment of new sewer treatment plants and the upgrading of existing sewage treatment plants would be a focus and would also include the removal of nitrogen and phosphorus. Therefore, control over the nutrient removal process will remain an important issue. There are still great expectations for cheaper and more durable online sensors, especially for the measurement of nutrients. Instrument redundancy and sensors with integrated data quality verification and a fault detection system are very important. Software-based monitoring and detection are other increasingly important areas. It is also reasonable to conclude that WWT control is moving towards the use of multi-input handling. In addition, the main concern seems to be related to the sludge problem. Reducing the production of sludge and treating excess sludge in an effective and environmentally friendly way, such as by recovering phosphorus, will be a major research area. Energy issues and previously unmonitored low concentrations of substances in water and sludge, for example micropollutants, endocrine disruptors, pathogens and new heavy metals, are other important themes. For new process technologies, the highest expectations are for membrane processes.[65]

Moreover, constructed wetlands are developing rapidly, and various hydrological models and structures can be used for design and construction to emphasize specific characteristics to improve treatment performance.

20.9 Conclusion

Textile industries, for example, are distributed all over the world. Wastewater from these industries directly or indirectly increases the threat to aquatic systems. Therefore, it is necessary to identify the impact of wastewater on aquatic ecosystems so that preventive measures can be taken before the wastewater is discharged into rivers. If the toxic substances present in wastewater are not treated before discharge, these toxic chemicals will be incorporated into marine animals, enter our food chain through these foods and cause adverse effects on human health.

It is estimated that 60% of Pakistan's fresh water is wasted because of mismanagement; just 40% of water is utilized for industrial, household and domestic tasks.[66] According to the National Bureau of Statistics, about 56% of the total world population consumes pure water.[67] However, taking into account the international standards for pure drinking water (WWF, 2007), only 25.6% of the population can use drinking water (WWF, 2007). The Pakistan Council of Research in Water Resources initiated a range of projects to investigate the quality of water in about 23 big cities in Pakistan from 2002 to 2006, which has expanded to more than 25 cities in 2015. Studies have shown that, on average, up to 89% of water resources in the country are lower than the recommended human consumption standards (PCRWR, 2008a).[68]

In order to improve the level of performance of removing biodegradable hard organic compounds in industrial wastewaters, for example in textile industries, oil fields/refineries, and pulp and paper and coffee-processing industries, strict strategies like aeration have increasingly been adopted. However, these strategies and their elimination performance and economic balance need further investigation. Moreover, the current industrial wastewater supervision and management systems need to be comprehensively improved depending on standards, settings, evaluation and implementation. As industrial waste behavior changes the standards, it is necessary to strengthen the water quality monitoring system, including water supplies coverage, parameters and timely reports to the public domain. It is time to move to ecosystem-specific emission standards to maintain the health and productivity of the natural resources people depend on. However, regulations are only effective if you have a strong monitoring component. The right path here is to ensure that the community has the right to a healthy environment and ultimately survival.

Ecosystem management is an adaptable process. Given our limited understanding of dynamic natural processes, there is at first no right answer, only the right direction. When no development strategies other than industrial growth are considered, environmental management, which is closely related to poverty alleviation, cannot be carried out. In addition to the corresponding policies to limit resource consumption and achieve intergenerational equity, restrictions on industrial emissions would be quite effective in order to carry out wastewater management. Based on technical capabilities, framework and the capacity to withstand industrial pressures, local communities have their initial limitations in terms of surveillance and law enforcement. However, as a comprehensive participatory approach, communities need opportunities and the time to consider and develop decisions that influence their subsistence, eventually ensuring its viability.

REFERENCES

1. A. Sonune and R. Ghate, *Desalination*, 2004, **167**, 55–63.
2. G. Crini and E. Lichtfouse, Green adsorbents for removal of antibiotics, pesticides and endocrine disruptors. In *Green Adsorbents for Pollutant Removal*, Springer Nature Switzerland, 2018, 327–351.
3. M. K. David, *A Review Paper on Industrial Waste Water Treatment Processes*, University of Nigeria, 2017.
4. N. H. Tran, M. Reinhard and K. Y. Gin, *Water Research*, 2018, **133**, 182–207.
5. M. Nafees, A. Nawab and W. Shah, *Journal of Engineering and Applied Sciences*, 2015, **34** (1), 1–9.
6. S. K. Koli and A. Hussain, Status of Electronic Waste Management in India: A Review. In *Advanced Treatment Techniques for Industrial Wastewater*, IGI Global, 2015, 238–250.
7. Y. J. Chan, M. F. Chong, C. L. Law and D. G. Hassell, *Chemical Engineering Journal*, 2009, **155** (1–2), 1–18.
8. S. Krishnan, H. Rawindran, C. M. Sinnathambi and J. W. Lim, Comparison of various advanced oxidation processes used in remediation of industrial wastewater laden with recalcitrant pollutants. *IOP Conference Series: Materials Science and Engineering*, 2015, IOP Publishing.
9. M. R. Al-Mamun, S. Kader, M. S. Islam and M. Z. H. Khan, *Journal of Environmental Chemical Engineering*, 2019, **7** (5), 103248.
10. K. Piaskowski, R. Swiderska-Dąbrowska and P. K, Zarzycki, *Journal of AOAC International*, 2018, **101** (5), 1371–1384.
11. M. Doble and A. Kumar, *Biotreatment of Industrial Effluents*, Elsevier Butterworth-Heinemann, 2005, 1–336.
12. C. Zaharia, D. Suteu, A. Muresan, R. Muresan and Alina Popescu, *Environmental Engineering and Management Journal*, 2009, **8** (6), 1359–1369.
13. F. H. Hussein and T. A. Abass, *International Journal of Chemical Sciences*, 2010, **8** (3), 1353–1364.
14. M. Kasina, M. Wendorff-Belon, P. R. Kowalski and M. Michalik, *Journal of Material Cycles and Waste Management*, 2019, **21**, 885–896.
15. M. Luan, G. Jing, Y. Piao, D. Liu and L. Jin, *Arabian Journal of Chemistry*, 2017, **10**, S769–S776.
16. A. S. Oliveira, J. A. Baeza, B. S. Miera, L. Calvo, J. J. Rodriguez and M. A. Gilarranz, *Journal of Environmental Management*, 2020, **274**, 111199.
17. N. Arumugam, S. Chelliapan, H. Kamyab, S. Thirugnana, N. Othman and N. S. Nasri, *International Journal of*

Environmental Research and Public Health, 2018, **15** (12), 2851.

18. Y. Deng and R. Zhao, *Current Pollution Reports*, 2015, **1**, 167–176.

19. K. Turhan and S. A. Ozturkcan, *Water, Air, & Soil Pollution*, 2013, **224** (1), 1353–1358.

20. M. Bourgin, E. Borowska, J. Helbing, J. Hollender, H. Kaiser, C. Kienle, C. S. McArdell, E. Simon and U. Gunten, *Water Research*, 2017, **122**, 234–245.

21. M. Hassan, X. Wang, F. Wang, D. Wu, A. Hussain and B. Xie, *Waste Management*, 2017, **63**, 292–298.

22. J. A. Giroto, A.C.S.C. Teixeira, C.A.O. Nascimento and R. Guardani, *Chemical Engineering and Processing: Process Intensification*, 2008, **47** (12), 2361–2369.

23. I. Tyagi, V. K. Gupta, H. Sadegh, R. S. Ghoshekandi and A. S. H. Makhlouf, *Science Technology and Development*, 2017, **34** (3), 195–214.

24. A. Pires, J. Morato, H. Peixoto, V. Botero, L. Zuluaga and A. Figueroa, *Science of the Total Environment*, 2017, **578**, 139–147.

25. S. N. Ahmed and W. Haider, *Nanotechnology*, 2018, **29** (34), 342001.

26. G. Vereb, L. Manczinger, A. Oszko, A. Sienkiewicz, L. Forro, K. Mogyorosi, A. Dombi and K. Hernadi, *Applied Catalysis B: Environmental*, 2013, **129**, 194–201.

27. S. K. Papageorgiou, F. K. Katsaros, E. P. Favvas, G. E. Romanos, C. P. Athanasekou, K. G. Beltsios, O. I. Tzialla and P. Falaras, *Water Research*, 2012, **46**, 1858–1872.

28. Y. Zhu, Y. Wang, Z. Chen, L. Qin, L. Yang, L. Zhu, P. Tang, T. Gao, Y. Huang, Z. Sha and G. Tang, *Applied Catalysis A: General*, 2015, **498**, 159–166.

29. T. Das, X. Rocquefelte, R. Laskowski, L. Lajaunie, S. Jobic, P. Blaha and K. Schwarz, *Chemistry of Materials*, 2017, **29** (8), 3380–3386.

30. J. Rodriguez, F. Paraguay-Delgado, A. Lopez, J. Alarcon and W. Estrada, *Thin Solid Films*, 2010, **519** (2), 729–735.

31. J. Shu, Z. Wang, G. Xia and Y. Zheng, *Chemical Engineering Journal*, 2014, **252**, 374–381.

32. R. J. Isaifan, A. Samara, W. Suwaileh, D. Johnson, W. Yiming, A. A. Abdallah and B. Aissa, *Scientific Reports*, 2017, **7**, 9466.

33. M. C. Mascolo, Y. Pei and T. A. Ring, *Mater*, 2013, **6** (12), 5549–5567.

34. L. Han, Q. Li, S. Chen, W. Xie, W. Bao, L. Chang and J. Wang, 2017, *Scientific Reports*, **7**, 11622.

35. D. Masih, Y. Ma and S. Rohani, *Applied Catalysis B: Environmental*, 2017, **206**, 556–588.

36. M. J. Sampaio, M. J. Lima, D. L. Baptista, A. M. T. Silva, C. G. Silva and J. L. Faria, *Chemical Engineering Journal*, 2017, **318**, 95–102.

37. K. S. Novoselov, A. K. Geim, S. V Morozov, D. Jiang, Y. Zhang, S. V. Dubonos, I. V. Grigorieva, and A. A. Firsov, *Science*, 2004, **306** (5696), 666–669.

38. W. Wang, Z. Wang, J. Liu, Z. Luo, S. L. Suib, P. He, G. Ding, Z. Zhang and L. Sun, *Scientific Reports*, 2017, **7** (1), 1–9.

39. A. Di Mauro, M. Cantarella, G. Nicotra, G. Pellegrino, A. Gulino, M. V. Brundo, V. Privitera and G. Impellizzeri, *Scientific Reports*, 2017, **7**, 1–12.

40. D. M. EL-Mekkawi, N. A. Abdelwahab, W A. A. Mohamed, N. A. Taha and M. S. A. Abdel-Mottaleb, *Journal of Cleaner Production*, 2020, **249**, 119430.

41. R. V. Solomon, I. S. Lydia, J. P. Merlin and P. Venuvanalingam, *Journal of the Iranian Chemical Society*, 2012, **9** (2), 101–109.

42. J. Li, S. Zhang, C. Chen, G. Zhao, X. Yang, J. Li, and X. Wang, *ACS Appl. Mater. Interfaces*, 2012, **4** (9), 4991–5000.

43. M. N. Chong, B. Jin, C. W. K. Chow and C. Saint, *Water Research*, 2010, **44** (10), 2997–3027.

44. A. Trapalis, N. Todorova, T. Giannakopoulou, N. Boukos, T. Speliotis, D. Dimotikali and J. Yu, *Applied Catalysis B: Environmental*, 2016, **180**, 637–647.

45. S. Malato, P. Fernandez-Ibanez, M. I. Maldonado, J. Blanco and W. Gernjak, *Catalysis Today*, 2009, **147** (1), 1–60.

46. J. Low, J. Yu, M. Jaroniec, S. Wageh and A. A. Al-Ghamdi, *Advanced Materials*, 2017, **29** (20), 1601694–1601714.

47. D. Spasiano, R. Marotta, S. Malato, P. Fernandez-Ibanez and I. Di Somma, *Applied Catalysis B: Environmental*, 2015, **170–217**, 90–123.

48. F. Shiraishi and T. Ishimatsu, *Chemical Engineering Science*, 2009, **64** (10), 2466–2472.

49. R. Chowdhury, H. Khan, E. Heydon, A. Shroufi, S. Fahimi, C. Moore, B. Stricker, S. Mendis, A. Hofman, J. Mant and O. H. Franco, *European Heart Journal*, 2013, **34** (38), 2940–2948.

50. A. Wosnie and A. Wondie, *African Journal of Environmental Science and Technology*, 2014, **8**, 312–318.

51. B. T. Sarfraz, M. Fazil, T. M. Malik, S. Bashir and H. Tariq, *PAFMJ*, 2013, **63** (1), 85–88.

52. H. Waqas, A. Shan, Y. G. Khan, R. Nawaz, M. Rizwan, M. Saif-Ur-Rehman, M. B. Shakoor, W. Ahmed and M. Jabeen, *Human and Ecological Risk Assessment: An International Journal*, 2017, **23** (4), 836–850.

53. N. Galic, A. Schmolke, V. Forbes, H. Baveco and P. J. Brink, *Science of the Total Environment*, 2012, **415**, 93–100.

54. A. V. B. Reddy, M. Moniruzzaman and T. M. Aminabhavi, *Chemical Engineering Journal*, 2019, **358**, 1186–1207.

55. P. S. Goh and A. F. Ismail, *Desalination*, 2018, **434**, 60–80.

56. M. Kamali, D. P. Suhas, M. E. Costa, I. Capela and T. M. Aminabhavi, *Chem. Eng. J.*, 2019, **368**, 474–494.

57. T. Melin, B. Jefferson, D. Bixio, C. Thoeye, W. D. Wilde, J. D. Koning, J. Graaf and T. Wintgens, *Desalination*, 2006, **187**, 271–282.

58. G. Sandin and G. M. Peters, *Journal of Cleaner Production*, 2018, **184**, 353–365.

59. N. Johansson, and H. Corvellec, *Waste Management*, 2018, **77**, 322–332.

60. T. Rajaram and A. Das, *Futures*, 2008, **40**, 56–69.

61. C. Prasse, J. Wenk, J. T. Jasper, T. A. Ternes and D. L. Sedlak, *Environmental Science & Technology*, 2015, **49** (24), 14136–14145.

62. M. P. Shah, Removal of heavy metals using bentonite clay and inorganic coagulants. In *Removal of Emerging Contaminants Through Microbial Processes*, Springer, 2021, 1–542.

63. M. P. Shah, Industrial wastewater and persistent organic pollutants. In *Advanced Oxidation Processes for Effluent Treatment Plants*, Elsevier, 2020, 1–342.

64. P. Kehrein, M. van-Loosdrecht, P. Osseweijer, M. Garfí, J. Dewulf and J. Posada, *Environmental Science-Water Research and Technology*, 2020, **6** (4), 877–910.

65. M. A. Kahlown and M. Azam, *Agricultural Water Management*, 2003, **62** (2), 127–138.

66. M. N. Bhutta, M. R. Chaudhry and A. H. Chaudhry, Groundwater quality and availability in Pakistan. *Proceedings of the Seminar on Strategies to Address the Present and Future Water Quality Issues*, 2005, **36** (26), 15–26.

67. S. Farooq, I. Hashmi, I. A. Qazi, S. Qaiser and S. Rasheed, *Environmental Monitoring and Assessment*, 2008, **140** (1–3), 339–347.

68. M. Raza, F. Hussain, J. Lee, M. B. Shakoor and K. D. Kwon, *Critical Reviews in Environmental Science and Technology*, 2017, **47** (18), 1713–1762.

21

Nanostructured Photocatalytic Materials for Water Purification

Jennyffer Martinez Quimbayo, Satu Ojala, Samuli Urpelainen, Mika
Huuhtanen, Wei Cao, Marko Huttula and Riitta L. Keiski

CONTENTS

List of abbreviations

AMR	Antimicrobial resistance
AOII	Acid orange II
APIs	Active pharmaceutical ingredients
BC	Biochar
BPA	Bisphenol-A
BTF	Benzotrifluoride
CB	Conduction band
CECs	Contaminants of emerging concern
CNT	Carbon nanotube
CQDs	Carbon quantum dots
DDE	Dichlorodiphenyldichloroethylene
DNP	Dinitrophenol
ECs	Emerging contaminants
E_f	Fermi level
Eg	Band gap energy
EU	European Union
fcc	Face-centered cubic
G	Graphene
g-C_3N_4	Graphitic carbon nitride
hcp	hexagonal close-packed
MB	Methylene blue
MIL	Materials of Institut Lavoisier
MO	Methyl orange
MOFs	Metal organic frameworks
mpg-C_3N_4	Mesoporous carbon nitride
PDOP	Photocatalytic decomposition of organic pollutants
Ph	Phenol
PHE	Photocatalytic hydrogen evolution
RhB	Rhodamine B
SC	Semiconductor
TC	Tetracycline
UN SDGs	United Nations Sustainable Development Goals
VB	Valence band
WHO	World Health Organization

21.1 Introduction

One of the largest problems humankind is facing is the availability of clean water. The world's freshwater sources are in threat as a result of industrialization, urbanization and climate change. According to the WHO report, half of the global population will face water stress in 2025 (WHO, 2019). Every day, two million tons of sewage and other polluted water are discharged with inadequate treatments. The wastewater, either used as an irrigation source or directly disposed to hydrologic systems, poses a significant health risk and sabotages the natural environment. Water has even become a financial asset to the point that it is being sold in the stock and futures markets.

Today, the global demand for water is intense because it is needed for drinking water, sanitation and many other activities such as agriculture and energy production. The quality of water is being affected by the amount of chemicals in the environment. The water is contaminated with pesticides, herbicides, fungicides and insecticides, commonly used in agriculture all around the globe. Water is also contaminated by waste disposed from industries, hospitals, domestic sanitation, and sewage and water treatment plants. Water is even contaminated by individuals through personal care products, pharmaceuticals, cyanotoxins, steroids, hormones, caffeine, detergents, fragrances and more. Several of these new water pollutants are considered emerging contaminants (ECs), since their concentrations are currently low but expected to increase in the future. Furthermore, their harmful effects are not yet known exactly, but they are expected to increase when concentrations are increased (Bedia, 2019).

ECs require special attention as they are commonly used, and water treatment plans are not able to degrade them efficiently. Many international directives and goals demand good quality and safe aquatic ecosystems, such as the European Water Framework Directive, the Marine Strategy Framework Directive and the United Nations Sustainable Development Goals (UN SDGs). The WHO and the EU have created an Action Plan on Antimicrobial Resistance to minimize the risks caused by antimicrobial resistant bacteria (AMR). However, there is no existing legislation limiting EC emissions. The EU has launched a dynamic watch list that contains a maximum of 14 ECs that will be monitored more carefully for increased information on their exposure and hazards (Carvalho et al., 2015). It can be expected that new restrictions will be implemented when more reliable information on EC emissions and their effects is available. To ensure clean water free from ECs, their utilization should be restricted as the primary pollution prevention method. Since in many cases restricted utilization is not possible (e.g., pharmaceuticals), their release to the environment should be avoided. One way of reaching this goal is to develop new and more effective water treatment technologies.

Decontamination of wastewater from organic pollutants using sunlight has been considered as one of the key candidates for sustainable water purification. Many chemicals can be directly decomposed under sunlight radiation or indirectly through bioprocessing. However, direct degradation of industrial pollutants by sunlight is almost impossible, mostly due to stable carbon bonds in organic pollutants. Semiconductor photocatalysis has been considered one of the most promising cleantech processes for eliminating organic and inorganic pollutants. As in photosynthesis, photocatalysts use light to convert unwanted chemical species to unharmful or usable chemicals. The process can be carried out with sunlight, making it sustainable, cheap and useful for the removal of toxic pollutants (Santhosh et al., 2019). This chapter focuses on different photocatalytic materials and their use in the treatment of ECs.

Photocatalysts are materials that convert light into chemical energy, which makes the material catalytically active. In other words, the light is used to activate the photocatalyst. When light (a photon) of suitable wavelength (energy) is absorbed by a photocatalyst, an electron-hole pair is formed. If the charges can migrate to the surface of the catalyst, they may be transferred to surface adsorbates participating in the redox reactions required for example in water treatment. The fundamental mechanism of heterogeneous photocatalysis is illustrated schematically in Figure 21.1.

The starting point of the photocatalytic reactions in water is rather well known. The water oxidization takes place after the creation of a photoinduced hole (h$^+$) on the semiconductor through the reaction $H_2O + h^+ \rightarrow \bullet OH + H^+$. The hole energy should be larger than water oxidation or hydrogen dissociation energy of 1.23 eV. In a rather straightforward manner, water purification is continued with the oxidation of the organic pollutants by possible oxidants of h$^+$, e$^-$, $\bullet OH$, H$^+$ or even the superoxide radicals $\bullet O_2$ (Kong et al., 2018).

Although sunlight is abundant on our planet, high-performance sunlight-activated photocatalysts are yet to be discovered to make them viable in an industrial and commercial scale: In contrast to light from laboratory UV sources and setups (used in pioneering experiments), harvesting sunlight remains challenging. When solar radiation passes from one medium to another (e.g., from air to water), unwanted reflection takes place at the interface. This significantly decreases the amount of light available for reactions. Low photon energies in the solar spectrum set up another barrier in applications of semiconductor catalysis. Typically, the most abundant spectral region of sunlight spectrum is around 300–1000 nm, corresponding to 4.13 eV–1.24 eV in photon energy. Efficient photocatalysts should have band gaps within this region. However, semiconductors with smaller band gaps suffer from fast recombination between photoinduced electrons and holes, resulting in lower photocatalytic efficiency. Besides the band structure, the effective contact area between a pollutant and a catalyst is important for catalytic reactions. Large surface areas increase contact probability between the reactive sites of photocatalysts and water/pollutants. For this reason, advanced engineering of photocatalysts is necessary,

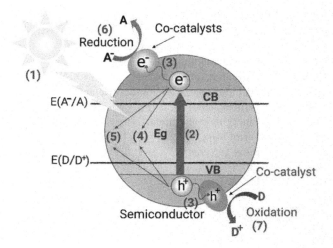

FIGURE 21.1 Mechanism of heterogeneous photocatalysis: (1) light harvesting; (2) charge excitation; (3) charge separation and charge transfer; (4) bulk charge recombination; (5) surface charge recombination; (6) surface reduction reaction; and (7) surface oxidation reaction.

and synthetic crystals are further engineered to nanoscale shapes increasing the surface area during modifying the band structure.

Despite many challenges, research concentrating on material development is very active, and numerous catalysts have already been developed for water purification applications. Examples of these materials are illustrated in Table 21.1. The next sections will discuss the physical and chemical properties of the different groups of materials used for photocatalytic water purification.

21.2 Structural, Physical and Chemical Properties

21.2.1 Description of Photocatalytic Materials

During the last decades, the study of photocatalytic materials used for water treatment has been a topic of great interest. This attention has resulted in the development and design of a vast number of photocatalytic materials and their modifications. To remove the contaminants from wastewater, a combination

TABLE 21.1

Examples of Photocatalytic Material Used for Water Treatment

Photocatalyst Type		Example	Method	Light Source	Pollutant	Reference
Metal oxides		TiO_2	Solvothermal	Solar light	Ofloxacin, sulfamethoxazole, carbamazepine, flumequine, ibuprofen	Carbajo, 2016
		WO_3	Hydrothermal	300W Xe lamp, $\lambda > 400$ nm	RhB	Murillo-Sierra, 2021
		$Nb-WO_3$	Hydrothermal	300 W high pressure lamp $\lambda = 248$ nm-1014 nm	MB	Murillo-Sierra, 2021
		$C-ZnO$	Hydrothermal	Solar light	RhB	Lee, 2016
Metal sulfides		MoS_2	Hydrothermal	Visible light	Thiobencarb	Huang, 2018
		CuS	Solvothermal	160 W, Hg vapor lamp – 12 W, UV lamp	MB, RhB, eosin Y, Congo red	Ayodhya et al., 2016
		$Cu-ZnS$	Coprecipitation	300 W Xe lamp, $\lambda > 420$ nm	Hexafluorobenzene	Kohtani et al., 2005
Nonmetal semiconductors		$Graphene-C_3N_4$	Impregnation-chemical reduction	350W Xe arc lamp, $\lambda > 400$nm	-	Xiang, 2011
		$N-CNT/mpg-C3N_4$	Thermal polycondensation	300 W Xe lamp, $\lambda > 400$ nm	RhB, MO	Liu, 2017
Magnetic composites		$ZnFe_2O_4$	Microwave	Simulated solar light	MB	Gómez-Pastora, 2017
		Fe_3O_4/TiO_2	Coprecipitation-sonochemical	UV light	RhB	Gómez-Pastora, 2017
		$TiO2/NiFe2O4$	Sol-gel	15W UV lamp, $\lambda = 253$ nm	MO	Jacinto, 2020
		$NiFe_2O_4/SiO_2/TiO2$	Solvothermal	UV light	Nitrobenzene	Gómez-Pastora, 2017
		$TiO_2/GO/CuFe_2O_4$	Ball-milling pathway	Visible light (300–400 nm)	Chlorinated pesticide	Ismael, 2020
Multicomponent materials	Binary compounds	CuS/TiO_2	Successive ionic layer adsorption and reaction (SILAR)	35 W Xe lamp	Enrofloxacin	(Jiang et al., 2019)
		$CoP/ZnSnO_3$	Hydrothermal	UV light	Tetracycline	(Chen et al., 2021)
	Ternary compounds	$TiO_2/WO_3/CQDs$	Hydrothermal-calcining	300 W Xe lamp, $\lambda > 420$ nm	Cephalexin	(Sun, 2021)
		$TiO_2/CdS/G$	Hummers	300 W Xe arc lamp	Benzyl alcohol	(Zhang, 2012)
		$Bi_2S_3/BiOBr/BC$	Solvothermal	50 W LED lamp, visible light	Diclofenac	(Li, 2020)
Metal organic frameworks (MOFs)		MIL-53(Al)	Hydrothermal	UV-Vis, visible light	MB	Gautam, 2020
		MIL-53(Cr)	Hydrothermal	UV-Vis, visible light	MB	Gautam, 2020
		MIL-68(In)-N H2/GO	Hummers-solvothermal	300 W Xe lamp, $\lambda > 420$ nm	Amoxiciline	Yang, 2017
		Fe-MIL-101	Hydrothermal	300 W Xe lamp, $\lambda > 420$ nm	Tetracycline	Wang, 2018

of different physical, biological and chemical methods is currently used. The photocatalytic degradation is interesting due to the effectiveness of the process. One of the advantages of this technique is that it uses O_2 as the main reactant, which is inexpensive and abundantly available. Photocatalysts are typically semiconductors, materials that have two separated energy bands for the electrons: The lower energy band occupied with electrons (valence band, VB) and the higher energy band free of electrons (conduction band, CB). The energetic barrier between these two bands is called band gap energy (Eg). When a photocatalyst receives energy equal to or higher than the bang gap energy, an electron (e^-) can be moved from the VB to the CB and a hole (h^+) is left behind in the VB. The photogenerated charges have two options: They can recombine and the energy is released as heat or light, or they can migrate to the surface of the semiconductor and generate reactive oxygen species. These oxygen species can couple to direct oxidation by h^+ and are able to mineralize the pollutant. During the process of recombination of e^-/h^+, other reactions can affect the global process, such as combination of HO_2^-, production of H_2O_2 and protonation of OH, resulting in the formation of water (Bedia, 2019; Kahng, 2020; Shah, 2020a, 2020b, 2021).

Effective photocatalysts have suitable CB and VB energy levels with good separation of charge carriers. Different materials can have these characteristics and are good candidates for photocatalysis. Here we classify the materials based on their chemical composition starting from simple binary metal oxides and sulfides, moving on to more complex ternary compounds, nonmetal semiconductors, multicomponent materials, magnetic composites and metal organic frameworks, or MOFs.

The design of photocatalytic materials aims at high photocatalytic activity, chemical stability and low toxicity. Another important factor is the radiation that can activate the principal photocatalytic phase of the material. For example, titanium dioxide (TiO_2), one of the most intensively studied photocatalysts, has a principal photocatalytic phase that can be activated just with UV radiation, covering only about 5% of the total electromagnetic solar spectrum. The present goal is to use solar energy to activate the photocatalytically active phase of the material. To reach this goal, modifications to the morphology and electronic structure of existing photocatalyst materials are frequently studied. The ultimate challenge will be to develop photocatalytic materials that can be used for water purification using visible and solar light instead of only UV light (Mamaghani, 2017; Shah, 2021).

21.2.2 Metal Oxides

Metal oxides have been recognized already for several decades as materials suitable for heterogeneous catalysis in the petrochemical industry, chemical synthesis, and more recently in environmental chemistry as depolluting catalysts. They are especially useful in oxidation and acid-base reactions, and their natural abundance, nontoxic nature and often suitable band gap make them interesting also for photocatalytic applications. Particularly interesting are the materials that have their band gap in the range of solar spectrum on earth, or in the visible range of the spectrum (ca. 400–800 nm, or 1.5–3 eV), and can therefore be activated by sunlight. Metal oxides

are either insulators or semiconductors, often having Eg in this range or close by. For example, TiO_2, one of the most widely studied photocatalysts, has Eg in the rage of 3.2–3.35 eV or about 380 nm (near UV) for its anatase and rutile forms.

An especially interesting class of materials is transition metal oxides and their nanostructures. For example, TiO_2 has been shown to be active for the removal of several ECs, such as ofloxacin, sulfamethoxazole, carbamezepine, flumequie and ibuprofen, to name a few (Carbajo, 2016). Several other transition metal oxide materials, such as WO_3 and ZnO, have been shown to work for rhodamine B and methylene blue (Murilllo-Siera, 2021; Lee, 2016). The research on photocatalytic activity and its origin has focused on different materials, and the photocatalysts are typically powders that have various exposed surface orientations. Research related to the photocatalytic activity of individual surface orientations has gained more momentum in the last decades through successful growth of nanomaterials and control of morphology (Batzill, 2011).

Metal oxides consist of metal (M) cations and oxygen (O) anions forming several types of oxides: suboxides, oxides, peroxides, superoxides and ozonides. Oxides have the oxide anion O^{2-} and peroxides the O_2^{2-} peroxide anion. Suboxides, on the other hand, have more metal cations than the regular oxides have. Superoxides and ozonides contain O_2^- superoxide ions and O_{3-} ozone ions, respectively. The oxidation state, stoichiometry and coordination of the metal cation play a key role in determining the structural and chemical properties of an oxide.

The metal ion typically has four or six oxygen ligands in its surroundings with binary oxides having the stoichiometries of MO, MO_2, M_2O_3 and M_3O_4. A metal ion with four oxygen ligands gives rise to a tetrahedral coordination, while six ligands result in an octahedral coordination. The crystalline structures can be then constructed by face-centered cubic (fcc) or hexagonal close-packed (hcp) oxygen ion lattices, with the metal ions occupying the empty interstitial sites. By filling these sites differently, several structures are obtained, the most common being rock salt, fluorite, spinel, perovskite, wurtzite, rutile and corundum (Sterrer & Freund, 2014). The chemical properties of a solid material are strongly coupled to the properties of its surface. The structure of the surface depends on the bulk structure, and in practice only certain cleavage planes can form stable structures on a metal oxide surface (Sterrer & Freund, 2014). Furthermore, the surface planes are rarely, if ever, perfectly flat. The surfaces of metal oxides consist of terraces with steps, kinks, corners and vacancies, and surface atoms have different coordination from bulk atoms. This affects the electronic structure and thus the chemical (and catalytic) properties of the surface. When the coordination of the metal atom changes, this will also influence the surrounding ligand field. For example, in transition metal oxides, such as NiO (Freund, 1995), the splitting of the d-orbitals and position and number of excited states are affected by reduced coordination at the surface. This reordering of the structure can extend several atomic layers deep into the material.

Typically, the bonding of adsorbates is weaker on an oxide surface (with weak multipole electrostatic forces) than on metal surfaces (with more covalent type bonding), although the electronic structure of the adsorbate obviously plays a role as well. The weaker bonding also affects the possible orientation of the adsorbates on the surface (Freund, 1995). In addition to adsorbate bonding, surface defects influence the chemistry of the surface. In TiO_2, for example, surface oxygen vacancies are non-dissociative binding sites for O_2, creating O_2^- superoxides on the surface. These superoxides act as oxidizing species for organic molecules or as hole scavengers on the surface. Molecularly adsorbed O_2 on the other hand can act as an electron scavenger of the photogenerated electrons, creating the reactive O_2^- superoxides that drive the photocatalytic oxidation processes (Thompson & Yates, 2005). The photocatalytic activity is also connected to the nature of the migration of the photoexcited electron and hole within the material. As the VB and CB, and thus also Eg, depend on the crystal facet that is exposed at the surface, the surface plays an important role in solar light absorption (Eg), charge transfer (band structure) and ability to catalyze redox reactions (band potentials). This has been shown to be true for TiO_2 (Liu, 2010) and WO_3 (Xie, 2012), for example. Furthermore, apart from band structure, the charge transfer may have a preferred direction due to polarization in ferroelectric photocatalysts, or electrostatic potentials induced by charged adsorbates, also resulting in different catalytic activity of different exposed surfaces and/or preference towards oxidation or reduction processes (Batzill, 2011). Together these properties make metal oxides good candidates for tailored nanomaterials for photocatalytic applications.

21.2.3 Metal Sulfides

Metal oxides described in the previous section are interesting because they have a wide band gap. Metal sulfides, on the other hand, are promising materials for catalytic and photocatalytic reactions due to their suitable electronic band gap, band position and exposed active sites. Metal sulfides generally have a shallow valence band, and they exhibit a robust quantum size effect due to their small effective mass. These types of materials are interesting because they are available in different shapes, sizes and crystalline forms. Their chemical compositions and the light response are excellent. For these reasons, they can contribute to harvesting the complete solar light spectrum (Chandrasekaran, 2019).

To create a metal sulfide, it is necessary to combine a metal or semi-metal cation with a sulfur anion. Mono-metal sulfides (e.g., FeS, Fe_3S_4, ZnS, CuS, CuS_2) can be represented as M_xS_y. Bi-metal sulfides also exist and follow the form $A_{1-x}B_xS_y$ (where x and y are integers). ZnY_2S_4, $CuSbS_2$ and Cu_5FeS_6 are some examples of bi-metal sulfides. Designed tri-metal sulfides are also possible, and some examples of common tri-metal sulfides are Cu_6WSnS_8, Cu_6GeWS_8 and $PbFeSb_6S_{14}$. Generally, the metal sulfides have a simple structure and high symmetry (Weber, 1998). Sulfur (powder), carbon disulfide, glutathione, sodium thiosulfate and dimethyl sulfoxide are some of the common sulfur source materials used to prepare a metal sulfide.

In a transition metal sulfide, the d electrons are part of a covalent bonding. This covalent bonding changes the charge of the transition metal, and the formation of metal-metal bonds are more probable. These interactions are important because

they play a fundamental role in determining the electronic structure of the material and the charge transfer properties. Metal sulfides are typically semiconducting as the bonding of the transition metal sulfides increases the semiconducting properties. However, metal sulfides can also exhibit metallic properties, and in some cases even superconducting properties are present (Rao, 1976).

The structure of metal sulfides is quite different from the structure of metal oxides. Some factors contributing to the differences between these materials are as follows (Weber, 1998):

- The molecular anions can present a rise due to the sulfur-sulfur bonding.
- The effective charge on the ions can be reduced because of the large covalency.
- The formation of a layered structure has more probability because the anions have a large polarizability. However, the possibility of a cubic spinel structure also exists (e.g., $CuCr_2S_4$, $LiTi_2S_4$ and $CdIn_2S_4$).

Generally, the transition metal mono-sulfides crystallize in an NiAs (B8) structure, with a space group $P6_3/mmc$. Here the Ni atoms have an fcc environment and the As atoms have an hcp environment. They can also crystallize in variations of the structure. Transition metal bi-sulfides, for example, form layered compounds using mono- and diatomic anions. These bi-sulfide materials can crystallize in diverse types of structures, and the cations in layered compounds can occupy the octahedral or the trigonal interstices. The crystalline structure can also be affected by the metal atom vacancies. Then the NiAs (B8) structure will become probable and monoclinic structure, or superstructures may occur (Rao, 1976). Depending on the solvent and method used to synthesize the material, there is evidence that the photocatalytic activity can change depending on these factors (Bedia, 2019).

In recent decades, the metal sulfides have been studied as hydrogen evolution photocatalysts. The mono-metal sulfides MoS_2 and ZnS have high crystallinity and defect morphologies that are crucial for the photocatalytic activity. MoS_2 has a high stability and excellent electronic arrangement and light absorption properties. However, there are some disadvantages as bulk MoS_2 has an indirect band gap of 1.2 eV that is unsuitable for initiating the photocatalytic reaction. To overcome this problem, different morphologies of the material can be explored to modify the band gap. For example, in the case of 2D MoS_2 nanosheets, the quantum confinement causes a different band gap (1.96 eV), which is suitable for the visible light absorption. In the case of the ZnS, it has been reported that without sacrificial electron donors the colloidal ZnS presents photocatalyst degradation, as the catalysts turned black after a small quantity of H_2 evolution. Nevertheless, when ZnS is in the presence of SO_3^{2-} in solutions and using UV radiation, a good quantum yield of approximately 90% and photocatalytic H_2 production are observed (Chandrasekaran, 2019).

With regards to the bi-metal sulfides, an example of a photocatalysts is $ZnIn_2S_4$, which is a semiconductor chalcogenide that is chemically stable and has an energy band gap suitable for the absorption of visible light. In addition, it is an eco-friendly photocatalyst that has been synthesized in different morphologies (nanoparticle, microspheres, nanoribbons, nanotubes). In the literature, three microstructures (such as flowering-cherry, microspheres and microclusters) are reported using different solvents such as water, methanol and ethylene glycol. The photocatalytic activity of these three different structures of ZnIn$_2$S$_4$ is the highest (hydrogen production rate 27.3 µmol h^{-1}) in an aqueous environment under visible light. In methanol, the photocatalytic activity has a decrease in the H$_2$ production rate of only 12.4 µmol h^{-1}. The ethylene glycol ZnIn$_2$S$_4$ mediated photocatalytic activity is even smaller with an H$_2$ formation rate of 9.1 µmol h^{-1} (Chen, 2016).

In comparison with metal oxides, metal sulfides are a new class of materials that can be used also for storage applications and energy conversion. They have a low cost with attractive properties such as redox reversibility, good optical properties, high conductivity and high electrochemical and photocatalytic activity. Many of the semiconducting metal sulfides have a lack of optical absorption, and for this reason cannot show photocatalytic activity in the range of the solar spectrum. To overcome these restrictions, the creation of heterojunctions may be useful. One example is WS$_2$ heterostructures with a mass ratio of 1.6:1 to fabricate WS$_2$-CdS. The synthesis resulted in nanorods of CdS with nanosheets of WS$_2$ attached. This heterostructure shows an excellent H$_2$ generation rate of 12.4 µmol h^{-1}g^{-1} in visible light for 20 mg catalysts (Chandrasekaran, 2019). Other types of heterojunctions and multicomponent materials will be described in section 21.2.4.

In summary, metal sulfides have good photocatalytic activity. Depending on the crystalline structure of these materials, the functional properties, amount and layer arrangement can change. For photocatalytic hydrogen generation applications,

the metal sulfides have an appropriate band gap, eco-friendly nature and low cost that make them suitable candidates for photo-responsive environmentally sustainable materials. Mono- and bi-metal sulfides have limitations due to photodegradation. Creating heterojunctions can improve the photocatalytic activity by exposing more active sites, facilitating charge carrier separation and boosting the electron transport. Heterojunctions of metal sulfides can also reduce the energy needed to activate the photocatalyst. The results obtained for these kinds of materials so far are intriguing, and metal sulfide heterojunctions should be studied further for increased photoactivity and whether they could replace other, more expensive, metal catalysts like Pt and Ru.

21.2.4 Multicomponent Materials

Ternary composites typically take three distinct components for forming a heterojunction system. Different from conventional homogeneous or binary systems where the charge transfer is relatively straightforward, in ternary composites the charge migrates between the heterosites through a buffer medium, thus delaying the electron-hole pair recombination. Strictly speaking, if interfaces of two different chemical sites A and B are counted as an individual composite AB, the binary system is also composed of three chemically distinct composites, while a binary system could be considered a pentanary system. Here we use the original definition of the ternary system where the nanojunctions *j* bridge the two components A and B forming an A*j*B system.

Different types of heterojunctions are illustrated in Figure 21.2. Two of the most common ternary systems

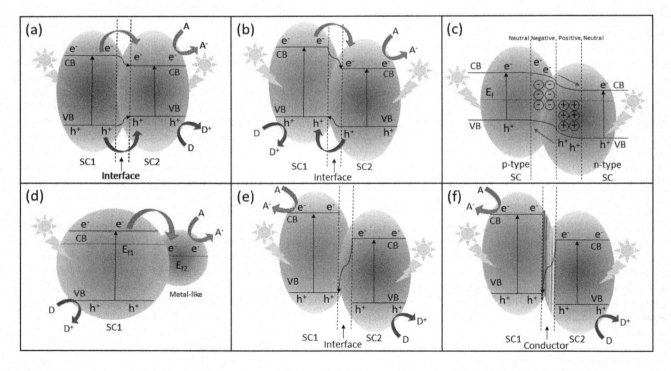

FIGURE 21.2 Different types of heterojunctions: (a) type-I heterojunction, (b) type-II heterojunction, (c) p-n junction, (d) Schottky junction, (e) Z-scheme heterojunction (without electron mediator) and (f) indirect Z-scheme (with electron mediator).

are categorized as Type-II (Figure 21.2b) and Z-scheme (Figure 21.2e) photocatalysts, with part in the S-scheme as frequently turned out in the heterodifunctional photocatalysts. In Type-II, photoexcited electrons or holes migrate from the CB or VB of A/B to the corresponding CB or VB of B/A via the *j* tunnel. Z-scheme systems can be further divided to the direct Z-type and the all-solid-state Z-type: When the recombination of an electron-hole pair happens between the electrons at the A site and holes at the B site, the holes at A and electrons at B sites are left available for redox reactions. This scenario is described as the direct Z-scheme, which is not limited to the ternary, but is also possible in the binary composites. The all-solid-state Z-type is defined in the case that the recombination happens on the joint side after charge migration. It is a unique feature mainly on the ternary system due to the irreplicable role of the interfacial joint. In both Z-schemes, these leftover charges at A and B sites are highly reactive and so the catalytic activities are enhanced (Xu, 2018). The multi-paths of type-II, S-scheme and Z-scheme have also been reported latterly following the complexity of development of the ternary systems (Shi, 2020). Owing to the dedicated design of charge migration channels, many ternary systems are capable of both photocatalytic water treatment and hydrogen evolution.

So far, many ternary systems have been developed for photocatalysis due to the vast number of possible combinations among compounds beside their unique catalytic activities. Thanks to its pioneering position in photocatalysis, TiO_2 has been used as the matrix and combined to various oxides, sulfides and nitrides. Despite many of its benefits, the large band gap substantially limits the visible light harvesting capabilities. During the past two decades, non-TiO_2 photocatalysts have been studied and developed as the matrix or composite in ternary systems (Marchelek, 2016). Among them, various Bi compounds are used in ternary systems. For example, bismuth oxide, zinc oxide, metal chloroxide, vanadate, phosphate and tungstenade have been combined among themselves or with other composites. The narrow band gap induced by Bi 6s and O 2p hybridization has enabled a wide range of solar energy harvesting. It is worth mentioning that many ternary systems with Bi utilizes the metallic Bi nanoparticles as a component due to the photoinduced surface plasmon resonance that creates hot electrons boosting photocatalytic activity (Wang, 2020). Metal sulfide matrices have been employed as a part of Z-scheme catalyst designs and connected to various oxides, metallic junctions and other compounds (Di, 2019). Among the compounds without metal involvement, graphitic carbon nitride (g-C_3N_4) has been identified as an important species, in parallel to its pioneering peer graphene. It has been reported to be a good matrix for developing the ternary systems with the Z-scheme, type-II, and combined type-II and Z-scheme due to its layered structure and easily modulated band gap. During the materials engineering, combinations with compositions have been performed directly with oxides, sulfides or through buffer layers of metallic nanoparticles (Liao, 2021).

During the past decade, 2D semiconductors have turned out to be another emerging type of host, or composite, in ternary systems thanks to their large surface area and tunable (by number of layers) band gap. Different from semi-bulk or nanoscale ternary systems, these rather thin slabs of 2D materials require

sophisticated materials engineering to ensure charge flow and migration between the slabs and other sites. In comparison to the 2D-nD (n = 0,1) counterparts, the 2D-2D heterojunctions have superior surface-surface charge migration channels and contact areas via interplane interfacial overlapping. Hence, the Z-schemes have been realized on slabs containing MoS_2, transition metal carbides and nitrides (MXenes), and black phosphorus, to name a few (Li, 2018). The predicted novel 2D systems are also proposed for photocatalysis, starting with adsorption and charge transfer interactions. However, these structures need further engineering efforts to be realized and combined in form of composites, and their functionalities remain yet to be verified (Kistanov, 2021).

The ternary systems are endowed with superior photocatalytic activities compared with those from their individual components and binary peers. For example, in water purification, the degradation rate of methylene blue (MB) was increased from ~17% on the flake MoS_2 to ~46% on MoS_2-Au composites, and to ~98% on the MoS_2-Au-Ni ternary composites under the same test conditions and 90 minutes of UV light radiation (Shi, 2018). Additionally, many ternary systems can remove persistent pollutants, which is hardly done with the individual components. An 85% removal of tetracycline (TC) and 65% of 2,4-Dinitrophenol (DNP) were reported by using the $(CuC_{10}H_{26}N_6)_3(PW_{12}O_{40})_2$/AgCl@Ag as photocatalysts under simulated sunlight in two hours (Chen, 2019). Degradation of ~88% TC was reported with the existence of Ag_3PO_4/$Co_3(PO4)_2$/g-C_3N_4 under visible-light irradiation ($\lambda > 420$ nm) in two hours. The rate is much larger than those from binary peers containing the g-C_3N_4 (Shi, 2020).

The ternary systems have also been found to be very catalytically active in hydrogen evolution. A high rate of 19.28 μmol g^{-1} h^{-1} was reached when using $(CuC_{10}H_{26}N_6)_3(PW_{12}O_{40})_2$/AgCl@Ag as the catalyst under the 300 W Xe irradiation and with 10 vol% triethanolamine (TEOA) as the sacrificial reagent. Such an evolution rate could be kept at least for 24 hours (Chen, 2019). The rate was boosted to 301 μmol g^{-1} h^{-1} with the ZnO/ZnS/g-C_3N_4 and the mixture sacrificing agents of Na_2S and Na_2SO_3, much higher than 4,1, and 192 μmol g^{-1} h^{-1} for the g-C_3N_4, ZnO nanoparticles and ZnO/ZnS peers (Dong, 2018). Similarly, the g-C_3N_4/Ag/MoS_2 ternary system owns 8.78-fold and 2.08-fold in the production of hydrogen (H_2) out of water compared to the binary peers of Ag/MoS_2 and g-C_3N_4/MoS_2 respectively under visible light (Lu, 2017).

As an emerging photocatalytic group, the ternary systems have been developing very rapidly during the past five years, and the trend is continuing. Large varieties in compositions, directional charge transfer schemes and stoichiometry of these combinations enrich the family of ternary compounds and enable their versatility in photochemical processes. Despite the striking functionality, controlled synthesis of the multicomponent heterojunctions is not an easy task. Being composites, the ternary catalysts suffer from relatively poor photocorrosion resistance in organic solutions and acidic conditions. Pure water splitting without a co-catalyst or sacrificing agents is scarcely reported on the systems. Furthermore, it is rather hard to directly probe the catalytic mechanism on these systems due to their complicated structural and electronic properties. Rational design and facile realization of the ternary systems is foreseen as challenges but also as opportunities to implement practical photocatalysis on these high-performance catalytic hosts.

21.2.5 Nonmetal Semiconductors

Nonmetal semiconductor photocatalysts comprise materials such as the graphitic carbon nitride (g-C_3N_4), G/C_3N_4 composites, mesoporous carbon nitride (mpg-C_3N_4), and its doped versions with nonmetal materials. This group of materials has attracted interest during the last years due to the versatile properties of the materials, including their crystal structure, surface area, stability, optical adsorption and electronic properties. The design efforts to improve these photocatalysts have focused especially on tuning the band gap and pore structure as well as surface sensitization and heterojunction construction. The photocatalytic activity of these materials has been tested in the degradation of different pollutants such as dyes and emerging pollutants (pharmaceuticals and pesticides) (Wen, 2017).

The C_3N_4 has seven different phases with different band gaps. From these seven phases just two have band gaps suitable for visible light absorption: g-h-triazine and g-heptazine with band gaps of 2.97 and 2.88 eV, respectively. The g-C_3N_4 is a material with a suitable band gap that can be used for photocatalytic applications. Due to its many advantages, such as its visible light response, easy availability, nontoxicity and thermodynamic stability, g-C_3N_4 is a promising photocatalyst (Fu, 2018; Wen, 2017). The good chemical stability of g-C_3N_4 originates from the optimized van der Waals interactions between single layers making it insoluble in water, acids, bases and some organic solvents such as ethanol, toluene and diethyl ether. The surface of polymeric g-C_3N_4 contains defects that can be useful for different functionalities such as the activation of CO_2 and, in comparison with a defect-free g-C_3N_4, the polymeric form is more interesting for catalytic applications. Additionally, the basic groups (-NH-, =N-, -NH$_2$ and -N-C=) that are on the surface can contribute to the removal of acidic toxic molecules using chemical adsorption (Wen, 2017). Also, it is possible to design heterostructures and composites with g-C_3N_4 to extend the light absorption and increase the specific surface area and active site density. The design of suitable heterostructures is a viable strategy for improving the separation efficiency of electron-hole pairs leading to better photocatalytic performance. The heterostructure can be realized in different ways, and metal oxides and metal sulfides can be used as parts of the heterostructure (Fu, 2018). One example is the composite of graphene/g-C_3N_4 reported in the literature by Xiang et al. (2011). They studied the optimal graphene content for the composite and found this material to be a promising visible light activated photocatalyst that is inexpensive and metal free (Xiang, 2011).

Interesting options for photocatalysts are mesoporous carbon nitrides (mpg-C_3N_4) with an estimated band gap of 2.73 eV. They have a considerably larger surface area that can provide more active sites compared to g-C_3N_4 but the disadvantage of low quantum yield. To overcome this, one option is coupling mpg-C_3N_4 with another co-catalyst to achieve higher photocatalytic activity. Liu et al. (2017) have reported

the use of nitrogen-doped carbon nanotubes as co-catalysts. The incorporation of a nitrogen atom into the graphitic structure of carbon nanotubes induces an excellent rate of electron transfer and a suitable electronic conductivity (Liu, 2017). The N-doped CNT has a more defective structure than the undoped CNT has. This characteristic of the structure can improve the photocatalytic activity. N-CNT reaches its maximum photocatalytic activity with visible light radiation, and the use of this material has been evaluated in the degradation of dyes and pharmaceutical waste in water with promising results (Liu, 2017).

Nonmetal semiconductors can be used to increase the light-harvesting capabilities of a photocatalyst in combination with, for example, metal oxides such as TiO_2. In the context of degradation of organic and inorganic pollutants, the use of heterojunctions made of g-C_3N_4/TiO_2 for the degradation of formaldehyde in gas phase has been demonstrated (Fu, 2018). Furthermore, it has been shown that doped-g-C_3N_4 can be used for the degradation of dyes such as rhodamine B (RhB) and methyl orange (MO) and that the materials based on g-C_3N_4 can be used as photocatalyst for water treatment under visible light (Xiang, 2011; Fu, 2018).

Nonmetal semiconductors based on g-C_3N_4 are attractive photocatalytic materials due to their nontoxic nature and low cost. Heterojunctions with graphene and carbon nanotubes have been studied demonstrating photocatalytic activity under visible light radiation. The activities of these materials have been evaluated with different pollutants and have shown results that make them a promising group of photocatalysts with good stability and surface properties that can be useful in water treatment. However, their design in different types of heterojunctions based on g-C_3N_4 with nonmetal options has not been explored extensively. In addition to the design of heterojunctions, the charge carrier transfer mechanism, thermodynamics and kinetics of the surface should be studied more deeply (Fu, 2018).

21.2.6 Magnetic Composites

During the last years, the development of magnetic materials as photocatalysts for water treatment has been explored. The most attractive property of these kind of materials is the possibility to recover the photocatalysts from a liquid by applying a magnetic field. This recovery process is easier than centrifugation at high speed. The efficient recovery of the catalyst is especially important when a heterogeneous catalyst is used as a powder suspended in the polluted water. The recovery of the catalyst from the liquid phase after the degradation of pollutants enables also recycling of the catalyst (Jacinto, 2020). Generally, magnetic composites and nanoparticles have a core-shell structure that is formed with magnetic and photocatalytic materials. The inner core can be composed of magnetic materials or oxides, such as magnetite or cobalt ferrite (Gómez-Pastora et al., 2017). The technique for creating a magnetic photocatalyst is based on the idea of coupling a known photocatalyst with a material that has magnetic properties. This way, the new photocatalyst can be more easily isolated from a solution and its reusability can be studied (Ismael, 2020). In the literature, there are reports of magnetic nanocomposites that have been used successfully for the degradation of organic pollutants. Even though research into these materials has increased during the last years, the design, separation and recovering systems for magnetic materials can be improved and explored.

To observe a general overview about the different characteristics of these material, it is possible to divide the magnetic photocatalysts into two groups: (1) magnetic photocatalysts based on TiO_2 and (2) other magnetic composite structures (Gómez-Pastora et al., 2017).

1. *Magnetic photocatalysts based on TiO_2*: In this group there are three main configurations of photocatalysts with titania.
   1. A magnetic core coated with a titania nanoshell (core/TiO_2). The core can be made of iron oxides or ternary oxides.
   2. Magnetic core with an interlayer and a titania outer shell (core/coating/TiO_2). For this configuration, the preferred material for the coating is SiO_2 due to its mesoporous characteristics.
   3. Configuration with a magnetic core and coating, but with doped titania (core/coating/TiO_2-doped). For doping, both metallic and nonmetallic materials can be used.

In this group, the most common configuration is the core-shell structure. The inner core supplies the magnetic properties to the composite. For this purpose, core nanocrystals based on nanocomposites are used. To control the size distribution thermal methods, microemulsion or sonochemical methods are studied. However, these methods suffer from low productivity and high reaction times. To avoid negative effects that can emerge from physical contact between the magnetic core and the TiO_2 shell, it is necessary to have a coating between them. Various materials for the coating have been reported in the literature. Among them, using the mesoporous silica is promising, because it provides a large surface area with well-defined pore sizes. Another advantage of the mesoporous silica is that the pores are useful in avoiding the magnetic attraction between the magnetic cores. For producing this type of coating, hydrolysis and the condensation of a sol-gel precursor are used. TiO_2 is then deposited on the silica coating contributing to the photocatalytic activity of the composite. The most common method for deposition is the sol-gel method (Gómez-Pastora et al., 2017; Jacinto, 2020).

There are several studies on magnetic photocatalysts based on TiO_2. One example is the incorporation of Fe_3O_4 and spinel ferrites ($CoFe_2O_4$, $NiFe_2O_4$) into the TiO_2, changing the magnetic properties of the photocatalysts (Ismael, 2020). Furthermore, the use of metal oxides such as magnetite and maghemite (Jing, 2013) and metals like Ni, Zn, Co and Ba have also been successfully employed as magnetic core materials of a TiO_2 composite (Fu, 2005). The photocatalytic activity of these materials has been studied mainly on the degradation of dyes and organic pollutants under UV irradiation. However, there are some studies that report the degradation of other kind of pollutants such as phenols and pesticides (Gómez-Pastora et al., 2017; Ismael, 2020).

Jing et al. (2013) have reported success using a magnetite-TiO_2 (Fe_3O_4) composite with a concentration of 1.5 photocatalysts gL^{-1} and a concentration of 71 mgL^{-1} for the removal of the carcinogenic quinoline from water. The solution was irradiated with UV for two hours and the photocatalysts reached a 90% of removal of the pollutant. The reusability of the photocatalyst was also studied, and after three cycles the photocatalysts showed a photodegradation of 84%, which is approximately 95% of the initial photocatalytic activity (Jing, 2013). The possibility of the core photo dissolution due to the contact between the magnetic core and the TiO_2 shell is a disadvantage. This phenomenon changes the properties of the core and deteriorates the photocatalytic activity of the TiO_2. An intermediate coating, such as SiO_2, is often used to avoid direct electrical contact between the core and photocatalysts and prevent charge transfer from TiO_2 to the magnetic core. SiO_2 also has absorbent properties that can improve the photocatalytic performance of the composite (Gómez-Pastora et al., 2017). This configuration of magnetic composites has shown a good photocatalytic performance on the degradation of different organic compounds under visible and UV light: Xue et al. (2013) have reported the degradation of MO using $Fe_3O_4/SiO_2/TiO_2$. After 180 minutes of irradiation with visible light, the degradation range was 93% (Xue, 2013).

The last configuration of TiO_2-based magnetic photocatalysts is the case of the core, intermediate coating and doped TiO_2 (core/coating/TiO_2-doped), where the dopants can be noble metals such as Ag, Au and Pt. The most studied dopant material is Ag, which improves the performance of the magnetic composite. For example, $Fe_3O_4/SiO_2/TiO_2$-Ag can achieve complete degradation of RhB under UV light after 10 minutes of irradiation, whereas without the dopant the photocatalyst can degrade only 80% of the pollutant (Chi, 2013). The use of other dopants is also possible. Rare earth metals and nonmetal groups have been reported to improve the photocatalytic activity with lower degradation times under visible and UV light.

2. *Other magnetic composite structures*: These magnetic composites are structures that are not based on TiO_2 and appear in three different configurations:
   1. Bare magnetic materials with photocatalytic properties, such as Zinc ferrites and other ternary oxide structures.
   2. Core shell structures with a magnetic core and a catalytically active nanoshell (core/coating/catalyst) consisting of a metal oxide or ternary compounds.
   3. Core shell structures using doped photocatalysts (core/coating/catalyst-doped). In this configuration the dopants can be based on polyoxometalate or metal-doped catalysts.

The most common configuration of the second group is the core/catalyst structure using other photocatalytic material (than TiO_2) such as ZnO. In this configuration, the core is responsible for the magnetic properties, and the external catalytic material provides the photocatalytic activity of the material. The magnetic core is synthesized in the same fashion as for TiO_2-based magnetic composites. For synthesizing the photocatalytic nanoshell, coprecipitation of metal salts is used, for example for ZnO (Gómez-Pastora et al., 2017; Jacinto, 2020).

Different magnetic composite photocatalysts have been successfully used for degradation of organic compounds. To analyze the photocatalytic activity of these materials, it is possible to continue with the same two groups of the previous section (magnetic composite structures based on titania and without titania).

The number of studies on materials that are not based on TiO_2 is small compared to those with TiO_2. Some of the studies report the use of $ZnFe_2O_4$ in the degradation of dyes achieving complete degradation under UV and visible light. Other organic compounds have also been tested, such as phenols under UV irradiation. The most studied configuration is the core/coating/photocatalyst configuration. This configuration has been reported to achieve high rate of degradation of the insecticide diazinon, bisphenol and different dyes. Also, the use of an external noble metal-doped photocatalyst layer is reported, and in some cases, it is possible to active the photocatalyst under visible light (Gómez-Pastora et al., 2017; Jacinto, 2020).

From the environmental point of view, the use of magnetic photocatalysts is interesting because recovering the photocatalyst is easy, and the reusability of the photocatalysts can be evaluated. Different types of configurations of these composites have been explored to find a way to improve the photocatalytic activity of these materials. However, there are some research challenges related to the magnetic composite photocatalysts such as complete recovery of the photocatalysts, design and synthesis methods. The ideas behind these challenges are achieving photocatalytic activity under visible light, assuring a better photocatalytic performance and avoiding the possibility of toxic secondary effects due to the discharge of the photocatalyst into the environment. These materials should be studied further to evaluate their efficiency in degradation of emerging pollutants.

21.2.7 Metal Organic Frameworks (MOFs)

During the last two decades, a certain group of materials have become interesting targets in photocatalysis research owing to their large and porous surface that is useful for water treatment and eliminating contaminants from wastewater. Porous solids have been playing an important role in catalysis as host materials useful in different applications. These materials can be divided into amorphous solid or structured (crystalline) solids (Gautam, 2020). An example of ordered solids is the metal organic frameworks. MOFs have a crystalline structure and a well-defined network, and different approaches are suitable for MOFs to modify their photocatalytic activity. These include using organic linkers with absorption bands in the visible region spectrum avoiding charge recombination, encapsulating chromophores in the internal structure of MOFs or promoting electron-hole separation in the metal structure like in metal oxides. Since MOFs have a wide variety of chemical and structural modifications, they have had a big impact on the

research of the photocatalytic treatment of pollutants in water during the last decades.

The structure of MOFs is achieved by the assembly of two components: the secondary building units (SBUs) that are clusters or metal ion nodes, and organic linkers between the SBUs. Usually, they have a crystalline structure with a porous texture (Bedia, 2019). MOFs have the necessary components for catalytic activity, because they are formed from transition metals (Mn, Fe, Co, Ni, Cu) and the organic linkers have C, H, O, N, S, and so on. Some examples of organic linkers that can be used are halides, anionic organic molecules (benzene dicarboxylic acid), neutral organic molecules (4,4-bipyridine) and cyanides. An interesting feature of MOFs is their variety of possible structures: linear, square planar, square pyramidal, tetrahedral, octahedral or trigonal-bipyramidal. The organic and inorganic components can form n-dimensional structures with n = 1, 2, 3. In comparison with zeolites, the bonds between the organic and inorganic connections of these materials are weaker, and these bonds are responsible for the structural stability of MOFs under external conditions (Gautam, 2020).

MOFs exhibit properties such as high porosity, large surface area and stability in different environments, and they have low cost. The MOFs can maintain their porous structure without guest molecules. The structure does not collapse under vacuum, and the ligands of the material play a crucial role in forming the symmetrical pore sizes and in the absorption capacity. MOFs generally have a large dimension due to the pore diameter that can be more than 10 Å in many MOF structures. To control the pore environment, the organic ligands can be changed to control the surface area of the MOF. The necessary characteristics of a MOF material are linking units, strong bonding providing robustness and high crystalline nature (or well-defined structure) (Gautam, 2020).

The combination of metal ions and organic linkers provides a huge number of possibilities of MOF materials. For example, it is possible to create different MOFs using the same SBU, but different organic linkers. For example, zeolite imidazolate frameworks (ZIF), ZIF-7 unit cell, ZIF-8 unit cell and ZIF-90 unit cell all use the same SBU (Zn), but the organic linker is different (Bedia, 2019).

MOFs have been studied for photocatalytic applications due to their semiconductor properties. In the case of wastewater treatment, the optimization of the size and surface activation are parameters to be considered. Studies on the photocatalytic degradation of organic pollutants and dyes and the photoreduction of heavy metals in water solutions have shown that the stability of MOFs in water for long periods of time and the pH ranges are crucial. One of the limitations of MOFs is their large band gap, capable of being activated only with UV light.

To enhance the photocatalytic activity, functionalization and heterostructures using semiconductors can be applied (Bedia, 2019). The MOFs can be functionalized during the synthesis step by using an organic linker that has the proper functional groups, such as replacing terephthalic acid by 2-aminoterephthalic acid to have amine groups in the MOF. They can also be modified using post-synthesis (post-synthetic modification, PSM). PSM consists of the functionalization of a previously synthetized MOF, and its main strain is to not disturb the initial structure during the process. One example of PSM is the deposition of metal nanoparticles in the MOF. MOFs can also be combined with common semiconductors, such as ZnO, TiO$_2$ and graphene. These heterostructures have been revealed to exhibit good catalytic performance, as well as changes in their luminescence and absorption properties. Furthermore, they can reduce the charge recombination with better charge separation.

MOFs can be used as efficient photocatalysts for eliminating organic dyes and treating wastewater. There is an important class of MOFs called MIL (Materials of Institut Lavoisier) that have been used as successfully as photocatalysts. The first attempt to use MOFs for methylene blue (MB) degradation was in 2011 with three different MIL (MIL-53(Al), MILS-53(Fe), and MIL-52(Cr)). The photocatalytic activity was studied with UV and visible lights, and after 40 min of irradiation, MIL-54(Fe) was able to degrade 11% of MB while the other two materials exhibited just a 3% efficiency, showing low photocatalytic activity. Since then, different experiments with MOFs have been performed to improve their photocatalytic activity in the degradation of dyes. A successful example is the UiO-66(Ti) nanocomposite for the degradation of MB dye using sunlight. The addition of Ti did not affect the framework and enhanced the photodegradation, resulting in a removal efficiency of 82.2% (Gautam, 2020).

Heavy metals such as Hg, Cd, Cr and As can be also found in wastewater. Heavy metal compounds are toxic for humans and aquatic organisms. In the literature, MOFs have been used for the removal of heavy metals from water (Bedia, 2019). In most of the studies found in the literature, the heavy metal is attempted to be removed by adsorption instead of trying to convert the toxic species into another with lower toxicity. Removing Cr(VI) from water is an important objective for researchers, as the use of Cr compounds has been increasing in various industries. Examples of MOFs that have achieved complete Cr(VI) degradation are the H$_2$-MIL-88 and NH$_2$-MIL-101 in a period between 40 min and 60 min. Other examples include the NH$_2$-MIL(Fe), which have reached complete degradation and can be reused without affecting their stability and performance. It is assumed that the high performance of these materials is caused by the incorporation of NH$_2$ in the linker promoting charge transfer between the organic linker and the iron cluster, thus improving the light absorption.

In the case of primary pollutants like phenol and related phenol compounds, the use of MOFs is unexplored. Only a handful of attempts have been reported on using MOFs for the degradation of phenol and phenol compounds. In contrast, there have been more efforts towards using MOFs as photocatalysts for the degradation of emerging pollutants. Some examples of these attempts include the MIL-101 that achieved 100% degradation of tetracycline, In$_2$S$_3$/MIL-125(Ti) heterostructure that under visible light can degrade 60% of the same pollutant, and the novel composite Ag/AgCl@MIL-88A(Fe) that after 3 hours with visible light can degrade 100% of ibuprofen (Bedia, 2019).

MOFs have versatile properties and a wide range of applications (Gautam, 2020). Some of the MOF and MOF-based composites have shown stability and reusability for the catalytic water treatment. However, MOFs have an inferior stability compared to other photocatalysts such as TiO$_2$, CdS and

BiVO$_4$. A reason for their instability can be the self-dissolution of some of the components into the medium. Indeed, the reduction in the photocatalytic activity after several cycles is caused by the dissolution of the MOF components and the deactivation of the surface generated for the absorption (Bedia, 2019). Some applications of MOFs, such as the removal of primary pollutants, have not been explored. Due to their versatile properties and the fact that they can be modified to absorb light in the visible range, MOFs can be considered an efficient type of photocatalysts with important applications in water treatment.

21.3 Preparation of Photocatalysts

As discussed in the previous section, various photocatalytic materials exist. The method used for the synthesis of a material is important because it can improve several characteristics, but it can also generate impurities that contaminate the catalyst. Each method will have its advantages and disadvantages and, depending on the targeted use of the material, one of them could be more suitable than the other. For example, the photocatalytic H$_2$ production rate depends on the density of the active sites, the exposed facets of the crystal and the surface areas that need to be taken into account when selecting the synthesis method (Kahng, 2020). The photocatalytic efficiency of a material can be altered by changing the morphology (nanorods, nanowires, urchin-like shape, nanoparticles, hollow spheres, etc.) of the material. In addition, besides the synthesis method, the addition of co-metal ions (e.g., Pt, Co, Fe, Ag) as dopants will enhance the catalytic activity and selectivity as well as the visible light absorption capacity (Ge, 2019). The next sections will present a few common synthesis methods and aspects related to the immobilization of the catalysts.

21.3.1 Most Common Synthesis Methods

Depending on the type of photocatalysts, some methods are more common than others. For metal oxide photocatalysts, solvothermal or hydrothermal synthesis and sol-gel synthesis are the most common methods used. For metal sulfide photocatalysts, in addition to the previous methods, precipitation and impregnation-sulfidation synthesis are used. For carbon materials, other methods available include the hummers method and thermal polymerization (Kahng, 2020). For the metal organic frameworks, slow evaporation, microwave-assisted synthesis and electrochemical synthesis also exist (Bedia, 2019). The next paragraphs will describe some of the most common methods of synthesis that are used for photocatalyst preparation.

Solvothermal or Hydrothermal Synthesis

The difference between solvothermal and hydrothermal synthesis is that for the first a solvent or a combination of solvents is used, and in the second the solvent used is water, which is cheaper and environmentally friendly. Generally, solvothermal synthesis occurs at low temperatures, no more than 200 °C, and is realized using closed vessels under pressure above the boiling point of the solvent with a preparation time from 48 to 96 hours (Mamaghani, 2019; Dey, 2014). TiO_2 nanofibers are synthesized, for example, by using aqueous NaOH solution with 24 h stirring at 175 °C. Hydrothermal syntheses are found to be good for simple production on a large scale (Wu, 2011).

Sol-Gel Synthesis

One of the characteristics of this method is that the texture and the surface of the material can be controlled. Sol-gel synthesis proceeds in five steps (Parashar, 2020):

a. Hydrolysis – mix the precursors (metal alkoxides) with the solvents (organic solvent or water).

b. Polycondensation – eliminate the water or alcohol while the metal oxides linkages are formed; then the viscosity of the solvent increases creating porous structures with liquid that are called gels.

c. Aging – change the structure of the gel, while the polycondensation still continues, decreasing the porosity.

d. Drying – eliminate the water and organic components from the structure of the gel.

e. Thermal decomposition – remove the remaining water molecules and their residues.

Slow Evaporation

Slow evaporation is a conventional and common method used for synthesis because it does not need external energy. Its advantage is that it generally can be performed at room temperature, and its disadvantage is that it requires more time than other methods do. The time frame for the synthesis can take from seven days to seven months (Parashar, 2020).

Microwave-Assisted Synthesis

The microwave-assisted method of synthesis is generally used with MOFs but can also be employed for metal oxides. The quality of the crystals produced with this method is almost the same as that obtained with solvothermal synthesis. In microwave-assisted synthesis, the solution is heated with microwave radiation for a period of approximately one hour. The time frame for the synthesis can vary from about four minutes to four hours (Parashar, 2020; Bedia, 2019).

Electrochemical Synthesis

The use of the electrochemical synthesis method is a good option for continuous production of MOFs, and for this reason is a good alternative for industrial production. Electrochemical synthesis reduces the film cracking that may occur with other methods and does not require salts in the solution. The principle consists of providing the metal ion by an anodic dissolution into mixtures that include organic linkers and electrolytes (Parashar, 2020; Bedia, 2019).

As an example, Table 21.2 shows how the preparation method affects the size of the photocatalyst.

21.3.2 Immobilization of Photocatalysts

One crucial issue in water purification is the form of the catalyst in the fluid. Powder-form or nano-sized particles are in many cases quite difficult and costly to recover and reuse. Various methods such as sol-gel, dip coating, sputtering, etching,

TABLE 21.2

Photocatalysts Prepared Using Different Synthesis Methods, Certain Characteristics, Example of Pollutants and Light Used in Activation (Gómez-Pastora et al., 2017)

Photocatalyst	Method	Characterization	Pollutant	Light
$ZnFe_2O_4$	Microwave-hydrothermal ionic liquid	Size: <50 nm	Ph	UV light
$ZnFe_2O_4$	Colloid mill and hydrothermal	Size: 9 nm, specific surf. area (SSA): 100.50 m^2g^{-1}	AOII	UV light
$ZnFe_2O_4$	Coprecipitation	Size: <500 nm	MO	UV light
$ZnFe_2O_4$	Solvothermal	Size: 212 nm, SSA: 51.81 m^2g^{-1}	RhB	UV light and visible light
$ZnFe_2O_4$	Microwave sintering method	SSA: 6 m^2g^{-1}	MB	Simulated solar light
$ZnFe_2O_4$	Coprecipitation	SSA: 35.4 m^2g^{-1}	Ph	UV light
Fe_3O_4/TiO_2	Coprecipitation and sonochemical	Size: 40 nm	RhB	UV light
Fe_3O_4/TiO_2	Coprecipitation and heteroagglomeration	Size: 100 nm, SSA: 63.5 m^2g^{-1}	MB	UV light and sunlight
Fe_3O_4/TiO_2	Precipitation and sol-gel	Size: 17 nm	Quinoline	UV light
Fe_3O_4/TiO_2	Sonochemical	Size. 50 nm	MB	Visible light
Fe_3O_4/TiO_2	Mechanical	Size: 35 nm, SSA: 117.7 m^2g^{-1}	MB	UV light
$ZnFe_2O_4/TiO_2$	Sol-gel	Size: 7.4 nm	MO	Visible light
$ZnFe_2O_4/TiO_2$	Solvothermal	Size: 20 nm	RhB	UV light

casting, chemical and physical vapor depositions among others have been studied in immobilization of photocatalysts on different supports (e.g., on glass, steel, carbon, zeolites, ceramics and polymers). The pre-treatment and modification of the support surface may contain various steps to make it attractive for catalyst layer immobilization. After coating, the produced catalyst/carrier complex may additionally need calcination, activation or post-treatment to achieve the highest possible surface density of photocatalyst in the final product (Borges et al., 2015; Srikanth, 2017; Zakria, 2021; Wood, 2020).

The aforementioned immobilization techniques could be applied for the treatment of aqueous and gaseous streams. However, none of these techniques are investigated widely in the continuous aqueous stream treatments, and thus more studies are needed to find out their suitability under real conditions. The immobilization of catalysts has also some chemical and physical drawbacks, especially when the catalysts are used in aqueous solutions, such as in water and wastewater treatment. The most common problems are the unwanted removal of the catalyst from the support (mechanical abrasion and friction), maintaining the photocatalytic material activity (there should be no binders covering the particles) and finding the proper binders to be used (e.g., ones that are photochemically inert) (Zakria, 2021).

Both slurry-type and immobilized catalyst photoreactors have certain advantages and disadvantages compared to one another. In a slurry reactor, the distribution of the catalyst throughout the reactor volume is better, deactivation rate is lower and the pressure drop in the reactor is smaller. However, there are also disadvantages, such as the scattering of light due to the particles and costly separation stages. The most important advantages of immobilized catalyst reactors are the continuous operation and ease of separation, but at the same time the catalyst washout, mass transfer limitations and lower activity of the material must be dealt with (Srikanth, 2017). One possibility to achieve photocatalytic degradation and separation in one unit is to use photocatalytic membranes. It is essential to understand the interaction between the catalyst and the membrane material as the chemical and physical properties (e.g., bonding, charges, porosity, permeability and durability) affect the activity, selectivity and separation efficiency. The immobilization method is dependent on the membrane material and whether the catalyst particles are to be on the membrane surface or inside the membrane material. Thus, the methods used are, for example, solvent casting, electrospinning, deposition, spinning and sputtering (Zakria, 2021; Salazar, 2020; Martins, 2019).

21.4 Photocatalysis in Water Treatment

The scientific research on the development of photocatalytic materials for the degradation of water pollutants has continued almost for 40 years. During this time several advancements have been made, leading to improvements in the efficiency of the removal of model pollutants and the decrease in energy consumption. As of today, very few reports exist related to upscaling the photocatalytic units to practical water treatment levels. Several limitations for practical application of photocatalytic processes are seen, and they remain understudied. The research lacks information, for example, on the long-term stability (deactivation) of the photocatalytic materials in real water treatment conditions, matrix effects appearing in the case of real waters (combination of several pollutants in different nature), toxicity of the possible intermediate degradation products, costs of production, robustness of the technology and separation of the photocatalyst from treated water (Loeb et al., 2018). According to current understanding, a photocatalytic treatment process could find its application either as a pre-treatment technology to convert recalcitrant compounds into biodegradable ones or as a post-treatment to improve the overall purification efficiencies in cases where other (or conventional) water treatment technologies fail (Rathnayake, 2020). In this perspective, the chapter will focus now on the photocatalytic treatment of emerging pollutants.

The contaminants of emerging concern (CECs) in water are a class of different chemical compounds and microbes (bacteria, viruses) that are very difficult to treat. Some of these contaminants are present in the ecosystem at quite low concentrations – some of them already at levels known to be harmful for humans. Their concentrations will increase due to their increased use and current improper water treatment, and they may end up in drinking water or in food through irrigation. CECs are released, for example, from industry, but also from hospitals or due to their use in daily life. Pharmaceuticals (antibiotics, pain killers, hormones, etc.), artificial sweeteners, personal care products, nanomaterials, persistent organic pollutants, endocrine disruptors, pesticides and herbicides end up into water bodies, since they are resistant against currently used primary (physicochemical) and secondary (biological) treatment techniques for wastewater purification (Fanourakis, 2020; Vilé, 2020). To reach the complete removal of these compounds, new technologies are developed, one of which is photocatalytic treatment.

One group of emerging contaminants in water is pharmaceuticals. Urban areas are the sources of complex mixtures of pharmaceuticals and biological pollutants. Part of the active pharmaceutical ingredients (APIs) and their metabolites are passed through toilets to wastewater treatment plants. The pharmaceutical industry and hospitals are adding to this load. While industrial contamination in high-income countries is at rather low levels due to strict legislation and good manufacturing practices, the situation in low-income countries may be much worse (Vilé, 2020). The different APIs have different effects on the environment. Diclofenac, a widely used anti-inflammatory drug, is one of the most frequently detected pharmaceuticals in wastewater plants. Diclofenac has been shown to cause neurological and oxidative stress for many aquatic animals and is a source of several diseases (Kimbi Yaah, 2021).

Endocrine-disrupting compounds are substances that interfere with the normal functioning of the human endocrine system, in which hormones control our metabolism, growth and development, and reproduction. Bisphenols, phthalates, pesticides, dioxins and estrogenic hormones, to mention a few, are detected in wastewater plant effluents more and more frequently, and they are known to have adverse endocrine-disrupting effects on fauna (Belgiorno, 2007). Bisphenol-A

(BPA), one of the endocrine-disrupting compounds, has been used in plastics and liners of beverage bottles since the 1940s. It is a contradictory chemical as it has been widely used even though its adverse effects were known even before its commercial production. Later, its use has been banned in several countries and products. BPA affects the structural development of the brain, disrupts estrogen regulation and has also effects on social behavior (Rathnayake, 2020). These are only a few examples of the emerging pollutants whose removal from the wastewaters needs to be resolved.

In photocatalytic treatment, the pollutants are degraded with the help of radical species (such as $\cdot O_2^-$, $\cdot HO_2$ and $\cdot OH$) or via direct oxidation with excited electrons and remaining positive holes generated in/on the photocatalyst during light irradiation. The aim is to reach the complete degradation of the pollutants and form mainly H_2O and CO_2 as the reaction products. This occurs through several elementary steps of reaction that depend on the molecule to be degraded (Vilé, 2020; Augugliaro, 2012). As an example, the photodegradation of benzene in the presence of TiO_2 starts with the reaction with an $\cdot OH$ radical and formation of phenol as the main reaction intermediate. In addition, muconaldehyde (hexa-2,4-dienedial, $C_6H_6O_2$) is formed. The oxygen for the hydroxylation may originate from both molecular oxygen and water. In the case of anatase, 70%–90% oxygen originates from water, while in the case of rutile, the amount remains at the level of 20%–40%. Further hydroxylation of phenol leads to the formation of catechol and hydroquinone together with oxidized benzoquinones that are further degraded to CO_2. The muconaldehyde route leads more easily to the formation of carboxylic acids by oxidation, and then to CO_2. It has been proven that during benzene oxidation, most of the CO_2 is formed via the muconaldehyde route (Augugliaro, 2012). This example illustrates well that both the properties of the catalyst and the pollutant affect the degradation. Quite often the last steps of degradation involve the formation of short organic acids. Therefore, simple pH measurement can be used in certain cases to follow the degree of mineralization.

Earlier in this chapter, different photocatalytic materials were categorized as metal oxides, metal sulfides, nonmetal semiconductors, multicomponent materials, magnetic composites and metal organic frameworks. Thus, we will focus first on the metal oxides used in water purification, and especially in the case of emerging pollutants.

Starting with *metal oxides*, *TiO₂*, and particularly P25 (aeroxide), is the most studied photocatalyst in water purification applications in general. TiO_2 is considered unselective in such a way that it can oxidize most of the pollutants (Augugliaro, 2012). TiO_2 is active in ambient conditions and its costs are low, promoting its application on a larger scale. However, TiO_2 has a wide band gap and requires UV irradiation for high performance. Since sunlight would be the most viable long-term source of energy, several modifications are proposed to reduce the TiO_2 band gap. There exists, however, certain very promising results related to the use of TiO_2 alone. For example, Carbajo et al. (2016) reported results on using the P25 (85% anatase/15% rutile) and self-made TiO_2 (100% anatase) photocatalysts in photocatalytic degradation of five pharmaceuticals. They found that sulfamethoxazole (antibiotic)

and carbamazepine (anticonvulsant) require five to six times' longer treatment for complete conversion than do ofloxacin (antibiotic), flumequine (antibiotic) and ibuprofen (anti-inflammatory). When these pollutants were combined with natural water matrix, the complete degradation of ofloxacin, flumequine and ibuprofen were affected significantly. In these pilot experiments, about 35 min was required for the removal of these five pollutants together in the presence of wastewater matrix. The photocatalytic experiments were realized using solar light irradiation and compound parabolic collector tubes, where the wastewater was in turbulent flow conditions. The total volume of the wastewater was 35 L and the irradiated volume 22 L. The total concentration of pharmaceuticals was 500 µg L^{-1} (Carbajo, 2016).

Tungsten trioxide (WO₃) is another interesting metal oxide for wastewater treatment, although it is much less studied than TiO_2. WO_3 has the ability to oxidize a wide range of organic pollutants that arises from its slightly higher CB edge than the H_2/H_2O potential and much positive VB edge than the H_2O/O_2 potential. The monoclinic phase of WO_3 has been observed to present the highest photocatalytic activity. Generally, WO_3 is less active than TiO_2 in degradation of organic pollutants when irradiated using UV light. However, doping of WO_3 with noble metal nanoparticles enhances its activity under visible light irradiation, making it more active than N-doped TiO_2. In this type of material, the noble metal (such as Pt) acts as a pool for the electrons injected from the WO_3 conduction band leading to more efficient reduction of the adsorbed oxygen molecules instead of significant recombination of electron-hole pairs typical for WO_3 (Dong, 2016; Murillo-Siera et al., 2021).

Metal sulfides, such as ZnS and CdS, are quite widely studied nanophotocatalysts. For example, nanoporous ZnS nanoparticles can be produced using a facile method and lead to a catalyst that is able to degrade eosin B dye more efficiently than commercial TiO_2 can. It has also been found that solid solution binary sulfides composed of crystalline ZnS and CdS are able to overcome the low photostability of the binary sulfides reported along the quick recombination rates of charge carriers (Xu, 2019). MoS_2 is another interesting material for photocatalytic applications. As bulk material, MoS_2 is not photocatalytically active due to its S-Mo-S coordination in the crystal lattice. However, when employed as layered nanomaterial, such as two-dimensional nanosheets, the indirect band gap of bulk MoS_2 is transformed to direct band gap (1.9 eV), which is in the level high enough for photocatalytic reactions and separation of charge carriers. In case of pollutant degradation, MoS_2 has found its role mainly as a co-catalyst that suppresses the recombination of charge carriers and improves the photostability of, for example, Ag-containing catalysts. Using MoS_2 alone in, for example, decomposition of organic pollutants is not effective, since the valence band edge potentials of bulk MoS_2 (1.23 eV) and monolayer MoS_2 (1.9 eV) are not effective enough for the production of radicals needed in pollutant oxidation. However, in combination with Ag_3PO_4 and TiO_2, MoS_2 effectively improves the photocatalyst's stability, making the material more promising for practical applications (Li, 2018). Interestingly, despite the conclusions made on bulk and layered MoS_2, information also exists on the activity of MoS_2 alone in the photocatalytic degradation of pollutants.

Huang et al. (2018) produced MoS_2 microspheres from sodium molybdate and ɪ-cysteine using the hydrothermal method. MoS_2 microspheres were used in the degradation of thiobencarb, a commonly used carbamate pesticide. The degradation was realized under visible light irradiation leading to 95% degradation during 12 h in pH range of 6–9. They found that the hydroxyl radicals and photogenerated holes were the main active species, and the degradation efficiency was maintained during three successive runs. As very often found for photocatalytic degradation, the real water matrix was decreasing the degradation of thiobencarb to around 70%. However, added inorganic ions had only a minor effect (Huang, 2018). These results shed some light for the possibilities of different types of MoS_2 structures in photocatalytic water treatment.

To improve the photocatalytic activity of the materials, *multicomponent materials* are an option that can combine different photocatalysts with diverse morphologies. These novel materials can be used for water treatment and can have a better performance than the single component materials. One example is the heterojunction made of Bi_2S_3/BiOBr/BC (where BC is denoting biochar) that can be synthetized using the solvothermal method. This multicomponent material shows a good performance in the degradation of diclofenac. Previous studies have reached less than 40% removal efficiency of diclofenac in water. The experiment with Bi_2S_3/BiOBr/BC used visible LED light to simulate solar light radiation to activate its photocatalytic activity. To evaluate the photocatalytic activity, a solution with a concentration of diclofenac of 10 mgL^{-1} in 50 ml of water together with 0.03 g of the photocatalysts was used. To reach absorption/desorption equilibrium, the solution was stirred for 30 min in the dark. For the irradiation, a 50 W visible LED light was used with a predominant wavelength of 475 nm for 40 min. The better performance was exhibited when the ratio of Bi_2S_3/BiOBr/BC was 1% Bi_2S_3 and 10% biochar that could reach a removal efficiency percent of 93.65%. This improvement on the photocatalytic activity can be attributed to the perfect selection of the heterostructure, the electron conductivity and the electron transfer ability of biochar (Li, 2020).

Inorganic semiconductors based on *ternary compounds* are also used for water treatment and H_2 production from water. The ternary photocatalysts contain active catalysts with electron transfer mediators. They can be in the forms of 2D materials, 1D nanorods, and 0D nanoparticles (Rani, 2020). Different studies have been published concerning the use of ternary nanocomposites for pollutant removal. One example is TiO_2/CdS/G (where G denotes graphene), which has good photocatalytic activity useful for the elimination of alcohols. The activity of TiO_2/CdS/G was evaluated in a 10 mL Pyrex glass container filled with molecular oxygen at 0.1 MPa with a mixture of 0.1 mmol alcohol and 8 mg catalysts with 1.5 ml of benzotrifluoride (BTF). The solvent was selected based on its solubility for molecular oxygen and inertness to oxidation. The well-agitated reaction mixture was irradiated for 5 h with a 300 W Xe arc lamp with a UV-Cut filter to cut the light with wavelength <420 nm. After the irradiation time, the solution was centrifuged and the catalysts particles were removed (Zhang, 2012). The oxidation results for different alcohols were benzyl alcohol (~100%), p-methyl benzyl alcohol (~85%), p-methoxy benzyl alcohol (~90%), p-nitro benzyl alcohol (~60%), p-fluoro benzyl alcohol (~95%), p-chloro benzyl alcohol (~85%), cinnamyl alcohol (~75%), 3-methyl-2-buten-1-ol (~60%) and 2-buten-1-ol (~60%). It was also reported that TiO_2/CdS/G maintained good stability during three cycles of experiments (Rani, 2020). Another example of ternary compounds for pollutant removal is the case of TiO_2/WO_3/CQDs (where CQDs denote carbon quantum dots) synthesized with the hydrothermal-calcination method. The photocatalytic activity of TiO_2/WO_3/CQDs was evaluated in the degradation of cephalexin, which is an antibiotic pharmaceutical used for the treatment of bacterial infections. To perform this experiment, 250 ml of cephalexin solution with a concentration of 10 mgL^{-1} and 0.125 g of catalysts were mixed in a photoreactor, the pH was adjusted to 7.0 and the solution was stirred for 30 min. The simulated solar irradiation was realized using 300 W Xe lamps with a filter to remove the UV irradiation. After 120 min, the material achieved a 76% degradation of cephalexin and the photocatalytic activity was stable for five repeated cycles (Sun, 2021).

Apart from metal compounds, *nonmetal semiconductors* are also potential candidates for photocatalytic water treatment. Graphitic carbon nitride (g-C_3N_4) is an n-type semiconductor composed of earth-abundant elements. This is the most stable form of carbon nitride in ambient conditions, and its light absorption properties as well as redox potential can be modified according to the requirements. The availability of different surface sites and tunable porosity make it interesting for several catalytic applications. It has been shown that g-C_3N_4 materials are also active in bacterial disinfection and degradation of pollutants. For example, tylosin (antibiotic) was degraded in 30 min under simulated solar light using novel crystal structure-modified g-C_3N_4 (Xu, 2019; Vilé, 2020).

Graphene is an allotrope of carbon consisting of a single layer of sp^2-bonded carbon networks. For this reason, it has a high thermal conductivity and excellent mobility of charge carriers already at room temperature (Fresno, 2014). Graphene can possess extremely high specific surface area (over 2500 m^{2}g^{-1}) allowing efficient adsorption of, for example, pharmaceutical compounds via π-π interaction. While graphene also exhibits some photocatalytic activity, it is rapidly deactivated due to agglomeration. However, graphene can be combined with other photocatalytically active materials. When incorporated with another photocatalyst, graphene can perform as an electron acceptor, minimizing the recombination of charge carriers and extending the light absorption range. Up to four-fold improvement to TiO_2 catalyst in organic pollutant degradation has been observed when a small amount of graphene is introduced to TiO_2 (Fresno, 2014; Li, 2019).

The use of *magnetic nanophotocatalysts* for water treatment is interesting because they have both physical and chemical properties that can be used to separate pollutants from water. The use of a magnetic material in the composite makes the separation easier because it can be performed applying an external magnetic field. As a result, these materials can be economic, more efficient and environmentally friendly for water treatment. For these reasons, they have been studied to degrade a variety of organic compounds (Gómez-Pastora et al., 2017). An example of a magnetic nanocomposite is TiO_2/GO/$CuFe_2O_4$ (where GO denotes graphene oxide), in which the combination of $CuFe_2O_4$ and TiO_2 in a heterojunction can improve the absorption of TiO_2 towards the visible light (300 nm–400 nm). To prepare this nanocomposite, Ismael et al. (2020) used 5 g of TiO_2 with 2.5 g of graphene oxide (GO) and 2.5 $CuFe_2O_4$, all in powder form, and blended the mixture. After that, 20 g of zirconia was added in a grinding bowl. To evaluate the photocatalytic activity, chlorinated pesticides were used as model pollutants in an aqueous solution. The photocatalyst was transferred to a quartz reactor and the solution was stirred for 10 min. The initial concentration of the pesticide was 0.5 mgL^{-1}. The UV radiation source was a 150 W Xe arc lamp with a wavelength of 365 nm and the solution was exposed for 20 min. The activity of the material was evaluated with 17 different pesticides, with and without UV radiation, and a sufficient removal of pesticides was observed. The best performance of the material was obtained for the DDE pesticide under UV irradiation, which removed 96.5% of the photocatalyst. An advantage of this nanocomposite is that it can be easily recovered for recycling purposes, and after five cycles there was no significant loss in the activity, since DDE removal remained at 95.05% (Ismael, 2020).

As discussed before, the *metal organic frameworks* have a photocatalytic activity that make them suitable for water treatment, due to their uniform distribution of the active sites. During the last decade, research on these materials has been focused on the development of MOF as photocatalyst with activity from ultraviolet light to visible light. Experiments using MOFs for photocatalysis in water purification are promising. MOFs can be used, for example, for the degradation of pharmaceutical and antibiotic residues. One example of these residues is amoxicillin, which is an antibiotic used all around the world. An attempt to degrade amoxicillin from water using MOFs was done with a composite of MIL-68(In)-N H_2/graphene oxide. The experiment was carried out at 23 ± 1 °C using 120 mg of the photocatalyst and 200 ml of 20 ppm amoxicillin aqueous solution in a 500 ml jacketed glass reactor. To achieve the adsorption-desorption equilibrium, the solution was stirred for 60 min in the dark. A 300 W Xe lamp with a 420 mn cutoff filter was used to irradiate the suspension (Yang, 2017). After 120 min of irradiation, the degradation was 93%. A possible explanation for the improvement of the photocatalytic activity could be that graphene oxide plays an important role in decreasing the recombination of the photogenerated electrons and holes and in increasing visible light absorption (Gautam, 2020). Tetracycline is other common antibiotic that can be degraded from water using MOFs. In the literature is reported the use of Fe-based MOFs, and especially Fe-MIL-101 has shown the best performance, with a degradation percentage of 96.6% (Gautam 2020). The light absorption experiments showed that it has a strong visible light absorption ability. For the photocatalytic experiment, the same light as in the previous case was used. The concentration of the tetracycline was 50 mgL^{-1}. The mixed solution was prepared with 50 mg of photocatalysts and 100 ml of the tetracycline aqueous solution, which was stirred in the dark for 60 min. After this, the solution was exposed to a visible light radiation for 3 hours. To prove the durability and recyclability of the material, the photocatalyst was recovered washed and dried at 60 °C to be used in other cycle, after three cycles of absorption photodegradation cycles, the material shows stability. With the used MOF, 96.6% removal of tetracycline was observed initially and after three cycles the decrease in total efficiency was ignorable (Wang, 2018).

The previous examples demonstrate the activities of different photocatalytic materials in water purification, especially in the treatment of emerging contaminants. Table 21.3 provides a summary of the photocatalysts shown as examples in the previous paragraphs. However, novel approaches and more research is needed for the practical application of the photocatalytic processes.

One practical concern hindering the application of photocatalytic processes for real wastewater purification is the

TABLE 21.3

Examples of Photocatalysts Used for the Degradation of ECs in Water

Photocatalyst	Method	Pollutant	Light	Time	Removal (%)	Cycles	Ref.
P25 (TiO_2)	-	Oxoflacin, flumequine, ibuprofen, sulfamethoxazole, carbamezepine	Sunlight	28 min	100	-	Carbajo, 2016
TiO_2 100% anatase	Hydrolysis	Oxoflacin, flumequine, ibuprofen, sulfamethoxazole, carbamezepine	Sunlight	35 min	100	-	Carbajo, 2016
MoS_2	Hydrothermal	Thiobencarb	Visible light	12 h	95	3	Huang, 2018
Bi_2S_3/BiOBr/BC	Solvothermal	Diclofenac	Visible LED light	30 min	94.65	5	Li, 2020
TiO_2/CdS/G	Solvothermal-hydrothermal	Alcohols	Visible light	5 h	60–100	3	Zhang, 2012
TiO_2/WO_3/CQDs	Hydrothermal-calcination	Cephalexin	Visible light	2 h	76	5	Sun et al., 2021
g-C_3N_4	Silica assisted	Tylosin	Simulated solar light	-	≈70	4	Xu, 2019
TiO_2/GO/$CuFe_2O_4$	Coprecipitation-hummers-hydrothermal	DDE	UV radiation	20 min	96.50	5	Ismael, 2020
MIL-68(In)-N H_2/graphene oxide	Solvothermal-hummers	Amoxicillin	Visible light	2 h	93	-	Yang, 2017
Fe-MIL-101	Hydrothermal	Tetracycline	Visible light	3 h	96.60	3	Wang, 2018

complicated mixture of different compounds present in the water. These species may be both organic and inorganic, and they may hinder the degradation of the target pollutant by adsorbing on the catalyst surface and preventing the expected reaction to occur. Inorganic compounds especially are considered harmful, since after adsorbing on the photocatalyst, they can scavenge the photogenerated holes and react with •OH radicals. Furthermore, for example CO_3^{2-}/HCO_3^-, NH_4^+, SO_4^{2-}, NO_3^- and Cl^- ions are often present in natural waters and can act as scavengers for the •OH radicals. The other pollutants present in water may also modify the catalyst-target pollutant interaction and contribute to the agglomeration of the catalyst leading to decrease in the removal efficiency (Carbajo, 2016). Furthermore, photoinduced radicals that are responsible for the decomposition of organic pollutants may also oxidize the catalysts themselves. These radicals are quite close to the catalysts, referring to a high probability of unwanted redox reactions. These issues are quite difficult to overcome completely, and if found impossible, the role of the photocatalysis will finally be in the after-treatment of the pollutants that are difficult to remove in conventional water treatment. An alternative is that new materials prove to be tolerant to this type of deactivation or inhibition and may maintain their activities during a reasonable time, after which the catalyst needs to be replaced. In this respect, the sustainability of the material and photocatalytic treatment process needs to be evaluated more thoroughly.

Another practical concern for the photocatalytic treatment of emerging pollutants is how to reach the complete mineralization of the pollutants. It has been observed in many cases that the degradation of the target pollutant leads to the appearance of one or several by-products that may be more toxic or harmful than the initial molecule. Furthermore, it is possible that these by-products are more difficult to degrade, such as the carbazole dimer in the case of diclofenac degradation (Iovino, 2017). The possibility of the increased toxicity of the treated water should be considered carefully, since in addition of possible toxic by-products, the concentration of reactive oxygen species in water increases during advanced oxidation treatments, which may lead to adverse effects at the cell level, and thus finding more information on the safety of the treated water is required (Rathnayake, 2020). Releasing unproperly treated water to nature should be avoided a priori.

21.5 Sustainability and Future Perspectives

Today, when the impacts of industrial and human actions on the environment and human well-being are more often considered in connection to production and consumption, the concept of sustainability has reached a significant and important role in steering the future decisions. In 2015, the United Nations member states accepted the 2030 Agenda for Sustainable Development. The Agenda includes 17 Sustainable Development Goals describing where urgent actions are needed. Two of these goals are directly related to water, and at least seven of them have an indirect impact on clean water (UN, 2021). From this perspective, when developing new processes, whether for production or environmental applications,

it is important to consider possible environmental and social impacts of the process in addition to the economy.

In the case of photocatalytic water treatment, in addition to improving the quality of water, we should also consider the questions related, for example, to the catalytic materials and their sources, catalyst preparation, energy consumption, the actual purification reactions and processes. While the complete removal of the harmful pollutants from water is the main target of water purification, the catalyst itself should not cause risks to the water. As the first approach, it should be possible to separate the catalyst efficiently from the cleaned water so that it will not cause pollution of the water bodies or the fresh water. One possibility is to immobilize the catalyst, which has been discussed briefly earlier in this chapter. Employing hybrid techniques such as photocatalytic membranes are also discussed. Magnetic separation would be an interesting alternative to filtration technologies. Filtration is a straightforward way to separate the catalyst from water, but questions arise about the efficiency and costs of separating the nano-sized catalytic particles and the environmental and health risks of nanoparticles that could escape the filtration. Doping the photocatalytic materials, such as TiO_2, has proven to be an efficient way to improve the performance of the catalyst and shift the light absorption towards the visible light region. However, leaching of dopants is possible, especially when the pH of the treated water is not close to neutral. The leaching of the dopants decreases the activity of the photocatalyst, but also poses a considerable health risk related to treated water (Vilé, 2020). That brings us to the selection of the photocatalytic material itself. Is the photocatalytic material sustainable? How can we choose the most sustainable materials? These questions should be considered more carefully when developing novel catalytic materials.

Related to photocatalytic materials, the lack of proper, cheap and feasible groups of hosting semiconductors makes the push for photocatalytic water treatment in real applications difficult. Most of the synthetic compounds are costly in production and energy consumption. Their preparation routes are normally accompanied by extreme reaction conditions, environmentally unfriendly reagents and low productivity of final products. On the other hand, treatments for naturally occurring semiconductors are easily lost due to selection difficulties among various species and methods of joining hetero-components together. It is noteworthy that an emerging group of inorganic layered crystals (ILCs) has been found useful as semiconductor hosts, thanks to their "unlimited" hostability. But their functionalities as photocatalysts are only seen as additives.

Photocatalysis requires light as an energy source. The sun would be the ultimate, clean and practically endless source of energy. Therefore, significant efforts are made to develop photocatalytic materials that would work efficiently under sunlight irradiation. The process using sunlight would also improve the sustainability of the treatment – it would make the process more environmentally friendly and probably more applicable in less-developed areas of the world. In addition to energy costs, the harmful environmental impact caused by energy production would be decreased. Using sunlight in water purification and clean fuel production would provide solutions for water purification, lowering CO_2 emissions and enabling green energy production. This will be only possible once proper photocatalytic media are developed.

Currently, insufficient physical knowledge sets up barriers to understand the whole sunlight-driven photocatalytic process and requirements for the materials. Molecular and electronic level information on the photocatalytic processes, where the determination of reaction sites is totally blurred in the solution environment, is required. Indeed, technically, *in situ* characterization tools are required to clarify these reactions. However, the aqueous medium will strongly hinder most experimental methods. While most of the studies related to the photocatalytic water treatment rely on the evaluation of activities of the materials and finding structure-activity relationships using *ex situ* characterization, detailed studies utilizing *in situ* and *operando* techniques are less available. *In situ* studies in photocatalysis means mainly studying the catalytic materials under light irradiation, that is, in an excited state. *Operando* studies are ideally carried out during the operation of the catalysis in real, or close to realistic, conditions of use. *Operando* studies especially require fast analysis capable of bringing out time-resolved information on the catalytic phenomena taking place. *Operando* infrared spectroscopy is probably the most common time-resolved approach, while X-ray spectroscopic techniques and Raman are increasingly used. Raman techniques are interesting also in water purification applications, since water does not interfere with the spectra in the same way that IR spectroscopy does. Improved techniques related to increased sensitivity and minimization of fluorescence interference start to be available. *In situ* and *operando* studies on photocatalytic materials and processes are expected to bring new valuable understanding that will help in practical application of photocatalytic processes in water treatment. This may also improve the sustainability of the photocatalysis-based water and wastewater treatment processes in the future.

The water treatments and hydrogen evolution have been studied separately on most of the photocatalysts so far, eventhough many of the photocatalysts show activity in both the fields (water treatment and hydrogen evolution), though many of photocatalysts own catalytic activities in both fields. In fact, many organic compounds are widely employed in photocatalytic hydrogen evolution as hole scavengers (Li, 2016) and commonly seen in the hydrogen conversion from biomass (Davis, 2021). The concept of simultaneous hydrogen evolution and pollutant removal from water has thus been developed. A recent study shows that simultaneous hydrogen evolution and water purification have been reached for sewage with specific contaminates (Rather, 2020). Though limited in specific dyes and dedicated catalysts, this track may implicitly become the future path to sustainability through sunlight photocatalysis (Rioja-Cabanillas, 2020).

21.6 Conclusions

This chapter summarizes different aspects related to photocatalytic phenomena, photocatalytic materials, how their properties can benefit water and wastewater treatment, and further development towards more sustainable water purification

solutions in the future. The purpose of this chapter is to provide the state of the art in photocatalytic materials and show some of the challenges that scientists have in the use of these materials for water treatment. Hydrogen production from water and treatment of organic load of waters are presented as examples in selected cases to boost sustainability of photocatalytic water purification systems. The principles of photocatalytic phenomena were described and examples of different photocatalytic materials and their application on the degradation of different pollutants were provided. In addition, the description of different photocatalytic materials with their characteristics, properties and the common synthesis methods were explained. At the end of the chapter, we discussed the use of different photocatalysts in water treatment with examples on the degradation of pollutants, concentrating on the degradation of ECs. During the survey, it was observed that the use of these materials for water treatment is promising, because according to the literature and our own research outcomes, they have shown efficiency in the degradation of dyes and ECs. There are, however, still several challenges in the development of these materials such as the design of heterostructures, reusability, stability and light harvesting. Finally, using sunlight to activate the photocatalytic reactions can be considered the ultimate goal for the development of sustainable photocatalytic water purification and hydrogen production catalysts and processes. For this purpose, novel materials based on scientific knowledge achieved from time-resolved operando experiments related to relevant phenomena, for example, should be developed to reach the real-life practical application of photocatalytic processes. The need to obtain *in situ* phenomenological information on materials' behavior, surface phenomena and reaction mechanisms, for example, would help to optimize the catalyst composition and its functioning in demanding real-life water purification and hydrogen production applications.

REFERENCES

Augugliaro, V., Bellardita, M., Loddo, V., Palmisano, G., Palmisano, L., & Yurdakal, S. (2012). Overview on oxidation mechanisms of organic compounds by TiO2 in heterogeneous photocatalysis. *Journal of Photochemistry and Photobiology C: Photochemistry Reviews*, 13(3), 224–245. DOI: 10.1016/j.jphotochemrev.2012.04.003

Ayodhya, D., Venkatesham, M., Santoshi Kumari, A., Reddy, G. B., Ramakrishna, D., & Veerabhadram, G. (2016). Photocatalytic degradation of dye pollutants under solar, visible and UV lights using green synthesised CuS nanoparticles. *Journal of Experimental Nanoscience*, 11(6), 418–432. DOI: 10.1080/17458080.2015.1070312

Batzill, M. (2011). Fundamental aspects of surface engineering of transition metal oxide photocatalysts. *Energy & Environmental Science*, 4(9), 3275–3286. DOI: 10.1039/C1EE01577J

Bedia, J., Muelas-Ramos, V., Peñas-Garzón, M., Gómez-Avilés, A., Rodríguez, J. J., & Belver, C. (2019). A review on the synthesis and characterization of metal organic frameworks for photocatalytic water purification. *Catalysts*, 9(1), 52. DOI: 10.3390/catal9010052

Belgiorno, V., Rizzo, L., Fatta, D., Della Rocca, C., Lofrano, G., Nikolaou, A., . . . Meric, S. (2007). Review on endocrine disrupting-emerging compounds in urban wastewater: Occurrence and removal by photocatalysis and ultrasonic irradiation for wastewater reuse. *Desalination*, 215(1–3), 166–176. DOI: 10.1016/j.desal.2006.10.035

Borges, M. E., García, D. M., Hernández, T., Ruiz-Morales, J. C., & Esparza, P. (2015). Supported photocatalyst for removal of emerging contaminants from wastewater in a continuous packed-bed photoreactor configuration. *Catalysts*, 5, 77–87. DOI:10.3390/catal5010077

Carbajo, J., Jiménez, M., Miralles, S., Malato, S., Faraldos, M., & Bahamonde, A. (2016). Study of application of titania catalysts on solar photocatalysis: Influence of type of pollutants and water matrices. *Chemical Engineering Journal*, 291, 64–73. DOI: 10.1016/j.cej.2016.01.092

Carvalho, R. N., Ceriani, L., Ippolito, A., & Lettieri, T. (2015) *Development of the first watch list under the environmental quality standards directive.* https://ec.europa.eu/jrc/en/publication/eur-scientific-and-technical-research-reports/development-first-watch-list-under-environmental-quality-standards-directive. DOI: 10.2788/101376

Chandrasekaran, S., Yao, L., Deng, L., Bowen, C., Zhang, Y., Chen, S., . . . Zhang, P. (2019). Recent advances in metal sulfides: From controlled fabrication to electrocatalytic, photocatalytic and photoelectrochemical water splitting and beyond. *Chemical Society Reviews*, 48(15), 4178–4280. DOI: 10.1039/C8CS00664D

Chen, S., Li, F., Li, T., & Cao, W. (2019). Loading AgCl@ Ag on phosphotungstic acid modified macrocyclic coordination compound: Z-scheme photocatalyst for persistent pollutant degradation and hydrogen evolution. *Journal of Colloid and Interface Science*, 547, 50–59. DOI: 10.1016/j.jcis.2019.03.092

Chen, Y., He, J., Li, J., Mao, M., Yan, Z., Wang, W., & Wang, J. (2016). Hydrilla derived $ZnIn_2S_4$ photocatalyst with hexagonal-cubic phase junctions: A bio-inspired approach for H_2 evolution. *Catalysis Communications*, 87, 1–5. DOI: 10.1016/j.catcom.2016.08.031

Chen, Z., Chu, X., Huang, X., Sun, H., Chen, L., & Guo, F. (2021). Fabrication of visible-light driven $CoP/ZnSnO_3$ composite photocatalyst for high-efficient photodegradation of antibiotic pollutant. *Separation and Purification Technology*, 257, 117900. DOI: 0.1016/j.seppur.2020.117900

Chi, Y., Yuan, Q., Li, Y., Zhao, L., Li, N., Li, X., & Yan, W. (2013). Magnetically separable $Fe3O4@ SiO2@ TiO2$-Ag microspheres with well-designed nanostructure and enhanced photocatalytic activity. *Journal of Hazardous Materials*, 262, 404–411. DOI: 10.1016/j.jhazmat.2013.08.077

Davis, K. A., Yoo, S., Shuler, E. W., Sherman, B. D., Lee, S., & Leem, G. (2021). Photocatalytic hydrogen evolution from biomass conversion. *Nano Convergence*, 8(1), 1–19. DOI: 10.1186/s40580-021-00256-9

Dey, C., Kundu, T., Biswal, B. P., Mallick, A., & Banerjee, R. (2014). Crystalline metal-organic frameworks (MOFs): synthesis, structure and function. *Acta Crystallographica Section B: Structural Science, Crystal Engineering and Materials*, 70(1), 3–10. DOI: 10.1107/S2052520613029557

Di, T., Xu, Q., Ho, W., Tang, H., Xiang, Q., & Yu, J. (2019). Review on metal sulphide-based Z-scheme photocatalysts. *ChemCatChem*, 11(5), 1394–1411. DOI: 10.1002/cctc.201802024

Dong, P., Xi, X., & Hou, G. (2016). Typical non-TiO_2-based visible-light photocatalysts, chapter 8 in semiconductor photocatalysis: Materials, mechanisms and applications. *Intech Open*. DOI: 10.5772/62889.

Dong, Z., Wu, Y., Thirugnanam, N., & Li, G. (2018). Double Z-scheme ZnO/ZnS/g-C3N4 ternary structure for efficient photocatalytic H_2 production. *Applied Surface Science*, 430, 293–300. DOI: 10.1016/j.apsusc.2017.07.186

Fanourakis, S. K., Peña-Bahamonde, J., Bandara, P. C., & Rodrigues, D. F. (2020). Nano-based adsorbent and photocatalyst use for pharmaceutical contaminant removal during indirect potable water reuse. *NPJ Clean Water*, 3(1), 1–15. DOI: 10.1038/s41545-019-0048-8

Fresno, F., Portela, R., Suárez, S., & Coronado, J. M. (2014). Photocatalytic materials: Recent achievements and near future trends. *Journal of Materials Chemistry A*, 2, 2863–2884. DOI: 10.1039/c3ta13793g.

Freund, H. J. (1995). Metal oxide surfaces: Electronic structure and molecular adsorption. *Physica Status Solidi (b)*, 192(2), 407–440. DOI: 10.1002/pssb.2221920214

Fu, J., Yu, J., Jiang, C., & Cheng, B. (2018). g-C3N4-Based heterostructured photocatalysts. *Advanced Energy Materials*, 8(3), 1701503. DOI: 10.1002/aenm.201701503

Fu, W., Yang, H., Li, M., Li, M., Yang, N., & Zou, G. (2005). Anatase TiO2 nanolayer coating on cobalt ferrite nanoparticles for magnetic photocatalyst. *Materials Letters*, 59(27), 3530–3534. DOI: 10.1016/j.matlet.2005.06.071

Gautam, S., Agrawal, H., Thakur, M., Akbari, A., Sharda, H., Kaur, R., & Amini, M. (2020). Metal oxides and metal organic frameworks for the photocatalytic degradation: A review. *Journal of Environmental Chemical Engineering*, 8(3), 103726. DOI: 10.1016/j.jece.2020.103726

Ge, J., Zhang, Y., Heo, Y. H., & Park, S. J. (2019). Advanced design and synthesis of composite photocatalysts for the remediation of wastewater: A review. *Catalysts*, 9, 122–154. DOI:10.3390/catal9020122

Gómez-Pastora, J., Dominguez, S., Bringas, E., Rivero, M. J., Ortiz, I., & Dionysiou, D. D. (2017). Review and perspectives on the use of magnetic nanophotocatalysts (MNPCs) in water treatment. *Chemical Engineering Journal*, 310, 407–427. DOI: 10.1016/j.cej.2016.04.140

Huang, S., Chen, C., Tsai, H., Shaya, J., & Lu, C. (2018). Photocatalytic degradation of thiobencarb by a visible light-driven MoS_2 photocatalyst. *Separation and Purification Technology*, 197, 147–155. DOI: 10.1016/j.seppur.2018.01.009.

Iovino, P., Chianese, S., Canzano, S., Prisciandaro, M., & Musmarra, M. (2017). Photodegradation of diclofenac in waste waters. *Desalination and Water Treatment*, 61, 293–297. DOI:10.5004/dwt.2016.11063

Ismael, A. M., El-Shazly, A. N., Gaber, S. E., Rashad, M. M., Kamel, A. H., & Hassan, S. S. M. (2020). Novel TiO 2/GO/CuFe$_2$O$_4$ nanocomposite: A magnetic, reusable and visible-light-driven photocatalyst for efficient photocatalytic removal of chlorinated pesticides from wastewater. *RSC Advances*, 10(57), 34806–34814. DOI: 10.1039/D0RA02874F

Jacinto, M. J., Ferreira, L. F., & Silva, V. C. (2020). Magnetic materials for photocatalytic applications – a review. *Journal of Sol-Gel Science and Technology*, 96, 1–14. DOI: 10.1007/s10971-020-05333-9

Jiang, Y., Zhang, M., Xin, Y., Chai, C., & Chen, Q. (2019). Construction of immobilized CuS/TiO2 nanobelts heterojunction photocatalyst for photocatalytic degradation of enrofloxacin: Synthesis, characterization, influencing factors and mechanism insight. *Journal of Chemical Technology & Biotechnology*, 94(7), 2219–2228. DOI: 10.1002/jctb.6006

Jing, J., Li, J., Feng, J., Li, W., & William, W. Y. (2013). Photodegradation of quinoline in water over magnetically separable Fe_3O_4/TiO_2 composite photocatalysts. *Chemical Engineering Journal*, 219, 355–360. DOI: 10.1016/j.cej.2012.12.058

Kahng, S., Yoo, H., & Kim, J.H. (2020). Recent advances in earth-abundant photocatalyst materials for solar H2 production. *Advanced Powder Technology*, 31(1), 11–28. DOI: 10.1016/j.apt.2019.08.035

Kimbi Yaah, V. B., Ojala, S., Khallok, H., Laitinen, T., Selent, M., Zhao, H., Sliz, R., & Botelho de Oliveira, S. (2021). Development and characterization of composite carbon adsorbents with photocatalytic regeneration ability. *Application to Diclofenac Removal from Water*, 11, 173. DOI: 10.3390/catal11020173

Kistanov, A. A., Shcherbinin, S. A., Ustiuzhanina, S. V., Huttula, M., Cao, W., Nikitenko, V. R., & Prezhdo, O. V. (2021). First-principles prediction of two-dimensional B3C2P3 and B2C4P2: Structural stability, fundamental properties, and renewable energy applications. *The Journal of Physical Chemistry Letters*, 12, 3436–3442. DOI: 10.1021/acs.jpclett.1c00411

Kohtani, S., Ohama, Y., Ohno, Y., Tsuji, I., Kudo, A., & Nakagaki, R. (2005). Photoreductive defluorination of hexafluorobenzene on metal-doped ZnS photocatalysts under visible light irradiation. *Chemistry Letters*, 34(7), 1056–1057. DOI: 10.1246/cl.2005.1056

Kong, D., Zheng, Y., Kobielusz, M., Wang, Y., Bai, Z., Macyk, W., Wang, X., & Tang, J. (2018). Recent advances in visible light-driven water oxidation and reduction in suspension systems. *Materials Today*, 21(8), 897–924. DOI: 10.1016/j.mattod.2018.04.009

Lee, K. M., Lai, C. W., Ngai, K. S., & Juan, J. C. (2016). Recent developments of zinc oxide based photocatalyst in water treatment technology: A review. *Water Research*, 88, 428–448. DOI: 10.1016/j.watres.2015.09.045

Li, J., Zhan, G., Yu, Y., & Zhang, L. (2016). Superior visible light hydrogen evolution of Janus bilayer junctions via atomic-level charge flow steering. *Nature Communications*, 7(1), 1–9. DOI: 10.1038/ncomms11480

Li, M., Liu, Y., Zeng, G., Liu, N., Liu, S. (2019) Graphene and graphene-based nanocomposites used for antibiotics removal in water treatment: A review. *Chemosphere*, 226, 360–380.

Li, S., Wang, Z., Xie, X., Liang, G., Cai, X., Zhang, X., & Wang, Z. (2020). Fabrication of vessel – like biochar – based heterojunction photocatalyst $Bi_2S_3/BiOBr/BC$ for diclofenac removal under visible LED light irradiation: Mechanistic investigation and intermediates analysis. *Journal of Hazardous Materials*, 391, 121407. DOI: 10.1016/j.jhazmat.2019.121407

Li, Z., Meng, X., & Zhang, Z. (2018). Recent development on MoS_2-based photocatalysis: A review. *Journal of Photochemistry and Photobiology C: Photochemistry Reviews*, 35, 39–55. DOI: 10.1016/j.jphotochemrev.2017.12.002

Liao, G., Li, C., Li, X., & Fang, B. (2021). Emerging polymeric carbon nitride Z-scheme systems for photocatalysis. *Cell Reports Physical Science*, 100355. DOI: 10.1016/j.xcrp.2021.100355

Liu, G., Sun, C., Yang, H. G., Smith, S. C., Wang, L., Lu, G. Q. (M.), & Cheng, H. M. (2010). Nanosized anatase TiO2 single crystals for enhanced photocatalytic activity. *Chem. Comm.* 46, 755–757. DOI: 10.1039/B919895D

Liu, J., Song, Y., Xu, H., Zhu, X., Lian, J., Xu, Y., . . . Li, H. (2017). Non-metal photocatalyst nitrogen-doped carbon nanotubes modified mpg-C3N4: Facile synthesis and the enhanced visible-light photocatalytic activity. *Journal of Colloid and Interface Science*, 494, 38–46. DOI: 10.1016/j.jcis.2017.01.010

Loeb, S. K., Alvarez, P. J., Brame, J. A., Cates, E. L., Choi, W., Crittenden, J., . . . Kim, J. H. (2018). The technology horizon for photocatalytic water treatment. *Sunrise or Sunset?*. DOI: 10.1021/acs.est.8b05041

Lu, D., Wang, H., Zhao, X., Kondamareddy, K. K., Ding, J., Li, C., & Fang, P. (2017). Highly efficient visible-light-induced photoactivity of Z-scheme $g-C_3N_4/Ag/MoS_2$ ternary photocatalysts for organic pollutant degradation and production of hydrogen. *ACS Sustainable Chemistry & Engineering*, 5(2), 1436–1445. DOI: 10.1021/acssuschemeng.6b02010

Mamaghani, A. H., Haghighat, F., & Lee, C. S. (2017). Photocatalytic oxidation technology for indoor environment air purification: the state-of-the-art. *Applied Catalysis B: Environmental*, 203, 247–269. DOI: 10.1016/j.apcatb.2016.10.037

Mamaghani, A. H., Haghighat, F., Lee, C. S. (2019). Hydrothermal/solvothermal synthesis and treatment of TiO2 for photocatalytic degradation of air pollutants: Preparation, characterization, properties, and performance. *Chemosphere*, 219, 804–825. DOI: 10.1016/j.chemosphere.2018.12.029

Marchelek, M., Diak, M., Kozak, M., Zaleska-Medynska, A., & Grabowska, E. (2016). Some unitary, binary, and ternary non-TiO2 photocatalysts. In *Semiconductor photocatalysis-materials, mechanisms and applications*. IntechOpen. DOI: 10.5772/62583

Martins, P. M., Ribeiro, J. M., Teizeira, S., Petrovykh, D. Y., Cuniberti, G., Pereira, L., & Lanceros-Méndez, S. (2019) Photocatalytic microporous membrane against the increasing problem of water emerging pollutants. *Materials*, 12, 1649. DOI: 10.3390/ma12101649

Murillo-Siera, J. C. M., Hernández-Ramírez, A., Hinojosa-Reyes, L., & Guzmán-Mar, J. L. (2021). A review on the development of visible light-responsive WO_3-based photocatalysts for environmental applications. *Chemical Engineering Journal Advances*, 5, 100070. DOI: 10.1016/j.ceja.2020.100070.

Parashar, M., Shukla, V. K., & Singh, R. (2020). Metal oxides nanoparticles via sol-gel method: A review on synthesis, characterization and applications. *Journal of Materials Science: Materials in Electronics*, 31(5), 3729–3749. Doi: 10.1007/s10854-020-02994-8

Rani, E., Talebi, P., Cao, W., Huttula, M., & Singh, H. (2020). Harnessing photo/electro-catalytic activity via nanojunctions in ternary nanocomposites for clean energy. *Nanoscale*. DOI: 10.1039/D0NR05782G

Rao, C. N. R., & Pisharody, K. P. R. (1976). Transition metal sulfides. *Progress in Solid State Chemistry*, 10, 207–270. DOI: 10.1016/0079-6786(76)90009-1

Rather, R. A., & Lo, I. M. (2020). Photoelectrochemical sewage treatment by a multifunctional g-C3N4/Ag/AgCl/BiVO4 photoanode for the simultaneous degradation of emerging pollutants and hydrogen production, and the disinfection of E. coli. *Water Research*, 168, 115166. DOI: 10.1016/j.watres.2019.115166

Rathnayake, B., Heponiemi, A., Huovinen, M., Ojala, S., Pirilä, M., Loikkanen, J., Azalim, S., Saouabe, M., Brahmi, R., Vähäkangas, K., Lassi, U., Keiski, R.L. (2020) Photocatalysis and catalytic wet air oxidation: Degradation and toxicity of bisphenol A containing wastewaters. *Environmental Technology*, 41, 25, 3272–3283.

Rioja-Cabanillas, A., Valdesueiro, D., Fernández-Ibáñez, P., & Byrne, J. A. (2020). Hydrogen from wastewater by photocatalytic and photoelectrochemical treatment. *Journal of Physics: Energy*, 3(1), 012006. DOI. 10.1088/2515-7655/abceab

Salazar, H., Martins, P. M., Santos, B., Fernandes, M. M., Reizabal, A., Sebastián, V., Botelho, G., Tavares, C. J., Vilas-Vileka, J. L., & Lanceros-Mendez, S. (2020). Photocatalytic and antimicrobial multifunctional nanocomposite membranes for emerging pollutants water treatment applications. *Chemosphere,* 250, 126299. DOI: 10.1016/j.chemosphere.2020.126299

Santhosh, C., Malathi, A., Dhaneshvar, E., Bhatnagar, A., Grace, A. N., & Madhavan, J. (2019). Iron oxide nanomaterials for water purification. In *Nanoscale materials in water purification* (pp. 431–446). Elsevier. DOI: 10.1016/B978-0-12-813926-4.00022-7

Shah, M. P. (2020a). *Advanced oxidation processes for effluent treatment plants*. Elsevier.

Shah, M. P. (2020b). *Microbial bioremediation and biodegradation*. Springer.

Shah, M. P. (2021). *Removal of emerging contaminants through microbial processes*. Springer.

Shi, W., Liu, C., Li, M., Lin, X., Guo, F., & Shi, J. (2020). Fabrication of ternary Ag3PO4/Co3 (PO4) 2/g-C3N4 heterostructure with following Type II and Z-Scheme dual pathways for enhanced visible-light photocatalytic activity. *Journal of Hazardous Materials*, 389, 121907. DOI: 10.1016/j.jhazmat.2019.121907

Shi, X., Posysaev, S., Huttula, M., Pankratov, V., Hoszowska, J., Dousse, J. C., . . . Cao, W. (2018). Metallic contact between MoS2 and Ni via Au nanoglue. *Small*, 14(22), 1704526. DOI: 10.1002/smll.201704526

Srikanth, B., Goutham, R., Badri Narayan, R., Ramprasath, A., Gopinath, K.P., & Sankaranarayanan, A.R. (2017). Recent advancements in supporting materials for immobilised photocatalytic applications in waste water treatment. *Journal of Environmental Engineering*, 200, 60–78. DOI: 10.1016/j.jenvman.2017.05.063

Sterrer, M., & Freund, H. J. (2014). Properties of oxide surfaces. In Wandelt, K. (ed.) *Surface and interface science: Properties of composite surfaces: Alloys, compounds, semiconductors*. Wiley-VCH Verlag GmbH & Co. KGaA, pp. 229–278. DOI: 10.1002/9783527680559

Sun, X., He, W., Yang, T., Ji, H., Liu, W., Lei, J., . . . Cai, Z. (2021). Ternary TiO2/WO3/CQDs nanocomposites for enhanced photocatalytic mineralization of aqueous cephalexin: Degradation mechanism and toxicity evaluation. *Chemical Engineering Journal*, 412, 128679. DOI: 10.1016/j.cej.2021.128679

Thompson, T. L., & Yates, J. T., Jr. (2005). TiO2-based photocatalysis: Surface defects, oxygen and charge transfer. *Topics in Catalysis*, 35, 197–210. DOI: 10.1007/s11244-005-3825-1

UN (2021). *The 17 goals: The UN department of economic and social affairs, sustainable development*. www.sdgs.un.org/goals

Vilé, G. (2020). Photocatalytic materials and light-driven continuous processes to remove emerging pharmaceutical pollutants from water and selectively close the carbon cycle. *Catalysis Science & Technology*. DOI: 10.1039/D0CY01713B

Wang, D., Jia, F., Wang, H., Chen, F., Fang, Y., Dong, W., . . . Yuan, X. (2018). Simultaneously efficient adsorption and photocatalytic degradation of tetracycline by Fe-based MOFs. *Journal of Colloid and Interface Science*, 519, 273–284. DOI: 10.1016/j.jcis.2018.02.067

Wang, S., Wang, L., & Huang, W. (2020). Bismuth-based photocatalysts for solar energy conversion. *Journal of Materials Chemistry A*, 8, 24307–24352. DOI: 10.1039/D0TA09729B

Weber, T., Prins, R., & van Santen, R. A. (eds.). (1998). *Transition metal sulphides: Chemistry and catalysis* (Vol. 60). Springer Science & Business Media.

Wen, J., Xie, J., Chen, X., & Li, X. (2017). A review on g-C3N4-based photocatalysts. *Applied Surface Science*, 391, 72–123. DOI /10.1016/j.apsusc.2016.07.030

WHO (2019). *Drinking water – key facts*. World Health Organization. www.who.int/mediacentre/factsheets/fs391/en/

Wood, D., Shaw, S., Cawte, T., Shanen, E., & Van Heyst, B. (2020). An overview of photocatalyst immobilization methods for air pollution remediation. *Chemical Engineering Journal*, 391, 123490. DOI: 10.1016/j.cej.2019.123490

Wu, M-C., Sápi, A., Avila, A., Szabó, M., Hiltunen, J., Huuhtanen, M., Tóth, G., Kukovecz, Á., Kónya, Z., Keiski, R., Su, W. F., Jantunen, H., & Kordás, K. (2011). Enhanced photocatalytic activity of TiO2 nanofibers and their flexible composite films: Decomposition of organic dyes and efficient H2 generation from ethanol-water mixtures. *Nano Research*, 4(4), 360–369. DOI: 10.1007/s12274-010-0090-9

Xiang, Q., Yu, J., & Jaroniec, M. (2011). Preparation and enhanced visible-light photocatalytic H2-production activity of graphene/C3N4 composites. *The Journal of Physical Chemistry C*, 115(15), 7355–7363. DOI: /10.1021/jp200953k

Xie, Y. P., Liu, G., Yin, L., & Cheng, H. M. (2012). Crystal facet-dependent photocatalytic oxidation and reduction reactivity of mononclinic WO3 for solar energy conversion. *Journal of Materials Chemistry*, 22, 6476. DOI: 10.1039/C2JM16178H

Xu, C., Anusuyadevi, P.R., Aymonier, C., Luque, R., & Marre, S. (2019). Nanostructured materials for photocatalysis. *Chemical Society Reviews*, 48, 3868–3902. DOI: 10.1039/c9cs00102f.

Xu, Q., Zhang, L., Yu, J., Wageh, S., Al-Ghamdi, A. A., & Jaroniec, M. (2018). Direct Z-scheme photocatalysts: Principles, synthesis, and applications. *Materials Today*, 21(10), 1042–1063. DOI: 10.1016/j.mattod.2018.04.008

Xue, C., Zhang, Q., Li, J., Chou, X., Zhang, W., Ye, H., . . . Dobson, P. J. (2013). High photocatalytic activity of Fe3O4-SiO2-TiO2 functional particles with core-shell structure. *Journal of Nanomaterials*. DOI: 10.1155/2013/762423

Yang, C., You, X., Cheng, J., Zheng, H., Chen, Y. (2017). A novel visible-light-driven In-based MOF/graphene oxide composite photocatalyst with enhanced photocatalytic activity toward the degradation of amoxicillin. *Applied Catalysis B: Environmental*, 200, 673–680. DOI: 10.1016/j.apcatb.2016.07.057

Zakria, H. S., Othman, M. H. D., Kamaludin, R., Kadir, S. H. S. A., Kurniawan, T. A., & Julani, A. (2021). Immobilization techniques of a photocatalyst into and onto a polymer membrane for photocatalytic activity. *RSC Advances*, 11, 6985. DOI: 10.1039/D0RA10964A

Zhang, N., Zhang, Y., Pan, X., Yang, M. Q., & Xu, Y. J. (2012). Constructing ternary CdS – graphene – TiO_2 hybrids on the flatland of graphene oxide with enhanced visible-light photoactivity for selective transformation. *The Journal of Physical Chemistry C*, 116(34), 18023–18031. DOI: 10.1021/jp303503c

22

Biological Based Methods for the Removal of VOCs and Heavy Metals

Amrin Pathan and Anupama Shrivastav

CONTENTS

22.1 Introduction

22.1.1 Heavy Metals in Industrial Wastewater

Heavy metals are components with nuclear weights between 63.5 and 200.6 and a particular gravity higher than 5.0. Living beings require a small amount of heavy metals, like vanadium, iron, cobalt, copper, manganese, molybdenum, zinc and strontium. Excessive levels of fundamental metals, however, can be detrimental to the life-form. Non-essential heavy metals of specific concern to surface water frameworks are cadmium, chromium, mercury, lead, arsenic and antimony. Heavy metals that are moderately copious within the earth and which are habitually utilized in mechanical forms or agribusiness are harmful to humans. These can make critical modifications to the biochemical cycles of living things.

Most of the point sources of heavy metal toxins include mechanical wastewater from mining, metal handling, tanneries, pharmaceuticals, pesticides, natural chemicals, elastics and plastics, and amble (tree logs) and wood items. The heavy metals are transported by runoff water and contaminate water sources downstream from the mechanical location. All living things, including microorganisms, plants and animals, depend on water for life. Heavy metals can bind to the surface of microorganisms and may enter the interior of the cell. Inside the microorganism, the heavy metals can be chemically changed as the microorganism employs chemical responses to process food.

The objective of this chapter is to analyze different organic strategies utilized for the treatment of heavy metals from mechanical wastewater and the changes to the productivity of the methods by rising strategies such as the application of genetic engineering for the treatment of heavy metal.

DOI: 10.1201/9781003165958-22

22.1.2 VOCs in Wastewater and Air

VOCs are omnipresent within ambient air, and they are vital antecedents to ozone arrangement and secondary organic pressurized canned products. VOCs are dangerous, and they are radiated from different sources such as the chemical industry, oil and gas abuse and so on. VOCs have different health impacts, including acute and chronic dangers depending on their chemical compositions. In a normal petroleum refinery, effluent treatment plant (ETP) is the last process of the pipe treatment set-up for different wastewater streams radiated from the processing units of refinery & offsite/utility sites. In this procedure, ETP gets diverse unstable natural compounds like VOCs and in-organics like NH3, H2S and so on related with the fluid effluents and in this way it becomes an outflow source for these gaseous substances. As the VOCs have lesser bubbling point or higher vapor weight at room temperature, they vanish effectively and a few are dangerous to environment and people. VOCs moreover create ozone after reacting with nitrogen oxides, which in turn creates an exhaust cloud that is destructive to people and vegetation. In petroleum refining, VOCs present in rough oil are disseminated to items, and some are discharged to the environment by means of leaks/venting, channels, fittings, types of gear, loading/unloading sites and so on, and a portion goes to fluid wastewater delivered from processing units and offsite facilities. Thus, VOC production in a refinery depends majorly upon the sort of crude handled, type of refining process, storage or capacity commitments, etc. Hence, the full VOC outflow regularly ranges from 50 to 1000 tons per million tons of rough oil handled [1]. In addition, hundreds of unstable natural compounds are found in a refinery and are classified as alkanes, alkenes, aromatics and cyclic hydrocarbons. Although various VOCs are present in ordinary petroleum refinery wastewater, common VOCs are methyl tertiary-butyl ether, benzene, biphenyl, cresols, xylene, cumene, ethylbenzene, hexane, naphthalene, phenol, styrene, toluene, 1,3-butadiene, 2,2,4-trimethylpentane [2, 3].

22.2 Biological Treatment for Heavy Metal Removal

Several physicochemical strategies have been broadly utilized for the evacuation of heavy metals from mechanical wastewater, such as particle trade, actuated charcoal, chemical precipitation, and chemical lessening and adsorption. The customary strategies utilized for the treatment of heavy metals from mechanical wastewater show a few restrictions. There are still a few common issues related to these strategies such as high cost and that they can create other waste issues, which have constrained their mechanical applications.

Among the accessible treatment forms, these days the application of natural forms is steadily gaining force due to the following reasons:

• Requirement of chemicals for the entire treatment is reduced
• Lower operating expenses

• Eco-friendly and cost-effective choice of method in comparison to old methods
• Effective for low level of contamination

22.2.1 Remediation of Heavy Metals

Heavy metals need to be remediated to prevent them from entering the human body. Industrial pollution is one of the issues that can be focused on. Other ways of heavy metals entering our body are through nourishment. Phytoremediation and intercropping are ways in which heavy metals can be retained and removed from the soils, silt and waters. Hyperaccumulator plants are planted within the soils to evacuate heavy metals. These sorts of plants have root frameworks that have a specific take-up, where the contaminant is translocated and bioaccumulated, and after that the plants are entirely corrupted to evacuate the heavy metals. Phytostabilization and phytoextraction can be used for inorganic compounds. In phytoextraction, heavy metals are translocated from the roots to the shoots. In Phytostabilization only parts of plant get involved in heavy metal removal. Organic compounds are expelled through phytodegradation, rhizofiltration and rhizodegradation. Rhizofiltration includes the adsorption or precipitation of the poisonous substances which are in soluble form and are surrounding the root or into the roots, once the land is damp. Rhizodegradation involves microbial movement where the contaminants are decomposed in the rhizosphere and which is enhanced by the plant roots. When choosing the plants, it is important to note that different plant species take up different heavy metals. Other factors include the soil's pH, natural matter display, the amount of phosphorous in the soil, the root zone, the chelating specialist added to the soil that influences the bioavailability of the metals such as ethylenediaminetetraacetic acid (EDTA), and temperature, which impacts the vegetative take-up through the root length. Hyperaccumulator plants don't show signs of toxicity when they assimilate heavy metals. The strategy of intercropping comprises growing two distinctive species of plants at the same time. Plant biomass is improved, and the collection of heavy metals is aided.

However, biofilters are the most recent and the foremost promising advancement in natural forms for the treatment of heavy metal contaminated mechanical wastewater.

22.2.2 Biofiltration Methods

Microorganisms settled to a permeable medium are utilized within biofiltration to break down poisonous substances found within the wastewater stream. The microorganisms grow in a biofilm on the surface of the medium or are suspended within the water stage encompassing the medium particles. The channel bed medium comprises generally inactive substances, which guarantee huge surface connection zones and extra supplement supply. The generally viability of a biofilter is determined by the properties and characteristics of the supportive medium, which incorporate porosity, degree of compaction, water maintenance capabilities and the capacity to have microbial populaces. Basic biofilter operational and execution parameters incorporate microbial vaccination, medium pH, temperature, the medium dampness and supplement substance.

In a biofiltration framework, the biodegradable poisons are expelled due to organic corruption instead of physical straining, as is the case in ordinary channels. With the advancement of filtration methods, microorganisms (oxygen consuming, anaerobic, facultative, microbes, parasites, green growth and protozoa) are steadily grown on the surface of the channel media and create an organic film or thin layer known as biofilm. The significant point for the fruitful operation of the biofilter is to control and keep up a solid biomass on the surface of the filter.

Since the execution of the biofilter generally depends on microbial activity, a steady source of substrates (natural substance and supplements) is required for its reliable and viable operation, although a few chemoautotrophic microbes may utilize inorganic chemicals as energy source. The expulsion productivity of biofilters is controlled by a few parameters like pH, temperature, O_2 substance and introductory concentration of poisonous poisons. The evacuation proficiency may be progressed by chemical alteration of the channel media or genetic alteration of microorganisms.

22.2.2.1 Components of Biofiltration

The basic instruments that permit biofilters to work and that must be controlled to guarantee success are complex. The biofilter contains a permeable medium whose surface is covered with water and microorganisms. The contaminant can make complexes with organic compounds within the water and may be adsorbed by the supporting medium. Eventually, biotransformation converts the contaminant to biomass, metabolic by-products or carbon dioxide and water. The biodegradation is carried out by a complex environment of degraders, competitors and predators that are mostly organized into a biofilm. There are three fundamental organic forms that can happen in a biofilter:

- Connection of microorganisms
- Development of microorganisms
- Rot and separation of microorganisms

The components by which microorganisms can join and colonize on the surface of the channel media of a biofilter are transportation of microorganisms, starting grip, firm connection and colonization. The transportation of microorganisms to the surface of the channel media is controlled by four fundamental forms:

- Dissemination (Brownian motion)
- Convection
- Sedimentation due to gravity
- Dynamic portability of the microorganisms

Before the microorganisms reach the surface, beginning attachment happens that may be reversible or irreversible depending upon the entire interaction energy, which is the whole of van der Waal's drive and electrostatic constraint. The forms of film connection and colonization of microorganisms depend on influent characteristics (such as natural sort and concentration) and surface properties of the channel media. The following parameters are considered to assess the connection of microorganisms on the surface of the channel media:

- Steric effect
- Hydrophobicity of the microorganisms
- Contact point
- Electrophoretic portability values

The components that impact the rate of substrate utilization inside a biofilm are substrate mass transport to the biofilm, dissemination of the substrate into the biofilm and utilization energy inside the biofilm. Adsorption and biodegradation perform at the same time in biofilters to evacuate biodegradable and water dissolvable perilous natural atoms, which occurs in synchronous adsorption and biodegradation. Although the component of heavy metals expulsion by biofilter contrasts that of natural chemicals expulsion, all the other conditions with respect to biomass development within the biofilm are the same. In this case, however, biodegradation does not occur in heavy metals expulsion. The instrument of heavy metals expulsion from contaminated water in biofilter is as follows. The nonbiodegradable water-dissolvable heavy metals are either oxidized or decreased by the microorganisms and less dissolvable species are formed. The less dissolvable frame of these metals that are shaped due to microbial responses are adsorbed or precipitated/co-precipitated on the surface of the adsorbent and the additional cellular protein (biocatalyst) of the microorganisms within the biolayer. The methylation of metals is additionally another vital course for the bioremediation of heavy metals in water. Although microbial activity on metal particle change is still a matter of inquiry, it is accepted that there are two ways. On one side oxidation or decrease of heavy metal particles is done by extracellular biocatalysts where the metal particles don't enter into the bacterial cell. On the otherside, the metal particles are transported into the microbial cells by transmembrane proteins and are transformed to other less soluble forms by metabolic activities of proteins within the cells followed by ensuing excretion from the cells, however in both ways the plasmid intervenes. Deciding which way is responsible for the microbial activity on a metal particle, could be a matter of research.

22.2.2.2 Advancement within Biofilter Effectiveness by Genetic Engineering of the Microorganisms

Bioremediation is the change or corruption of contaminants into nonhazardous or less dangerous chemicals. Microscopic organisms are by and large utilized for bioremediation, but parasites, green growth and plants have also been used. There are three classifications of bioremediation:

- Biotransformation: The change of contaminant particles into less or nonhazardous molecules.

- Biodegradation: The breakdown of natural substances in smaller natural or inorganic molecules.
- Mineralization: The total biodegradation of natural materials into inorganic constituents.

These three types of bioremediation can happen either *ex situ* or *in situ*, which both have advantages and disadvantages. In *ex situ* bioremediation, the contaminants are expelled and set in a contained environment, which makes the remediation handle quicker by permitting simpler observing and maintaining conditions and advances. In any case, the expulsion of the contaminant from the contaminated location is time consuming, expensive and possibly unsafe, which are the major impediments of the method. In contrast, the *in situ* prepare does not require the expulsion of the contaminant from the contaminated location. Instep, either biostimulation or bioaugmentation is connected. The previous is the expansion of nutrients, oxygen or other electron givers or acceptors to the facilitated location in arrange to extend the populace or movement of normally happening microorganisms accessible for remediation, whereas the last mentioned is the expansion of microorganisms that can biotransform or biodegrade contaminants.

Bioremediation innovation includes the use of microorganisms to diminish, kill, contain or change to generous items contaminants show in soils, silt, water and atmosphere. *Staphylococcus, Bacillus, Alcaligens, Escherichia, Pseudomonas, Citrobacteria, Klebsilla* and *Rhodococcus* are biological organisms that are commonly utilized in bioremediation. This handle includes biochemical responses or pathways in a living being that results in movement, development and generation of that life-form. Chemical forms included in microbial digestion systems comprise reactants, contaminants, oxygen or other electron acceptors that convert metabolites to well characterized items. Key to the remediation of metals is that metals are nonbiodegradable but can be changed through sorption, methylation and complexation and changes in valence state. Although utilizing bioremediation could be effective, often the contaminants are also toxic to the dynamic organisms included within the bioremediation preparation. This issue can make it exceptionally troublesome to keep the rate of bioremediation high. An alternative is to genetically design organisms that are safe to the extraordinary conditions of the contaminated location and conjointly have bioremediation properties. Bioremediation has gained significance recently as an interchange innovation for the evacuation of essential toxins in soil and water, which require compelling strategies of purification.

The designed microorganisms are more specific. The change in expulsion productivity for all the cases is recognizable. It is trusted that in combination, these methods would advance channel proficiency.

22.2.3 Bioelectrochemical System

A unique approach based on the integration of electrochemistry and microbiology has developed of late. It is commonly referred to as electrode-assisted bioremediation or electrobioremediation or bioelectrochemical remediation of heavy metal contamination. The reactor frameworks utilized to attain this preparation are referred to as bioelectrochemical systems (BESs). BESs combine microbial and electrochemical forms to change over the chemical energy stored in biodegradable natural matter into power, hydrogen or other valuable chemicals or to drive methods such as saltwater desalination and the expulsion of different contaminants.

Electrochemical treatment methods work on the rule of electrolysis in which, by adding energy from a connected source, metal oxidation at the anode and metal reduction at the cathode are encouraged at particular electric possibilities. Although this technique is successful, it requires support, care and high energy input and therefore incurs significant operational costs. To overcome some of the disadvantages of this approach, analysts began investigating BESs, which utilize microorganisms to intervene, encourage or catalyze the redox responses at both or either of the cathodes [4]. A few ponders have archived that the microorganisms are compelling biocatalysts to impact the portability of metal particles in natural and built or planned frameworks beneath controlled conditions [5, 6]. BESs open up an opportunity to improve a novel and valuable biotechnology based on the use of microorganisms in electrochemical frameworks for the evacuation and recovery of heavy metals from contaminated soil, dregs and water environments.

22.2.3.1 Concept and Principle

BESs comprise the anode and cathode that are ordinarily isolated by an ion-selective layer. Microorganisms that have extracellular electron exchange capabilities are for the most part utilized to catalyze either the substrate oxidation or diminishment responses at the anodes in BESs. Such microorganisms are referred to as electroactive microorganisms. They have the capacity to connect and intervene with electron exchange to solid-state electron acceptors and from donors such as cathodes to support their respiratory or metabolic activities. BESs can be worked in two diverse modes, microbial fuel cell (MFC) and microbial electrolysis cell (MEC), depending upon the target handle. Microbial substrate oxidation response at the anode is ordinarily utilized for the treatment of natural matter containing wastewaters. While substrate diminishment responses at the cathode are particularly utilized for the generation of decreased items such as hydrogen and biochemicals [7], or the remediation of obstinate contaminants and decrease of oxidized shapes of heavy metals display in several sorts of wastes, prepare streams, wastewaters, and leachate arrangements. It can moreover be utilized for the lessening of other electron acceptors such as perchlorate, nitrobenzene, azo colors, and nitrate [8, 9]. In BESs, the electrons delivered by the microbial oxidation of natural matter at the anode can be utilized to drive or encourage the diminishment of heavy metals at the cathode (Figure 22.1). There are two conceivable outcomes for the 56 coordinates microbial fuel cells for wastewater treatment lessening of heavy metals at the cathode in BESs. To begin with, in the event that the lessening potential of the heavy metal at the cathode is higher (or more positive) than the oxidation potential of the electron giver at the anode, the response will continue suddenly and create a net positive cell voltage in MFC mode operation. The metals that can be decreased at the cathode of MFCs incorporate Cu^{2+}/Cu^0, $Se^{4+}/$

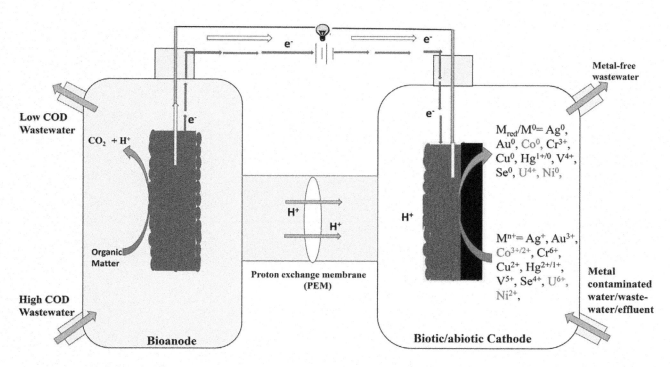

FIGURE 22.1 Schematic of a two-chambered bioelectrochemical system (BES) showing the principle of heavy metals removal [10].

Se^0, Cr^{6+}/Cr^{3+} and V^{5+}/V^{4+}. The moment plausibility is through the utilizing of MECs when the diminishment potential of the metal is lower (or more negative) than the oxidation potential of electron donors; outside energy should be added to encourage the metal diminishment response at the cathode.

22.2.4 Biosorption

The component of biosorption of heavy metals happens in two steps: the primary step occurs on the cell surface, counting physical adsorption, particle trade and surface complexation. This step is driven by the chemical osmotic slope of the natural cytoplasmic layer [11], and it can also be done by a dead natural cell. The moment steps are the inside aggregation of cells and surface microprecipitation, both which expend cellular energy and require the interest of metabolic processes, but as it were happening within the uncommon state of organic cells [12]. The intuitive adsorption mechanism is shown in Figure 22.2.

Although biosorption of heavy metals by numerous natural materials has been researched, a single organic adsorbent is difficult to meet the needs of practical applications. In this way, after a long time, modern heavy metals treatment strategies based on biomass have been developing. Among them, genetically built microorganisms are highly anticipated. Be that as it may, a genetically engineered microorganism is difficult to realize, due to the huge sum of building and the demanding working conditions it requires. Most importantly, the existing genetic information is not sufficient to completely foresee how the genetically engineered microorganisms will influence the human living environment in the future. In this manner, agreeing to the current natural, financial and innovative status, the use of generally straight forward chemical

or physical strategies to advance the adsorption effectiveness of biosorbents is necessary for the treatment of heavy metal wastewater. Chemical adjustment may be largely separated into two categories: surface modification and inner adjustment of the natural cell. The main reason to alter the surface is to expel the debasements on the cell surface, and increment the heavy metal official location on the surface of the natural cell, or alter the charge on the cell surface. Inner adjustment is more complex and includes changes within the inside structure or composition of natural cells, counting protein expression, enzymatic movement or self-generated nanomaterials inside the cell. Invigorating or restraining the movement of chemicals, which includes heavy metal transport or collection inside the cell, will have a great impact on the decontamination of heavy metals.

The combination of biomass and other materials cannot as it was fathoming the issue with the lacking of mechanical quality of free suspended biomass in building application and the trouble of solid-liquid division, but moreover at the same time apply two sorts of materials perform on the adsorption [14, 15]. There may moreover be synergistic impacts within the assimilation of heavy metals beneath fitting fabric choice and working conditions. The commonly utilized biologically bound materials are alginate, enacted carbon and different polymer materials. For a long time, only a couple of studies have combined biomass with nanoparticles to form an unused fabric that has tremendous adsorption properties for heavy metals [16, 17].

In expansion, multibiological combinations have moreover been demonstrated to have high heavy metal expulsion and steadiness in complex heavy metal contaminated situations. The use of multibiological combinations frequently has an

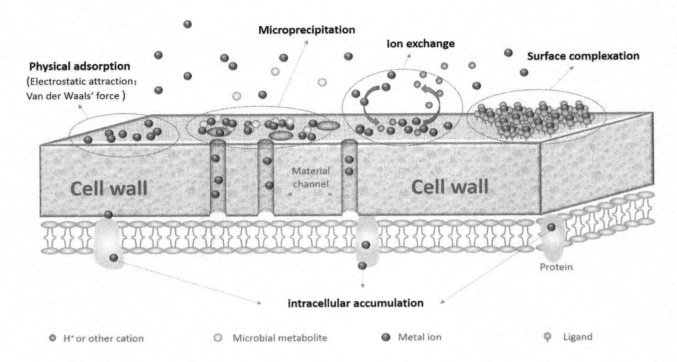

FIGURE 22.2 Adsorption mechanism of biological cells for heavy metals removal [13].

advantageous and synergistic impact. A strain in a multibiological combination can alter the natural oxygen substance and pH level through an organic digestion system that supplies superior adsorption conditions for the other strain. The point of this work is to show the state of the craftsmanship of three strategies on heavy metals biosorption. Fabric planning, strategy investigation, treatment impact and expulsion component have already been discussed, and their potential application and assist improvement have been anticipated. Methodically comparing these strategies, we given a point of view approximately biosorption for heavy metals expulsion. It is hoped that this audit will help advance investigation in this field.

22.3 Biological Treatment for VOCs Removal

Despite the viability of physicochemical strategies in destroying VOCs and rotten gases, the drawbacks of tall taken a toll and auxiliary contamination are characteristic of these techniques.

22.3.1 Organically Actuated Froth for VOCs Removal

Prepare designers at the Orange County Water District (OCWD) in Wellspring Valley, CA, USA, have created a ceaselessly renewable biologically actuated foam (BAF) reactor that quickly "eats" VOCs, converting the contaminants into carbon dioxide, water and characteristic biological buildup. As the late protected handle employments a vapor-phase reactor and BAF to annihilate the contaminants. Engineers state that the manufactured "biofilm" delivered amid this handle boosts

transport of VOCs and oxygen and kills microbial fouling issues customarily related with settled film methods. A major advantage of the BAF reactor is that unlike a fixed-film reactor, the microbe's concentration is controllable. In case microscopic organisms within the framework do multiply out of control, it is simple to turn the framework "off", close it down and start the reactor up once more. The downtime is brief, as there should be no reason to extricate the strong pressing, clean it, put it back and reinoculate the packing.

22.3.2 Hollow-Fiber Film Bioreactors

A hollow-fiber film employed novel bioreactor was created for the coupled transcription/translation framework utilizing T7RNA polymcrase and *Escherichia coli*S30 extricate. The expansive surface zone per the response volume of the reactor guarantees quick mass exchanges of substrates into the response blend and of wastes out from it over the layer by its atomic dissemination. The flux was sufficient to preserve nucleotide concentrations for more than 3 h, which expanded the protein union greatly. In expansion, the T; eliminator grouping, downstream from the columnist qualities, was found to extend the integrated protein significantly, especially when the item of polymerase chain response (PCR) was utilized as a template. Implementation of this finding and utilize of the bioreactor created, increased the efficiency of protein by the in *z&v* coordinate expression from the PCR template.

22.3.3 Biotrickling Filters

In BTF frameworks, wastegas being treated is carried through a pressed bed, which is persistently or irregularly watered with

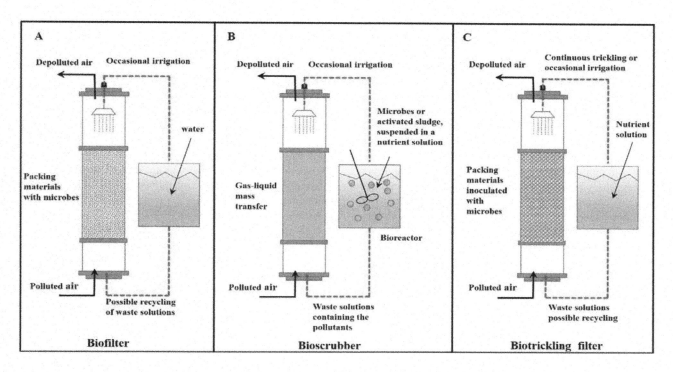

FIGURE 22.3 Schematic view of conventional bioreactors: (A) biofilter (BF); (B) bioscrubber; (C) biotrickling filter (BTF) [19].

an aqueous arrangement containing fundamental supplements required by the natural framework. Microorganisms from an outside source are immunized on the surface of the pressing fabric and shape a biofilm. Toxins are at first retained by a fluid film that encompasses the biofilm and are at that point corrupted by the biofilm. The BTF framework has the points of interest of low working and capital costs, lower weight drop amid long-term operation, and ability to treat acidic corruption items of VOCs and acidic rotten gases [18]. In any case, BTFs are moderately complex in development and operation and amass an overabundance of biomass. Comparison of BFs, BTFs and bioscrubbers reveal that BTFs have much higher capacity in treating obstinate VOCs and acidic or soluble compounds. Additionally, water administration is the major advantage of BTFs, when compared with the BF frameworks, permitting characterized control of pH, supplement supply and expulsion of poisonous metabolites, hence accomplishing higher toxins end rates. Otherwise, when compared with bioscrubbers, BTFs show quick biodegradation of toxins owing to enhancement of poisons with extracellular polymeric substances acting as surfactants, resulting in expansive numbers of immobilized microorganisms coming into contact with the pollutants.

Although the development and operation of a BTF may be complex, these frameworks as a normal frame of natural treatment innovation have succeeded in treating numerous sorts of toxins. Hence, when combining this advantage with its taken a toll adequacy, BTF innovation is an appealing choice for controlling VOC emanation from different mechanical methods. In any case, a few confinements in BTF methods (such as abundance biomass amassing, lesser mass exchange rate, etc.) may influence the toxin expulsion execution of BTFs. In a

BTF framework, the plan, working condition, mass exchange, pressing materials and microorganisms are basic impacts amid BTF methods and can altogether influence the expulsion of vaporous toxins.

Numerous strategies have been examined for upgrading the expulsion execution of BTFs, such as poison pre-treatment, parasitic BTF, surfactant expansion and other strategies. Basically, these strategies are outlined to optimize the basic parameters of BTFs. In this way, examination of these basic parameters is fundamental, since they are the establishment for utilizing BTFs and moving forward the toxin evacuation execution.

22.3.3.1 BTF Instruments and Limitations

A BTF framework for VOCs and musty gas evacuation is ordinarily a complex combination of distinctive physicochemical and organic marvels. As the vapor passes through the BTF bed, contaminants and supplements are exchanged to the microorganisms, where a cellular digestion system breaks down the chemicals into less difficult components. Microorganisms will be immobilized on the surface of the pressing fabric and the surface of biofilms is secured by a water stage. The VOCs and musty toxins already within the contaminated discus have to be transferred to the water stage layer, and after that moved to microbial cells within the biofilm where they are eventually debased by the microorganisms.

At first, the vapor-phase contaminants and any supplements required for microbial development (oxygen, nitrogen and phosphorus, etc.) must be broken up within the water stage encompassing the biofilm. The poison vapors and oxygen are transported in the muggy discus or specifically into the water stage by concentration angle contrasts, and this handle is shown as mass exchange. Along these lines, biodegradation responses happen, amid which the broken down contaminants are acclimatized and corrupted by the biofilm through change to CO_2 or other ineffectively unstable intermediates with simple chemical structures. Near the supplements, the energy discharged by the oxidation of these contaminants is utilized by the microorganisms to create a biofilm. In a BTF, the biodegradation impact is controlled both by mass exchange of vaporous toxins from discus to biofilm (dissemination restriction) and by the biodegradation responses (response impediment). In a past study [20], the relationship between end capacities (EC) and channel stack (IL) was discussed at three purge bed maintenance times (EBRT). When IL was settled, longer EBRT implied the vaporous poison characteristics were a high channel concentration and less add up to stream rate. In this case, the water stage encompassing the biofilm was immersed by vaporous toxins; in this way, the debasement response was controlled by the response restriction. At shorter EBRT, the vaporous poison characteristics added up to high stream rate and less gulf concentration. Under these conditions, the vaporous poison left the BTF without satisfactory contact with the water stage; in this way, the debasement was controlled by dissemination impediment [20].

The real biodegradation of target contaminants happens inside the biofilm, and the arrangement of auxiliary metabolites is conceivable. In the event that shaped, auxiliary metabolites will experience the same concurrent dissemination, biodegradation and sorption methods as the essential toxins. At long last, the conclusion items (such as CO_2) after microbial corruption are exchanged to gas stage or shapes carbonate in fluid stage by VOC oxidation. In any case, in the event that the auxiliary metabolites appear to be more harmful than the first poisons, their arrangement will halt the biodegradation forms, for example, generation of p-benzoquinone amid the biodegradation of chlorophenol or 4-chloronitrobenzene, and collection of tert-butyl liquor amid the change of methyl tert-butyl ether (MTBE) [21, 22].

For viable BTF treatment, the contaminants must be biodegradable and nontoxic or have lesser poisonous quality to BTF. The foremost fruitful toxin expulsion utilizing BTFs has been accomplished for less atomic weight and exceedingly solvent natural compounds with straight forward bond structures. Compounds with complex bond structures more often than not require more energy, which occasionally are not continuously accessible to the microorganisms, and as a result, small or no biodegradation of these compounds happens. It is well known that natural compounds such as alcohols, aldehydes, ketones and a few basic aromatics are highly biodegradable; phenols, chlorinated hydrocarbons, PAHs and exceedingly halogenated hydrocarbons appear to have moderate debasement; and certain anthropogenic compounds may not biodegrade at all unless a few extra components, such as vital proteins, are included.

22.3.3.2 Microorganisms in BTF Systems

The adequacy of a BTF depends on the capacity of the microorganisms to biodegrade toxins. Microorganisms vary in their corruption execution in treating a particular toxin. The biofilms in BTFs often comprise high numbers of microbes and less numbers of organisms; hence, most articles focus on the investigation of bacterial communities [23]. Microbes of the genera *Pseudomonas*, *Bacillus*, *Staphylococcus* and *Rhodococcus* are most regularly found in BTFs. *Pseudomonas* has been recognized as the dominant species of the bacterial populace in several bioreactors used to expel nitrogen, H_2S and numerous VOCs [24–26]. Bacillus can happen simultaneously under high-impact nitrification-denitrification conditions, *Staphylococcus* can decrease nitrate to nitrite [27], and *Rhodococcus* has the capacity to metabolize destructive natural toxins, including toluene, naphthalene, herbicides and other compounds.

Although toxin corruption in BTFs is often ascribed to bacteria, occasionally organisms may play a vital part. A few studies have demonstrated that organisms display higher VOC evacuation performance than microscopic organisms. Moreover, fungi permit simple assimilation of numerous VOCs from the bulk gas phase because of their filamentous structures with ethereal mycelia and a large surface range. Meanwhile, the resistance of fungi to lesser stickiness favors mass exchange of hydrophobic VOCs from the gas phase to parasitic surfaces. Compared with microbes, parasites can better resist the situations of less stickiness and high sharpness. Therefore, utilizing parasites as the most corrupt microorganism in BTFs has awesome potential.

Filamentous organisms having a place to the genera *Scedosporium*, *Paecilomyces*, *Cladosporium* and *Cladophialophora*, white-rot parasites and yeasts of the sort *Exophiala* are regularly gotten in BTFs [28]. Jin et al. [29] demonstrated that a fungi-dominated BTF showed a great potential to resist stun loads within the treatment of alpha-pinene and might quickly accomplish its full execution after a three- to seven-day starvation period. Besides, organisms can evacuate profoundly concentrated VOCs in exceedingly acidic conditions. However, parasites have a few downsides. Compared to microscopic organisms, fungi have overall lower metabolic rates that lengthen the start-up periods of BTFs, and their filamentous structure frequently leads to clogging in BTFs.

To advance the evacuation execution of BTF, bacteria-dominated BTF and fungi-dominated BTF can be combined. Cheng et al. [30] set up three distinctive BTFs (bacterial BTF, parasitic BTF and bacterial-and-fungal BTF) to study their differences in treating toluene. The bacterial-and-fungi BTF come to an RE of 90%, whereas the RE of the contagious BTF and the bacterial BTF were 60% and 20%, respectively.

22.3.3.3 Other Procedures Combined with BTF Systems

The combination of a BTF framework with other sorts of treatment can achieve superior evacuation execution than the use of BTF system alone. A few microorganisms can create distinctive sorts of enzymes such as monooxygenase, which can upgrade the debasement of refractory VOCs to straightforward atoms. Co-metabolism can assist the microorganisms in invigorating the discharge of substrate-oxidizing proteins. Quan et al. [31] utilized an inactive attractive field combined BTF to treat TCE waste gas and compared this combined technique to a single BTF. The comes about demonstrated progressed TCE removal performance and change within the bacterial community beneath 60 mT of magnetic field escalated. Attractive field fortifying can create similar effects of poison evacuation with the expansion of phenol and sodium acetate as co-metabolic substrate. Besides, distinctive magnetic field power altogether influenced the contagious community in the BTF frameworks and progressed the wealth of the phylum Ascomycota, thus expanding the TCE expulsion rate[32]. Compared to using proteins, pre-treatment of persistent VOCs may advance the metabolism of toxins by microorganisms. For example, in the treatment of a few fragrant compounds such as styrene and chlorobenzene, UV pre-treatment is ordinarily utilized earlier to a BTF framework. The UV energy can straightforwardly change over the hydrophobic and stable VOCs into water-soluble and effectively biodegradable intermediates, making it a productive pre-treatment step to BTF treatment. As a result, both mass exchange and response rate are improved by UV pre-treatment since it changes toxins into more solvent and biodegradable compounds. Other than UV, plasma or other photolysis pre-treatment methodologies can also be connected.

22.3.4 Bioscrubbers

In bioscrubbers, the contaminated gas is treated in two steps. In with the first step, the contaminated discus stream is committed with water in a reactor pressed with inactive media, coming about in assimilation of contaminant to the fluid stage. The fluid is at that point coordinated into an enacted slime reactor or any organic unit where the contaminants are organically corrupted. The treated liquid effluent from the moment bioreactor (after clarification) is recirculated to the primary reactor [33]. Subsequently, the moment reactor permits the bioscrubber to treat higher concentrations of VOCs than biofilters. Since retention and biodegradation happen in two different reactors, optimization of these reactors can be accomplished. Though biotrickling channels and bioscrubbers offer more control over pH, nutrient and wash out of corruption by-products since of nearness of fluid stage.

The main disadvantage of bioscrubbers over biofilters is transfer of abundant slime/profluent. As the system depends on retention, exceedingly water dissolvable poisons can be effectively treated. Since a large portion of the VOCs appearing in ETP or STP emanations are modestly hydrophobic in nature, bioscrubbing at that point ends up less prevalent [33]. In bioscrubbers, the inlet VOC or odor stacking for the most part is <5 g m^{-3}[34].

Rene at al. [35] examined the BTEX expulsion in a fungi-dominated biofilter, and in general evacuation productivity was found to be 35%–97% under different operating conditions. Zamir et al. [36] conducted a compost biofilter trial with toluene after immunization with an extraordinary sort of white-rot organism, *Phanerochaetechrysosporium*, and found 92% effective diminishment. Li et al. [37] created a styrene biofilter test containing PU froth as media and the evacuation effectiveness was >96%. Chen et al. [38] planned a biofilter for toluene evacuation obtaining suspended biofilm. The expulsion proficiency was >90.2% with a start-up time of under 14

days and 128 days of working time. Rene et al. [39] conducted a compost biofilter test with benzene and toluene stacking by shifting channel concentrations, and the expulsion was 72.7% for benzene and 81.1% for toluene individually. Natarajan et al. [40]treated an ethylbenzene and xylene blend in biofilter having blended microbial culture with tree bark media. The expulsion effectiveness of 58%–78% and 68%–89% were recorded for xylene and ethylbenzene for a nonstop 96 days of biofilter operation. Rahul et al. [41] conducted a biofilter operation with corncob as channel media with BTEX and obtained more than 99% removal. Gallastegui et al. [42] tested a toluene and p-xylene biofilter pressed with inert material. They watched hindrance of p-xylene in nearness of toluene whereas the nearness of pxylene upgraded the toluene evacuation. Li et al. [37] outlined a coordinate bioreactor system —a gas division film module introduced after a control biofilter. Due to combining, the expulsion of styrene was upgraded and in general the framework productively handled for the fluctuating channel stack. Moreover, within the film module, styrene was condensed and recovered back to the biofilter, which in turn amplified the maintenance time. Schiavon et al. [43] recently conducted a trial with non-warm plasma (NTP) upstream of a biofilter for expulsion of blended VOCs. The NTP utilizing the diverse particular energy densities reduced the VOC concentrations down to an ideal level. Also, plasma treatment converted water insoluble VOCs to more dissolvable compounds. They utilized NTP effectively for pre-treatment some time recently biofiltration.

22.3.5 Biofiltration

Biofiltration for the most part refers to natural treatment or change of organics or inorganic contaminants into safe compounds, whether at discus or gas stage. Although biodegradation in range of wastewater treatment and bioremediation methods are broadly connected for treatment of soil and groundwater, in recent decades biofiltration has been developed as treatment for VOC expulsion procedure for VOC expulsion and mechanical application. Most of the biofilters built as an open single bed frameworks for treating compost gas or odor and continuously developing as VOC expulsion procedure for mechanical application. Within the biofiltration, contaminated stream is dampened and pumped to a bio channel bed. Whereas the stream gradually runs through the filter media, the contaminants within the discus stream are retained and metabolized. Organic oxidation by microorganisms can be expressed as follows:

$$\text{Organic pollutant} + O_2 + \text{Microorganisms} \rightarrow O_2 + H_2O + \text{Heat} + \text{Biomass}$$

Biofiltration is favored over ordinary control strategies because of its lower capital and working costs, less chemical utilization, adaptability in plan, expelling of a wide run of contaminants, high treatment productivity and customizability. Nevertheless, as this can be living handle it is exceptionally touchy to environment and cannot handle high vacillation in stack or extraordinary climate. A few components contribute to the generally biofiltration handle like channel media, temperature, stickiness, pH, home time, weight drop and supplement. Biofiltration is effective for the evacuation or annihilation of much sort of off-gas poisons, especially natural compounds counting inorganic compounds such as H_2S and NH_3 [44]. Biofiltration is exceptionally much compelling and prudent when the discus volume is high and contaminant is less.

Essentially, contaminants that can be treated by biofiltration might have the following characteristics [34, 44]:

1. High water solvency–Compounds having highwater solvency diffuse to water layer at biofilm more effectively, which is required in advance for its biodegradation. Compounds having higher esteem of Henry's steady (H, mol m^{-3} Pa^{-1}) and lower vapor weight display higher water solubility.

2. Ready biodegradability–After the compounds are ingested into the water film (biofilm), the same is required to be biodegraded, something else the concentration of the same will increment in biofilm which may be inconvenient to microorganisms additionally the encourage dissemination into biofilm may be detrimental to microorganisms additionally the encourage dissemination into biofilm may be diminished. Compounds having lower water solvency and lower biodegradability like halogenated hydrocarbons can moreover be treated in biofilter. Nevertheless, the treatment proficiency depends on planning specific sort and environment for the same.

Evacuation of contaminants in a biofilter may be a multistep process beginning from dissemination of contaminants from gas stage to fluid phase/biofilm and after that cell surface of microorganisms and finally the biodegradation by microorganisms (basically microscopic organisms an parasites). Major forms included in biofiltration are outlined as follows [34, 45]:

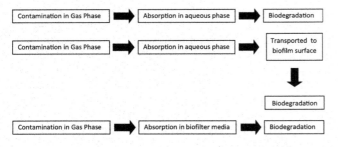

As the discus passes through the pressing bed, contaminants are exchanged from the gas stream to the water within the biofilm. Once retained within the biofilm layer or broken up within the water layer around the biofilm, the natural compound is accessible as nourishment for the microorganism's digestion system, serving as carbon and energy source to maintain microbial life and development transported to microbial cells and corrupted. For exceedingly dissolvable compounds, major expulsion happens in water broken down frame, though for hydrophobic contaminants, the major expulsion component may be adsorption on the surface of the medium [34] and

ensuing biodegradation inside the microorganism cells. The organic deterioration rate depends on concentration of biomass and particular development rate coefficient. Overseeing control parameters in bioreactors are nature of pressing fabric, gulf stream rate, temperature, pH, mugginess and inorganic supplements. Although in the biofiltration framework both microscopic organisms and parasites utilized and occasionally specialized bacterial or parasitic culture are connected, most biofiltration considerations are based on microbial duplication, which is commonly found in biofilters [33].

Biofilters are the best and most seasoned reactor framework among the three vapor-phase bioreactors for the treatment of VOCs. VOC-loaded wastegas is passed through a biofilm immobilized back media and are changed over by the microorganisms into carbon dioxide, water and extra biomass. Waste gas is humidified prior to being directed to a bioreactor that is critical for treatment. By and large media bed consists of organic packing fabric like compost, peat, wood or bark chips or a manufactured material like plastic or ceramic. Microorganisms create a biofilm on this pressing media bolster, which (organic media) can moreover be a source for additional supplements to back microbial development. When a synthetic media bed is utilized, supplements ought to be included for microbial development. For organic packing, supplements are required after a while as the media supplements are exhausted. Supplements are mostly splashed irregularly onto the pressing material. Sprinkling of water over the channel media is additionally connected irregularly to guarantee suitable moisture substance inside the pressing media and empower the wash-out of debasement by-products [33]. In biofilters, channel stacking of VOC or odor for the most part is <1 g m^{-3}[34]. Although biofilters continuously show odor diminishment efficiencies that are higher than 80%, VOC expulsion efficiencies are 20%–90% [33]. Subsequently, the operational parameters ought to be kept up securely to guarantee compelling expulsion of VOCs.

22.3.6 NTP and UV-BTF Coordinate Innovation for Cl-VOCs Removal

For less biodegradable and less water-soluble CleVOCs, NTP oxidation and UV photo-oxidation are included as pretreatment steps to change over CleVOCs to biodegradable intermediates, which may hold guarantee for the intensive treatment of CleVOCs waste air [46, 47]. Jiang et al. [47] inquired about a NTP-BTF framework (RE 97.6%), which had stronger ability to evacuate CB than did single BTF (RE 68.7%, EBRT 60 s) and single NTP (RE around 59%, EBRT 4.32 s) at low-energy conditions (SIE [particular input energy] 3500 J/L). Besides, the NTP pre-treatment may be a commonsense strategy for restraining abundance development of biomass with small impact on the BTF execution due to the biomass development in NTP-BTF beings lower than within the single BTF. Jiang et al. and Zhu et al. [48, 49] combined DBD (dielectric barrier discharge, one of the ordinary frameworks for NTP) with BTF for purification of waste discus containing 1,2-DCA and CB individually. The results appeared that the nearness of the catalyst significantly reduced the number of by-products, and the water solvency and biodegradability of the intermediates were altogether improved. In expansion, the test results showed that the biodiversity and plenitude of microbial community within the NTP-BTF system were lower than that in single BTF framework. This may be since the intermediates shaped by NTP pre-treatment influence the development of microorganisms in BTF, or the ozone created by NTP pre-treatment kills the microorganisms. Yu et al. [50] coupled UV photooxidation with BTF to expel DCM from polluted air. Compared with single BTF, it displayed higher RE of DCM (81.78%) with inlet concentration of 750 mg/m³, and UV-BTF had higher microbial diversity. UV photo-oxidation might change over hydrophobic VOCs to water-soluble and biodegradable intermediates, which may be carbonyl compounds and unstable greasy acids. These components may have a positive impact on mass exchange of the parent compound and microorganism development, which could greatly decrease the time for biofilm arrangement [46]. The homogeneity of the bacterial community helps to maintain the high performance of the bioreactor.

22.4 Conclusion

The literature indicates that the release of heavy metals and volatile organic compounds into the environment is unavoidable. Conversely, their ample treatment is essential to save the environment. The release of these compounds into the environment causes disturbance of aquatic flora and fauna, a risk to human health and also air pollution. Published investigation reveals that biological methods in comparison to other processes could be an efficient treatment process for their removal. Based on the present study, bioreactors (biofilter, biotrickling filter and bioscrubber) can be used as a potential technology for removal of VOCs as compared to the other available elimination techniques. However, use of biological systems to remove VOCs still has limitations and challenges. Additionally, biofilters have emerging applications for the treatment of heavy metal contaminated wastewater, whereas BES for heavy metal eradication is still in the development stage, and consequently, more research is required to understand microbe-metal interactions, electron transfer mechanisms, and electrode materials to enhance the efficiency of bioelectrochemical processes. Hence, more research in the field is required for the development of optimized eco-friendly as well as economically feasible technology to protect the planet for future generations.

REFERENCES

[1] Barthe P, Chaugny M, Roudier S, Delgado Sancho L. *Best available techniques (BAT)reference document for the refining of mineral oil and gas.* European Commission. 2015;754.

[2] Version –*Corrected FI.* Emission Estimation Protocol for Petroleum Refineries.

[3] Malakar S, Saha PD. Estimation of VOC emission in petroleum refinery ETP and comparative analysis with measured VOC emission rate. *The IJES.* 2015;4(10):20–29.

[4] Dermentzis K, Christoforidis A, Valsamidou E, Lazaridou A, Kokkinos N. Removal of hexavalent chromium from electroplating wastewater by electrocoagulation with ironelectrodes. *Global Nest Journal.* 2011 Dec 1;13(4):412–418.

[5] van der Maas P, Peng S, Klapwijk B, Lens P. Enzymatic versus nonenzymatic conversions during the reduction of EDTA-chelated Fe (III) in BioDeNO x reactors. *Environmental Science &Technology.* 2005 Apr 15;39(8):2616–2623.

[6] Francis AJ, Nancharaiah YV. In situ and ex situ bioremediation of radionuclide-contaminated soils at nuclear and norm sites. In *Environmental remediation and restoration of contaminated nuclear and norm sites* (pp. 185–236). Woodhead Publishing. 2015 Jan 1.

[7] Patil SA, Arends JB, Vanwonterghem I, Van Meerbergen J, Guo K, Tyson GW, Rabaey K. Selective enrichment establishes a stable performing community for microbial electrosynthesis of acetate from CO2. *Environmental Science &Technology.* 2015 July 21; 49(14):8833–8843.

[8] Ter Heijne A, Hamelers HV, De Wilde V, Rozendal RA, Buisman CJ. A bipolar membrane combined with ferric iron reduction as an efficient cathode system in microbial fuel cells. *Environmental Science &Technology.* 2006 Sept 1;40(17):5200–5205.

[9] Strycharz-Glaven SM, Glaven RH, Wang Z, Zhou J, Vora GJ, Tender LM. Electrochemical investigation of a microbial solar cell reveals a nonphotosynthetic biocathode catalyst. *Applied and Environmental Microbiology.* 2013 July 1;79(13):3933–3942.

[10] Malyan SK, Kumar SS, Singh L, Singh R, Jadhav DA, Kumar V. Bioelectrochemical systems for removal and recovery of heavy metals. In *Bioremediation, nutrients, and other valuable product recovery.* 2021 Jan 1 (pp. 185–203). Elsevier.

[11] Nies DH. Microbial heavy-metal resistance. *Applied Microbiology and Biotechnology.* 1999 June 1;51(6):730–750.

[12] Javanbakht V, Alavi SA, Zilouei H. Mechanisms of heavy metal removal using microorganisms as biosorbent. *Water Science and Technology.* 2014 May 1;69(9):1775–1787.

[13] Qin H, Hu T, Zhai Y, Lu N, Aliyeva J. The improved methods of heavy metals removal by biosorbents: A review. *Environmental Pollution.* 2020 Mar 1;258:113777.

[14] Arıca MY, Kacar Y, Genç Ö. Entrapment of white-rot fungus Trametes versicolor in Ca-alginate beads: preparation and biosorption kinetic analysis for cadmium removal from an aqueous solution. *Bioresource Technology.* 2001 Nov 1;80(2):121–129.

[15] Arıca MY, Arpa C, Ergene A, Bayramoğlu G, Genç Ö. Ca-alginate as a support for Pb (II) and Zn (II) biosorption with immobilized Phanerochaetechrysosporium. *Carbohydrate Polymers.* 2003 May 1;52(2):167–174.

[16] Xu P, Zeng GM, Huang DL, Feng CL, Hu S, Zhao MH, Lai C, Wei Z, Huang C, Xie GX, Liu ZF. Use of iron oxide nanomaterials in wastewater treatment: A review. *Science of the Total Environment.* 2012 May 1;424:1.

[17] Xu P, Zeng GM, Huang DL, Lai C, Zhao MH, Wei Z, Li NJ, Huang C, Xie GX. Adsorption of Pb (II) by iron oxide nanoparticles immobilized Phanerochaetechrysosporium: Equilibrium, kinetic, thermodynamic and mechanisms analysis. *Chemical Engineering Journal.* 2012 Sept 1;203:423–431.

[18] Lebrero R, Estrada JM, Muñoz R, Quijano G. Toluene mass transfer characterization in a biotrickling filter. *Biochemical Engineering Journal*. 2012 Jan 15;60:44–49.

[19] Wu H, Yan H, Quan Y, Zhao H, Jiang N, Yin C. Recent progress and perspectives in biotrickling filters for VOCs and odorous gases treatment. *Journal of Environmental Management*. 2018 Sept 15;222:409–419.

[20] Wu H, Yin Z, Quan Y, Fang Y, Yin C. Removal of methyl acrylate by ceramic-packed biotrickling filter and their response to bacterial community. *Bioresource Technology*. 2016 June 1;209:237–245.

[21] Purswani J, Juárez B, Rodelas B, Gónzalez-López J, Pozo C. Biofilm formation and microbial activity in a biofilter system in the presence of MTBE, ETBE and TAME. *Chemosphere*. 2011 Oct 1;85(4):616–624.

[22] Skiba A, Hecht V, Pieper DH. Formation of protoanemonin from 2-chloro-cis, cis-muconate by the combined action of muconatecycloisomerase and muconolactone isomerase. *Journal of Bacteriology*. 2002 Oct 1;184(19):5402–5409.

[23] Zhao L, Huang S, Wei Z. A demonstration of biofiltration for VOC removal in petrochemical industries. *Environmental Science: Processes & Impacts*. 2014;16(5):1001–1007.

[24] Giri BS, Kim KH, Pandey RA, Cho J, Song H, Kim YS. Review of biotreatment techniques for volatile sulfur compounds with an emphasis on dimethyl sulfide. *Process Biochemistry*. 2014 Sept 1;49(9):1543–1554.

[25] Li Y, Zhang W, Xu J. Siloxanes removal from biogas by a lab-scale biotrickling filter inoculated with Pseudomonas aeruginosa S240. *Journal of Hazardous Materials*. 2014 Jun 30;275:175–184.

[26] Zheng M, Li C, Liu S, Gui M, Ni J. Potential application of aerobic denitrifying bacterium Pseudomonas aeruginosa PCN-2 in nitrogen oxides (NOx) removal from flue gas. *Journal of Hazardous Materials*. 2016 Nov 15;318:571–578.

[27] Cheng CY, Mei HC, Tsao CF, Liao YR, Huang HH, Chung YC. Diversity of the bacterial community in a bioreactor during ammonia gas removal. *Bioresource Technology*. 2010 Jan 1;101(1):434–437.

[28] Repečkienė J, Švedienė J, Paškevičius A, Tekorienė R, Raudonienė V, Gudeliūnaitė E, Baltrėnas P, Misevičius A. Succession of microorganisms in a plate-type air treatment biofilter during filtration of various volatile compounds. *Environmental Technology*. 2015 Apr 3;36(7):881–889.

[29] Jin Y, Guo L, Veiga MC, Kennes C. Fungal biofiltration of α-pinene: Effects of temperature, relative humidity, and transient loads. *Biotechnology and Bioengineering*. 2007 Feb 15;96(3):433–443.

[30] Cheng Z, Lu L, Kennes C, Yu J, Chen J. Treatment of gaseous toluene in three biofilters inoculated with fungi/bacteria: microbial analysis, performance and starvation response. *Journal of Hazardous Materials*. 2016 Feb 13;303:83–93.

[31] Quan Y, Wu H, Yin Z, Fang Y, Yin C. Effect of static magnetic field on trichloroethylene removal in a biotrickling filter. *Bioresource Technology*. 2017 Sept 1;239:7–16.

[32] Quan Y, Wu H, Guo C, Han Y, Yin C. Enhancement of TCE removal by a static magnetic field in a fungal biotrickling filter. *Bioresource Technology*. 2018 July 1;259:365–372.

[33] Omil F. *Biological technologies for the removal of VOCs, odours and greenhouse gases*. Departmente of Chemical Engineering, University of Santiago de Compostela. 2014.

[34] Frederickson J, Boardman CP, Gladding TL, Simpson AE, Howell G, Sgouridis F. *Evidence: Biofilter performance and operation as related to commercial composting*. Environment Agency. 2013.

[35] Rene ER, Mohammad BT, Veiga MC, Kennes C. Biodegradation of BTEX in a fungal biofilter: influence of operational parameters, effect of shock-loads and substrate stratification. *Bioresource Technology*. 2012 July 1;116:204–213.

[36] Zamir SM, Halladj R, Nasernejad B. Removal of toluene vapors using a fungal biofilter under intermittent loading. *Process Safety and Environmental Protection*. 2011 Jan 1;89(1):8–14.

[37] Li L, Lian J, Han Y, Liu J. A biofilter integrated with gas membrane separation unit for the treatment of fluctuating styrene loads. *Bioresource Technology*. 2012 May 1;111:76–83.

[38] Chen X, Qian W, Kong L, Xiong Y, Tian S. Performance of a suspended biofilter as a new bioreactor for removal of toluene. *Biochemical Engineering Journal*. 2015 June 15;98:56–62.

[39] Rene ER, Kar S, Krishnan J, Pakshirajan K, López ME, Murthy DV, Swaminathan T. Start-up, performance and optimization of a compost biofilter treating gas-phase mixture of benzene and toluene. *Bioresource Technology*. 2015 Aug 1;190:529–535.

[40] Natarajan R, Al-Sinani J, Viswanathan S, Manivasagan R. Biodegradation of ethyl benzene and xylene contaminated air in an up flow mixed culture biofilter. *International Biodeterioration & Biodegradation*. 2017 Apr 1;119:309–315.

[41] Mathur AK, Balomajumder C. Biological treatment and modeling aspect of BTEX abatement process in a biofilter. *Bioresource Technology*. 2013 Aug 1;142:9–17.

[42] Gallastegui G, Ramirez AÁ, Elías A, Jones JP, Heitz M. Performance and macrokinetic analysis of biofiltration of toluene and p-xylene mixtures in a conventional biofilter packed with inert material. *Bioresource Technology*. 2011 Sept 1;102(17):7657–7665.

[43] Schiavon M, Schiorlin M, Torretta V, Brandenburg R, Ragazzi M. Non-thermal plasma assisting the biofiltration of volatile organic compounds. *Journal of Cleaner Production*. 2017 Apr 1;148:498–508.

[44] Adler SF. Biofiltration- aprimer. *Chemical Engineering Progress*. 2001 Apr 1;97(4):33–41.

[45] Berenjian A, Chan N, Malmiri HJ. Volatile organic compounds removal methods: A review. *American Journal of Biochemistry and Biotechnology*. 2012;8(4):220–229.

[46] Den W, Ravindran V, Pirbazari M. Photooxidation and biotrickling filtration for controlling industrial emissions of trichloroethylene and perchloroethylene. *Chemical Engineering Science*. 2006 Dec 1;61(24):7909–7923.

[47] Jiang L, Li H, Chen J, Zhang D, Cao S, Ye J. Combination of non-thermal plasma and biotrickling filter for chlorobenzene removal. *Journal of Chemical Technology & Biotechnology*. 2016 Dec;91(12):3079–3087.

[48] Jiang L, Li S, Cheng Z, Chen J, Nie G. Treatment of 1, 2-dichloroethane and n-hexane in a combined system of non-thermal plasma catalysis reactor coupled with a biotrickling filter. *Journal of Chemical Technology & Biotechnology*. 2018 Jan;93(1):127–137.

[49] Zhu R, Mao Y, Jiang L, Chen J. Performance of chlorobenzene removal in a nonthermal plasma catalysis reactor and evaluation of its byproducts. *Chemical Engineering Journal.* 2015 Nov 1;279:463–471.

[50] Jianming Y, Wei L, Zhuowei C, Yifeng J, Wenji C, Jianmeng C. Dichloromethane removal and microbial variations in a combination of UV pretreatment and biotrickling filtration. *Journal of Hazardous Materials.* 2014 Mar 15;268:14–22.